OPTIMAL QUADRATIC PROGRAMMING ALGORITHMS

Springer Optimization and Its Applications

VOLUME 23

Aims and Scope
Optimization has been expanding in all directions at an astonishing rate during the last few decades. New algorithmic and theoretical techniques have been developed, the diffusion into other disciplines has proceeded at a rapid pace, and our knowledge of all aspects of the field has grown even more profound. At the same time, one of the most striking trends in optimization is the constantly increasing emphasis on the interdisciplinary nature of the field. Optimization has been a basic tool in all areas of applied mathematics, engineering, medicine, economics and other sciences.

The series *Springer Optimization and Its Applications* publishes undergraduate and graduate textbooks, monographs and state-of-the-art expository works that focus on algorithms for solving optimization problems and also study applications involving such problems. Some of the topics covered include nonlinear optimization (convex and nonconvex), network flow problems, stochastic optimization, optimal control, discrete optimization, multi-objective programming, description of software packages, approximation techniques and heuristic approaches.

OPTIMAL QUADRATIC PROGRAMMING ALGORITHMS

With Applications to Variational Inequalities

By

ZDENĚK DOSTÁL
VŠB - Technical University of Ostrava, Czech Republic

Springer

Zdeněk Dostál
Department of Applied Mathematics
VŠB - Technical University of Ostrava
70833 Ostrava
Czech Republic
zdenek.dostal@vsb.cz

ISSN 1931-6828
ISBN 978-1-4419-4648-5 e-ISBN 978-0-387-84806-8
DOI 10.1007/978-0-387-84806-8

Mathematics Subject Classification (2000): 90C20, 90C06, 65K05, 65N55

Cover illustration: "Decomposed cubes with the trace of decomposition" by Marta Domoraádová

Printed on acid-free paper

springer.com

To Maruška, Matěj, and Michal, the dearest ones

Preface

The main purpose of this book is to present some recent results concerning the development of in a sense optimal algorithms for the solution of large bound and/or equality constrained quadratic programming (QP) problems. The unique feature of these algorithms is the rate of convergence in terms of the bounds on the spectrum of the Hessian matrix of the cost function. If applied to the class of QP problems with the cost functions whose Hessian has the spectrum confined to a given positive interval, the algorithms can find approximate solutions in a uniformly bounded number of simple iterations, such as the matrix–vector multiplications. Moreover, if the class of problems admits a sparse representation of the Hessian, it simply follows that the cost of the solution is proportional to the number of unknowns.

Notice also that the cost of duplicating the solution is proportional to the number of variables. The only difference is a constant coefficient. But the constants are important; people are interested in their salaries, as Professor Babuška nicely points out. We therefore tried hard to present a *quantitative* theory of convergence of our algorithms wherever possible. In particular, we tried to give realistic bounds on the rate of convergence, usually in terms of the extreme nonzero eigenvalues of the matrices involved in the definition of the problem. The theory covers also the problems with dependent constraints.

The presentation of each new algorithm is complete in the sense that it starts from its classical predecessors, describes their drawbacks, introduces modifications that improve their performance, and documents the improvements by numerical experiments. Since the exposition is self-contained, the book can serve as an introductory text for anybody interested in QP. Moreover, since the solution of a number of more general nonlinear problems can be reduced to the solution of a sequence of QP problems, the book can also serve as a convenient introduction to nonlinear programming. Such presentation has also a considerable methodological appeal as it enables us to separate the simple geometrical ideas, which are behind many theoretical results and algorithms, from the technical difficulties arising in the analysis of more general nonlinear optimization problems.

Our algorithms are based on modifications of the active set strategy that is optionally combined with variants of the augmented Lagrangian method. Small observations and careful analysis resulted in their qualitatively improved performance. Surprisingly, these methods can solve some large QP problems with less effort than a single step of the popular interior point methods. The reason is that the standard implementation of the interior point methods can hardly use a favorable distribution of the spectrum of the Hessian due to the barrier function. On the other hand, the standard implementations of interior point methods do not rely on the conditioning of the Hessian and can exploit efficiently its sparsity pattern to simplify LU decomposition. Hence there are also many problems that can be solved more efficiently by the interior point methods, and our approach may be considered as complementary to them.

Contact Problems and Scalable Algorithms

The development of the algorithms presented in this book was motivated by an effort to solve the large sparse problems arising from the discretization of elliptic variational inequalities, such as those describing the equilibrium of elastic bodies in mutual contact. A simple academic example is the contact problem of elasticity to describe the deformation and contact pressure due to volume forces of the cantilever cube over the obstacle in Fig. 0.1.

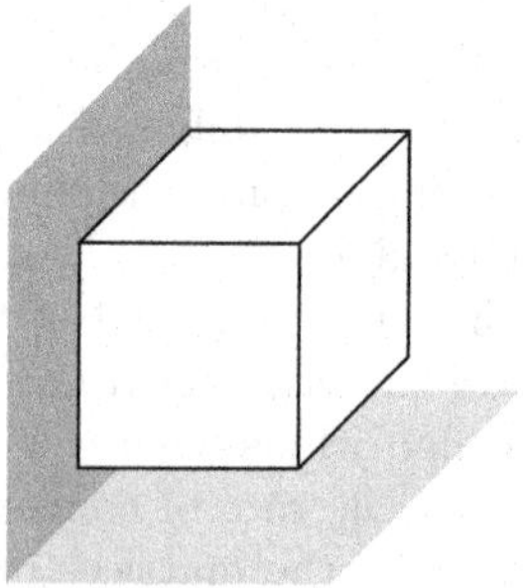

Fig. 0.1. Cantilever cube over the obstacle

The class of problems arising from various discretizations of a given variational inequality by the finite element or boundary element method can be reduced to the class of QP problems with a uniformly bounded spectrum by an application of the FETI/BETI (Finite/Boundary Element Tearing and Interconnecting)-based domain decomposition methods. Let us recall that the basic idea of these methods is to decompose the domain into subdomains as in Fig. 0.2 and then "glue" them by the Lagrange multipliers that are found by an iterative procedure.

Combination of the results on scalability of variants of the FETI methods for unconstrained problems with the algorithms presented in this book resulted in development of *scalable* algorithms for elliptic boundary variational inequalities. Let us recall that an algorithm is numerically scalable if the cost of the solution is nearly proportional to the number of unknowns, and it enjoys the parallel scalability if the time required for the solution can be reduced nearly proportionally to the number of available processors. For example, the solution of our toy problem required from 111 to 133 sparse matrix multiplications for varying discretizations with the number of nodes on the surface ranging from 417 to 163275.

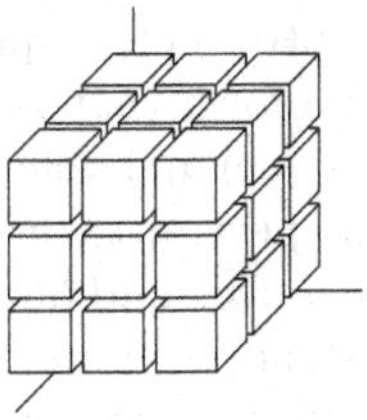

Fig. 0.2. Decomposed cube

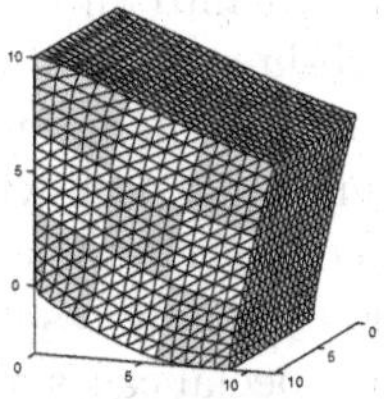

Fig. 0.3. Solution

As a more realistic example, let us consider the problem to describe the deformation and contact pressure in the ball bearings in Fig. 0.4. We can easily recognize that it comprises several bodies – balls, rings, and cages. The balls are not fixed in their cages, so that their stiffness matrices are necessarily singular and the discretized nonpenetration conditions can be described naturally by dependent constraints. Though the displacements and forces are typically given on parts of the surfaces of some bodies, exact places where the deformed balls come into contact with the cages or the rings are known only after the problem is solved.

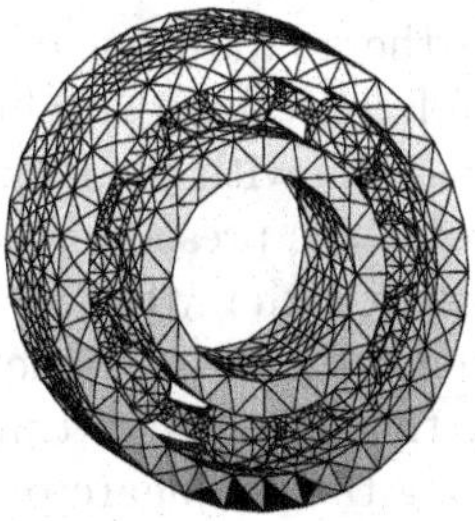

Fig. 0.4. Ball bearings

It should not be surprising that the duality-based methods can be more successful for the solution of variational inequalities than for the linear problems. The duality turns the general inequality constraints into bound constraints for free; the aspect not exploited in the solution of linear problems. The first fully scalable algorithm for numerical solution of linear problems, FETI, was introduced only in the early 1990s. It was quite challenging to get similar results for variational inequalities. Since the cost of the solution of a linear problem is proportional to the number of variables, a scalable algorithm must identify the active constraits in a sense for free!

Synopsis of the Book

The book is arranged into three parts. We start the introductory part by reviewing some well-known facts on linear algebra in the form that is useful in the following analysis, including less standard estimates, matrix decompositions of semidefinite matrices with known kernel, and spectral theory. The results are then used in the review of standard results on convex and quadratic programming. Though many results concerning the existence and uniqueness of QP problems are special cases of more general theory of nonlinear programming, it is often possible to develop more straightforward proofs that exploit specific structure of the QP problems, in particular the three-term Taylor's expansion, and sometimes to get stronger results. We paid special attention to the results for dependent constraints and/or positive semidefinite Hessian, including the sensitivity analysis and the duality theory in Sect. 2.6.5.

The second part is the core of the book and comprises four sections on the algorithms for specific types of constraints. It starts with Chap. 3 which summarizes the basic facts on the application of the conjugate gradient method to unconstrained QP problems. The material included is rather standard, possibly except Sect. 3.7 on the preconditioning by a conjugate projector.

Chapter 4 reviews in detail the Uzawa-type algorithms. A special attention is paid to the quantitative analysis of the penalty method and of an inexact solution of auxiliary unconstrained problems. The standard results on exact algorithms are also included. A kind of optimality is proved for a variant of the inexact penalty method and for the semimonotonic augmented Lagrangian algorithm SMALE. A bound on the penalty parameter which guarantees the linear convergence is also presented.

Chapter 5 describes the adaptations of the conjugate gradient algorithm to the solution of bound constrained problems. The algorithms include the variants of Polyak's algorithm with the inexact solution of auxiliary problems and the precision control which preserves the finite termination property. The main result of this chapter is the MPRGP algorithm with the linear rate of convergence which depends on the extreme eigenvalues of the Hessian of the cost function. We show that the rate of convergence can be improved by the preconditioning exploiting the conjugate projectors.

The last chapter of the second part combines Chaps. 4 and 5 to obtain optimal convergence results for the SMALBE algorithm for the solution of bound and equality constrained QP problems.

The performance of the representative algorithms of the second part is illustrated in each chapter by numerical experiments. We chose the benchmarks arising from the discretization of the energy functions associated with the Laplace operator to mimic typical applications. The benchmarks involve in each chapter one ill-conditioned problem to illustrate the typical performance of our algorithms in such situation and the class of well-conditioned problems to demonstrate the optimality of the best algorithms. Using the same cost functions in all benchmarks of the second part in combination with the boundary inequalities and multipoint constraints enables additional comparison. For convenience of the reader, Chaps. 3–5 are introduced by an overview of the algorithms presented there.

The concept of optimality is fully exploited in the last part of our book, where the algorithms of Chaps. 5 and 6 are combined with the FETI–DP (Dual–Primal FETI) and TFETI (Total FETI) methods to develop theoretically supported scalable algorithms for numerical solution of the classes of problems arising from the discretization of elliptic boundary variational inequalities. The numerical and parallel scalability is demonstrated on the solution of a coercive boundary variational inequality and on the solution of a semicoercive multidomain problem with more then two million nodal variables. The application of the algorithms presented in the last part of our book to the solution of contact problems of elasticity in two and three dimensions, including the contact problems with friction, is straightforward. The same is true for the applications of the algorithms to the development of scalable BETI-based algorithms for the solution of contact problems discretized by the direct boundary element methods. An interested reader can find the references at the end of the last two chapters.

Acknowledgments

Most of the nonstandard results presented in this book have been found by the author over the last fifteen years, often in cooperation with other colleagues. I would like to acknowledge here my thanks especially to Ana Friedlander and Mario Martínez for proper assessment of the efficiency of the augmented Lagrangian methods, to Sandra A. Santos and F.A.M. Gomes for their share in early development of our algorithms for the solution of variational inequalities, to Joachim Schöberl for sharing his original insight into the gradient projection method, to Dan Stefanica for joint work on scalable FETI–DP methods, especially for the proofs of optimal estimates without preconditioners, and to Charbel Farhat for drawing attention to practical aspects of our algorithms and an inspiration for thinking twice about simple topics.

The first results on optimal algorithms were presented at the summer schools organized by Ivo Marek, whose encouragement was essential in the decision to write this book.

My thanks go also to my colleagues and students from the Faculty of Electrical Engineering and Computer Science of VŠB–Technical University of Ostrava. Vít Vondrák implemented the first versions of the algorithms to the solution of contact problems of mechanics and shape optimization, David Horák first implemented many variants of the algorithms that appear in this book, and Dalibor Lukáš adapted the algorithms of Chap. 4 to the solution of the Stokes problem. My thanks go also to Marta Domorádová for her share in research of conjugate projectors and assistance with figures, to Radek Kučera who adapted the algorithms for bound constrained QP to the solution of more general problems with separated constraints and carried out a lot of joint work, and to Petr Beremlijski, Tomáš Kozubek, and Oldřich Vlach who applied at least some of these algorithms to the solution of engineering benchmarks. The book would be much worse without critical reading of its early versions by the colleagues mentioned above and especially by Marie Sadowská, who also participated with Jiří Bouchala in development of scalable algorithms for the problems discretized by boundary elements. Marta Domorádová, Marie Sadowská, Dalibor Lukáš, and David Horák kindly assisted with numerical experiments. There would be more errors in English if it were not for Barunka Dostálová.

It was a pleasure to work on the book with the publication staff at Springer. I am especially grateful to Elizabeth Loew and Frank Ganz for their share in final refinements of the book.

I gratefully acknowledge the support by the grants of the Ministry of Education of the Czech Republic No. MSM6198910027, GA CR 201/07/0294, and AS CR 1ET400300415. Last, but not least, my thanks go to the VŠB-Technical University of Ostrava and to the Faculty of Electrical Engineering and Computer Science for supporting the research of the whole group when needed.

The book is inscribed to my closest family, who have never complained much when my mind switched to quadratic forms. I would have hardly finished the book without the kind support of my wife Maruška.

Ostrava and Dolní Bečva

Zdeněk Dostál
August 2008

Contents

Part I

Background

1

Linear Algebra

The purpose of this chapter is to briefly review definitions, notations, and results of linear algebra that are used in the rest of our book. A few results especially developed for analysis of our algorithms are also included. There is no claim of completeness as the reader is assumed to be familiar with basic concepts of the college linear algebra such as vector spaces, linear mappings, matrix decompositions, etc. More systematic exposition and additional material can be found in the books by Strang [171], Hager [112], Demmel [31], Golub and Van Loan [103], Saad [163], and Axelsson [4]. We use without any reference basic concepts and standard results of analysis as they are reviewed in the books by Bertsekas [12] or Conn, Gould, and Toint [28].

1.1 Vectors

In this book we work with *n-dimensional arithmetic vectors* $\mathbf{v} \in \mathbb{R}^n$, where $\mathbb{R}$ denotes the set of real numbers. The only exception is Sect. 1.8, where vectors with complex entries are considered. We denote the ith component of an arithmetic vector $\mathbf{v} \in \mathbb{R}^n$ by $[\mathbf{v}]_i$. Thus $[\mathbf{v}]_i = v_i$ if $\mathbf{v} = [v_i]$ is defined by its components v_i. All the arithmetic vectors are considered by default to be column vectors. The relations between vectors $\mathbf{u}, \mathbf{v} \in \mathbb{R}^n$ are defined componentwise. Thus $\mathbf{u} = \mathbf{v}$ is equivalent to $[\mathbf{u}]_i = [\mathbf{v}]_i$, $i = 1, \dots, n$, and $\mathbf{u} \le \mathbf{v}$ is equivalent to $[\mathbf{u}]_i \le [\mathbf{v}]_i$, $i = 1, \dots, n$. We sometimes call the elements of $\mathbb{R}^n$ *points* to indicate that the concepts of length and direction are not important.

Having arithmetic vectors $\mathbf{u}, \mathbf{v} \in \mathbb{R}^n$ and a scalar $\alpha \in \mathbb{R}$, we define the *addition* and *multiplication by scalar* componentwise by

$$[\mathbf{u} + \mathbf{v}]_i = [\mathbf{u}]_i + [\mathbf{v}]_i \text{ and } [\alpha\mathbf{v}]_i = \alpha[\mathbf{v}]_i, \quad i = 1, \dots, n.$$

The rules that govern these operations, such as associativity, may be easily deduced from the related rules for computations with real numbers.

Zdeněk Dostál, *Optimal Quadratic Programming Algorithms*,
Springer Optimization and Its Applications, DOI 10.1007/978-0-387-84806-8_1,

The vector analog of $0 \in \mathbb{R}$ is the *zero vector* $\mathbf{o}_n \in \mathbb{R}^n$ with all the entries equal to zero. When the dimension can be deduced from the context, possibly using the assumption that all the expressions in our book are well defined, we often drop the subscript and write simply $\mathbf{o}$.

A nonempty set $\mathcal{V} \subseteq \mathbb{R}^n$ with the operations defined above is a *vector space* if $\alpha \in \mathbb{R}$ and $\mathbf{u}, \mathbf{v} \in \mathcal{V}$ imply $\mathbf{u} + \mathbf{v} \in \mathcal{V}$ and $\alpha\mathbf{u} \in \mathcal{V}$. In particular, both $\mathbb{R}^n$ and $\{\mathbf{o}\}$ are vector spaces. Given vectors $\mathbf{v}_1, \dots, \mathbf{v}_k \in \mathbb{R}^n$, the set

$$\mathrm{Span}\{\mathbf{v}_1, \dots, \mathbf{v}_k\} = \{\mathbf{v} \in \mathbb{R}^n : \ \mathbf{v} = \alpha_1\mathbf{v}_1 + \dots + \alpha_k\mathbf{v}_k, \ \alpha_i \in \mathbb{R}\}$$

is a vector space called the *linear span* of $\mathbf{v}_1, \dots, \mathbf{v}_k$. If $\mathcal{U}$ and $\mathcal{V}$ are vector spaces, then the sets $\mathcal{U} \cap \mathcal{V}$ and

$$\mathcal{U} + \mathcal{V} = \{\mathbf{x} + \mathbf{y} : \ \mathbf{x} \in \mathcal{U} \ \text{ and } \ \mathbf{y} \in \mathcal{V}\}$$

are also vector spaces. If $\mathcal{W} = \mathcal{U} + \mathcal{V}$ and $\mathcal{U} \cap \mathcal{V} = \{\mathbf{o}\}$, then $\mathcal{W}$ is said to be the *direct sum* of $\mathcal{U}$ and $\mathcal{V}$. We denote it by

$$\mathcal{W} = \mathcal{U} \oplus \mathcal{V}.$$

If $\mathcal{U}, \mathcal{V} \subseteq \mathbb{R}^n$ are vector spaces and $\mathcal{U} \subseteq \mathcal{V}$, then $\mathcal{U}$ is a *subspace* of $\mathcal{V}$.

A vector space $\mathcal{V} \subseteq \mathbb{R}^n$ can be spanned by different sets of vectors. A finite set of vectors $\mathcal{E} \subset \mathbb{R}^n$ that spans a given vector space $\mathcal{V} \neq \{\mathbf{o}\}$ is called a *basis* of $\mathcal{V}$ if no proper subset of $\mathcal{E}$ spans $\mathcal{V}$. For example, the set of vectors

$$\mathcal{S} = \{\mathbf{s}_1, \dots, \mathbf{s}_n\}, \quad [\mathbf{s}_i]_j = \delta_{ij}, \quad i, j = 1, \dots, n,$$

where δ_{ij} denotes the *Kronecker symbol* defined by $\delta_{ij} = 1$ for $i = j$ and $\delta_{ij} = 0$ for $i \neq j$, is the *standard basis* of $\mathbb{R}^n$. If $\mathcal{E} = \{\mathbf{e}_1, \dots, \mathbf{e}_d\}$ is a basis of a vector space $\mathcal{V}$, then $\mathcal{E}$ is *independent*, that is,

$$\alpha_1\mathbf{e}_1 + \dots + \alpha_d\mathbf{e}_d = \mathbf{o}$$

implies

$$\alpha_1 = \dots = \alpha_d = 0.$$

Any two bases of a vector space $\mathcal{V}$ have the same number of vectors. We call it the *dimension* of $\mathcal{V}$ and denote it $\dim\mathcal{V}$. Obviously $\dim\mathbb{R}^n = n$ and $\dim\mathcal{V} \leq n$ for any subspace $\mathcal{V} \subseteq \mathbb{R}^n$. For convenience, we define $\dim\{\mathbf{o}\} = 0$.

We sometimes use the componentwise extensions of scalar functions to vectors. Thus if $\mathbf{v} \in \mathbb{R}^n$, then $\mathbf{v}^+$ and $\mathbf{v}^-$ are the vectors whose ith components are $\max\{[\mathbf{v}]_i, 0\}$ and $\min\{[\mathbf{v}]_i, 0\}$, respectively. Similarly, if $\mathbf{u}, \mathbf{v} \in \mathbb{R}^n$, then $\max\{\mathbf{u}, \mathbf{v}\}$ and $\min\{\mathbf{u}, \mathbf{v}\}$ denote the vectors whose ith components are $\max\{[\mathbf{u}]_i, [\mathbf{v}]_i\}$ and $\min\{[\mathbf{u}]_i, [\mathbf{v}]_i\}$, respectively.

If $\mathcal{I}$ is a nonempty subset of $\{1, \dots, n\}$ and $\mathbf{v} \in \mathbb{R}^n$, then we denote by $[\mathbf{v}]_\mathcal{I}$ or simply $\mathbf{v}_\mathcal{I}$ the subvector of $\mathbf{v}$ with components $[\mathbf{v}]_i$, $i \in \mathcal{I}$. Thus if $\mathcal{I}$ has m elements, then $\mathbf{v}_\mathcal{I} \in \mathbb{R}^m$, so that we can refer to the components of $\mathbf{v}_\mathcal{I}$ either by the *global indices* $i \in \mathcal{I}$ or by the *local indices* $j \in \{1, \dots, m\}$. We usually rely on the reader's judgment to recognize the appropriate type of indexing.

1.2 Matrices and Matrix Operations

Throughout the whole book, all the matrices are assumed to be real except Sect. 1.8, where also complex matrices are considered. Similarly to the related convention for vectors, the (i,j)th component of a matrix $\mathsf{A} \in \mathbb{R}^{m\times n}$ is denoted by $[\mathsf{A}]_{ij}$, so that $[\mathsf{A}]_{ij} = a_{ij}$ for $\mathsf{A} = [a_{ij}]$ which is defined by its entries a_{ij}. A matrix $\mathsf{A} \in \mathbb{R}^{m\times n}$ is called an (m,n)-*matrix*; a matrix $\mathsf{A} \in \mathbb{R}^{n\times n}$ is called a *square matrix* of the order n.

Having (m,n)-matrices A, B and a scalar $\alpha \in \mathbb{R}$, we define *addition* and *multiplication by a scalar* by

$$[\mathsf{A}+\mathsf{B}]_{ij} = [\mathsf{A}]_{ij} + [\mathsf{B}]_{ij} \quad \text{and} \quad [\alpha\mathsf{A}]_{ij} = \alpha[\mathsf{A}]_{ij}.$$

The rules that govern the addition of matrices and their multiplication by scalars are the same as those for corresponding vector operations.

The matrix analog of 0 is the *zero matrix* $\mathsf{O}_{mn} \in \mathbb{R}^{m\times n}$ with all the entries equal to zero. When the dimension is clear from the context, we often drop the subscripts and write simply O.

Having matrices $\mathsf{A} \in \mathbb{R}^{m\times k}$ and $\mathsf{B} \in \mathbb{R}^{k\times n}$, we define their *product* $\mathsf{AB} \in \mathbb{R}^{m\times n}$ by

$$[\mathsf{AB}]_{ij} = \sum_{l=1}^{k} [\mathsf{A}]_{il}[\mathsf{B}]_{lj}.$$

Matrix multiplication is associative, therefore we do not need to use brackets to specify the order of multiplication. In particular, given a positive integer k and a square matrix A, we can define the kth *power of a square matrix* A by

$$\mathsf{A}^k = \underbrace{\mathsf{AA}\ldots\mathsf{A}}_{k\text{-times}}.$$

Matrix multiplication is not commutative.

The matrix counterpart of $1 \in \mathbb{R}$ in $\mathbb{R}^{n\times n}$ is the *identity matrix* $\mathsf{I}_n = [\delta_{ij}]$ of the order n. When the dimension may be deduced from the context, we often drop the subscripts and write simply I. Thus we can write

$$\mathsf{A} = \mathsf{IA} = \mathsf{AI}$$

for any matrix A, having in mind that the order of I on the left may be different from that on the right.

Given $\mathsf{A} \in \mathbb{R}^{m\times n}$, we define the *transposed matrix* $\mathsf{A}^T \in \mathbb{R}^{n\times m}$ to A by $[\mathsf{A}^T]_{ij} = [\mathsf{A}]_{ji}$. Having matrices $\mathsf{A} \in \mathbb{R}^{m\times k}$ and $\mathsf{B} \in \mathbb{R}^{k\times n}$, it may be checked that

$$(\mathsf{AB})^T = \mathsf{B}^T\mathsf{A}^T. \tag{1.1}$$

A square matrix A is *symmetric* if $\mathsf{A} = \mathsf{A}^T$.

A matrix A is *positive definite* if $\mathbf{x}^T\mathsf{A}\mathbf{x} > 0$ for any $\mathbf{x} \neq \mathbf{o}$, *positive semidefinite* if $\mathbf{x}^T\mathsf{A}\mathbf{x} \geq 0$ for any $\mathbf{x}$, and *indefinite* if neither A nor $-\mathsf{A}$ is positive definite or semidefinite. We are especially interested in symmetric positive definite (SPD) matrices.

If $\mathsf{A} \in \mathbb{R}^{m \times n}$, $\mathcal{I} \subseteq \{1, \dots, m\}$, and $\mathcal{J} \subseteq \{1, \dots, n\}$, $\mathcal{I}$ and $\mathcal{J}$ nonempty, we denote by $\mathsf{A}_{\mathcal{I}\mathcal{J}}$ the submatrix of A with the components $[\mathsf{A}]_{ij}$, $i \in \mathcal{I}$, $j \in \mathcal{J}$. The local indexing of the entries of $\mathsf{A}_{\mathcal{I}\mathcal{J}}$ is used whenever it is convenient in a similar way as the local indexing of subvectors which was introduced in Sect. 1.1. The full set of indices may be replaced by $*$ so that $\mathsf{A} = \mathsf{A}_{**}$ and $\mathsf{A}_{\mathcal{I}*}$ denotes the submatrix of A with the row indices belonging to $\mathcal{I}$.

Sometimes it is useful to rearrange the matrix operations into manipulations with submatrices of given matrices called blocks. A *block matrix* $\mathsf{A} \in \mathbb{R}^{m \times n}$ is defined by its blocks $\mathsf{A}_{ij} = \mathsf{A}_{\mathcal{I}_i \mathcal{J}_j}$, where $\mathcal{I}_i$ and $\mathcal{J}_j$ denote nonempty contiguous sets of indices decomposing $\{1, \dots, m\}$ and $\{1, \dots, n\}$, respectively. We can use the block structure to implement matrix operations only when the block structure of the involved matrices matches.

Very large matrices are often *sparse* in the sense that they have a small number of nonzero entries distributed in a pattern which can be exploited to the efficient implementation of matrix operations, to the reduction of storage requirements, or to the effective solution of standard problems of linear algebra. Such matrices arise, e.g., from the discretization of problems described by differential operators. The matrices with a large number of nonzero entries are often called *full* or *dense matrices.*

1.3 Matrices and Mappings

Each matrix $\mathsf{A} \in \mathbb{R}^{m \times n}$ defines the mapping which assigns to each $\mathbf{x} \in \mathbb{R}^n$ the vector $\mathsf{A}\mathbf{x} \in \mathbb{R}^m$. Two important subspaces associated with this mapping are its *range* or *image space* $\mathrm{Im}\mathsf{A}$ and its *kernel* or *null space* $\mathrm{Ker}\mathsf{A}$; they are defined by

$$\mathrm{Im}\mathsf{A} = \{\mathsf{A}\mathbf{x} : \ \mathbf{x} \in \mathbb{R}^n\} \quad \text{and} \quad \mathrm{Ker}\mathsf{A} = \{\mathbf{x} \in \mathbb{R}^n : \ \mathsf{A}\mathbf{x} = \mathbf{o}\}.$$

The range of A is the span of its columns.

If f is a mapping defined on $\mathcal{D} \subseteq \mathbb{R}^n$ and $\Omega \subseteq \mathcal{D}$, then $f|\Omega$ denotes the *restriction* of f to Ω, that is, the mapping defined on Ω which assigns to each $\mathbf{x} \in \Omega$ the value $f(\mathbf{x})$. If $\mathsf{A} \in \mathbb{R}^{m \times n}$ and $\mathcal{V}$ is a subspace of $\mathbb{R}^n$, we define $\mathsf{A}|\mathcal{V}$ as a restriction of the mapping associated with A to $\mathcal{V}$. The restriction $\mathsf{A}|\mathcal{V}$ is said to be positive definite if $\mathbf{x}^T\mathsf{A}\mathbf{x} > 0$ for $\mathbf{x} \in \mathcal{V}$, $\mathbf{x} \neq \mathbf{o}$, and positive semidefinite if $\mathbf{x}^T\mathsf{A}\mathbf{x} \geq 0$ for $\mathbf{x} \in \mathcal{V}$.

The mapping associated with A is *injective* if $\mathsf{A}\mathbf{x} = \mathsf{A}\mathbf{y}$ implies $\mathbf{x} = \mathbf{y}$. It is easy to check that the mapping associated with A is injective if and only if $\mathrm{Ker}\mathsf{A} = \{\mathbf{o}\}$. More generally, it may be proved that

$$\dim \mathrm{Im}\mathsf{A} + \dim \mathrm{Ker}\mathsf{A} = n \tag{1.2}$$

for any $\mathsf{A} \in \mathbb{R}^{m \times n}$. If $m = n$, then A is injective if and only if $\mathrm{Im}\mathsf{A} = \mathbb{R}^n$.

The *rank* or *column rank* of a matrix A is equal to the dimension of the range of A. The *column rank* is known to be equal to the *row rank*, the number of linearly independent rows. A matrix is of *full row rank* or *full column rank* when its rank is equal to the number of its rows or columns, respectively. A matrix $\mathsf{A} \in \mathbb{R}^{m\times n}$ is of *full rank* when its rank is the smaller of m and n.

A subspace $\mathcal{V} \subseteq \mathbb{R}^n$ which satisfies

$$\mathsf{A}\mathcal{V} = \{\mathsf{A}\mathbf{x} : \ \mathbf{x} \in \mathcal{V}\} \subseteq \mathcal{V}$$

is an *invariant subspace* of A. Obviously

$$\mathsf{A}(\mathrm{Im}\mathsf{A}) \subseteq \mathrm{Im}\mathsf{A},$$

so that $\mathrm{Im}\mathsf{A}$ is an invariant subspace of A.

A *projector* is a square matrix P that satisfies

$$\mathsf{P}^2 = \mathsf{P}.$$

Such a matrix is also said to be *idempotent.* A vector $\mathbf{x} \in \mathrm{Im}\mathsf{P}$ if and only if there is $\mathbf{y} \in \mathbb{R}^n$ such that $\mathbf{x} = \mathsf{P}\mathbf{y}$, so that

$$\mathsf{P}\mathbf{x} = \mathsf{P}(\mathsf{P}\mathbf{y}) = \mathsf{P}\mathbf{y} = \mathbf{x}.$$

If P is a projector, then $\mathsf{Q} = \mathsf{I} - \mathsf{P}$ and P^T are also projectors as

$$(\mathsf{I} - \mathsf{P})^2 = \mathsf{I} - 2\mathsf{P} + \mathsf{P}^2 = \mathsf{I} - \mathsf{P} \quad \text{and} \quad \left(\mathsf{P}^T\right)^2 = \left(\mathsf{P}^2\right)^T = \mathsf{P}^T.$$

Since for any $\mathbf{x} \in \mathbb{R}^n$

$$\mathbf{x} = \mathsf{P}\mathbf{x} + (\mathsf{I} - \mathsf{P})\mathbf{x},$$

it simply follows that $\mathrm{Im}\mathsf{Q} = \mathrm{Ker}\mathsf{P}$,

$$\mathbb{R}^n = \mathrm{Im}\mathsf{P} + \mathrm{Ker}\mathsf{P}, \quad \text{and} \quad \mathrm{Ker}\mathsf{P} \cap \mathrm{Im}\mathsf{P} = \{\mathbf{o}\}.$$

We say that P is a projector onto $\mathcal{U} = \mathrm{Im}\mathsf{P}$ along $\mathcal{V} = \mathrm{Ker}\mathsf{P}$ and Q is a complementary projector onto $\mathcal{V}$ along $\mathcal{U}$. The above relations may also be rewritten as

$$\mathrm{Im}\mathsf{P} \oplus \mathrm{Ker}\mathsf{P} = \mathbb{R}^n. \tag{1.3}$$

Let $(\pi(1), \ldots, \pi(n))$ be a permutation of numbers $1, \ldots, n$. Then the mapping which assigns to each $\mathbf{v} = [v_i] \in \mathbb{R}^n$ a vector $[v_{\pi(1)}, \ldots, v_{\pi(n)}]^T$ is associated with the *permutation matrix*

$$\mathsf{P} = [\mathbf{s}_{\pi(1)}, \ldots, \mathbf{s}_{\pi(n)}],$$

where $\mathbf{s}_i$ denotes the ith column of the identity matrix I_n. If P is a permutation matrix, then

$$\mathsf{P}\mathsf{P}^T = \mathsf{P}^T\mathsf{P} = \mathsf{I}.$$

Notice that if B is a matrix obtained from a matrix A by reordering of the rows of A, then there is a permutation matrix P such that $\mathsf{B} = \mathsf{P}\mathsf{A}$. Similarly, if B is a matrix obtained from A by reordering of the columns of A, then there is a permutation matrix P such that $\mathsf{B} = \mathsf{A}\mathsf{P}$.

1.4 Inverse and Generalized Inverse Matrices

If A is a square full rank matrix, then there is the unique *inverse matrix* A^{-1} such that

$$\mathsf{AA}^{-1} = \mathsf{A}^{-1}\mathsf{A} = \mathsf{I}. \tag{1.4}$$

The mapping associated with A^{-1} is inverse to that associated with A.

If A^{-1} exists, we say that A is *nonsingular*. A square matrix is *singular* if its inverse matrix does not exist. Any positive definite matrix is nonsingular. If P is a permutation matrix, then P is nonsingular and

$$\mathsf{P}^{-1} = \mathsf{P}^T.$$

If A is a nonsingular matrix, then $\mathsf{A}^{-1}\mathbf{b}$ is the unique solution of the system of linear equations $\mathsf{A}\mathbf{x} = \mathbf{b}$.

If A is a nonsingular matrix, then we can transpose (1.4) and use (1.1) to get

$$(\mathsf{A}^{-1})^T\mathsf{A}^T = \mathsf{A}^T(\mathsf{A}^{-1})^T = \mathsf{I},$$

so that

$$(\mathsf{A}^T)^{-1} = (\mathsf{A}^{-1})^T. \tag{1.5}$$

It follows that if A is symmetric, then A^{-1} is symmetric.

If $\mathsf{A} \in \mathbb{R}^{n\times n}$ is positive definite, then A^{-1} is also positive definite, as any vector $\mathbf{x} \neq \mathbf{o}$ can be expressed as $\mathbf{x} = \mathsf{A}\mathbf{y}$, $\mathbf{y} \neq \mathbf{o}$, and

$$\mathbf{x}^T\mathsf{A}^{-1}\mathbf{x} = (\mathsf{A}\mathbf{y})^T\mathsf{A}^{-1}\mathsf{A}\mathbf{y} = \mathbf{y}^T\mathsf{A}^T\mathbf{y} = \mathbf{y}^T\mathsf{A}\mathbf{y} > 0.$$

If A and B are nonsingular matrices, then it is easy to check that AB is also nonsingular and

$$(\mathsf{AB})^{-1} = \mathsf{B}^{-1}\mathsf{A}^{-1}.$$

If $\mathsf{U}, \mathsf{V} \in \mathbb{R}^{m\times n}$, $m < n$, and A, $\mathsf{A} + \mathsf{U}^T\mathsf{V}$ are nonsingular, then it can be verified directly that

$$(\mathsf{A} + \mathsf{U}^T\mathsf{V})^{-1} = \mathsf{A}^{-1} - \mathsf{A}^{-1}\mathsf{U}^T(\mathsf{I} + \mathsf{VA}^{-1}\mathsf{U}^T)^{-1}\mathsf{VA}^{-1}. \tag{1.6}$$

The formula (1.6) is known as *Sherman–Morrison–Woodbury's formula* (see [103, p. 51]). The formula is useful in theory and for evaluation of the inverse matrix to a low rank perturbation of A provided A^{-1} is known.

If $\mathsf{B} \in \mathbb{R}^{m\times n}$ denotes a full row rank matrix, $\mathsf{A} \in \mathbb{R}^{n\times n}$ is positive definite, and $\mathbf{y} \neq \mathbf{o}$, then $\mathbf{z} = \mathsf{B}^T\mathbf{y} \neq \mathbf{o}$ and

$$\mathbf{y}^T\mathsf{BA}^{-1}\mathsf{B}^T\mathbf{y} = \mathbf{z}^T\mathsf{A}^{-1}\mathbf{z} > 0.$$

Thus if A is positive definite and B is a full row rank matrix such that $\mathsf{BA}^{-1}\mathsf{B}^T$ is well defined, then the latter matrix is also positive definite.

A real matrix $\mathsf{A} = [a_{ij}] \in \mathbb{R}^{n\times n}$ is called a (nonsingular) *M-matrix* if $a_{ij} \leq 0$ for $i \neq j$ and if all entries of A^{-1} are nonnegative. If

$$a_{ii} > \sum_{j \neq i}^{n} |a_{ij}|, \quad i = 1, \ldots, n,$$

then A is an M-matrix (see Fiedler and Pták [88] or Axelsson [4, Chap. 6]).

If $\mathsf{A} \in \mathbb{R}^{m \times n}$ and $\mathbf{b} \in \mathrm{Im}\mathsf{A}$, then we can express a solution of the system of linear equations $\mathsf{A}\mathbf{x} = \mathbf{b}$ by means of a *left generalized inverse matrix* $\mathsf{A}^+ \in \mathbb{R}^{n \times m}$ which satisfies $\mathsf{A}\mathsf{A}^+\mathsf{A} = \mathsf{A}$. Indeed, if $\mathbf{b} \in \mathrm{Im}\mathsf{A}$, then there is $\mathbf{y}$ such that $\mathbf{b} = \mathsf{A}\mathbf{y}$ and $\overline{\mathbf{x}} = \mathsf{A}^+\mathbf{b}$ satisfies

$$\mathsf{A}\overline{\mathbf{x}} = \mathsf{A}\mathsf{A}^+\mathbf{b} = \mathsf{A}\mathsf{A}^+\mathsf{A}\mathbf{y} = \mathsf{A}\mathbf{y} = \mathbf{b}.$$

Thus A^+ acts on the range of A like the inverse matrix. If A is a nonsingular square matrix, then obviously

$$\mathsf{A}^+ = \mathsf{A}^{-1}.$$

Moreover, if $\mathsf{S} \in \mathbb{R}^{n \times p}$ is such that $\mathsf{AS} = \mathsf{O}$ and $\mathsf{N} \in \mathbb{R}^{n \times p}$, then $(\mathsf{A}^+) + \mathsf{SN}^T$ is also a left generalized inverse as

$$\mathsf{A}\left((\mathsf{A}^+) + \mathsf{SN}^T\right)\mathsf{A} = \mathsf{A}\mathsf{A}^+\mathsf{A} + \mathsf{ASN}^T\mathsf{A} = \mathsf{A}.$$

If A is a symmetric singular matrix, then there is a permutation matrix P such that

$$\mathsf{A} = \mathsf{P}^T \begin{bmatrix} \mathsf{B} & \mathsf{C}^T \\ \mathsf{C} & \mathsf{C}\mathsf{B}^{-1}\mathsf{C}^T \end{bmatrix} \mathsf{P},$$

where B is a nonsingular matrix whose dimension is equal to the rank of A. It may be verified directly that the matrix

$$\mathsf{A}^{\#} = \mathsf{P}^T \begin{bmatrix} \mathsf{B}^{-1} & \mathsf{O}^T \\ \mathsf{O} & \mathsf{O} \end{bmatrix} \mathsf{P} \tag{1.7}$$

is a left generalized inverse of A. If A is symmetric positive semidefinite, then $\mathsf{A}^{\#}$ is also symmetric positive semidefinite. Notice that if $\mathsf{AS} = \mathsf{O}$, then $\mathsf{A}^+ = \mathsf{A}^{\#} + \mathsf{SS}^T$ is also a symmetric positive semidefinite generalized inverse.

1.5 Direct Methods for Solving Linear Equations

The inverse matrix is a useful tool for theoretical developments, but not for computations, especially when sparse matrices are involved. The reason is that the inverse matrix is usually full, so that its evaluation results in large storage requirements and high computational costs. It is often much more efficient to implement the multiplication of a vector by the inverse matrix by solving the related system of linear equations. We recall here briefly the *direct methods*, which reduce solving of the original system of linear equations to solving of a system or systems of linear equations with triangular matrices.

A matrix $\mathsf{L} = [l_{ij}]$ is *lower triangular* if $l_{ij} = 0$ for $i < j$. It is easy to solve a system $\mathsf{L}\mathbf{x} = \mathbf{b}$ with the nonsingular lower triangular matrix $\mathsf{L} \in \mathbb{R}^n$. As there is only one unknown in the first equation, we can find it and then substitute it into the remaining equations to obtain a system with the same structure, but with only $n-1$ remaining unknowns. We can repeat the procedure until we find all the components of $\mathbf{x}$.

A similar procedure, but starting from the last equation, can be applied to a system with the nonsingular *upper triangular matrix* $\mathsf{U} = [u_{ij}]$ with $u_{ij} = 0$ for $i > j$.

The solution costs of a system with triangular matrices is proportional to the number of its nonzero entries. In particular, the solution of a system of linear equations with a *diagonal matrix* $\mathsf{D} = [d_{ij}]$, $d_{ij} = 0$ for $i \neq j$, reduces to the solution of a sequence of linear equations with one unknown.

If we are to solve the system of linear equations with a nonsingular matrix, we can use systematically *equivalent transformations* that do not change the solution in order to modify the original system to that with an upper triangular matrix. It is well-known that the solutions of a system of linear equations are the same as the solutions of a system of linear equations obtained from the original system by interchanging two equations, replacing an equation by its nonzero multiple, or adding a multiple of one equation to another equation. The *Gauss elimination* for the solution of a system of linear equations with a nonsingular matrix thus consists of two steps: the *forward reduction*, which exploits equivalent transformations to reduce the original system to the system with an upper triangular matrix, and the *backward substitution*, which solves the resulting system with the upper triangular matrix.

Alternatively, we can use suitable matrix factorizations. For example, it is well-known that any positive definite matrix A can be decomposed into the product

$$\mathsf{A} = \mathsf{L}\mathsf{L}^T, \tag{1.8}$$

where L is a nonsingular lower triangular matrix with positive diagonal entries. Having the decomposition, we can evaluate $\mathbf{z} = \mathsf{A}^{-1}\mathbf{x}$ by solving the systems

$$\mathsf{L}\mathbf{y} = \mathbf{x} \quad \text{and} \quad \mathsf{L}^T\mathbf{z} = \mathbf{y}.$$

The factorization-based solvers may be especially useful when we are to solve several systems of equations with the same coefficients but different right-hand sides coming one after another.

The method of evaluation of the factor L is known as the *Cholesky factorization*. The Cholesky factor L can be computed in a number of equivalent ways. For example, we may compute it column by column. Suppose that

$$\mathsf{A} = \begin{bmatrix} a_{11} & \mathbf{a}_1^T \\ \mathbf{a}_1 & \mathsf{A}_{22} \end{bmatrix} \quad \text{and} \quad \mathsf{L} = \begin{bmatrix} l_{11} & \mathbf{o} \\ \mathbf{l}_1 & \mathsf{L}_{22} \end{bmatrix}.$$

Substituting for A and L into (1.8) and comparing the corresponding terms immediately reveals that

$$l_{11} = \sqrt{a_{11}}, \qquad \mathbf{l}_1 = l_{11}^{-1}\mathbf{a}_1, \qquad \mathsf{L}_{22}\mathsf{L}_{22}^T = \mathsf{A}_{22} - \mathbf{l}_1\mathbf{l}_1^T. \tag{1.9}$$

This gives us the first column of L, and the remaining factor L_{22} is simply the Cholesky factor of the Schur complement $\mathsf{A}_{22} - \mathbf{l}_1\mathbf{l}_1^T$ which is known to be positive definite, so we can find its first column by the above procedure. The algorithm can be implemented to exploit a sparsity pattern of A, e.g., when $\mathsf{A} = [a_{ij}] \in \mathbb{R}^{n\times n}$ is a *band matrix* with $a_{ij} = 0$ for $|i-j| > b$, $b \ll n$.

If $\mathsf{A} \in \mathbb{R}^{n\times n}$ is only positive semidefinite, it can happen that $a_{11} = 0$. Then

$$0 \le \mathbf{x}^T\mathsf{A}\mathbf{x} = \mathbf{y}^T\mathsf{A}_{22}\mathbf{y} + 2x_1\mathbf{a}_1^T\mathbf{y}$$

for any vector $\mathbf{x} = \left[x_1, \mathbf{y}^T\right]^T$. The inequality implies that $\mathbf{a}_1 = \mathbf{o}$, as otherwise we could take $\mathbf{y} = -\mathbf{a}_1$ and large x_1 to get

$$\mathbf{y}^T\mathsf{A}_{22}\mathbf{y} + 2x_1\mathbf{a}_1^T\mathbf{y} = \mathbf{a}_1^T\mathsf{A}_{22}\mathbf{a}_1 - 2x_1\|\mathbf{a}_1\|^2 < 0.$$

Thus for A symmetric positive semidefinite and $a_{11} = 0$, (1.9) reduces to

$$l_{11} = 0, \qquad \mathbf{l}_1 = \mathbf{o}, \qquad \mathsf{L}_{22}\mathsf{L}_{22}^T = \mathsf{A}_{22}. \tag{1.10}$$

Of course, this simple modification assumes exact arithmetics. In the computer arithmetics, the decision whether a_{11} is to be treated as zero depends on some small $\varepsilon > 0$.

In some important applications, it is possible to exploit additional information. In mechanics, e.g., the basis of the kernel of the stiffness matrix of a floating body is formed by three (2D) or six (3D) known and independent rigid body motions. Any basis of the kernel of a matrix can be used to identify the zero rows (and columns) of a Cholesky factor by means of the following lemma.

Lemma 1.1. *Let* $\mathsf{A} = \mathsf{L}\mathsf{L}^T$ *denote a triangular decomposition of a symmetric positive semidefinite matrix* A*, let* $\mathsf{A}\mathbf{e} = \mathbf{o}$*, and let* $l(\mathbf{e})$ *denote the largest index of a nonzero entry of* $\mathbf{e} \in \mathrm{Ker}\mathsf{A}$*, so that*

$$[\mathbf{e}]_{l(\mathbf{e})} \neq 0 \quad \text{and} \quad [\mathbf{e}]_j = 0 \quad \text{for} \quad j > l(\mathbf{e}).$$

Then

$$[\mathsf{L}]_{l(\mathbf{e})l(\mathbf{e})} = 0.$$

Proof. If $\mathsf{A}\mathbf{e} = \mathbf{o}$, then

$$\mathbf{e}^T\mathsf{A}\mathbf{e} = \mathbf{e}^T\mathsf{L}\mathsf{L}^T\mathbf{e} = (\mathsf{L}^T\mathbf{e})^T(\mathsf{L}^T\mathbf{e}) = 0.$$

Thus $\mathsf{L}^T\mathbf{e} = \mathbf{o}$ and in particular

$$[\mathsf{L}^T\mathbf{e}]_{l(\mathbf{e})} = [\mathsf{L}]_{l(\mathbf{e})l(\mathbf{e})}[\mathbf{e}]_{l(\mathbf{e})} = 0.$$

Since $[\mathbf{e}]_{l(\mathbf{e})} \neq 0$, we have $[\mathsf{L}]_{l(\mathbf{e})l(\mathbf{e})} = 0$. □

Let $\mathsf{A} \in \mathbb{R}^{n\times n}$ be positive semidefinite and let $\mathsf{R} \in \mathbb{R}^{n\times d}$ denote a full column rank matrix such that $\mathrm{Ker}\mathsf{A} = \mathrm{Im}\mathsf{R}$. Observing that application of equivalent transformations to the columns of R preserves the image space and the rank of R, we can modify the forward reduction to find R which satisfies

$$l(\mathsf{R}_{*1}) < \cdots < l(\mathsf{R}_{*d}).$$

The procedure can be described by the following transformations of R: transpose R, reverse the order of columns, apply the forward reduction, reverse the order of columns back, and transpose the resulting matrix back. Then $l(\mathsf{R}_{*1}), \dots, l(\mathsf{R}_{*d})$ are by Lemma 1.1 the indices of zero columns of a factor of the modified Cholesky factorization; the factor cannot have any other zero columns due to the rank argument. The procedure has been described and tested in Menšík [151]. Denoting by the crosses and dots the nonzero and undetermined entries, respectively, the relations between the pivots of R and the zero columns of the Cholesky factor L can be illustrated by

$$\mathsf{R} = \begin{bmatrix} . & . \\ . & . \\ \times & . \\ 0 & . \\ 0 & \times \end{bmatrix} \quad \Rightarrow \quad \mathsf{L} = \begin{bmatrix} \times & 0 & 0 & 0 & 0 \\ . & \times & 0 & 0 & 0 \\ . & . & 0 & 0 & 0 \\ . & . & 0 & \times & 0 \\ . & . & 0 & . & 0 \end{bmatrix}.$$

Alternatively, we can combine the basic algorithm with a suitable rank revealing decomposition, such as the singular value decomposition (SVD) introduced in Sect. 1.9. For example, Frahat and Gérardin [82] proposed to start with the Cholesky decomposition and to switch to SVD in case of doubts.

1.6 Norms

General concepts of size and distance in a vector space are expressed by norms. A *norm* on $\mathbb{R}^n$ is a function which assigns to each $\mathbf{x} \in \mathbb{R}^n$ a number $\|\mathbf{x}\| \in \mathbb{R}$ in such a way that for any vectors $\mathbf{x}$, $\mathbf{y} \in \mathbb{R}^n$ and any scalar $\alpha \in \mathbb{R}$, the following three conditions are satisfied:

(i) $\|\mathbf{x}\| \geq 0$, and $\|\mathbf{x}\| = 0$ if and only if $\mathbf{x} = \mathbf{o}$.
(ii) $\|\mathbf{x} + \mathbf{y}\| \leq \|\mathbf{x}\| + \|\mathbf{y}\|$.
(iii) $\|\alpha\mathbf{x}\| = |\alpha|\ \|\mathbf{x}\|$.

It is easy to check that the functions

$$\|\mathbf{x}\|_1 = |\mathbf{x}_1| + \cdots + |\mathbf{x}_n| \quad \text{and} \quad \|\mathbf{x}\|_\infty = \max\{|\mathbf{x}_1|, \dots, |\mathbf{x}_n|\}$$

are norms. They are called ℓ_1 and ℓ_∞ norms, respectively. We often use the Euclidean norm defined by

$$\|\mathbf{x}\|_2 = \sqrt{\mathbf{x}_1^2 + \cdots + \mathbf{x}_n^2}.$$

The norms on $\mathbb{R}^n$ introduced above satisfy the inequalities

$$\|\mathbf{x}\|_\infty \le \|\mathbf{x}\|_2 \le \|\mathbf{x}\|_1 \le \sqrt{n}\|\mathbf{x}\|_2 \le n\|\mathbf{x}\|_\infty.$$

Given a norm defined on the domain and the range of a matrix A, we can define the *induced norm* $|\!|\!|\mathsf{A}|\!|\!|$ of A by

$$|\!|\!|\mathsf{A}|\!|\!| = \sup_{|\!|\!|\mathbf{x}|\!|\!|=1} |\!|\!|\mathsf{A}\mathbf{x}|\!|\!| = \sup_{\mathbf{x}\ne\mathbf{o}} \frac{|\!|\!|\mathsf{A}\mathbf{x}|\!|\!|}{|\!|\!|\mathbf{x}|\!|\!|}.$$

If $\mathsf{B} \ne \mathsf{O}$, then

$$|\!|\!|\mathsf{AB}|\!|\!| = \sup_{\mathbf{x}\ne\mathbf{o}} \frac{|\!|\!|\mathsf{AB}\mathbf{x}|\!|\!|}{|\!|\!|\mathbf{x}|\!|\!|} = \sup_{\mathsf{B}\mathbf{x}\ne\mathbf{o}} \frac{|\!|\!|\mathsf{AB}\mathbf{x}|\!|\!|}{|\!|\!|\mathsf{B}\mathbf{x}|\!|\!|}\frac{|\!|\!|\mathsf{B}\mathbf{x}|\!|\!|}{|\!|\!|\mathbf{x}|\!|\!|} \le \sup_{\substack{\mathbf{y}\in\mathrm{Im}\mathsf{B},\\ \mathbf{y}\ne\mathbf{o}}} \frac{|\!|\!|\mathsf{A}\mathbf{y}|\!|\!|}{|\!|\!|\mathbf{y}|\!|\!|} \sup_{\mathbf{x}\ne\mathbf{o}} \frac{|\!|\!|\mathsf{B}\mathbf{x}|\!|\!|}{|\!|\!|\mathbf{x}|\!|\!|}.$$

It follows easily that the induced norm is *submultiplicative*, i.e.,

$$|\!|\!|\mathsf{AB}|\!|\!| \le |\!|\!|\mathsf{A}|\mathrm{Im}\mathsf{B}|\!|\!|\ |\!|\!|\mathsf{B}|\!|\!| \le |\!|\!|\mathsf{A}|\!|\!|\ |\!|\!|\mathsf{B}|\!|\!|. \tag{1.11}$$

If $\mathsf{A} = [a_{ij}] \in \mathbb{R}^{m\times n}$ and $\mathbf{x} = [x_i] \in \mathbb{R}^n$, then

$$\|\mathsf{A}\mathbf{x}\|_\infty = \max_{i=1,\dots,m} \Big|\sum_{j=1}^n a_{ij}x_j\Big| \le \max_{i=1,\dots,m} \sum_{j=1}^n |a_{ij}||x_j| \le \|\mathbf{x}\|_\infty \max_{i=1,\dots,m} \sum_{j=1}^n |a_{ij}|,$$

that is, $\|\mathsf{A}\|_\infty \le \max_{i=1,\dots,m} \sum_{j=1}^n |a_{ij}|$. Since the last inequality turns into the equality for a vector $\mathbf{x}$ with suitably chosen entries $x_i \in \{1, -1\}$, we have

$$\|\mathsf{A}\|_\infty = \max_{i=1,\dots,m} \sum_{j=1}^n |a_{ij}|. \tag{1.12}$$

Similarly

$$\|\mathsf{A}\mathbf{x}\|_1 = \sum_{i=1}^m \Big|\sum_{j=1}^n a_{ij}x_j\Big| \le \sum_{j=1}^n |x_j| \sum_{i=1}^m |a_{ij}| \le \|\mathbf{x}\|_1 \max_{j=1,\dots,n} \sum_{i=1}^m |a_{ij}|,$$

that is, $\|\mathsf{A}\|_1 \le \max_{i=1,\dots,n} \sum_{i=1}^m |a_{ij}|$. Taking for the vector $\mathbf{x}$ a suitably chosen column of the identity matrix I_n, we get

$$\|\mathsf{A}\|_1 = \max_{j=1,\dots,n} \sum_{i=1}^m |a_{ij}| = \|\mathsf{A}^T\|_\infty. \tag{1.13}$$

The matrix norms induced by ℓ_1 and ℓ_∞ norms are relatively inexpensive to compute. If $\mathsf{A} \in \mathbb{R}^{m\times n}$, they may be used to estimate the typically expensive Euclidean norm $\|\mathsf{A}\|_2$ by means of the inequalities

$$\|\mathsf{A}\|_\infty \le \sqrt{n}\|\mathsf{A}\|_2 \le n\|\mathsf{A}\|_1 \le n\sqrt{m}\|\mathsf{A}\|_2 \le nm\|\mathsf{A}\|_\infty.$$

Another useful inequality is

$$\|\mathsf{A}\|_2 \le \sqrt{\|\mathsf{A}\|_1\|\mathsf{A}\|_\infty}. \tag{1.14}$$

1.7 Scalar Products

General concepts of length and angle in a vector space are introduced by means of a *scalar product*; it is the mapping which assigns to each couple $\mathbf{x},\ \mathbf{y} \in \mathbb{R}^n$ a number $\langle \mathbf{x}, \mathbf{y} \rangle \in \mathbb{R}$ in such a way that for any vectors $\mathbf{x},\ \mathbf{y},\ \mathbf{z} \in \mathbb{R}^n$ and any scalar $\alpha \in \mathbb{R}$, the following four conditions are satisfied:

(i) $\langle \mathbf{x}, \mathbf{y} + \mathbf{z} \rangle = \langle \mathbf{x}, \mathbf{y} \rangle + \langle \mathbf{x}, \mathbf{z} \rangle$.
(ii) $\langle \alpha\mathbf{x}, \mathbf{y} \rangle = \alpha \langle \mathbf{x}, \mathbf{y} \rangle$.
(iii) $\langle \mathbf{x}, \mathbf{y} \rangle = \langle \mathbf{y}, \mathbf{x} \rangle$.
(iv) $\langle \mathbf{x}, \mathbf{x} \rangle > 0$ for $\mathbf{x} \neq \mathbf{o}$.

We often use the *Euclidean scalar product* or *Euclidean inner product* which assigns to each couple of vectors $\mathbf{x},\ \mathbf{y} \in \mathbb{R}^n$ a number defined by

$$(\mathbf{x}, \mathbf{y}) = \mathbf{x}^T \mathbf{y}.$$

If A is a symmetric positive definite matrix, then we can define the more general A-*scalar product* on $\mathbb{R}^n$ by

$$(\mathbf{x}, \mathbf{y})_{\mathsf{A}} = \mathbf{x}^T \mathsf{A} \mathbf{y}.$$

Using a scalar product, we can define the norm $\|\mathbf{x}\|$ of $\mathbf{x}$ and the angle α between $\mathbf{x}$ and $\mathbf{y}$ by

$$\|\mathbf{x}\|^2 = \langle \mathbf{x}, \mathbf{x} \rangle, \quad \cos \alpha = \frac{\langle \mathbf{x}, \mathbf{y} \rangle}{\|\mathbf{x}\| \|\mathbf{y}\|}.$$

We denote for any $\mathbf{x} \in \mathbb{R}^n$ its *Euclidean norm* and A-*norm* by

$$\|\mathbf{x}\| = (\mathbf{x}, \mathbf{x})^{1/2}, \quad \|\mathbf{x}\|_{\mathsf{A}} = (\mathbf{x}, \mathbf{x})_{\mathsf{A}}^{1/2}.$$

It is easy to see that any norm induced by a scalar product satisfies the properties (i) and (iii) of the norm. The property (ii) follows from the *Cauchy–Schwarz inequality*

$$\langle \mathbf{x}, \mathbf{y} \rangle^2 \leq \|\mathbf{x}\|^2 \, \|\mathbf{y}\|^2, \tag{1.15}$$

which is valid for any $\mathbf{x},\ \mathbf{y} \in \mathbb{R}^n$ and any scalar product. The bound is tight in the sense that the inequality becomes the equality when $\mathbf{x},\ \mathbf{y}$ are dependent. The property (ii) of the norm then follows by

$$\|\mathbf{x} + \mathbf{y}\|^2 = \|\mathbf{x}\|^2 + 2\langle \mathbf{x}, \mathbf{y} \rangle + \|\mathbf{y}\|^2 \leq \|\mathbf{x}\|^2 + 2\|\mathbf{x}\|\|\mathbf{y}\| + \|\mathbf{y}\|^2 = (\|\mathbf{x}\| + \|\mathbf{y}\|)^2.$$

A pair of vectors $\mathbf{x}$ and $\mathbf{y}$ is *orthogonal* (with respect to a given scalar product) if

$$\langle \mathbf{x}, \mathbf{y} \rangle = 0.$$

If the scalar product is not specified, then we assume by default the Euclidean scalar product. The vectors $\mathbf{x}$ and $\mathbf{y}$ that are orthogonal in A-scalar product are also called A-*conjugate* or briefly *conjugate*.

Two *sets of vectors* $\mathcal{E}$ *and* $\mathcal{F}$ *are orthogonal* (also stated "$\mathcal{E}$ orthogonal to $\mathcal{F}$") if every $\mathbf{x} \in \mathcal{E}$ is orthogonal to any $\mathbf{y} \in \mathcal{F}$. The set $\mathcal{E}^\perp$ of all the vectors of $\mathbb{R}^n$ that are orthogonal to $\mathcal{E} \subseteq \mathbb{R}^n$ is a vector space called the *orthogonal complement* of $\mathcal{E}$. If $\mathcal{E} \subseteq \mathbb{R}^n$, then

$$\mathbb{R}^n = \operatorname{Span} \mathcal{E} \oplus \mathcal{E}^\perp.$$

A *set of vectors* $\mathcal{E}$ *is orthogonal* if its elements are pairwise orthogonal, i.e., any $\mathbf{x} \in \mathcal{E}$ is orthogonal to any $\mathbf{y} \in \mathcal{E}$, $\mathbf{y} \neq \mathbf{x}$. A set of vectors $\mathcal{E}$ is *orthonormal* if it is orthogonal and $\langle \mathbf{x}, \mathbf{x} \rangle = 1$ for any $\mathbf{x} \in \mathcal{E}$.

Any orthogonal set $\mathcal{E} = \{\mathbf{e}_1, \ldots, \mathbf{e}_n\}$ of nonzero vectors $\mathbf{e}_i$ is independent. Indeed, if

$$\alpha_1 \mathbf{e}_1 + \cdots + \alpha_n \mathbf{e}_n = \mathbf{o},$$

then we can take the scalar product of both sides of the equation with $\mathbf{e}_i$ and use the assumption on orthogonality of $\mathcal{E}$ to get that

$$\alpha_i \langle \mathbf{e}_i, \mathbf{e}_i \rangle = 0,$$

so that $\alpha_i = 0$.

If $\mathcal{E}$ is an orthonormal basis of a vector space $\mathcal{V} \subseteq \mathbb{R}^n$, then the same procedure as above may be used to get conveniently the coordinates ξ_i of any $\mathbf{x} \in \mathcal{V}$. For example, if $\mathcal{E}$ is orthonormal with respect to the Euclidean scalar product, it is enough to multiply

$$\mathbf{x} = \xi_1 \mathbf{e}_1 + \cdots + \xi_n \mathbf{e}_n$$

on the left by $\mathbf{e}_i^T$ to get

$$\xi_i = \mathbf{e}_i^T \mathbf{x}.$$

A square matrix U *is orthogonal* if $\mathsf{U}^T \mathsf{U} = \mathsf{I}$, that is, $\mathsf{U}^{-1} = \mathsf{U}^T$. Multiplication by an orthogonal matrix U preserves both the angles between any two vectors and the Euclidean norm of any vector as

$$(\mathsf{U}\mathbf{x})^T \mathsf{U}\mathbf{y} = \mathbf{x}^T \mathsf{U}^T \mathsf{U}\mathbf{y} = \mathbf{x}^T \mathbf{y}.$$

A matrix $\mathsf{P} \in \mathbb{R}^{n \times n}$ is an *orthogonal projector* if P is a projector, i.e., $\mathsf{P}^2 = \mathsf{P}$, and $\operatorname{Im}\mathsf{P}$ is orthogonal to $\operatorname{Ker}\mathsf{P}$. The latter condition can be rewritten equivalently as

$$\mathsf{P}^T (\mathsf{I} - \mathsf{P}) = \mathsf{O}.$$

It simply follows that

$$\mathsf{P}^T = \mathsf{P}^T \mathsf{P} = \mathsf{P},$$

so that orthogonal projectors are symmetric matrices and symmetric projectors are orthogonal projectors. If P is an orthogonal projector, then $\mathsf{I} - \mathsf{P}$ is also an orthogonal projector as

$$(\mathsf{I} - \mathsf{P})^2 = I - 2\mathsf{P} + \mathsf{P}^2 = \mathsf{I} - \mathsf{P} \quad \text{and} \quad (\mathsf{I} - \mathsf{P})^T \mathsf{P} = (\mathsf{I} - \mathsf{P})\mathsf{P} = \mathsf{O}.$$

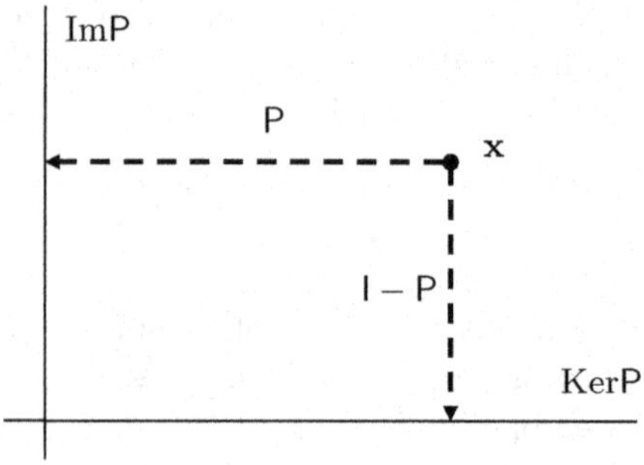

Fig. 1.1. Orthogonal projector

See Fig. 1.1 for a geometric interpretation.

If $\mathcal{U} \subseteq \mathbb{R}^n$ is the subspace spanned by the columns of a full column rank matrix $\mathsf{U} \in \mathbb{R}^{m\times n}$, then

$$\mathsf{P} = \mathsf{U}(\mathsf{U}^T\mathsf{U})^{-1}\mathsf{U}^T$$

is an orthogonal projector as

$$\mathsf{P}^2 = \mathsf{U}(\mathsf{U}^T\mathsf{U})^{-1}\mathsf{U}^T\mathsf{U}(\mathsf{U}^T\mathsf{U})^{-1}\mathsf{U}^T = \mathsf{P} \quad \text{and} \quad \mathsf{P}^T = \mathsf{P}.$$

Since any vector $\mathbf{x} \in \mathcal{U}$ may be written in the form $\mathbf{x} = \mathsf{U}\mathbf{y}$ and

$$\mathsf{P}\mathbf{x} = \mathsf{U}(\mathsf{U}^T\mathsf{U})^{-1}\mathsf{U}^T\mathsf{U}\mathbf{y} = \mathsf{U}\mathbf{y} = \mathbf{x},$$

it follows that

$$\mathcal{U} = \mathrm{Im}\mathsf{P}.$$

Observe that $\mathsf{U}^T\mathsf{U}$ is nonsingular; since $\mathsf{U}^T\mathsf{U}\mathbf{x} = \mathbf{o}$ implies

$$\|\mathsf{U}\mathbf{x}\|^2 = \mathbf{x}^T(\mathsf{U}^T\mathsf{U}\mathbf{x}) = 0,$$

it follows that $\mathbf{x} = \mathbf{o}$ by the assumption on the full column rank of U.

Let $\mathsf{B} \in \mathbb{R}^{m\times n}$ and $\mathbf{x} \in \mathrm{Im}\mathsf{B}^T$, so that there is $\mathbf{y}$ such that $\mathbf{x} = \mathsf{B}^T\mathbf{y}$. Then for any $\mathbf{z} \in \mathrm{Ker}\mathsf{B}$

$$\mathbf{x}^T\mathbf{z} = (\mathsf{B}^T\mathbf{y})^T\mathbf{z} = \mathbf{y}^T(\mathsf{B}\mathbf{z}) = 0,$$

so that $\mathrm{Ker}\mathsf{B}$ is orthogonal to $\mathrm{Im}\mathsf{B}^T$. An important result of linear algebra is that

$$(\mathrm{Ker}\mathsf{B})^{\perp} = \mathrm{Im}\mathsf{B}^T, \tag{1.16}$$

thus

$$\mathbb{R}^n = \mathrm{Ker}\mathsf{B} \oplus \mathrm{Im}\mathsf{B}^T. \tag{1.17}$$

The orthogonal projectors and their generalization, the conjugate projectors that we introduce in Sect. 3.7.1, are useful computational tools for manipulations with subspaces.

1.8 Eigenvalues and Eigenvectors

If $\mathsf{A} \in \mathbb{R}^{n \times n}$ is a square matrix, then it may happen that there is a vector $\mathbf{e} \in \mathbb{R}^n$ such that $\mathsf{A}\mathbf{e}$ is just a scalar multiple of $\mathbf{e}$. Such vectors turned out to be useful for analysis of problems described by matrices. Since the theory has been developed for complex matrices, we consider in this section the vectors and matrices with the entries belonging to the set of complex numbers $\mathbb{C}$ in order to simplify our exposition.

Let $\mathsf{A} \in \mathbb{C}^{n \times n}$ denote a square matrix with complex entries. If a vector $\mathbf{e} \in \mathbb{C}^n$ and a scalar $\lambda \in \mathbb{C}$ satisfy

$$\mathsf{A}\mathbf{e} = \lambda \mathbf{e},$$

then $\mathbf{e}$ is said to be an *eigenvector* of A associated with an *eigenvalue* λ. A vector $\mathbf{e}$ is an eigenvector of A if and only if $\mathrm{Span}\{\mathbf{e}\}$ is an invariant subspace of A; the restriction $\mathsf{A}|\mathrm{Span}\{\mathbf{e}\}$ reduces to the multiplication by λ. If $\{\mathbf{e}_1, \ldots, \mathbf{e}_k\}$ are eigenvectors of a symmetric matrix A, then it is easy to check that $\mathrm{Span}\{\mathbf{e}_1, \ldots, \mathbf{e}_k\}$ and $\mathrm{Span}\{\mathbf{e}_1, \ldots, \mathbf{e}_k\}^{\perp}$ are invariant subspaces.

The set of all eigenvalues of A is called the *spectrum* of A; we denote it by $\sigma(\mathsf{A})$. Obviously $\lambda \in \sigma(\mathsf{A})$ if and only if $\mathsf{A} - \lambda \mathsf{I}$ is singular, and $0 \in \sigma(\mathsf{A})$ if and only if A is singular.

If $\mathcal{U} \subseteq \mathbb{C}^n$ is an invariant subspace of $\mathsf{A} \in \mathbb{C}^{n \times n}$, then we denote by $\sigma(\mathsf{A}|\mathcal{U})$ the eigenvalues of A that correspond to the eigenvectors belonging to $\mathcal{U}$.

Fig. 1.2. The spectrum of a matrix

Many important relations in this book are proved by analysis of the spectrum of a given matrix. The information about the spectrum is typically obtained indirectly, in a similar way as information about a ghost in a spirit session, when the participants are not assumed to see the ghost, but to observe that the table is moving.

The eigenvalues can be characterized algebraically by means of the *determinant* which can be defined by induction in two steps:

(i) A matrix $[a_{11}] \in \mathbb{C}^{1 \times 1}$ is assigned the value $\det[a_{11}] = a_{11}$.

(ii) Assuming that the determinant of a matrix $\mathsf{A} \in \mathbb{C}^{(n-1) \times (n-1)}$ is already defined, the determinant of $\mathsf{A} = [a_{ij}] \in \mathbb{C}^{n \times n}$ is assigned the value

$$\det(\mathsf{A}) = \sum_{j=1}^{n} (-1)^{j+1} a_{1j} \det(\mathsf{A}_{1j}),$$

where A_{1j} is the square matrix of the order $n-1$ obtained from A by deleting its first row and jth column.

Since it is well-known that a matrix is singular if and only if its determinant is equal to zero, it follows that the eigenvalues of A are the roots of the *characteristic equation*

$$\det(\mathsf{A} - \lambda \mathsf{I}) = 0. \tag{1.18}$$

The *characteristic polynomial* $p_{\mathsf{A}}(\lambda) = \det(\mathsf{A} - \lambda \mathsf{I})$ is of the degree n. Thus there are at most n distinct eigenvalues and $\sigma(\mathsf{A})$ is not the empty set.

In what follows, we associate with each matrix $\mathsf{A} \in \mathbb{C}^{n \times n}$ a sequence $\lambda_1, \dots, \lambda_n$ of the eigenvalues of A labeled as the roots of the characteristic polynomial. Each distinct eigenvalue appears in this sequence as many times as corresponds to its *algebraic multiplicity* as a root of the characteristic polynomial. Using the factorization

$$p_{\mathsf{A}}(\lambda) = (\lambda_1 - \lambda) \cdots (\lambda_n - \lambda),$$

we get

$$\det(\mathsf{A}) = p_{\mathsf{A}}(0) = \lambda_1 \cdots \lambda_n. \tag{1.19}$$

Comparing the coefficients in the factorization of $p_{\mathsf{A}}(\lambda)$ with those arising from evaluation of $\det(\mathsf{A} - \lambda \mathsf{I})$, we get

$$\mathrm{trace}(\mathsf{A}) = \lambda_1 + \cdots + \lambda_n, \tag{1.20}$$

where $\mathrm{trace}(\mathsf{A})$ denotes the sum of all diagonal entries of A.

The dimension of $\mathrm{Ker}(\mathsf{A} - \lambda \mathsf{I})$ is called the *geometric multiplicity* of λ. If the algebraic multiplicity of an eigenvalue λ is k, then the number of the independent eigenvectors corresponding to λ is an integer between 1 and k.

Even though it is in general difficult to evaluate the eigenvalues of a given matrix A, it is still possible to get nontrivial information about $\sigma(\mathsf{A})$ without heavy computations. Useful information about the location of eigenvalues can be obtained by *Gershgorin's theorem*, which guarantees that every eigenvalue of $\mathsf{A} = [a_{ij}] \in \mathbb{C}^{n \times n}$ is located in at least one of the n circular disks in the complex plane with the centers a_{ii} and radii $r_i = \sum_{j \neq i} |a_{ij}|$.

The *eigenvalues of a real symmetric matrix are real.* Since it is easy to check whether a matrix is symmetric, this gives us useful information about the location of eigenvalues.

Let $\mathsf{A} \in \mathbb{R}^{n \times n}$ denote a real symmetric matrix, let $\mathcal{I} = \{1, \dots, n-1\}$, and let $\mathsf{A}^1 = \mathsf{A}_{\mathcal{I}\mathcal{I}}$. Let $\lambda_1 \geq \cdots \geq \lambda_n$ and $\lambda_1^1 \geq \cdots \geq \lambda_{n-1}^1$ denote the eigenvalues of A and A^1, respectively. Then by the *Cauchy interlacing theorem*

$$\lambda_1 \geq \lambda_1^1 \geq \lambda_2 \geq \lambda_2^1 \geq \cdots \geq \lambda_{n-1}^1 \geq \lambda_n. \tag{1.21}$$

1.9 Matrix Decompositions

If $\mathsf{A} \in \mathbb{R}^{n\times n}$ is a symmetric matrix, then it is possible to find n orthonormal eigenvectors $\mathbf{e}_1, \dots, \mathbf{e}_n$ that form the basis of $\mathbb{R}^n$. Moreover, the corresponding eigenvalues are real. Denoting by $\mathsf{U} = [\mathbf{e}_1, \dots, \mathbf{e}_n] \in \mathbb{R}^{n\times n}$ an orthogonal matrix whose columns are the eigenvectors, we may write the *spectral decomposition* of A as

$$\mathsf{A} = \mathsf{U}\mathsf{D}\mathsf{U}^T, \tag{1.22}$$

where $\mathsf{D} = \mathrm{diag}(\lambda_1, \dots, \lambda_n) \in \mathbb{R}^{n\times n}$ is the diagonal matrix whose diagonal entries are the eigenvalues corresponding to the eigenvectors $\mathbf{e}_1, \dots, \mathbf{e}_n$. Reordering the columns of U, we can achieve that $\lambda_1 \geq \dots \geq \lambda_n$.

The spectral decomposition reveals close relations between the properties of a symmetric matrix and its eigenvalues. Thus a symmetric matrix is positive definite if and only if all its eigenvalues are positive, and it is positive semidefinite if and only if they are nonnegative. It is easy to check that the rank of a symmetric matrix is equal to the number of nonzero entries of D.

If A is symmetric, then we can use the spectral decomposition (1.22) to check that for any nonzero $\mathbf{x}$

$$\lambda_1 = \lambda_{\max} \geq \|\mathbf{x}\|^{-2}\mathbf{x}^T\mathsf{A}\mathbf{x} \geq \lambda_{\min} = \lambda_n. \tag{1.23}$$

Thus for any symmetric positive definite matrix A

$$\|\mathsf{A}\| = \lambda_{\max}, \ \|\mathsf{A}^{-1}\| = \lambda_{\min}^{-1}, \ \|\mathbf{x}\|_{\mathsf{A}} \leq \lambda_{\max}\|\mathbf{x}\|, \ \|\mathbf{x}\|_{\mathsf{A}^{-1}} \leq \lambda_{\min}^{-1}\|\mathbf{x}\|. \tag{1.24}$$

The *spectral condition number* $\kappa(\mathsf{A}) = \|\mathsf{A}\|\|\mathsf{A}^{-1}\|$, which is a measure of departure from the identity, can be expressed for real symmetric matrix by

$$\kappa(\mathsf{A}) = \lambda_{\max}/\lambda_{\min}.$$

Another consequence of the spectral decomposition theorem is the *Courant–Fischer minimax principle*, (see, e.g., [103]) which states that if $\lambda_1 \geq \dots \geq \lambda_n$ are the eigenvalues of a real symmetric matrix $\mathsf{A} \in \mathbb{R}^{n\times n}$, then

$$\lambda_k = \max_{\substack{\mathcal{V}\subseteq\mathbb{R}^n,\\ \dim\mathcal{V}=k}} \ \min_{\substack{\mathbf{x}\in\mathcal{V},\\ \|\mathbf{x}\|=1}} \mathbf{x}^T\mathsf{A}\mathbf{x}, \quad k = 1, \dots, n. \tag{1.25}$$

If A is a real symmetric matrix and f is a real function defined on $\sigma(\mathsf{A})$, we can use the spectral decomposition to define the *scalar function* by

$$f(\mathsf{A}) = \mathsf{U}f(\mathsf{D})\mathsf{U}^T,$$

where $f(\mathsf{D}) = \mathrm{diag}\left(f(\lambda_1), \dots, f(\lambda_n)\right)$. It is easy to check that if a is the identity function on $\mathbb{R}$ defined by $a(x) = x$, then

$$a(\mathsf{A}) = \mathsf{A},$$

and if f and g are real functions defined on $\sigma(\mathsf{A})$, then

$$(f+g)(\mathsf{A}) = f(\mathsf{A}) + g(\mathsf{A}) \quad \text{and} \quad (f \cdot g)(\mathsf{A}) = f(\mathsf{A})g(\mathsf{A}).$$

Moreover, if $f(x) \geq 0$ for $x \in \sigma(\mathsf{A})$, then $f(\mathsf{A})$ is positive semidefinite, and if $f(x) > 0$ for $x \in \sigma(\mathsf{A})$, then $f(\mathsf{A})$ is positive definite. For example, if A is symmetric positive semidefinite, then the square root of A is well defined and

$$\mathsf{A} = \mathsf{A}^{1/2}\mathsf{A}^{1/2}.$$

Obviously

$$\sigma(f(\mathsf{A})) = f(\sigma(\mathsf{A})), \tag{1.26}$$

and if $\mathbf{e}_i$ is an eigenvector corresponding to $\lambda_i \in \sigma(\mathsf{A})$, then it is also an eigenvector of $f(\mathsf{A})$ corresponding to $f(\lambda_i)$. It follows easily that for any symmetric positive semidefinite matrix

$$\mathrm{Im}\mathsf{A} = \mathrm{Im}\mathsf{A}^{1/2} \quad \text{and} \quad \mathrm{Ker}\mathsf{A} = \mathrm{Ker}\mathsf{A}^{1/2}. \tag{1.27}$$

A key to understanding nonsymmetric matrices is the *singular value decomposition* (SVD). If $\mathsf{B} \in \mathbb{R}^{m\times n}$, then SVD of B is given by

$$\mathsf{B} = \mathsf{U}\mathsf{S}\mathsf{V}^T, \tag{1.28}$$

where $\mathsf{U} \in \mathbb{R}^{m\times m}$ and $\mathsf{V} \in \mathbb{R}^{n\times n}$ are orthogonal, and $\mathsf{S} \in \mathbb{R}^{m\times n}$ is a diagonal matrix with nonnegative diagonal entries $\sigma_1 \geq \cdots \geq \sigma_{\min\{m,n\}} = \sigma_{\min}$ called *singular values* of B. If $\mathsf{A} \neq \mathsf{O}$, it is often more convenient to use the *reduced singular value decomposition* (RSVD)

$$\mathsf{B} = \widehat{\mathsf{U}}\widehat{\mathsf{S}}\widehat{\mathsf{V}}^T, \tag{1.29}$$

where $\widehat{\mathsf{U}} \in \mathbb{R}^{m\times r}$ and $\widehat{\mathsf{V}} \in \mathbb{R}^{n\times r}$ are matrices with orthonormal columns, $\widehat{\mathsf{S}} \in \mathbb{R}^{r\times r}$ is a nonsingular diagonal matrix with positive diagonal entries $\sigma_1 \geq \cdots \geq \sigma_r = \overline{\sigma}_{\min}$, and $r \leq \min\{m,n\}$ is the rank of B. The matrices $\widehat{\mathsf{U}}$ and $\widehat{\mathsf{V}}$ are formed by the first r columns of U and V. If $\mathbf{x} \in \mathbb{R}^m$, then

$$\mathsf{B}\mathbf{x} = \widehat{\mathsf{U}}\widehat{\mathsf{S}}\widehat{\mathsf{V}}^T\mathbf{x} = (\widehat{\mathsf{U}}\widehat{\mathsf{S}}\widehat{\mathsf{V}}^T)(\widehat{\mathsf{V}}\widehat{\mathsf{S}}\widehat{\mathsf{U}}^T)(\widehat{\mathsf{U}}\widehat{\mathsf{S}}^{-1}\mathbf{x}) = \mathsf{B}\mathsf{B}^T\mathbf{y},$$

so that

$$\mathrm{Im}\mathsf{B} = \mathrm{Im}\mathsf{B}\mathsf{B}^T. \tag{1.30}$$

The singular value decomposition reveals close relations between the properties of a matrix and its singular values. Thus the rank of $\mathsf{B} \in \mathbb{R}^{m\times n}$ is equal to the number of its nonzero singular values,

$$\|\mathsf{B}\| = \|\mathsf{B}^T\| = \sigma_1, \tag{1.31}$$

and for any vector $\mathbf{x} \in \mathbb{R}^n$

$$\sigma_{\min}\|\mathbf{x}\| \leq \|\mathsf{B}\mathbf{x}\| \leq \|\mathsf{B}\|\|\mathbf{x}\|. \tag{1.32}$$

Let $\overline{\sigma}_{\min}$ denote the least nonzero singular value of $\mathsf{B} \in \mathbb{R}^{m\times n}$, let $\mathbf{x} \in \mathrm{Im}\mathsf{B}^T$, and consider a reduced singular value decomposition $\mathsf{B} = \widehat{\mathsf{U}}\widehat{\mathsf{S}}\widehat{\mathsf{V}}^T$ with $\widehat{\mathsf{U}} \in \mathbb{R}^{m\times r}$, $\widehat{\mathsf{V}} \in \mathbb{R}^{n\times r}$, and $\widehat{\mathsf{S}} \in \mathbb{R}^{r\times r}$. Then there is $\mathbf{y} \in \mathbb{R}^r$ such that $\mathbf{x} = \widehat{\mathsf{V}}\mathbf{y}$ and

$$\|\mathsf{B}\mathbf{x}\| = \|\widehat{\mathsf{U}}\widehat{\mathsf{S}}\widehat{\mathsf{V}}^T\widehat{\mathsf{V}}\mathbf{y}\| = \|\widehat{\mathsf{U}}\widehat{\mathsf{S}}\mathbf{y}\| = \|\widehat{\mathsf{S}}\mathbf{y}\| \geq \overline{\sigma}_{\min}\|\mathbf{y}\|.$$

Since

$$\|\mathbf{x}\| = \|\widehat{\mathsf{V}}\mathbf{y}\| = \|\mathbf{y}\|,$$

we conclude that

$$\overline{\sigma}_{\min}\|\mathbf{x}\| \leq \|\mathsf{B}\mathbf{x}\| \quad \text{for any} \quad \mathbf{x} \in \mathrm{Im}\mathsf{B}^T, \tag{1.33}$$

or, equivalently,

$$\overline{\sigma}_{\min}\|\mathbf{x}\| \leq \|\mathsf{B}^T\mathbf{x}\| \quad \text{for any} \quad \mathbf{x} \in \mathrm{Im}\mathsf{B}. \tag{1.34}$$

The singular value decomposition (1.28) can be used to introduce the *Moore–Penrose generalized inverse* of an $m \times n$ matrix B by

$$\mathsf{B}^\dagger = \mathsf{V}\mathsf{S}^\dagger\mathsf{U}^T = \widehat{\mathsf{V}}\widehat{\mathsf{S}}^\dagger\widehat{\mathsf{U}}^T,$$

where $\mathsf{S}^\dagger$ is the diagonal matrix with the entries $[\mathsf{S}^\dagger]_{ii} = 0$ if $\sigma_i = 0$ and $[\mathsf{S}^\dagger]_{ii} = \sigma_i^{-1}$ otherwise. It is easy to check that

$$\mathsf{B}\mathsf{B}^\dagger\mathsf{B} = \widehat{\mathsf{U}}\widehat{\mathsf{S}}\widehat{\mathsf{V}}^T\widehat{\mathsf{V}}\widehat{\mathsf{S}}^\dagger\widehat{\mathsf{U}}^T\widehat{\mathsf{U}}\widehat{\mathsf{S}}\widehat{\mathsf{V}}^T = \widehat{\mathsf{U}}\widehat{\mathsf{S}}\widehat{\mathsf{V}}^T = \mathsf{B}, \tag{1.35}$$

so that the Moore–Penrose generalized inverse is a generalized inverse. If B is a full row rank matrix, then it may be checked directly that

$$\mathsf{B}^\dagger = \mathsf{B}^T(\mathsf{B}\mathsf{B}^T)^{-1}.$$

If B is a singular matrix and $\mathbf{c} \in \mathrm{Im}\mathsf{B}$, then $\mathbf{x}_{\mathrm{LS}} = \mathsf{B}^\dagger\mathbf{c}$ is a solution of the system of linear equations $\mathsf{B}\mathbf{x} = \mathbf{c}$, i.e.,

$$\mathsf{B}\mathbf{x}_{\mathrm{LS}} = \mathbf{c}.$$

Notice that $\mathbf{x}_{\mathrm{LS}} \in \mathrm{Im}\mathsf{B}^T$, so that if $\overline{\mathbf{x}}$ is any other solution, then $\overline{\mathbf{x}} = \mathbf{x}_{\mathrm{LS}} + \mathbf{d}$, where $\mathbf{d} \in \mathrm{Ker}\mathsf{B}$, $\mathbf{x}_{\mathrm{LS}}^T\mathbf{d} = 0$ by (1.16), and

$$\|\mathbf{x}_{\mathrm{LS}}\|^2 \leq \|\overline{\mathbf{x}}_{\mathrm{LS}}\|^2 + \|\mathbf{d}\|^2 = \|\overline{\mathbf{x}}\|^2. \tag{1.36}$$

The vector $\mathbf{x}_{\mathrm{LS}}$ is called the *least square solution* of $\mathsf{B}\mathbf{x} = \mathbf{c}$.

Obviously

$$\|\mathsf{B}^\dagger\| = \overline{\sigma}_{\min}^{-1}, \tag{1.37}$$

where $\overline{\sigma}_{\min}$ denotes the least nonzero singular value of B, so that

$$\|\mathbf{x}_{\mathrm{LS}}\| = \|\mathsf{B}^\dagger\mathbf{c}\| \leq \overline{\sigma}_{\min}^{-1}\|\mathbf{c}\|. \tag{1.38}$$

It can be verified directly that

$$\left(\mathsf{B}^\dagger\right)^T = \left(\mathsf{B}^T\right)^\dagger.$$

1.10 Penalized Matrices

We often use the matrices

$$\mathsf{A}_\varrho = \mathsf{A} + \varrho \mathsf{B}^T \mathsf{B},$$

where $\mathsf{A} \in \mathbb{R}^{n\times n}$ is a symmetric matrix, $\mathsf{B} \in \mathbb{R}^{m\times n}$, and $\varrho \geq 0$. The matrix A_ϱ is called the *penalized matrix* as it is closely related to the penalty method described in Sect. 4.2. Let us first give a simple sufficient condition that enforces A_ϱ to be positive definite.

Lemma 1.2. *Let* $\mathsf{A} \in \mathbb{R}^{n\times n}$ *be a symmetric positive semidefinite matrix, let* $\mathsf{B} \in \mathbb{R}^{m\times n}$, $\varrho > 0$, *and let* $\mathrm{Ker}\mathsf{A} \cap \mathrm{Ker}\mathsf{B} = \{\mathbf{o}\}$. *Then* A_ϱ *is positive definite.*

Proof. If $\mathbf{x} \neq \mathbf{o}$ and $\mathrm{Ker}\mathsf{A} \cap \mathrm{Ker}\mathsf{B} = \{\mathbf{o}\}$, then either $\mathsf{A}\mathbf{x} \neq \mathbf{o}$ or $\mathsf{B}\mathbf{x} \neq \mathbf{o}$. Since $\mathsf{A}\mathbf{x} \neq \mathbf{o}$ is by (1.27) equivalent to $\mathsf{A}^{1/2}\mathbf{x} \neq \mathbf{o}$, we get for $\varrho > 0$

$$\mathbf{x}^T \mathsf{A}_\varrho \mathbf{x} = \mathbf{x}^T \mathsf{A}\mathbf{x} + \varrho \|\mathsf{B}\mathbf{x}\|^2 = \|\mathsf{A}^{1/2}\mathbf{x}\|^2 + \varrho\|\mathsf{B}\mathbf{x}\|^2 > 0.$$

Thus A_ϱ is positive definite. □

If A is positive definite, then it can be verified either directly or by Lemma 1.2, using $\mathrm{Ker}\mathsf{A} = \{\mathbf{o}\}$, that A_ϱ is also positive definite. It follows that we can use the Sherman–Morrison–Woodbury formula (1.6) to get

$$\mathsf{A}_\varrho^{-1} = \mathsf{A}^{-1} - \mathsf{A}^{-1}\mathsf{B}^T(\varrho^{-1}\mathsf{I} + \mathsf{B}\mathsf{A}^{-1}\mathsf{B}^T)^{-1}\mathsf{B}\mathsf{A}^{-1}. \tag{1.39}$$

The following lemma shows that A_ϱ can be positive definite even when A is indefinite.

Lemma 1.3. *Let* $\mathsf{A} \in \mathbb{R}^{n\times n}$ *denote a symmetric matrix, let* $\mathsf{B} \in \mathbb{R}^{m\times n}$, *and let there be* $\mu > 0$ *such that*

$$\mathbf{x}^T \mathsf{A}\mathbf{x} \geq \mu\|\mathbf{x}\|^2, \quad \mathbf{x} \in \mathrm{Ker}\mathsf{B}.$$

Then A_ϱ *is positive definite for sufficiently large* ϱ.

Proof. Let $\lambda_{\min}$ and $\overline{\sigma}_{\min}$ denote the least eigenvalue of A and the least positive singular value of B, respectively, and recall that by (1.17) any $\mathbf{x} \in \mathbb{R}^n$ can be written in the form

$$\mathbf{x} = \mathbf{y} + \mathbf{z}, \quad \mathbf{y} \in \mathrm{Ker}\mathsf{B}, \quad \mathbf{z} \in \mathrm{Im}\mathsf{B}^T.$$

Using the definition of A_ϱ, the assumptions, and (1.33), we get

$$\begin{aligned}
\mathbf{x}^T \mathsf{A}_\varrho \mathbf{x} &= \mathbf{y}^T \mathsf{A}\mathbf{y} + 2\mathbf{y}^T\mathsf{A}\mathbf{z} + \mathbf{z}^T\mathsf{A}\mathbf{z} + \varrho\|\mathsf{B}\mathbf{z}\|^2 \\
&\geq \mu\|\mathbf{y}\|^2 - 2\|\mathsf{A}\|\|\mathbf{y}\|\|\mathbf{z}\| + (\lambda_{\min} + \varrho\overline{\sigma}_{\min}^2)\|\mathbf{z}\|^2 \\
&= \left[\, \|\mathbf{y}\|, \|\mathbf{z}\| \,\right] \begin{bmatrix} \mu, & -\|\mathsf{A}\| \\ -\|\mathsf{A}\|, & \lambda_{\min} + \varrho\overline{\sigma}_{\min}^2 \end{bmatrix} \begin{bmatrix} \|\mathbf{y}\| \\ \|\mathbf{z}\| \end{bmatrix}.
\end{aligned} \tag{1.40}$$

We shall complete the proof by showing that the matrix

$$\mathsf{H}_\varrho = \begin{bmatrix} \mu, & -\|\mathsf{A}\| \\ -\|\mathsf{A}\|, & \lambda_{\min} + \varrho\overline{\sigma}_{\min}^2 \end{bmatrix}$$

is positive definite for sufficiently large values of ϱ. To this end, it is enough to show that the eigenvalues λ_1, λ_2 of H_ϱ are positive for sufficiently large ϱ.

First observe that H_ϱ is symmetric, so that λ_1, λ_2 are real. Moreover, both the determinant and the trace of H_ϱ are obviously positive for a sufficiently large ϱ, so that by (1.19) and (1.20) both $\lambda_1\lambda_2 > 0$ and $\lambda_1 + \lambda_2 > 0$ for sufficiently large values of ϱ. Since the latter implies that at least one of the eigenvalues of H_ϱ is positive for sufficiently large ϱ, it follows from $\lambda_1\lambda_2 > 0$ that $\lambda_1 > 0$ and $\lambda_2 > 0$ provided ϱ is sufficiently large. Thus H_ϱ is positive definite for sufficiently large ϱ and the statement of our lemma follows by(1.40). □

We often use bounds on the spectrum of some matrix expressions with penalized matrices that are based on the following lemma.

Lemma 1.4. *Let $m < n$ be given positive integers, let $\mathsf{A} \in \mathbb{R}^{n\times n}$ denote a symmetric positive definite matrix, and let $\mathsf{B} \in \mathbb{R}^{m\times n}$.*

Then

$$\mathsf{BA}_\varrho^{-1} = (\mathsf{I} + \varrho\mathsf{BA}^{-1}\mathsf{B}^T)^{-1}\mathsf{BA}^{-1} \tag{1.41}$$

and the eigenvalues μ_i of $\mathsf{BA}_\varrho^{-1}\mathsf{B}^T$ are related to the eigenvalues β_i of $\mathsf{BA}^{-1}\mathsf{B}^T$ by

$$\mu_i = \beta_i/(1+\varrho\beta_i),\ i = 1,\dots,n. \tag{1.42}$$

Proof. Following [56], we can use a special form (1.39) of the Sherman–Morrison–Woodbury formula to get

$$\begin{aligned}
\mathsf{B}(\mathsf{A} + \varrho\mathsf{B}^T\mathsf{B})^{-1} &= \mathsf{BA}^{-1} - \mathsf{BA}^{-1}\mathsf{B}^T(\varrho^{-1}\mathsf{I} + \mathsf{BA}^{-1}\mathsf{B}^T)^{-1}\mathsf{BA}^{-1} \\
&= (\mathsf{I} - \varrho\mathsf{BA}^{-1}\mathsf{B}^T(\mathsf{I} + \varrho\mathsf{BA}^{-1}\mathsf{B}^T)^{-1})\mathsf{BA}^{-1} \\
&= \Big(\mathsf{I} - ((\mathsf{I} + \varrho\mathsf{BA}^{-1}\mathsf{B}^T) - \mathsf{I})(\mathsf{I} + \varrho\mathsf{BA}^{-1}\mathsf{B}^T)^{-1}\Big)\mathsf{BA}^{-1} \\
&= \left(\mathsf{I} + \varrho\mathsf{BA}^{-1}\mathsf{B}^T\right)^{-1}\mathsf{BA}^{-1}.
\end{aligned}$$

To prove (1.42), notice that by (1.41)

$$\mathsf{BA}_\varrho^{-1}\mathsf{B}^T = (\mathsf{I} + \varrho\mathsf{BA}^{-1}\mathsf{B}^T)^{-1}\mathsf{BA}^{-1}\mathsf{B}^T$$

and apply (1.26) with $f(x) = x/(1+\varrho x)$. □

The eigenvalues of the restriction of $\mathsf{BA}^{-1}\mathsf{B}^T$ to its invariant subspace ImB are given by the following lemma.

Lemma 1.5. *Let $m < n$ be positive integers, let $\mathsf{A} \in \mathbb{R}^{n\times n}$ denote a symmetric positive definite matrix, let $\mathsf{B} \in \mathbb{R}^{m\times n}$, and let r denote the rank of the matrix B.*

Then

$$\mathrm{Im}\mathsf{B} = \mathrm{Im}\mathsf{B}\mathsf{A}^{-1} = \mathrm{Im}\mathsf{B}\mathsf{A}^{-1}\mathsf{B}^T \tag{1.43}$$

and the eigenvalues $\overline{\beta}_i$ of $\mathsf{B}\mathsf{A}_\varrho^{-1}\mathsf{B}^T|\mathrm{Im}\mathsf{B}$ of the restriction of $\mathsf{B}\mathsf{A}_\varrho^{-1}\mathsf{B}^T$ to $\mathrm{Im}\mathsf{B}$ are related to the positive eigenvalues $\beta_1 \geq \beta_2 \geq \cdots \geq \beta_r$ of $\mathsf{B}\mathsf{A}^{-1}\mathsf{B}^T$ by

$$\overline{\beta}_i = \beta_i/(1+\varrho\beta_i), \; i = 1,\ldots,r. \tag{1.44}$$

Proof. First observe that if $\mathsf{C} \in \mathbb{R}^{n\times n}$ is nonsingular and $\mathsf{B} \in \mathbb{R}^{m\times n}$, then $\mathsf{B}\mathbf{x} = \mathsf{B}\mathsf{C}(\mathsf{C}^{-1}\mathbf{x})$, so that

$$\mathrm{Im}\mathsf{B} = \mathrm{Im}\mathsf{B}\mathsf{C}.$$

It follows that if $\mathsf{A} \in \mathbb{R}^{n\times n}$ is positive definite, so that $\mathsf{A}^{1/2}$ is a well-defined nonsingular matrix, then we can use (1.27) and (1.30) to get

$$\mathrm{Im}\mathsf{B} = \mathrm{Im}\mathsf{B}\mathsf{A}^{-1} = \mathrm{Im}\mathsf{B}\mathsf{A}^{-1/2} = \mathrm{Im}\mathsf{B}\mathsf{A}^{-1/2}(\mathsf{B}\mathsf{A}^{-1/2})^T = \mathrm{Im}\mathsf{B}\mathsf{A}^{-1}\mathsf{B}^T.$$

We have thus proved (1.43).

To prove (1.44), notice that we can use (1.43) with $\mathsf{A} = \mathsf{A}_\varrho$ to get $\mathrm{Im}\mathsf{B} = \mathrm{Im}\mathsf{B}\mathsf{A}_\varrho^{-1}\mathsf{B}^T$. Since $\mathrm{Im}\mathsf{B}\mathsf{A}_\varrho^{-1}\mathsf{B}^T$ is spanned by the eigenvectors of $\mathsf{B}\mathsf{A}_\varrho^{-1}\mathsf{B}^T$ which correspond to the positive eigenvalues of $\mathsf{B}\mathsf{A}_\varrho^{-1}\mathsf{B}^T$, we can use (1.42) to finish the proof. □

The following lemma gives the estimates that are useful in the analysis of the Uzawa-type algorithms.

Lemma 1.6. *Let $m < n$ be given positive integers, let $\mathsf{A} \in \mathbb{R}^{n\times n}$ denote a symmetric positive definite matrix, and let $\mathsf{B} \in \mathbb{R}^{m\times n}$. Let $\overline{\beta}_{\min} > 0$ and $\beta_{\max} \geq \overline{\beta}_{\min}$ denote the least nonzero eigenvalue and the largest eigenvalue of the matrix $\mathsf{B}\mathsf{A}^{-1}\mathsf{B}^T$, respectively.*

Then for any $\varrho > 0$

$$\|\mathsf{B}\mathsf{A}_\varrho^{-1}\| \leq \|\mathsf{B}\mathsf{A}^{-1}\|/\left(1+\varrho\overline{\beta}_{\min}\right), \tag{1.45}$$

$$\|(\mathsf{I}+\varrho\mathsf{B}\mathsf{A}^{-1}\mathsf{B}^T)^{-1}|\mathrm{Im}\mathsf{B}\| = 1/\left(1+\varrho\overline{\beta}_{\min}\right), \tag{1.46}$$

and

$$\|\mathsf{B}\mathsf{A}_\varrho^{-1}\mathsf{B}^T\| = \beta_{\max}/\left(1+\varrho\beta_{\max}\right) < \varrho^{-1}. \tag{1.47}$$

Moreover

$$\lim_{\varrho\to\infty} \kappa(\mathsf{B}\mathsf{A}_\varrho^{-1}\mathsf{B}^T|\mathrm{Im}\mathsf{B}) = 1. \tag{1.48}$$

Proof. Applying submultiplicativity of the matrix norms (1.11) to (1.41), we get that

$$\|\mathsf{BA}_\varrho^{-1}\| \leq \|(\mathsf{I} + \varrho\mathsf{BA}^{-1}\mathsf{B}^T)^{-1}|\mathrm{Im}\mathsf{BA}^{-1}\|\|\mathsf{BA}^{-1}\|.$$

To evaluate the first factor, notice that by (1.43) $\mathrm{Im}\mathsf{B} = \mathrm{Im}\mathsf{BA}^{-1} = \mathrm{Im}\mathsf{BA}^{-1}\mathsf{B}^T$, so that our task reduces to the evaluation of

$$\|(\mathsf{I} + \varrho\mathsf{BA}^{-1}\mathsf{B}^T)^{-1}|\mathrm{Im}\mathsf{BA}^{-1}\mathsf{B}^T\|.$$

Since $\mathrm{Im}\mathsf{BA}^{-1}\mathsf{B}^T$ is spanned by the eigenvectors of $\mathsf{BA}^{-1}\mathsf{B}^T$ which correspond to the positive eigenvalues, and the eigenvectors of $\mathsf{I} + \varrho\mathsf{BA}^{-1}\mathsf{B}^T$ are just the eigenvectors of $\mathsf{BA}^{-1}\mathsf{B}^T$, it follows by (1.26) with $f(x) = 1/(1+\varrho x)$ that

$$\|(\mathsf{I} + \varrho\mathsf{BA}^{-1}\mathsf{B}^T)^{-1}|\mathrm{Im}\mathsf{BA}^{-1}\mathsf{B}^T\| = \max_{\substack{i=1,\dots,m\\ \beta_i>0}} 1/\left(1+\varrho\beta_i\right) = 1/\left(1+\varrho\overline{\beta}_{\min}\right).$$

This completes the proof of (1.45) and (1.46).

To prove (1.47), recall that the eigenvalues μ_i of $\mathsf{BA}_\varrho^{-1}\mathsf{B}^T$ are related to the eigenvalues β_i of $\mathsf{BA}^{-1}\mathsf{B}^T$ by (1.42), so that

$$\|\mathsf{BA}_\varrho^{-1}\mathsf{B}^T\| = \max_{i=1,\dots,m} \mu_i = \max_{i=1,\dots,m} \beta_i\left(1+\varrho\beta_i\right) = \beta_{\max}/\left(1+\varrho\beta_{\max}\right). \quad (1.49)$$

Finally, using Lemma 1.5, we get that

$$\min_{\substack{\|\mathbf{x}\|=1\\ \mathbf{x}\in\mathrm{Im}\mathsf{B}}} \|\mathsf{BA}^{-1}\mathsf{B}^T\mathbf{x}\| = \min_{\substack{i=1,\dots,m\\ \beta_i>0}} \beta_i/\left(1+\varrho\beta_i\right) = \overline{\beta}_{\min}/\left(1+\varrho\overline{\beta}_{\min}\right),$$

so that

$$\lim_{\varrho\to\infty} \kappa(\mathsf{BA}_\varrho^{-1}\mathsf{B}^T|\mathrm{Im}\mathsf{B}) = \lim_{\varrho\to\infty} \frac{\beta_{\max}}{1+\varrho\beta_{\max}} \frac{1+\varrho\overline{\beta}_{\min}}{\overline{\beta}_{\min}} = 1.$$

□

The last result to be presented here concerns the distribution of the eigenvalues of the penalized matrices. To simplify its formulation, let us denote for each symmetric matrix M of the order n by $\lambda_i(\mathsf{M})$ the ith eigenvalue of M in the decreasing order, so that

$$\lambda_1(\mathsf{M}) \geq \lambda_2(\mathsf{M}) \geq \dots \geq \lambda_n(\mathsf{M}).$$

Lemma 1.7. *Let* $\mathsf{A} \in \mathbb{R}^{n\times n}$ *denote a symmetric matrix, let* $\mathsf{B} \in \mathbb{R}^{m\times n}$ *denote a matrix of the rank* r, $0 < r \leq m < n$, *and* $\varrho > 0$.

Then $\lambda_r(\mathsf{B}^T\mathsf{B}) > 0$ *and*

$$\lambda_r(\mathsf{A}_\varrho) \geq \lambda_n(\mathsf{A}) + \varrho\lambda_r(\mathsf{B}^T\mathsf{B}), \quad (1.50)$$

$$\lambda_{r+1}(\mathsf{A}_\varrho) \leq \lambda_1(\mathsf{A}). \quad (1.51)$$

Proof. Using the spectral decomposition (1.22), we can find an orthogonal matrix U such that

$$\mathsf{U}^T\mathsf{B}^T\mathsf{B}\mathsf{U} = \operatorname{diag}(\gamma_1, \dots, \gamma_n)$$

with

$$\gamma_i = \lambda_i(\mathsf{B}^T\mathsf{B}) \quad \text{and} \quad \gamma_1 \geq \dots \geq \gamma_r > \gamma_{r+1} = \gamma_n = 0.$$

Thus

$$\mathsf{U}^T(\mathsf{A} + \varrho\mathsf{B}^T\mathsf{B})\mathsf{U} = \mathsf{U}^T\mathsf{A}\mathsf{U} + \varrho\mathsf{U}^T\mathsf{B}^T\mathsf{B}\mathsf{U} = \begin{bmatrix} \mathsf{E} + \varrho\mathsf{G} & \mathsf{F} \\ \mathsf{F}^T & \mathsf{H} \end{bmatrix}$$

with

$$\mathsf{G} = \operatorname{diag}(\gamma_1, \dots, \gamma_r) \quad \text{and} \quad \mathsf{U}^T\mathsf{A}\mathsf{U} = \begin{bmatrix} \mathsf{E} & \mathsf{F} \\ \mathsf{F}^T & \mathsf{H} \end{bmatrix},$$

where H is a square matrix of the order $n - r$. The eigenvalues of $\mathsf{U}^T\mathsf{A}\mathsf{U}$ and A are identical.

To prove (1.50), notice that the elementary properties of the spectrum of symmetric matrices and the Cauchy interlacing property of bordering matrices (1.21) imply that

$$\begin{aligned}\lambda_r(\mathsf{A}_\varrho) &\geq \lambda_r(\mathsf{E} + \varrho\mathsf{G}) \geq \lambda_r(\lambda_n(\mathsf{A})\mathsf{I} + \varrho\mathsf{G}) = \lambda_n(\mathsf{A}) + \varrho\lambda_r(\mathsf{G}) \\ &= \lambda_n(\mathsf{A}) + \varrho\gamma_r = \lambda_n(\mathsf{A}) + \varrho\lambda_r(\mathsf{B}^T\mathsf{B}).\end{aligned}$$

To prove (1.51), observe that by the Courant–Fischer min-max principle (1.25)

$$\begin{aligned}\lambda_{r+1}(\mathsf{A}_\varrho) &= \max_{\substack{\mathcal{V}\subseteq\mathbb{R}^n \\ \dim\mathcal{V}=r+1}} \min_{\substack{\mathbf{x}\in\mathcal{V} \\ \|\mathbf{x}\|=1}} \mathbf{x}^T(\mathsf{A} + \varrho\mathsf{B}^T\mathsf{B})\mathbf{x} \\ &\leq \max_{\substack{\mathcal{V}\subseteq\mathbb{R}^n \\ \dim\mathcal{V}=r+1}} \min_{\substack{\mathbf{x}\in\mathcal{V}\cap\mathrm{Ker}\mathsf{B} \\ \|\mathbf{x}\|=1}} \mathbf{x}^T(\mathsf{A} + \varrho\mathsf{B}^T\mathsf{B})\mathbf{x} \\ &= \max_{\substack{\mathcal{V}\subseteq\mathbb{R}^n \\ \dim\mathcal{V}=r+1}} \min_{\substack{\mathbf{x}\in\mathcal{V}\cap\mathrm{Ker}\mathsf{B} \\ \|\mathbf{x}\|=1}} \mathbf{x}^T\mathsf{A}\mathbf{x} \leq \lambda_1(\mathsf{A}),\end{aligned} \tag{1.52}$$

so that the inequality (1.51) is proved. We used the fact that $\dim\mathcal{V} = r + 1$ implies

$$\mathcal{V} \cap \mathrm{Ker}\mathsf{B} \neq \{\mathbf{o}\}.$$

□

2

Optimization

In this chapter we briefly review the results of optimization to the extent that is sufficient for understanding of the rest of our book. Since we are interested mainly in quadratic programming, we present most results with specialized arguments, typically algebraic, that exploit the specific structure of these problems. Such approach not only simplifies the analysis, but sometimes enables us to obtain stronger results. Since the results of this section are useful also in the analysis of more general optimization problems that are locally approximated by quadratic problems, this chapter may also serve as a simple introduction to nonlinear optimization. Systematic exposition of the optimization theory in the framework of nonlinear optimization may be found in the books by Bertsekas [12], Nocedal and Wright [155], Conn, Gould, and Toint [28], or Bazaraa, Sherali, and Shetty [8].

2.1 Optimization Problems and Solutions

Optimization problems considered in this book are described by a *cost (objective) function* f defined on a subset $\mathcal{D} \subseteq \mathbb{R}^n$ and by a *constraint set* $\Omega \subseteq \mathcal{D}$. The elements of the constraint set Ω are called *feasible vectors.* The main topic of this book is development of efficient algorithms for the solution of *quadratic programming (QP) problems* with a quadratic cost function f and a constraint set $\Omega \subseteq \mathbb{R}^n$ described by linear equalities and inequalities.

We look either for a solution $\overline{\mathbf{x}} \in \mathbb{R}^n$ of the *unconstrained minimization problem* which satisfies

$$f(\overline{\mathbf{x}}) \le f(\mathbf{x}), \ \ \mathbf{x} \in \mathbb{R}^n, \tag{2.1}$$

or for a solution $\overline{\mathbf{x}} \in \Omega$ of the *constrained minimization problem*

$$f(\overline{\mathbf{x}}) \le f(\mathbf{x}), \ \ \mathbf{x} \in \Omega. \tag{2.2}$$

A solution of the minimization problem is called its *minimizer* or *global minimizer.* The value of f corresponding to a minimizer is the *minimum.*

Zdeněk Dostál, *Optimal Quadratic Programming Algorithms*,
Springer Optimization and Its Applications, DOI 10.1007/978-0-387-84806-8_2,

As a characterization of global minimizers of the inequality constrained problems may be a too ambitious goal, we consider also *local minimizers* that satisfy for some $\delta > 0$

$$f(\overline{\mathbf{x}}) \le f(\mathbf{x}), \quad \mathbf{x} \in \Omega, \quad \|\mathbf{x} - \overline{\mathbf{x}}\| \le \delta. \tag{2.3}$$

Obviously each global minimizer is a local minimizer.

A nonzero vector $\mathbf{d} \in \mathbb{R}^n$ is a *feasible direction* of Ω at a feasible point $\mathbf{x}$ if $\mathbf{x} + \varepsilon\mathbf{d} \in \Omega$ for all sufficiently small $\varepsilon > 0$. Feasible directions are useful in analysis of local minimizers.

A nonzero vector $\mathbf{d} \in \mathbb{R}^n$ is a *recession direction*, or simply a *direction*, of Ω if for each $\mathbf{x} \in \Omega$, $\mathbf{x} + \alpha\mathbf{d} \in \Omega$ for all $\alpha > 0$.

2.2 Unconstrained Quadratic Programming

Let us first recall some simple results which concern unconstrained quadratic programming.

2.2.1 Quadratic Cost Functions

We consider the cost functions in the form

$$f(\mathbf{x}) = \frac{1}{2}\mathbf{x}^T \mathsf{A}\mathbf{x} - \mathbf{b}^T\mathbf{x}, \tag{2.4}$$

where $\mathsf{A} \in \mathbb{R}^{n\times n}$ denotes a given symmetric matrix of order n and $\mathbf{b} \in \mathbb{R}^n$.

If $\mathbf{x}, \mathbf{d} \in \mathbb{R}^n$, then using elementary computations and $\mathsf{A} = \mathsf{A}^T$, we get

$$f(\mathbf{x} + \mathbf{d}) = f(\mathbf{x}) + (\mathsf{A}\mathbf{x} - \mathbf{b})^T\mathbf{d} + \frac{1}{2}\mathbf{d}^T\mathsf{A}\mathbf{d}. \tag{2.5}$$

The formula (2.5) is *Taylor's expansion* of f at $\mathbf{x}$, so that the *gradient* of f at $\mathbf{x}$ is given by

$$\nabla f(\mathbf{x}) = \mathsf{A}\mathbf{x} - \mathbf{b}, \tag{2.6}$$

and the *Hessian* of f at $\mathbf{x}$ is given by

$$\nabla^2 f(\mathbf{x}) = \mathsf{A}.$$

Taylor's expansion will be our simple but powerful tool in what follows.

A vector $\mathbf{d}$ is a *decrease direction* of f at $\mathbf{x}$ if

$$f(\mathbf{x} + \varepsilon\mathbf{d}) < f(\mathbf{x})$$

for all sufficiently small values of $\varepsilon > 0$. Using Taylor's expansion (2.5) in the form

$$f(\mathbf{x} + \varepsilon\mathbf{d}) = f(\mathbf{x}) + \varepsilon(\mathsf{A}\mathbf{x} - \mathbf{b})^T\mathbf{d} + \frac{\varepsilon^2}{2}\mathbf{d}^T\mathsf{A}\mathbf{d},$$

we get that $\mathbf{d}$ is a decrease direction if and only if

$$(\mathsf{A}\mathbf{x} - \mathbf{b})^T\mathbf{d} < 0.$$

2.2.2 Unconstrained Minimization of Quadratic Functions

The following proposition gives algebraic conditions that are satisfied by the solutions of the unconstrained QP problem to find

$$\min_{\mathbf{x}\in\mathbb{R}^n} f(\mathbf{x}), \tag{2.7}$$

where f is a quadratic function defined by (2.4).

Proposition 2.1. *Let the quadratic function f be defined by a symmetric matrix $\mathsf{A} \in \mathbb{R}^{n\times n}$ and $\mathbf{b} \in \mathbb{R}^n$. Then the following statements hold:*
(i) A vector $\overline{\mathbf{x}}$ is a solution of the unconstrained minimization problem (2.1) if and only if A is positive semidefinite and

$$\nabla f(\overline{\mathbf{x}}) = \mathsf{A}\overline{\mathbf{x}} - \mathbf{b} = \mathbf{o}. \tag{2.8}$$

(ii) The unconstrained minimization problem (2.1) has a unique solution if and only if A is positive definite.

Proof. (i) If $\overline{\mathbf{x}}$ and $\mathbf{d}$ denote arbitrary n-vectors and $\alpha \in \mathbb{R}$, then we can use Taylor's expansion (2.5) to get

$$f(\overline{\mathbf{x}} + \alpha\mathbf{d}) - f(\overline{\mathbf{x}}) = \alpha(\mathsf{A}\overline{\mathbf{x}} - \mathbf{b})^T\mathbf{d} + \frac{\alpha^2}{2}\mathbf{d}^T\mathsf{A}\mathbf{d}.$$

Let us first assume that $\overline{\mathbf{x}}$ is a solution of (2.1), so that the right-hand side of the above equation is nonnegative for any α and $\mathbf{d}$. For α sufficiently large and $\mathbf{d} \in \mathbb{R}^n$ arbitrary but fixed, the nonnegativity of the right-hand side implies that $\mathbf{d}^T\mathsf{A}\mathbf{d} \geq 0$; thus A is positive semidefinite. On the other hand, for α sufficiently small, the sign of the right-hand side is determined by the linear term, so the nonnegativity of the right-hand side implies that $(\mathsf{A}\overline{\mathbf{x}} - \mathbf{b})^T\mathbf{d} = 0$ for any $\mathbf{d} \in \mathbb{R}^n$. Thus $\mathsf{A}\overline{\mathbf{x}} - \mathbf{b} = \mathbf{o}$.

If A is positive semidefinite and $\overline{\mathbf{x}}$ satisfies (2.8), then for any $\mathbf{d} \in \mathbb{R}^n$

$$f(\overline{\mathbf{x}} + \mathbf{d}) - f(\overline{\mathbf{x}}) = \frac{1}{2}\mathbf{d}^T\mathsf{A}\mathbf{d} \geq 0;$$

therefore $\overline{\mathbf{x}}$ is a solution of (2.1).

(ii) If $\widehat{\mathbf{x}}$ is the unique solution of the unconstrained minimization problem (2.1), then by (i) A is positive semidefinite and $\widehat{\mathbf{x}}$ is the only vector which satisfies $\mathsf{A}\widehat{\mathbf{x}} = \mathbf{b}$. Thus A is nonsingular and positive semidefinite, i.e., positive definite. On the other hand, if A is positive definite, then it is nonsingular and the gradient condition (2.8) has the unique solution. □

Examining the gradient condition (2.8), we get that problem (2.1) has a solution if and only if A is positive semidefinite and

$$\mathbf{b} \in \mathrm{Im}\mathsf{A}. \tag{2.9}$$

Denoting by R a matrix whose columns span the kernel of A, we can rewrite the latter condition as

$$\mathbf{b}^T\mathsf{R} = \mathbf{o}.$$

This condition has a simple mechanical interpretation: if a mechanical system is in equilibrium, the external forces must be orthogonal to the rigid body motions.

If $\mathbf{b} \in \mathrm{Im}\mathsf{A}$, a solution of (2.7) is given by

$$\overline{\mathbf{x}} = \mathsf{A}^+\mathbf{b},$$

where A^+ is a left generalized inverse introduced in Sect. 1.4. After substituting into f and simple manipulations, we get

$$\min_{\mathbf{x}\in\mathbb{R}^n} f(\mathbf{x}) = -\frac{1}{2}\mathbf{b}^T\mathsf{A}^+\mathbf{b}. \tag{2.10}$$

In particular, if A is positive definite, then

$$\min_{\mathbf{x}\in\mathbb{R}^n} f(\mathbf{x}) = -\frac{1}{2}\mathbf{b}^T\mathsf{A}^{-1}\mathbf{b}. \tag{2.11}$$

The formulae for the minimum of the unconstrained minimization problems can be used to develop useful estimates. Indeed, if (2.9) holds and $\mathbf{x} \in \mathbb{R}^n$, we can use (2.10), properties of generalized inverses, and (1.37) to get

$$f(\mathbf{x}) \geq -\frac{1}{2}\mathbf{b}^T\mathsf{A}^+\mathbf{b} = -\frac{1}{2}\mathbf{b}^T\mathsf{A}^\dagger\mathbf{b} \geq -\frac{1}{2}\|\mathsf{A}^\dagger\|\|\mathbf{b}\|^2 = -\|\mathbf{b}\|^2/(2\overline{\lambda}_{\min}),$$

where $\mathsf{A}^\dagger$ denotes the Moore–Penrose generalized inverse and $\overline{\lambda}_{\min}$ denotes the least nonzero eigenvalue of A. In particular, it follows that if A is positive definite and $\lambda_{\min}$ denotes the least eigenvalue of A, then for any $\mathbf{x} \in \mathbb{R}^n$

$$f(\mathbf{x}) \geq -\frac{1}{2}\mathbf{b}^T\mathsf{A}^{-1}\mathbf{b} \geq -\frac{1}{2}\|\mathsf{A}^{-1}\|\|\mathbf{b}\|^2 = -\|\mathbf{b}\|^2/(2\lambda_{\min}). \tag{2.12}$$

If the dimension n of the unconstrained minimization problem (2.7) is large, then it can be too ambitious to look for a solution which satisfies the gradient condition (2.8) exactly. A natural idea is to consider the weaker condition

$$\|\nabla f(\mathbf{x})\| \leq \varepsilon \tag{2.13}$$

with a small epsilon. If $\mathbf{x}$ satisfies the latter condition with ε sufficiently small and A nonsingular, then $\mathbf{x}$ is near the unique solution $\widehat{\mathbf{x}}$ as

$$\|\mathbf{x} - \widehat{\mathbf{x}}\| = \|\mathsf{A}^{-1}\mathsf{A}\left(\mathbf{x} - \widehat{\mathbf{x}}\right)\| = \|\mathsf{A}^{-1}(\mathsf{A}\mathbf{x} - \mathbf{b})\| \leq \|\mathsf{A}^{-1}\|\|\nabla f(\mathbf{x})\|. \tag{2.14}$$

The typical "solution" returned by an iterative solver is just $\mathbf{x}$ that satisfies the condition (2.13). Using the Taylor expansion (2.5), we can obtain

$$\begin{aligned} f(\mathbf{x}) - f(\widehat{\mathbf{x}}) &= f(\widehat{\mathbf{x}} + (\mathbf{x} - \widehat{\mathbf{x}})) - f(\widehat{\mathbf{x}}) \\ &= f(\widehat{\mathbf{x}}) + \mathbf{g}(\widehat{\mathbf{x}})^T(\mathbf{x} - \widehat{\mathbf{x}}) + \frac{1}{2}\|\mathbf{x} - \widehat{\mathbf{x}}\|_{\mathsf{A}}^2 - f(\widehat{\mathbf{x}}) \\ &= \frac{1}{2}\|\mathbf{x} - \widehat{\mathbf{x}}\|_{\mathsf{A}}^2. \end{aligned}$$

2.3 Convexity

Many strong results can be proved when the problem obeys convexity assumptions. Intuitively, convexity is a property of the sets that contain with any two points the joining segment as in Fig. 2.1. More formally, a subset Ω of $\mathbb{R}^n$ is *convex* if for any $\mathbf{x}$ and $\mathbf{y}$ in Ω and $\alpha \in (0,1)$, the vector $\mathbf{s} = \alpha\mathbf{x} + (1-\alpha)\mathbf{y}$ is also in Ω.

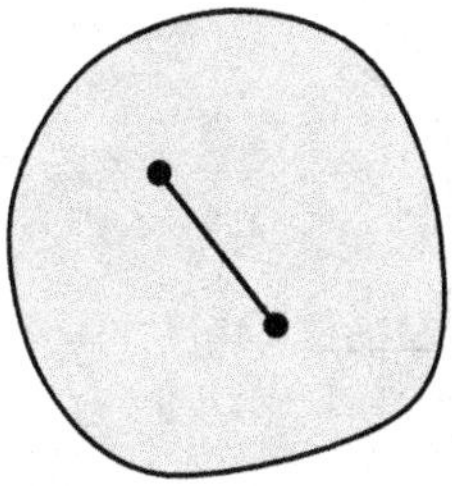

Fig. 2.1. Convex set

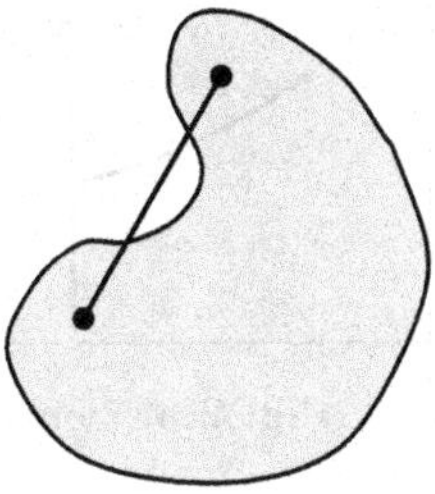

Fig. 2.2. Nonconvex set

Let $\mathbf{x}_1, \dots, \mathbf{x}_k$ be vectors of $\mathbb{R}^n$. If $\alpha_1, \dots, \alpha_k$ are scalars such that

$$\alpha_i \geq 0, \quad i = 1, \dots, k, \qquad \sum_{i=1}^{k} \alpha_i = 1,$$

then the vector $\mathbf{v} = \sum_{i=1}^{k} \alpha_i \mathbf{x}_i$ is said to be a *convex combination* of vectors $\mathbf{x}_1, \dots, \mathbf{x}_k$. The *convex hull* of $\mathbf{x}_1, \dots, \mathbf{x}_k$, denoted $\mathrm{Conv}\{\mathbf{x}_1, \dots, \mathbf{x}_k\}$, is the set of all convex combinations of $\mathbf{x}_1, \dots, \mathbf{x}_k$. The convex hull of $\mathbf{x}_1, \dots, \mathbf{x}_k$ is the smallest convex set to which $\mathbf{x}_1, \dots, \mathbf{x}_k$ belong. *Caratheodory's theorem* guarantees that $\mathrm{Conv}\{\mathbf{x}_1, \dots, \mathbf{x}_k\}$ can be represented as a convex combination of no more than $n+1$ elements of $\{\mathbf{x}_1, \dots, \mathbf{x}_k\}$. The *convex boundary* of a convex set Ω is a set of vectors that cannot be expressed as a convex combination of any other vectors of Ω. Thus the convex boundary of a square is formed by its four corners, while the convex boundary of a circle is formed by its boundary. The intersection of two or more convex sets is also convex. In this book, we consider minimization over the convex sets defined by a finite set of linear equations like $\mathbf{b}^T\mathbf{x} = c$ or inequalities like $\mathbf{b}^T\mathbf{x} \leq c$.

2.3.1 Convex Quadratic Functions

Given a convex set $\Omega \in \mathbb{R}^n$, a mapping $h : \Omega \rightarrow \mathbb{R}$ is said to be a *convex function* if its epigraph is convex, that is, if

$$h(\alpha\mathbf{x} + (1-\alpha)\mathbf{y}) \le \alpha h(\mathbf{x}) + (1-\alpha)h(\mathbf{y})$$

for all $\mathbf{x}, \mathbf{y} \in \Omega$ and $\alpha \in (0,1)$, and it is *strictly convex* if

$$h(\alpha\mathbf{x} + (1-\alpha)\mathbf{y}) < \alpha h(\mathbf{x}) + (1-\alpha)h(\mathbf{y})$$

for all $\mathbf{x}, \mathbf{y} \in \Omega$, $\mathbf{x} \ne \mathbf{y}$, and $\alpha \in (0,1)$. The concept of convex function is illustrated in Fig. 2.3.

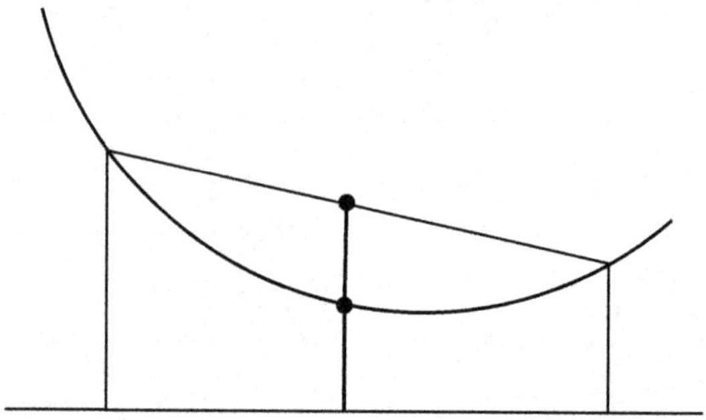

Fig. 2.3. Convex function

The following proposition offers an algebraic characterization of convex functions.

Proposition 2.2. *Let $\mathcal{V}$ be a subspace of $\mathbb{R}^n$. The restriction $f|\mathcal{V}$ of a quadratic function f with the Hessian matrix A to $\mathcal{V}$ is convex if and only if $\mathsf{A}|\mathcal{V}$ is positive semidefinite, and $f|\mathcal{V}$ is strictly convex if and only if $\mathsf{A}|\mathcal{V}$ is positive definite.*

Proof. Let $\mathcal{V}$ be a subspace, let $\mathbf{x}, \mathbf{y} \in \mathcal{V}$, $\alpha \in (0,1)$, and $\mathbf{s} = \alpha\mathbf{x} + (1-\alpha)\mathbf{y}$. Then by Taylor's expansion (2.5) of f at $\mathbf{s}$

$$f(\mathbf{s}) + \nabla f(\mathbf{s})^T(\mathbf{x}-\mathbf{s}) + \frac{1}{2}(\mathbf{x}-\mathbf{s})^T\mathsf{A}(\mathbf{x}-\mathbf{s}) = f(\mathbf{x}),$$
$$f(\mathbf{s}) + \nabla f(\mathbf{s})^T(\mathbf{y}-\mathbf{s}) + \frac{1}{2}(\mathbf{y}-\mathbf{s})^T\mathsf{A}(\mathbf{y}-\mathbf{s}) = f(\mathbf{y}).$$

Multiplying the first equation by α, the second equation by $1-\alpha$, and summing up, we get

$$f(\mathbf{s}) + \frac{\alpha}{2}(\mathbf{x}-\mathbf{s})^T\mathsf{A}(\mathbf{x}-\mathbf{s}) + \frac{1-\alpha}{2}(\mathbf{y}-\mathbf{s})^T\mathsf{A}(\mathbf{y}-\mathbf{s}) \\ = \alpha f(\mathbf{x}) + (1-\alpha)f(\mathbf{y}). \tag{2.15}$$

It follows that if $\mathsf{A}|\mathcal{V}$ is positive semidefinite, then $f|\mathcal{V}$ is convex. Moreover, since $\mathbf{x} = \mathbf{y}$ is equivalent to $\mathbf{x} = \mathbf{s}$ and $\mathbf{y} = \mathbf{s}$, it follows that if $\mathsf{A}|\mathcal{V}$ is positive definite, then $f|\mathcal{V}$ is strictly convex.

Let us now assume that $f|\mathcal{V}$ is convex, let $\mathbf{z} \in \mathcal{V}$, set $\alpha = \frac{1}{2}$, and denote $\mathbf{x} = 2\mathbf{z}$, $\mathbf{y} = \mathbf{o}$. Then $\mathbf{s} = \mathbf{z}$, $\mathbf{x} - \mathbf{s} = \mathbf{z}$, $\mathbf{y} - \mathbf{s} = -\mathbf{z}$, and substituting into (2.15) results in

$$f(\mathbf{s}) + \frac{1}{2}\mathbf{z}^T\mathsf{A}\mathbf{z} = \alpha f(\mathbf{x}) + (1-\alpha)f(\mathbf{y}).$$

Since $\mathbf{z} \in \mathcal{V}$ is arbitrary and $f|\mathcal{V}$ is assumed to be convex, it follows that

$$\frac{1}{2}\mathbf{z}^T\mathsf{A}\mathbf{z} = \alpha f(\mathbf{x}) + (1-\alpha)f(\mathbf{y}) - f\left(\alpha\mathbf{x} + (1-\alpha)\mathbf{y}\right) \geq 0.$$

Thus $\mathsf{A}|\mathcal{V}$ is positive semidefinite. Moreover, if $f|\mathcal{V}$ is strictly convex, then $\mathsf{A}|\mathcal{V}$ is positive definite. □

The following simple corollary is useful in the analysis of equality constrained problems.

Corollary 2.3. *Let f denote a quadratic function with the Hessian $\mathsf{A} \in \mathbb{R}^{n\times n}$, let $\mathsf{B} \in \mathbb{R}^{m\times n}$, $\mathbf{c} \in \mathbb{R}^m$, and $\Omega = \{\mathbf{x} \in \mathbb{R}^n : \mathsf{B}\mathbf{x} = \mathbf{c}\}$. Then $f|\Omega$ is convex if and only if $f|\mathrm{Ker}\mathsf{B}$ is convex, and $f|\Omega$ is strictly convex if and only if $f|\mathrm{Ker}\mathsf{B}$ is strictly convex.*

Proof. First observe that Ω is convex. If $\overline{\mathbf{x}} \in \Omega$, then $\Omega = \{\overline{\mathbf{x}} + \mathbf{d} : \mathbf{d} \in \mathrm{Ker}\mathsf{B}\}$. It follows that the restriction of $f(\mathbf{x})$ to Ω has the same graph as the restriction of $f(\overline{\mathbf{x}} + \mathbf{d})$ to $\mathrm{Ker}\mathsf{B}$. The statement then follows by Proposition 2.2 and

$$\nabla^2 f(\mathbf{x}) = \nabla^2_{\mathbf{dd}} f(\overline{\mathbf{x}} + \mathbf{d}) = \mathsf{A}.$$

□

The strictly convex functions have a nice property that $f(\mathbf{x}) \to \infty$ when $\|\mathbf{x}\| \to \infty$. The functions with this property are called *coercive functions.* More generally, a function $f : \mathbb{R}^n \mapsto \mathbb{R}$ is said to be *coercive on* $\Omega \subseteq \mathbb{R}^n$ if

$$f(\mathbf{x}) \to \infty \quad \text{for} \quad \|\mathbf{x}\| \to \infty, \ \mathbf{x} \in \Omega.$$

A quadratic function need not be strictly convex to be coercive on a given set, as in the case of $f(x,y) = x^2 - y$, which is coercive on $\Omega = \mathbb{R} \times (-\infty, 1]$. More generally, a quadratic function f with a semidefinite Hessian A is coercive on a convex set Ω if $\mathbf{d}^T\mathbf{b} < 0$ for any recession direction $\mathbf{d}$ of Ω which belongs to $\mathrm{Ker}\mathsf{A}$. The coercive quadratic function with a semidefinite Hessian matrix is also called a *semicoercive function.* For example, the function $f(x,y) = x^2 - y$ is semicoercive on $\Omega = \mathbb{R} \times (-\infty, 1]$.

2.3.2 Local and Global Minimizers of Convex Function

Under the convexity assumptions, each local minimizer is a global minimizer. We shall formulate this result together with some observations concerning the set of solutions.

Proposition 2.4. *Let f and $\Omega \subseteq \mathbb{R}^n$ be a quadratic function defined by (2.4) and a closed convex set, respectively. Then the following statements hold:*
(i) If f is convex, then each local minimizer of f subject to $\mathbf{x} \in \Omega$ is a global minimizer of f subject to $\mathbf{x} \in \Omega$.
(ii) If f is convex on a subspace $\mathcal{V} \supseteq \Omega$ and $\overline{\mathbf{x}}$, $\overline{\mathbf{y}}$ are two minimizers of f subject to $\mathbf{x} \in \Omega$, then

$$\overline{\mathbf{x}} - \overline{\mathbf{y}} \in \mathrm{Ker}\mathsf{A}.$$

(iii) If f is strictly convex on Ω and $\overline{\mathbf{x}}$, $\overline{\mathbf{y}}$ are two minimizers of f subject to $\mathbf{x} \in \Omega$, then $\overline{\mathbf{x}} = \overline{\mathbf{y}}$.

Proof. (i) Let $\overline{\mathbf{x}} \in \Omega$ and $\overline{\mathbf{y}} \in \Omega$ be local minimizers of f subject to $\mathbf{x} \in \Omega$, $f(\overline{\mathbf{x}}) < f(\overline{\mathbf{y}})$. Denoting $\mathbf{y}_\alpha = \alpha\overline{\mathbf{x}} + (1-\alpha)\overline{\mathbf{y}}$ and using that f is convex, we get

$$f(\mathbf{y}_\alpha) = f(\alpha\overline{\mathbf{x}} + (1-\alpha)\overline{\mathbf{y}}) \leq \alpha f(\overline{\mathbf{x}}) + (1-\alpha)f(\overline{\mathbf{y}}) < f(\overline{\mathbf{y}})$$

for every $\alpha \in (0,1)$. Since

$$\|\overline{\mathbf{y}} - \mathbf{y}_\alpha\| = \alpha\|\overline{\mathbf{y}} - \overline{\mathbf{x}}\|,$$

the inequality contradicts the assumption that $\overline{\mathbf{y}}$ is a local minimizer.
(ii) Let $\overline{\mathbf{x}}$ and $\overline{\mathbf{y}}$ be global minimizers of f on Ω. Then for any $\alpha \in [0,1]$

$$\overline{\mathbf{x}} + \alpha(\overline{\mathbf{y}} - \overline{\mathbf{x}}) = (1-\alpha)\overline{\mathbf{x}} + \alpha\overline{\mathbf{y}} \in \Omega, \quad \overline{\mathbf{y}} + \alpha(\overline{\mathbf{x}} - \overline{\mathbf{y}}) = (1-\alpha)\overline{\mathbf{y}} + \alpha\overline{\mathbf{x}} \in \Omega.$$

Moreover, using Taylor's formula, we get

$$0 \leq f(\overline{\mathbf{x}} + \alpha(\overline{\mathbf{y}} - \overline{\mathbf{x}})) - f(\overline{\mathbf{x}}) = \alpha(\mathsf{A}\overline{\mathbf{x}} - \mathbf{b})^T(\overline{\mathbf{y}} - \overline{\mathbf{x}}) + \frac{\alpha^2}{2}(\overline{\mathbf{y}} - \overline{\mathbf{x}})^T\mathsf{A}(\overline{\mathbf{y}} - \overline{\mathbf{x}}),$$

$$0 \leq f(\overline{\mathbf{y}} + \alpha(\overline{\mathbf{x}} - \overline{\mathbf{y}})) - f(\overline{\mathbf{y}}) = \alpha(\mathsf{A}\overline{\mathbf{y}} - \mathbf{b})^T(\overline{\mathbf{x}} - \overline{\mathbf{y}}) + \frac{\alpha^2}{2}(\overline{\mathbf{x}} - \overline{\mathbf{y}})^T\mathsf{A}(\overline{\mathbf{x}} - \overline{\mathbf{y}}).$$

Since the latter inequalities hold for arbitrarily small α, it follows that

$$(\mathsf{A}\overline{\mathbf{x}} - \mathbf{b})^T(\overline{\mathbf{y}} - \overline{\mathbf{x}}) \geq 0 \quad \text{and} \quad (\mathsf{A}\overline{\mathbf{y}} - \mathbf{b})^T(\overline{\mathbf{x}} - \overline{\mathbf{y}}) \geq 0.$$

After summing up the latter inequalities and simple manipulations, we have

$$-(\overline{\mathbf{x}} - \overline{\mathbf{y}})^T\mathsf{A}(\overline{\mathbf{x}} - \overline{\mathbf{y}}) \geq 0.$$

Since the convexity of $f|\mathcal{V}$ implies by Proposition 2.2 that $\mathsf{A}|\mathcal{V}$ is positive semidefinite, it follows that $\overline{\mathbf{x}} - \overline{\mathbf{y}} \in \mathrm{Ker}\mathsf{A}$.
(iii) Let f be strictly convex and let $\overline{\mathbf{x}} \in \Omega$ and $\overline{\mathbf{y}} \in \Omega$ be different global minimizers of f on Ω, so that $f(\overline{\mathbf{x}}) = f(\overline{\mathbf{y}})$. Then $\mathrm{Ker}\mathsf{A} = \{\mathbf{o}\}$ and by (ii) $\overline{\mathbf{x}} - \overline{\mathbf{y}} = \mathbf{o}$. Alternatively, taking $\alpha \in (0,1)$, we get

$$f(\alpha\overline{\mathbf{x}} + (1-\alpha)\overline{\mathbf{y}}) < \alpha f(\overline{\mathbf{x}}) + (1-\alpha)f(\overline{\mathbf{y}}) = f(\overline{\mathbf{x}}),$$

which contradicts the assumption that $\overline{\mathbf{x}}$ is a global minimizer of f on Ω. □

2.3.3 Existence of Minimizers

Since quadratic functions are continuous, existence of at least one minimizer is guaranteed by the Weierstrass theorem provided Ω is compact, that is, closed and bounded. We can also use the following standard results which do not assume that Ω is bounded.

Proposition 2.5. *Let f be a quadratic function defined on a nonempty closed convex set $\Omega \subseteq \mathbb{R}^n$. Then the following statements hold:*
(i) If f is strictly convex, then a global minimizer of f subject to $\mathbf{x} \in \Omega$ exists and is necessarily unique.
(ii) If f is coercive on Ω, then a global minimizer of f subject to $\mathbf{x} \in \Omega$ exists.
(iii) A global minimizer of f subject to $\mathbf{x} \in \Omega$ exists if and only if f is bounded from below on Ω.

Proof. (i) If f is strictly convex, it follows by Proposition 2.2 that its Hessian A is positive definite, and $\mathbf{z} = \mathsf{A}^{-1}\mathbf{b}$ is by Proposition 2.1 the unique minimizer of f on $\mathbb{R}^n$. Thus for any $\mathbf{x} \in \mathbb{R}^n$

$$f(\mathbf{x}) \geq f(\mathbf{z}).$$

It follows that the infimum of $f(\mathbf{x})$ subject to $\mathbf{x} \in \Omega$ exists, and there is a sequence of vectors $\mathbf{x}^k \in \Omega$ such that

$$\lim_{k \to \infty} f(\mathbf{x}^k) = \inf_{\mathbf{x} \in \Omega} f(\mathbf{x}).$$

The sequence $\{\mathbf{x}^k\}$ is bounded as

$$f(\mathbf{x}^k) - f(\mathbf{z}) = \frac{1}{2}(\mathbf{x}^k - \mathbf{z})^T \mathsf{A}(\mathbf{x}^k - \mathbf{z}) \geq \frac{\lambda_{\min}}{2} \|\mathbf{x}^k - \mathbf{z}\|^2,$$

where $\lambda_{\min}$ denotes the least eigenvalue of A. It follows that $\{\mathbf{x}^k\}$ has at least one cluster point $\overline{\mathbf{x}} \in \Omega$. Since f is continuous, we get

$$f(\overline{\mathbf{x}}) = \inf_{\mathbf{x} \in \Omega} f(\mathbf{x}).$$

The uniqueness follows by Proposition 2.4.
(ii) The proof is similar to that of (i). See, e.g., Bertsekas [12, Proposition A.8].
(iii) The statement is the well-known Frank–Wolfe theorem [93]. See also Eaves [79] or Blum and Oettli [15]. □

Using a special structure of the feasible set, it is possible to get stronger existence results. For example, it is known that the quadratic function with a positive semidefinite Hessian attains its minimum on a polyhedral cone if and only if its linear term satisfies $\mathbf{b}^T\mathbf{d} \leq 0$ for any recession direction $\mathbf{d} \in \mathrm{Ker}\mathsf{A}$ (see Zeidler [184, pp. 553–556]). See also Sect. 2.5.4.

2.3.4 Projections to Convex Sets

Having the results of the previous subsection, we can naturally define the *projection P_Ω to the (closed) convex set $\Omega \subset \mathbb{R}^n$* as a mapping which assigns to each $\mathbf{x} \in \mathbb{R}^n$ its nearest vector $\widehat{\mathbf{x}} \in \Omega$ as in Fig. 2.4. The distance can be measured by the norm induced by any scalar product. The following proposition concerns the projection induced by the Euclidean scalar product.

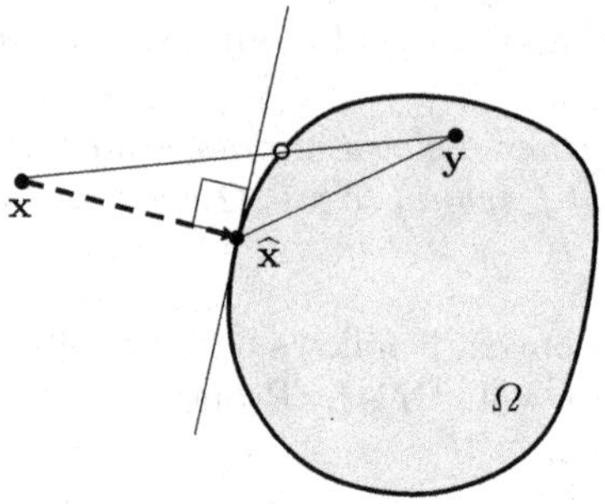

Fig. 2.4. Projection to the convex set

Proposition 2.6. *Let $\Omega \subseteq \mathbb{R}^n$ be a nonempty closed convex set and $\mathbf{x} \in \mathbb{R}^n$. Then there is a unique point $\widehat{\mathbf{x}} \in \Omega$ with the minimum Euclidean distance from $\mathbf{x}$, and for any $\mathbf{y} \in \Omega$*

$$(\mathbf{x} - \widehat{\mathbf{x}})^T(\mathbf{y} - \widehat{\mathbf{x}}) \le 0. \tag{2.16}$$

Proof. Since the proof is trivial for $\mathbf{x} \in \Omega$, let us assume that $\mathbf{x} \notin \Omega$ is arbitrary but fixed and observe that the function f defined on $\mathbb{R}^n$ by

$$f(\mathbf{y}) = \|\mathbf{x} - \mathbf{y}\|^2 = \mathbf{y}^T\mathbf{y} - 2\mathbf{y}^T\mathbf{x} + \|\mathbf{x}\|^2$$

has the Hessian

$$\nabla^2 f(\mathbf{y}) = 2\mathsf{I}.$$

The identity matrix being positive definite, it follows by Proposition 2.2 that f is strictly convex, so that the unique minimizer $\widehat{\mathbf{x}} \in \Omega$ of $f(\mathbf{y})$ with respect to $\mathbf{y} \in \Omega$ exists by Proposition 2.5(i).

If $\mathbf{y} \in \Omega$ and $\alpha \in (0, 1)$, then by convexity of Ω

$$(1 - \alpha)\widehat{\mathbf{x}} + \alpha\mathbf{y} = \widehat{\mathbf{x}} + \alpha(\mathbf{y} - \widehat{\mathbf{x}}) \in \Omega,$$

so that for any $\mathbf{x} \in \mathbb{R}^n$

$$\|\mathbf{x} - \widehat{\mathbf{x}}\|^2 \le \|\mathbf{x} - \widehat{\mathbf{x}} - \alpha(\mathbf{y} - \widehat{\mathbf{x}})\|^2.$$

Using simple manipulations and the latter inequality, we get

$$\|\mathbf{x}-\widehat{\mathbf{x}}-\alpha(\mathbf{y}-\widehat{\mathbf{x}})\|^2 = \|\widehat{\mathbf{x}}-\mathbf{x}\|^2+\alpha^2\|\mathbf{y}-\widehat{\mathbf{x}}\|^2-2\alpha(\mathbf{x}-\widehat{\mathbf{x}})^T(\mathbf{y}-\widehat{\mathbf{x}})$$
$$\leq \|\mathbf{x}-\widehat{\mathbf{x}}-\alpha(\mathbf{y}-\widehat{\mathbf{x}})\|^2$$
$$+\alpha^2\|\mathbf{y}-\widehat{\mathbf{x}}\|^2-2\alpha(\mathbf{x}-\widehat{\mathbf{x}})^T(\mathbf{y}-\widehat{\mathbf{x}}).$$

Thus

$$2\alpha(\mathbf{x}-\widehat{\mathbf{x}})^T(\mathbf{y}-\widehat{\mathbf{x}}) \leq \alpha^2\|\mathbf{y}-\widehat{\mathbf{x}}\|^2$$

for any $\alpha \in (0,1)$. To obtain (2.16), just divide the last inequality by $\alpha > 0$ and observe that α may be arbitrarily small. □

Using Proposition 2.6, it is not difficult to show that the mapping P_Ω which assigns to each $\mathbf{x} \in \mathbb{R}^n$ its projection to Ω is *nonexpansive* as in Fig. 2.5.

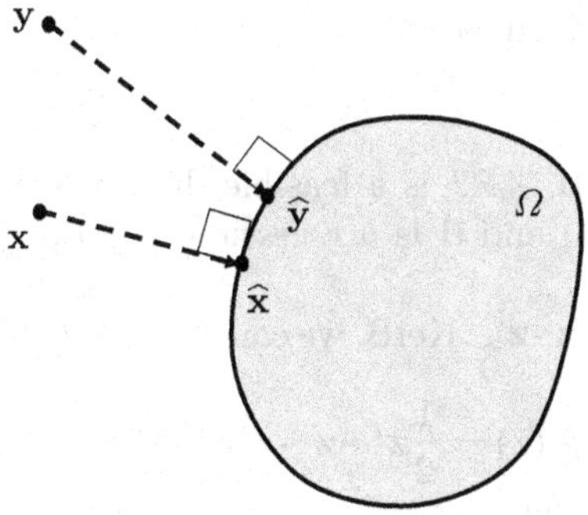

Fig. 2.5. Projection P_Ω is nonexpansive

Corollary 2.7. *Let $\Omega \subseteq \mathbb{R}^n$ be a nonempty closed convex set, and for any $\mathbf{x} \in \mathbb{R}^n$, let $\widehat{\mathbf{x}} \in \Omega$ denote the projection of $\mathbf{x}$ to Ω. Then for any $\mathbf{x}, \mathbf{y} \in \Omega$*

$$\|\widehat{\mathbf{x}}-\widehat{\mathbf{y}}\| \leq \|\mathbf{x}-\mathbf{y}\|. \tag{2.17}$$

Proof. If $\mathbf{x}, \mathbf{y} \in \mathbb{R}$, then by Proposition 2.6 their projections $\widehat{\mathbf{x}}, \widehat{\mathbf{y}}$ to Ω satisfy

$$(\mathbf{x}-\widehat{\mathbf{x}})^T(\mathbf{z}-\widehat{\mathbf{x}}) \leq 0 \quad \text{and} \quad (\mathbf{y}-\widehat{\mathbf{y}})^T(\mathbf{z}-\widehat{\mathbf{y}}) \leq 0$$

for any $\mathbf{z} \in \Omega$. Substituting $\mathbf{z} = \widehat{\mathbf{y}}$ into the first inequality, $\mathbf{z} = \widehat{\mathbf{x}}$ into the second inequality, and summing up, we get

$$(\mathbf{x}-\widehat{\mathbf{x}}-\mathbf{y}+\widehat{\mathbf{y}})^T(\widehat{\mathbf{y}}-\widehat{\mathbf{x}}) \leq 0.$$

After rearranging the entries and using the Schwarz inequality, we get

$$\|\widehat{\mathbf{x}}-\widehat{\mathbf{y}}\|^2 \leq (\mathbf{x}-\mathbf{y})^T(\widehat{\mathbf{x}}-\widehat{\mathbf{y}}) \leq \|\mathbf{x}-\mathbf{y}\|\|\widehat{\mathbf{x}}-\widehat{\mathbf{y}}\|,$$

showing that the projection to the convex set is nonexpansive and proving (2.17). □

2.4 Equality Constrained Problems

We shall now consider the problems with the constraint set described by a set of linear equations. More formally, we shall look for

$$\min_{\mathbf{x}\in\Omega_E} f(\mathbf{x}), \tag{2.18}$$

where f is a quadratic function defined by (2.4), $\Omega_E = \{\mathbf{x} \in \mathbb{R}^n : \mathsf{B}\mathbf{x} = \mathbf{c}\}$, $\mathsf{B} \in \mathbb{R}^{m\times n}$, and $\mathbf{c} \in \mathrm{Im}\mathsf{B}$. We assume that $\mathsf{B} \neq \mathsf{O}$ is not a full column rank matrix, so that $\mathrm{Ker}\mathsf{B} \neq \{\mathbf{o}\}$, but we admit dependent rows of B. It is easy to check that Ω_E is a nonempty closed convex set.

A feasible set Ω_E is a *linear manifold* of the form

$$\Omega_E = \overline{\mathbf{x}} + \mathrm{Ker}\mathsf{B},$$

where $\overline{\mathbf{x}}$ is any vector which satisfies

$$\mathsf{B}\overline{\mathbf{x}} = \mathbf{c}.$$

Thus a nonzero vector $\mathbf{d} \in \mathbb{R}^n$ is a feasible direction of Ω_E at any $\mathbf{x} \in \Omega_E$ if and only if $\mathbf{d} \in \mathrm{Ker}\mathsf{B}$, and $\mathbf{d}$ is a recession direction of Ω_E if and only if $\mathbf{d} \in \mathrm{Ker}\mathsf{B}$.

Substituting $\mathbf{x} = \overline{\mathbf{x}}+\mathbf{z}$, $\mathbf{z} \in \mathrm{Ker}\mathsf{B}$, we can reduce (2.18) to the minimization of

$$f_{\overline{x}}(\mathbf{z}) = \frac{1}{2}\mathbf{z}^T\mathsf{A}\mathbf{z} - (\mathbf{b} - \mathsf{A}\overline{\mathbf{x}})^T\mathbf{z} \tag{2.19}$$

over the subspace $\mathrm{Ker}\mathsf{B}$. Thus we can assume, without loss of generality, that $\mathbf{c} = \mathbf{o}$ in the definition of Ω_E. We shall occasionally use this assumption to simplify our exposition.

A useful tool for the analysis of equality constrained problems is the *Lagrangian function* $L_0 : \mathbb{R}^{n+m} \to \mathbb{R}$ defined by

$$L_0(\mathbf{x}, \boldsymbol{\lambda}) = f(\mathbf{x}) + \boldsymbol{\lambda}^T(\mathsf{B}\mathbf{x} - \mathbf{c}) = \frac{1}{2}\mathbf{x}^T\mathsf{A}\mathbf{x} - \mathbf{b}^T\mathbf{x} + (\mathsf{B}\mathbf{x} - \mathbf{c})^T\boldsymbol{\lambda}. \tag{2.20}$$

Obviously

$$\nabla^2_{\mathbf{xx}}L_0(\mathbf{x}, \boldsymbol{\lambda}) = \nabla^2 f(\mathbf{x}) = \mathsf{A}, \tag{2.21}$$

$$\nabla_{\mathbf{x}}L_0(\mathbf{x}, \boldsymbol{\lambda}) = \nabla f(\mathbf{x}) + \mathsf{B}^T\boldsymbol{\lambda} = \mathsf{A}\mathbf{x} - \mathbf{b} + \mathsf{B}^T\boldsymbol{\lambda}, \tag{2.22}$$

$$L_0(\mathbf{x} + \mathbf{d}, \boldsymbol{\lambda}) = L_0(\mathbf{x}, \boldsymbol{\lambda}) + (\mathsf{A}\mathbf{x} - \mathbf{b} + \mathsf{B}^T\boldsymbol{\lambda})^T\mathbf{d} + \frac{1}{2}\mathbf{d}^T\mathsf{A}\mathbf{d}. \tag{2.23}$$

The Lagrangian function is defined in such a way that if considered as a function of $\mathbf{x}$, then its Hessian and its restriction to Ω_E are exactly those of f, but its gradient $\nabla_{\mathbf{x}}L_0(\mathbf{x}, \boldsymbol{\lambda})$ varies depending on the choice of $\boldsymbol{\lambda}$. It simply follows that if f is convex, then L_0 is convex for any fixed $\boldsymbol{\lambda}$, and the global minimizer of L_0 with respect to $\mathbf{x}$ also varies with $\boldsymbol{\lambda}$. We shall see that it is possible to give conditions on A, B, and $\mathbf{b}$ such that with a suitable choice $\boldsymbol{\lambda} = \widehat{\boldsymbol{\lambda}}$, the solution of the constrained minimization problem (2.18) reduces to the unconstrained minimization of L_0 as in Fig. 2.6.

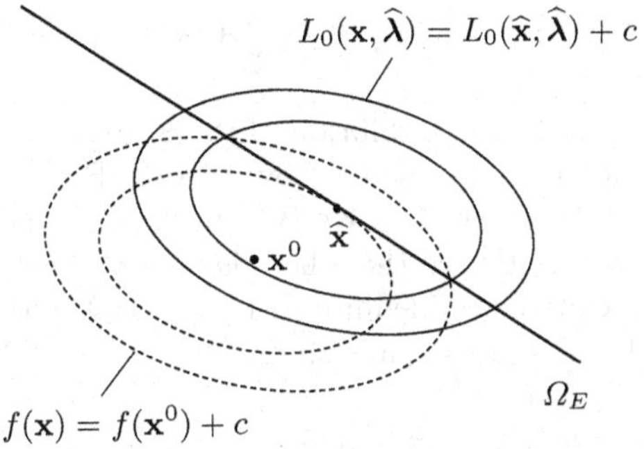

Fig. 2.6. Geometric illustration of the Lagrangian function

2.4.1 Optimality Conditions

The main questions concerning the optimality and solvability conditions of (2.18) are answered by the next proposition.

Proposition 2.8. *Let the equality constrained problem (2.18) be defined by a symmetric matrix* $\mathsf{A} \in \mathbb{R}^{n\times n}$, *a constraint matrix* $\mathsf{B} \in \mathbb{R}^{m\times n}$ *whose column rank is less than* n, *and vectors* $\mathbf{b} \in \mathbb{R}^n$, $\mathbf{c} \in \mathrm{Im}\mathsf{B}$. *Then the following statements hold:*
(i) A vector $\overline{\mathbf{x}} \in \Omega_E$ *is a solution of (2.18) if and only if* $\mathsf{A}|\mathrm{Ker}\mathsf{B}$ *is positive semidefinite and*

$$(\mathsf{A}\overline{\mathbf{x}} - \mathbf{b})^T\mathbf{d} = 0 \tag{2.24}$$

for any $\mathbf{d} \in \mathrm{Ker}\mathsf{B}$.
(ii) A vector $\overline{\mathbf{x}} \in \Omega_E$ *is a solution of (2.18) if and only if* $\mathsf{A}|\mathrm{Ker}\mathsf{B}$ *is positive semidefinite and there is a vector* $\overline{\boldsymbol{\lambda}} \in \mathbb{R}^m$ *such that*

$$\mathsf{A}\overline{\mathbf{x}} - \mathbf{b} + \mathsf{B}^T\overline{\boldsymbol{\lambda}} = \mathbf{o}. \tag{2.25}$$

Proof. (i) Let $\overline{\mathbf{x}}$ be a solution of the equality constrained minimization problem (2.18), so that for any $\mathbf{d} \in \mathrm{Ker}\mathsf{B}$ and $\alpha \in \mathbb{R}$

$$0 \le f(\overline{\mathbf{x}} + \alpha\mathbf{d}) - f(\overline{\mathbf{x}}) = \alpha(\mathsf{A}\overline{\mathbf{x}} - \mathbf{b})^T\mathbf{d} + \frac{\alpha^2}{2}\mathbf{d}^T\mathsf{A}\mathbf{d}. \tag{2.26}$$

Fixing $\mathbf{d} \in \mathrm{Ker}\mathsf{B}$ and taking α sufficiently large, we get that the nonnegativity of the right-hand side of (2.26) implies $\mathbf{d}^T\mathsf{A}\mathbf{d} \ge 0$. Thus $\mathsf{A}|\mathrm{Ker}\mathsf{B}$ must be positive semidefinite. On the other hand, for sufficiently small values of α and $(\mathsf{A}\overline{\mathbf{x}} - \mathbf{b})^T\mathbf{d} \ne 0$, the sign of the right-hand side of (2.26) is determined by the sign of $\alpha(\mathsf{A}\overline{\mathbf{x}} - \mathbf{b})^T\mathbf{d}$. Since we can choose the sign of α arbitrarily and the right-hand side of (2.26) is nonnegative, we conclude that (2.24) holds for any $\mathbf{d} \in \mathrm{Ker}\mathsf{B}$.

Let us now assume that (2.24) holds for a vector $\overline{\mathbf{x}} \in \Omega_E$ and $\mathsf{A}|\mathrm{Ker}\mathsf{B}$ is positive semidefinite. Then

$$f(\overline{\mathbf{x}}+\mathbf{d}) - f(\overline{\mathbf{x}}) = \frac{1}{2}\mathbf{d}^T\mathsf{A}\mathbf{d} \geq 0$$

for any $\mathbf{d} \in \mathrm{Ker}\mathsf{B}$, so that $\overline{\mathbf{x}}$ is a solution of (2.18).

(ii) Let $\overline{\mathbf{x}}$ be a solution of (2.18), so that by (i) $\mathsf{A}|\mathrm{Ker}\mathsf{B}$ is positive semidefinite and $\overline{\mathbf{x}}$ satisfies (2.24) for any $\mathbf{d} \in \mathrm{Ker}\mathsf{B}$. The latter condition is by (1.16) equivalent to $\mathsf{A}\overline{\mathbf{x}} - \mathbf{b} \in \mathrm{Im}\mathsf{B}^{\mathrm{T}}$, so that there is $\overline{\boldsymbol{\lambda}} \in \mathbb{R}^m$ such that (2.25) holds.

Let $\mathsf{A}|\mathrm{Ker}\mathsf{B}$ be positive semidefinite. If there are $\overline{\boldsymbol{\lambda}}$ and $\overline{\mathbf{x}} \in \Omega_E$ such that (2.25) holds, then by Taylor's expansion (2.23)

$$f(\overline{\mathbf{x}}+\mathbf{d}) - f(\overline{\mathbf{x}}) = L_0(\overline{\mathbf{x}}+\mathbf{d}, \overline{\boldsymbol{\lambda}}) - L_0(\overline{\mathbf{x}}, \overline{\boldsymbol{\lambda}}) = \frac{1}{2}\mathbf{d}^T\mathsf{A}\mathbf{d} \geq 0$$

for any $\mathbf{d} \in \mathrm{Ker}\mathsf{B}$, so that $\overline{\mathbf{x}}$ is a solution of the equality constrained problem (2.18). □

Proposition 2.8(i) may be easily modified to characterize the solutions of the minimization problem whose feasible set is a manifold defined by a vector and a subspace; this modification is often useful in what follows.

Corollary 2.9. *Let f be a convex quadratic function on $\mathbb{R}^n$, let $\mathcal{S}$ be a subspace of $\mathbb{R}^n$, and let $\mathbf{x}^0 \in \mathbb{R}^n$. Then $\overline{\mathbf{x}}$ is a solution of*

$$\min_{\mathbf{x}\in\Omega_S} f(\mathbf{x}), \quad \Omega_S = \mathbf{x}^0 + \mathcal{S} \tag{2.27}$$

if and only if

$$\nabla f(\overline{\mathbf{x}})^T\mathbf{d} = 0 \quad \text{for any} \quad \mathbf{d} \in \mathcal{S}.$$

Proof. Let $\mathcal{S}$ be a subspace of $\mathbb{R}^n$, $\mathcal{S} = \mathrm{Im}\mathsf{S}$, where $\mathsf{S} \in \mathbb{R}^{n\times m}$ is a full column rank matrix, $\mathbf{x}^0 \in \mathbb{R}^n$, and let $\mathsf{B} = \mathsf{I} - \mathsf{S}(\mathsf{S}^T\mathsf{S})^{-1}\mathsf{S}^T$, so that

$$\mathcal{S} = \mathrm{Ker}\mathsf{B} \quad \text{and} \quad \Omega_S = \{\mathbf{x} \in \mathbb{R}^n : \ \mathsf{B}\mathbf{x} = \mathsf{B}\mathbf{x}^0\}.$$

Using Proposition 2.8, we get that $\overline{\mathbf{x}} \in \Omega_S$ is the minimizer of a convex quadratic function f on Ω_S if and only if $\nabla f(\overline{\mathbf{x}})$ is orthogonal to $\mathcal{S}$. □

The conditions (ii) of Proposition 2.8 are known as the *Karush–Kuhn–Tucker (KKT) conditions* for the solution of the equality constrained problem (2.18). If $\overline{\mathbf{x}} \in \Omega_E$ and $\overline{\boldsymbol{\lambda}} \in \mathbb{R}^m$ satisfy (2.25), then $(\overline{\mathbf{x}}, \overline{\boldsymbol{\lambda}})$ is called a *KKT pair* of problem (2.18). Its second component $\overline{\boldsymbol{\lambda}}$ is called a vector of *Lagrange multipliers* or simply a *multiplier*. We shall often use the notation $\widehat{\mathbf{x}}$ or $\widehat{\boldsymbol{\lambda}}$ to denote the components of a KKT pair that are uniquely determined.

Proposition 2.8 has a simple geometrical interpretation. The condition (2.24) requires that the gradient of f at a solution $\overline{\mathbf{x}}$ is orthogonal to $\mathrm{Ker}\mathsf{B}$, the set of feasible directions of Ω_E, so that there is no feasible decrease direction as illustrated in Fig. 2.7. Since $\mathbf{d}$ is by (1.16) orthogonal to $\mathrm{Ker}\mathsf{B}$ if and only if $\mathbf{d} \in \mathrm{Im}\mathsf{B}^T$, it follows that (2.24) is equivalent to the possibility to choose $\overline{\boldsymbol{\lambda}}$

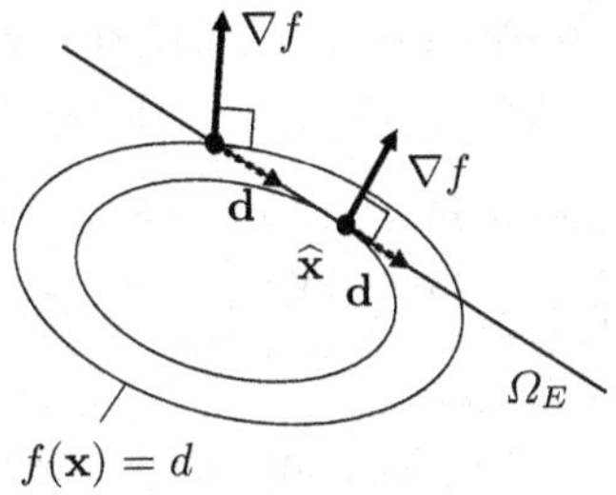

Fig. 2.7. Solvability condition (i)

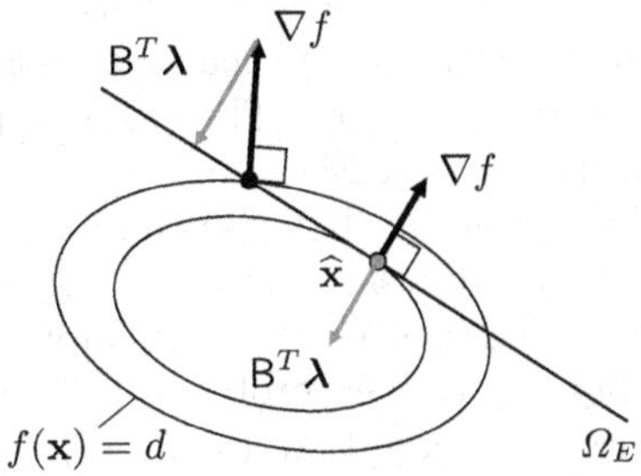

Fig. 2.8. Solvability condition (ii)

so that $\nabla_{\mathbf{x}} L_0(\overline{\mathbf{x}}, \overline{\boldsymbol{\lambda}}) = \mathbf{o}$. If f is convex, then the latter condition is equivalent to the condition for the unconstrained minimizer of L_0 with respect to $\mathbf{x}$ as illustrated in Fig. 2.8.

Notice that if f is convex, then the vector of Lagrange multipliers which is the component of a KKT pair modifies the linear term of the original problem in such a way that the solution of the unconstrained modified problem is exactly the same as the solution of the original constrained problem. In terms of mechanics, if the original problem describes the equilibrium of a constrained elastic body subject to traction, then the modified problem is unconstrained with the constraints replaced by the reaction forces.

2.4.2 Existence and Uniqueness

Using the optimality conditions of Sect. 2.4.1, we can formulate the conditions that guarantee the existence or uniqueness of a solution of (2.18).

Proposition 2.10. *Let the equality constrained problem (2.18) be defined by a symmetric matrix* $\mathsf{A} \in \mathbb{R}^{n\times n}$, *a constraint matrix* $\mathsf{B} \in \mathbb{R}^{m\times n}$ *whose column rank is less than* n, *and vectors* $\mathbf{b} \in \mathbb{R}^n, \mathbf{c} \in \mathrm{Im}\mathsf{B}$. *Let* R *denote a matrix whose columns span* $\mathrm{Ker}\mathsf{A}$ *and let* $\mathsf{A}|\mathrm{Ker}\mathsf{B}$ *be positive semidefinite. Then the following statements hold:*
(i) Problem (2.18) has a solution if and only if

$$\mathsf{R}^T\mathbf{b} \in \mathrm{Im}(\mathsf{R}^T\mathsf{B}^T). \tag{2.28}$$

(ii) If $\mathsf{A}|\mathrm{Ker}\mathsf{B}$ *is positive definite, then problem (2.18) has a unique solution.*
(iii) If $(\overline{\mathbf{x}}, \overline{\boldsymbol{\lambda}})$ *and* $(\overline{\mathbf{y}}, \overline{\boldsymbol{\mu}})$ *are KKT couples for problem (2.18), then*

$$\overline{\mathbf{x}} - \overline{\mathbf{y}} \in \mathrm{Ker}\mathsf{A} \quad \text{and} \quad \overline{\boldsymbol{\lambda}} - \overline{\boldsymbol{\mu}} \in \mathrm{Ker}\mathsf{B}^T.$$

In particular, if problem (2.18) has a solution and

$$\mathrm{Ker}\mathsf{B}^T = \{\mathbf{o}\},$$

then there is a unique Lagrange multiplier $\widehat{\boldsymbol{\lambda}}$.

Proof. (i) Using Proposition 2.8(ii), we have that problem (2.18) has a solution if and only if there is $\boldsymbol{\lambda}$ such that $\mathbf{b}-\mathsf{B}^T\boldsymbol{\lambda} \in \mathrm{Im}\mathsf{A}$, or, equivalently, that $\mathbf{b}-\mathsf{B}^T\boldsymbol{\lambda}$ is orthogonal to $\mathrm{Ker}\mathsf{A}$. The latter condition reads $\mathsf{R}^T\mathbf{b} - \mathsf{R}^T\mathsf{B}^T\boldsymbol{\lambda} = \mathbf{o}$ and can be rewritten as (2.28).
(ii) First observe that if $\mathsf{A}|\mathrm{Ker}\mathsf{B}$ is positive definite, then $f|\mathrm{Ker}\mathsf{B}$ is strictly convex by Proposition 2.2 and $f|\Omega_E$ is strictly convex by Corollary 2.3. Since Ω_E is closed, convex, and nonempty, it follows by Proposition 2.5(i) that the equality constrained problem (2.18) has a unique solution.
(iii) First observe that $\mathrm{Ker}\mathsf{B} = \{\mathbf{x} - \mathbf{y} : \mathbf{x}, \mathbf{y} \in \Omega_E\}$ and that f is convex on $\mathrm{Ker}\mathsf{B}$ by the assumption and Proposition 2.2. Thus if $\overline{\mathbf{x}}$ and $\overline{\mathbf{y}}$ are any solutions of (2.18), then by Proposition 2.4(ii) $\mathsf{A}\overline{\mathbf{x}} = \mathsf{A}\overline{\mathbf{y}}$. The rest follows by a simple analysis of the KKT conditions (2.25). □

If B is not a full row rank matrix and $\overline{\boldsymbol{\lambda}}$ is a Lagrange multiplier for (2.18), then by Proposition 2.10(iii) any Lagrange multiplier $\boldsymbol{\lambda}$ can be expressed in the form

$$\boldsymbol{\lambda} = \overline{\boldsymbol{\lambda}} + \boldsymbol{\delta}, \quad \boldsymbol{\delta} \in \mathrm{Ker}\mathsf{B}^T. \tag{2.29}$$

The Lagrange multiplier $\boldsymbol{\lambda}_{\mathrm{LS}}$ which minimizes the Euclidean norm is called the *least square Lagrange multiplier*; it is a unique multiplier which belongs to $\mathrm{Im}\mathsf{B}$. If $\overline{\boldsymbol{\lambda}}$ is a vector of Lagrange multipliers, then $\boldsymbol{\lambda}_{\mathrm{LS}}$ can be evaluated by

$$\boldsymbol{\lambda}_{\mathrm{LS}} = \left(\mathsf{B}^\dagger\right)^T \mathsf{B}^T\overline{\boldsymbol{\lambda}} \tag{2.30}$$

and

$$\overline{\boldsymbol{\lambda}} = \boldsymbol{\lambda}_{\mathrm{LS}} + \boldsymbol{\delta}, \ \boldsymbol{\delta} \in \mathrm{Ker}\mathsf{B}^T.$$

2.4.3 KKT Systems

If A is positive definite, then the unique solution $\widehat{\mathbf{x}}$ of (2.18) is by Proposition 2.8 fully determined by the matrix equation

$$\begin{bmatrix} \mathsf{A} & \mathsf{B}^T \\ \mathsf{B} & \mathsf{O} \end{bmatrix} \begin{bmatrix} \mathbf{x} \\ \boldsymbol{\lambda} \end{bmatrix} = \begin{bmatrix} \mathbf{b} \\ \mathbf{c} \end{bmatrix}, \tag{2.31}$$

which is known as the *Karush–Kuhn–Tucker system*, briefly *KKT system* or *KKT conditions for the equality constrained problem* (2.18). Proposition 2.8 does not require that the related KKT system is nonsingular, in agreement with observation that the solution of the equality constrained problem should not depend on the description of Ω_E.

An alternative proof of the uniqueness of the component $\widehat{\mathbf{x}}$ of the solution of the KKT system (2.31) for A positive definite and B with dependent rows can be obtained by analysis of the solutions of the homogeneous system

$$\begin{bmatrix} \mathsf{A} & \mathsf{B}^T \\ \mathsf{B} & \mathsf{O} \end{bmatrix} \begin{bmatrix} \mathbf{d} \\ \boldsymbol{\mu} \end{bmatrix} = \begin{bmatrix} \mathbf{o} \\ \mathbf{o} \end{bmatrix}. \tag{2.32}$$

Indeed, after multiplying the first block row of (2.32) by $\mathbf{d}^T$ on the left, $\mathbf{d} \in \mathrm{Ker}\mathsf{B}$, we get

$$\mathbf{d}^T\mathsf{A}\mathbf{d} + \mathbf{d}^T\mathsf{B}^T\boldsymbol{\mu} = 0.$$

Since $\mathsf{B}\mathbf{d} = \mathbf{o}$, it follows that $\mathbf{d}^T\mathsf{B}^T\boldsymbol{\mu} = (\mathsf{B}\mathbf{d})^T\boldsymbol{\mu} = 0$, so that $\mathbf{d}^T\mathsf{A}\mathbf{d} = 0$ and, due to the positive definiteness of A, also $\mathbf{d} = \mathbf{o}$. The same argument is valid for A positive semidefinite provided

$$\mathrm{Ker}\mathsf{A} \cap \mathrm{Ker}\mathsf{B} = \{\mathbf{o}\}.$$

If A and B are respectively positive definite and full row rank matrices, then we can directly evaluate the inverse of the matrix of the KKT system (2.31) to get

$$\begin{bmatrix} \mathsf{A} & \mathsf{B}^T \\ \mathsf{B} & \mathsf{O} \end{bmatrix}^{-1} = \begin{bmatrix} \mathsf{A}^{-1} - \mathsf{A}^{-1}\mathsf{B}^T\mathsf{S}^{-1}\mathsf{B}\mathsf{A}^{-1}, & \mathsf{A}^{-1}\mathsf{B}^T\mathsf{S}^{-1} \\ \mathsf{S}^{-1}\mathsf{B}\mathsf{A}^{-1}, & -\mathsf{S}^{-1} \end{bmatrix}, \tag{2.33}$$

where $\mathsf{S} = \mathsf{B}\mathsf{A}^{-1}\mathsf{B}^T$ denotes the *Schur complement matrix.*

Even though not very useful computationally, the inverse matrix is useful in analysis. In the following lemma, we use it to get information on the distribution of the eigenvalues of the spectrum of a matrix of the KKT system.

Lemma 2.11. *Let $\mathcal{A} \in \mathbb{R}^{(n+m)\times(n+m)}$ denote the matrix of the KKT system (2.31) with a positive definite matrix* A *and a full rank matrix* B. *Then $\mathcal{A}$ is nonsingular and its eigenvalues $\alpha_1 \geq \cdots \geq \alpha_{n+m}$ satisfy*

$$\alpha_1 \geq \cdots \geq \alpha_n \geq \lambda_{\min}(\mathsf{A}) > 0 > \alpha_{n+1} \geq \cdots \geq \alpha_{n+m},$$

where

$$\lambda_{\min}(\mathsf{A}) = \|\mathsf{A}^{-1}\|^{-1}$$

denotes the smallest eigenvalue of A.

Proof. Using repeatedly the Cauchy interlacing inequalities (1.21) to $\mathcal{A}$, we get

$$\alpha_1 \geq \lambda_1(\mathsf{A}),\ \alpha_2 \geq \lambda_2(\mathsf{A}), \ldots,\ \alpha_n \geq \lambda_n(\mathsf{A}) = \lambda_{\min}(\mathsf{A}),$$

where

$$\lambda_1(\mathsf{A}) \geq \cdots \geq \lambda_n(\mathsf{A})$$

denote the eigenvalues of A.

Now observe that if $\sigma_1 \geq \cdots \geq \sigma_m > 0$ denote the eigenvalues of the Schur complement $\mathsf{S} = \mathsf{B}\mathsf{A}^{-1}\mathsf{B}^T$, then by (1.26) $0 > -\sigma_1^{-1} \geq \cdots \geq -\sigma_m^{-1}$ are the eigenvalues of $-\mathsf{S}^{-1}$. Thus we can apply the Cauchy interlacing inequalities (1.21) to the formula (2.33) for $\mathcal{A}^{-1}$ to get the inequalities

$$-\sigma_1^{-1} \geq \mu_{n+1},\ -\sigma_2^{-1} \geq \mu_{n+2},\ \ldots,\ -\sigma_m^{-1} \geq \mu_{n+m}$$

for the m smallest eigenvalues $\mu_{n+1} \geq \cdots \geq \mu_{n+m}$ of $\mathcal{A}^{-1}$. We have thus proved that $\mathcal{A}^{-1}$ has at least m negative eigenvalues. Since $\mu_{n+1}^{-1}, \ldots, \mu_{n+m}^{-1}$

are the eigenvalues of $\mathcal{A}$, it follows that $\mathcal{A}$ has at least m negative eigenvalues. As $\mathcal{A}$ has altogether $n+m$ eigenvalues counting multiplicity and includes at least n positive ones, we conclude that $0 > \alpha_{n+1}$. □

Using the extreme singular values of B, Rusten and Winther [162] established stronger bounds on the eigenvalues of $\mathcal{A}$ including

$$\frac{1}{2}\left(\lambda_1(\mathsf{A}) - \sqrt{\lambda_1(\mathsf{A})^2 + 4\sigma_{\min}(\mathsf{B})^2}\right) \geq \alpha_{n+1},$$

where $\sigma_{\min}(\mathsf{B})$ denotes the smallest singular value of B. More results concerning the spectrum of $\mathcal{A}$ may be found also in Benzi, Golub, and Liesen [10].

2.4.4 Min-max, Dual, and Saddle Point Problems

To simplify our exposition, we shall assume in this subsection that A and B are positive definite and full row rank matrices, respectively, postponing the analysis of more general convex problems to Sects. 2.6.4 and 2.6.5. The assumptions imply that the related KKT system

$$\nabla_{\mathbf{x}} L_0(\mathbf{x}, \boldsymbol{\lambda}) = \mathsf{A}\mathbf{x} - \mathbf{b} + \mathsf{B}^T\boldsymbol{\lambda} = \mathbf{o}, \tag{2.34}$$

$$\nabla_{\boldsymbol{\lambda}} L_0(\mathbf{x}, \boldsymbol{\lambda}) = \mathsf{B}\mathbf{x} - \mathbf{c} = \mathbf{o} \tag{2.35}$$

has a unique solution $(\widehat{\mathbf{x}}, \widehat{\boldsymbol{\lambda}})$, which can be found by first solving (2.34) with respect to $\mathbf{x}$, and then substituting for $\mathbf{x}$ into (2.35) to get an equation for $\widehat{\boldsymbol{\lambda}}$. We shall now associate these two steps with optimization problems.

First observe that by the gradient argument of Proposition 2.1, equation (2.34) is just the condition for $\mathbf{x}$ to be the unconstrained minimizer of L_0 with respect to $\mathbf{x}$. Thus for a given $\boldsymbol{\lambda} \in \mathbb{R}^m$, the first step is equivalent to evaluating the minimizer

$$\mathbf{x} = \mathbf{x}(\boldsymbol{\lambda}) = \mathsf{A}^{-1}(\mathbf{b} - \mathsf{B}^T\boldsymbol{\lambda})$$

of $L_0(\mathbf{x}, \boldsymbol{\lambda})$ with respect to $\mathbf{x}$. We can use this observation to express explicitly the *dual function*

$$\begin{aligned}\Theta(\boldsymbol{\lambda}) &= \inf_{\mathbf{x}\in\mathbb{R}^n} L_0(\mathbf{x}, \boldsymbol{\lambda}) = \min_{\mathbf{x}\in\mathbb{R}^n} L_0(\mathbf{x}, \boldsymbol{\lambda}) = L_0(\mathbf{x}(\boldsymbol{\lambda}), \boldsymbol{\lambda}) \\ &= -\frac{1}{2}\boldsymbol{\lambda}^T\mathsf{B}\mathsf{A}^{-1}\mathsf{B}^T\boldsymbol{\lambda} + (\mathsf{B}\mathsf{A}^{-1}\mathbf{b} - \mathbf{c})^T\boldsymbol{\lambda} - \frac{1}{2}\mathbf{b}^T\mathsf{A}^{-1}\mathbf{b}\end{aligned} \tag{2.36}$$

and its gradient

$$\nabla\Theta(\boldsymbol{\lambda}) = -\mathsf{B}\mathsf{A}^{-1}\mathsf{B}^T\boldsymbol{\lambda} + (\mathsf{B}\mathsf{A}^{-1}\mathbf{b} - \mathbf{c}). \tag{2.37}$$

We can also substitute for $\mathbf{x}$ into (2.35) to get

$$-\mathsf{BA}^{-1}\mathsf{B}^T\boldsymbol{\lambda} + (\mathsf{BA}^{-1}\mathbf{b} - \mathbf{c}) = \mathbf{o}.$$

Comparing the left-hand side of the last equation with the explicit expression (2.37) for $\nabla\Theta(\boldsymbol{\lambda})$, we get that the last equation can be written in the form

$$\nabla\Theta(\boldsymbol{\lambda}) = \mathbf{o}.$$

As $\mathsf{BA}^{-1}\mathsf{B}^T$, the Hessian of $-\Theta$, is positive definite, we conclude that the latter is equivalent to the condition (2.8) for the minimizer of $-\Theta$ or, equivalently, for the maximizer of Θ. Therefore the KKT couple $(\widehat{\mathbf{x}}, \widehat{\boldsymbol{\lambda}})$ for the equality constrained problem (2.18) solves the *min-max problem*

$$L_0(\widehat{\mathbf{x}}, \widehat{\boldsymbol{\lambda}}) = \max_{\boldsymbol{\lambda}\in\mathbb{R}^m} \min_{\mathbf{x}\in\mathbb{R}^n} L_0(\mathbf{x}, \boldsymbol{\lambda}), \tag{2.38}$$

$\widehat{\boldsymbol{\lambda}}$ solves the *dual problem*

$$\Theta(\widehat{\boldsymbol{\lambda}}) = \max_{\boldsymbol{\lambda}\in\mathbb{R}^m} \Theta(\boldsymbol{\lambda}), \tag{2.39}$$

and, since $\widehat{\mathbf{x}}$ is feasible, we get

$$f(\widehat{\mathbf{x}}) = L_0(\widehat{\mathbf{x}}, \boldsymbol{\lambda}) = L_0(\widehat{\mathbf{x}}, \widehat{\boldsymbol{\lambda}}) = \Theta(\widehat{\boldsymbol{\lambda}}). \tag{2.40}$$

Moreover, it follows that

$$f(\widehat{\mathbf{x}}) = L_0(\widehat{\mathbf{x}}, \widehat{\boldsymbol{\lambda}}) = \min_{\mathbf{x}\in\mathbb{R}^n} L_0(\mathbf{x}, \widehat{\boldsymbol{\lambda}}) \le L_0(\mathbf{x}, \widehat{\boldsymbol{\lambda}}), \quad \mathbf{x} \in \mathbb{R}^n. \tag{2.41}$$

There is yet another equivalent problem which is related to the penalty method. Since

$$\sup_{\boldsymbol{\lambda}\in\mathbb{R}^m} L_0(\mathbf{x}, \boldsymbol{\lambda}) = \infty \text{ for } \mathbf{x} \notin \Omega_E \quad \text{and} \quad \sup_{\boldsymbol{\lambda}\in\mathbb{R}^m} L_0(\mathbf{x}, \boldsymbol{\lambda}) = f(\mathbf{x}) \text{ for } \mathbf{x} \in \Omega_E,$$

it follows that the solution $\widehat{\mathbf{x}}$ of the KKT system (2.32) satisfies

$$f(\widehat{\mathbf{x}}) = \min_{\mathbf{x}\in\mathbb{R}^n} \sup_{\boldsymbol{\lambda}\in\mathbb{R}^m} L_0(\mathbf{x}, \boldsymbol{\lambda}). \tag{2.42}$$

Comparing (2.42) with (2.38) and (2.40), we get the well-known *duality relation*

$$\max_{\boldsymbol{\lambda}\in\mathbb{R}^m} \min_{\mathbf{x}\in\mathbb{R}^n} L_0(\mathbf{x}, \boldsymbol{\lambda}) = \min_{\mathbf{x}\in\mathbb{R}^n} \sup_{\boldsymbol{\lambda}\in\mathbb{R}^m} L_0(\mathbf{x}, \boldsymbol{\lambda}). \tag{2.43}$$

Using (2.40) and (2.41), we get that $(\widehat{\mathbf{x}}, \widehat{\boldsymbol{\lambda}})$ solves the *saddle point problem* to find $(\widehat{\mathbf{x}}, \widehat{\boldsymbol{\lambda}})$ so that for any $\mathbf{x} \in \mathbb{R}^n$ and $\boldsymbol{\lambda} \in \mathbb{R}^m$

$$L_0(\widehat{\mathbf{x}}, \boldsymbol{\lambda}) \le L_0(\widehat{\mathbf{x}}, \widehat{\boldsymbol{\lambda}}) \le L_0(\mathbf{x}, \widehat{\boldsymbol{\lambda}}). \tag{2.44}$$

We have thus obtained two *unconstrained* problems which are equivalent to the original equality constrained problem (2.18). The saddle point formulation enhances explicitly the Lagrange multipliers and is unconstrained at the cost of two sets of variables, while the dual formulation may enjoy a small dimension at the cost of dealing with more complex matrices. The dual problem may be also better conditioned. Notice that the left inequality in (2.44) can be replaced by the equality.

2.4.5 Sensitivity

The Lagrange multipliers emerged in Proposition 2.8 as auxiliary variables which nobody had asked for, but which turned out to be useful in alternative formulations of the optimality conditions. However, it turns out that the Lagrange multipliers frequently have an interesting interpretation in specific practical contexts, as we have mentioned at the end of Sect. 2.4.1, where we briefly described their mechanical interpretation. Here we show that if they are uniquely determined by the KKT conditions (2.31), then they are related to the rates of change of the optimal cost due to the violation of constraints.

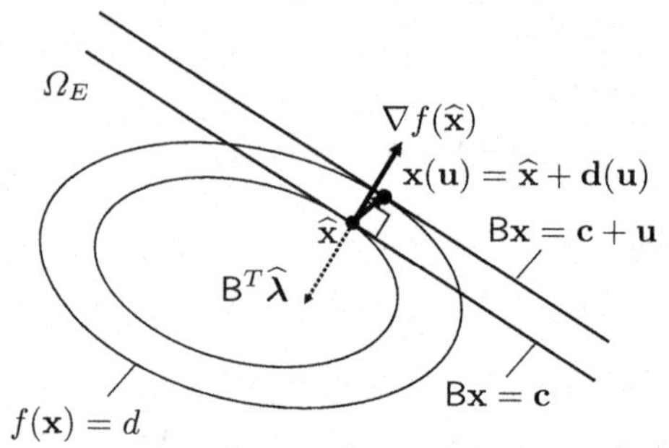

Fig. 2.9. Minimization with perturbed constraints

Let us assume that A and B are positive definite and full rank matrices, respectively, so that there is a unique KKT couple $(\widehat{\mathbf{x}}, \widehat{\boldsymbol{\lambda}})$ of the equality constrained problem (2.18). For $\mathbf{u} \in \mathbb{R}^m$, let us consider also the perturbed problem

$$\min_{\mathsf{B}\mathbf{x}=\mathbf{c}+\mathbf{u}} f(\mathbf{x})$$

as in Fig. 2.9. Its solution $\mathbf{x}(\mathbf{u})$ and the corresponding vector of Lagrange multipliers $\boldsymbol{\lambda}(\mathbf{u})$ are fully determined by the KKT conditions

$$\begin{bmatrix} \mathsf{A} & \mathsf{B}^T \\ \mathsf{B} & \mathsf{O} \end{bmatrix} \begin{bmatrix} \mathbf{x}(\mathbf{u}) \\ \boldsymbol{\lambda}(\mathbf{u}) \end{bmatrix} = \begin{bmatrix} \mathbf{b} \\ \mathbf{c}+\mathbf{u} \end{bmatrix},$$

so that

$$\begin{bmatrix} \mathbf{x}(\mathbf{u}) \\ \boldsymbol{\lambda}(\mathbf{u}) \end{bmatrix} = \begin{bmatrix} \mathsf{A} & \mathsf{B}^T \\ \mathsf{B} & \mathsf{O} \end{bmatrix}^{-1} \begin{bmatrix} \mathbf{b} \\ \mathbf{c}+\mathbf{u} \end{bmatrix} = \begin{bmatrix} \mathsf{A} & \mathsf{B}^T \\ \mathsf{B} & \mathsf{O} \end{bmatrix}^{-1} \begin{bmatrix} \mathbf{b} \\ \mathbf{c} \end{bmatrix} + \begin{bmatrix} \mathsf{A} & \mathsf{B}^T \\ \mathsf{B} & \mathsf{O} \end{bmatrix}^{-1} \begin{bmatrix} \mathbf{o} \\ \mathbf{u} \end{bmatrix}.$$

First observe that $\mathbf{d}(\mathbf{u}) = \mathbf{x}(\mathbf{u}) - \widehat{\mathbf{x}}$ satisfies

$$\mathsf{B}\mathbf{d}(\mathbf{u}) = \mathsf{B}\mathbf{x}(\mathbf{u}) - \mathsf{B}\widehat{\mathbf{x}} = \mathbf{u},$$

so that we can use $\nabla f(\widehat{\mathbf{x}}) = -\mathsf{B}^T\widehat{\boldsymbol{\lambda}}$ to approximate the change of optimal cost by

$$\nabla f(\widehat{\mathbf{x}})^T \mathbf{d}(\mathbf{u}) = -(\mathsf{B}^T\widehat{\boldsymbol{\lambda}})^T \mathbf{d}(\mathbf{u}) = -\widehat{\boldsymbol{\lambda}}^T \mathsf{B}\mathbf{d}(\mathbf{u}) = -\widehat{\boldsymbol{\lambda}}^T \mathbf{u}.$$

It follows that $-[\widehat{\boldsymbol{\lambda}}]_i$ can be used to approximate the change of the optimal cost due to the violation of the ith constraint by $[\mathbf{u}]_i$.

To give more detailed analysis of the sensitivity of the optimal cost with respect to the violation of constraints, let us define for each $\mathbf{u} \in \mathbb{R}^m$ the *primal function*

$$p(\mathbf{u}) = f\left(\mathbf{x}(\mathbf{u})\right).$$

Observing that $\widehat{\mathbf{x}} = \mathbf{x}(\mathbf{o})$ and using the explicit formula (2.33) to evaluate the inverse of the KKT system, we get

$$\mathbf{x}(\mathbf{u}) = \widehat{\mathbf{x}} + \mathsf{A}^{-1}\mathsf{B}^T\mathsf{S}^{-1}\mathbf{u},$$

where $\mathsf{S} = \mathsf{B}\mathsf{A}^{-1}\mathsf{B}^T$ denotes the Schur complement matrix. Thus

$$\mathbf{x}(\mathbf{u}) - \widehat{\mathbf{x}} = \mathsf{A}^{-1}\mathsf{B}^T\mathsf{S}^{-1}\mathbf{u},$$

so that

$$\begin{aligned} p(\mathbf{u}) - p(\mathbf{o}) &= f\left(\mathbf{x}(\mathbf{u})\right) - f(\widehat{\mathbf{x}}) \\ &= \nabla f(\widehat{\mathbf{x}})^T\left(\mathbf{x}(\mathbf{u}) - \widehat{\mathbf{x}}\right) + \frac{1}{2}\left(\mathbf{x}(\mathbf{u}) - \widehat{\mathbf{x}}\right)^T \mathsf{A}\left(\mathbf{x}(\mathbf{u}) - \widehat{\mathbf{x}}\right) \\ &= \nabla f(\widehat{\mathbf{x}})^T\mathsf{A}^{-1}\mathsf{B}^T\mathsf{S}^{-1}\mathbf{u} + \frac{1}{2}\mathbf{u}^T\mathsf{S}^{-1}\mathsf{B}\mathsf{A}^{-1}\mathsf{B}^T\mathsf{S}^{-1}\mathbf{u}. \end{aligned}$$

It follows that the gradient of the primal function p at $\mathbf{o}$ is given by

$$\nabla p(\mathbf{o}) = \left(\nabla f(\widehat{\mathbf{x}})^T\mathsf{A}^{-1}\mathsf{B}^T\mathsf{S}^{-1}\right)^T = \mathsf{S}^{-1}\mathsf{B}\mathsf{A}^{-1}\nabla f(\widehat{\mathbf{x}}).$$

Recalling that $\nabla f(\widehat{\mathbf{x}}) = -\mathsf{B}^T\widehat{\boldsymbol{\lambda}}$, we get

$$\nabla p(\mathbf{o}) = -\mathsf{S}^{-1}\mathsf{B}\mathsf{A}^{-1}\mathsf{B}^T\widehat{\boldsymbol{\lambda}} = -\widehat{\boldsymbol{\lambda}}. \tag{2.45}$$

Our analysis shows that if the total differential of f at $\widehat{\mathbf{x}}$ decreases outside Ω_E, then this decrease is compensated by the increase of $\widehat{\boldsymbol{\lambda}}^T(\mathsf{B}\mathbf{x} - \mathbf{c})$. See also Fig. 2.6. The components of $\widehat{\boldsymbol{\lambda}}$ are also called the *shadow prices* after their meaning in the applications in economics.

For the sensitivity analysis of the solution of more general equality constrained problems, we refer to the book by Bertsekas [12].

2.4.6 Error Analysis

We shall now give the bounds on the error of the solution of the KKT system (2.31) in terms of perturbation of the right-hand side. As we do not assume here that the constraints are necessarily defined by a full rank matrix B, we shall use bounds on the nonzero singular values of the constraint matrix.

Proposition 2.12. *Let matrices* A, B *and vectors* $\mathbf{b}, \mathbf{c}$ *be those from the definition of problem (2.18) with A SPD and* $\mathsf{B} \in \mathbb{R}^{m \times n}$ *not necessarily a full rank matrix of the column rank less than n. Let* $\lambda_{\min}(\mathsf{A})$ *denote the least eigenvalue of* A*, let* $\overline{\sigma}_{\min}(\mathsf{B})$ *denote the least nonzero singular value of* B*, let* $(\widehat{\mathbf{x}}, \overline{\boldsymbol{\lambda}})$ *denote any KKT pair for the equality constrained problem (2.18), let* $\mathbf{g} \in \mathbb{R}^n$*,* $\mathbf{e} \in \mathbb{R}^m$*, and let* $(\mathbf{x}, \boldsymbol{\lambda})$ *denote an approximate KKT pair which satisfies*

$$\begin{aligned} \mathsf{A}\mathbf{x} + \mathsf{B}^T\boldsymbol{\lambda} &= \mathbf{b} + \mathbf{g}, \\ \mathsf{B}\mathbf{x} \qquad &= \mathbf{c} + \mathbf{e}. \end{aligned} \tag{2.46}$$

Then

$$\|\mathsf{B}^T(\boldsymbol{\lambda} - \overline{\boldsymbol{\lambda}})\| \leq \kappa(\mathsf{A})\|\mathbf{g}\| + \frac{\|\mathsf{A}\|}{\overline{\sigma}_{\min}(\mathsf{B})}\|\mathbf{e}\| \tag{2.47}$$

and

$$\|\mathbf{x} - \widehat{\mathbf{x}}\| \leq \frac{\kappa(\mathsf{A}) + 1}{\lambda_{\min}(\mathsf{A})}\|\mathbf{g}\| + \frac{\kappa(\mathsf{A})}{\overline{\sigma}_{\min}(\mathsf{B})}\|\mathbf{e}\|. \tag{2.48}$$

Moreover, if $\boldsymbol{\lambda}_{\mathrm{LS}}$ *denotes the least square Lagrange multiplier for (2.18) and* $\boldsymbol{\lambda} \in \mathrm{Im}\mathsf{B}$*, then*

$$\|\boldsymbol{\lambda} - \boldsymbol{\lambda}_{\mathrm{LS}}\| \leq \frac{1}{\overline{\sigma}_{\min}(\mathsf{B})}\left(\kappa(\mathsf{A})\|\mathbf{g}\| + \frac{\|\mathsf{A}\|}{\overline{\sigma}_{\min}(\mathsf{B})}\|\mathbf{e}\|\right). \tag{2.49}$$

Proof. Let us recall that we assume $\mathbf{c} \in \mathrm{Im}\mathsf{B}$, so that also $\mathbf{e} \in \mathrm{Im}\mathsf{B}$. If $\mathsf{B}^\dagger$ denotes the Moore–Penrose pseudoinverse of B, it follows that $\boldsymbol{\delta} = \mathsf{B}^\dagger\mathbf{e}$ satisfies $\mathsf{B}\boldsymbol{\delta} = \mathbf{e}$ and $\|\boldsymbol{\delta}\| \leq \overline{\sigma}_{\min}(\mathsf{B})^{-1}\|\mathbf{e}\|$ (see (1.38)). Moreover, $\mathbf{y} = \mathbf{x} - \boldsymbol{\delta}$ satisfies

$$\begin{aligned} \mathsf{A}(\mathbf{y} - \widehat{\mathbf{x}}) + \mathsf{B}^T(\boldsymbol{\lambda} - \overline{\boldsymbol{\lambda}}) &= \mathbf{g} + \mathsf{A}\boldsymbol{\delta}, \\ \mathsf{B}(\mathbf{y} - \widehat{\mathbf{x}}) \qquad &= \mathbf{o}. \end{aligned}$$

After eliminating $\mathbf{y} - \widehat{\mathbf{x}}$ from the first equation, we get

$$\mathsf{B}\mathsf{A}^{-1}\mathsf{B}^T(\boldsymbol{\lambda} - \overline{\boldsymbol{\lambda}}) = \mathsf{B}(\mathsf{A}^{-1}\mathbf{g} + \boldsymbol{\delta}),$$

so that, after multiplication on the left by $(\boldsymbol{\lambda} - \overline{\boldsymbol{\lambda}})^T$ and taking norms, we get

$$\|\mathsf{A}\|^{-1}\|\mathsf{B}^T(\boldsymbol{\lambda} - \overline{\boldsymbol{\lambda}})\|^2 \leq \|\mathsf{B}^T(\boldsymbol{\lambda} - \overline{\boldsymbol{\lambda}})\|\|\mathsf{A}^{-1}\mathbf{g} + \boldsymbol{\delta}\|.$$

Thus

$$\|\mathsf{B}^T(\boldsymbol{\lambda} - \overline{\boldsymbol{\lambda}})\| \leq \kappa(\mathsf{A})\|\mathbf{g}\| + \|\mathsf{A}\|\|\boldsymbol{\delta}\| \leq \kappa(\mathsf{A})\|\mathbf{g}\| + \frac{\|\mathsf{A}\|}{\overline{\sigma}_{\min}(\mathsf{B})}\|\mathbf{e}\|.$$

After subtracting $\mathsf{A}\widehat{\mathbf{x}} + \mathsf{B}^T\overline{\boldsymbol{\lambda}} = \mathbf{b}$ from the first equation of (2.46), multiplying the result on the left by A^{-1}, and taking the norms, we get

$$\|\mathbf{x} - \widehat{\mathbf{x}}\| = \|\mathsf{A}^{-1}(\mathbf{g} - \mathsf{B}^T(\boldsymbol{\lambda} - \overline{\boldsymbol{\lambda}}))\| \leq \frac{\kappa(\mathsf{A}) + 1}{\lambda_{\min}(\mathsf{A})}\|\mathbf{g}\| + \frac{\kappa(\mathsf{A})}{\overline{\sigma}_{\min}(\mathsf{B})}\|\mathbf{e}\|.$$

If $\boldsymbol{\lambda} \in \mathrm{Im}\mathsf{B}$, then $\boldsymbol{\lambda} - \boldsymbol{\lambda}_{\mathrm{LS}} \in \mathrm{Im}\mathsf{B}$ and

$$\boldsymbol{\lambda} - \boldsymbol{\lambda}_{\mathrm{LS}} = \left(\mathsf{B}^\dagger\right)^T\mathsf{B}^T(\boldsymbol{\lambda} - \boldsymbol{\lambda}_{\mathrm{LS}}).$$

The last inequality then follows by (1.37) and (2.47). □

2.5 Inequality Constrained Problems

Let us now consider the problems whose feasible sets are described by linear inequalities. Such sets are also called the *polyhedral sets.* More formally, we look for

$$\min_{\mathbf{x}\in\Omega_I} f(\mathbf{x}), \tag{2.50}$$

where f is a quadratic function defined by (2.4), $\Omega_I = \{\mathbf{x} \in \mathbb{R}^n : \mathsf{B}\mathbf{x} \le \mathbf{c}\}$, $\mathsf{B} = [\mathbf{b}_1, \dots, \mathbf{b}_m]^T \in \mathbb{R}^{m\times n}$, and $\mathbf{c} = [c_i] \in \mathbb{R}^m$. We assume that $\Omega_I \neq \emptyset$.

At any feasible point $\mathbf{x}$, we define the *active set*

$$\mathcal{A}(\mathbf{x}) = \{i \in \{1, \dots, m\} : \mathbf{b}_i^T\mathbf{x} = c_i\}.$$

In particular, if $\overline{\mathbf{x}}$ is a local solution of (2.50), then each feasible direction of $\overline{\Omega}_E = \{\mathbf{x} \in \mathbb{R}^n : [\mathsf{B}\mathbf{x}]_{\mathcal{A}(\overline{\mathbf{x}})} = \mathbf{c}_{\mathcal{A}(\overline{\mathbf{x}})}\}$ at $\overline{\mathbf{x}}$ is a feasible direction of Ω_I at $\overline{\mathbf{x}}$. Using the arguments of Sect. 2.4.1, we get that $\overline{\mathbf{x}}$ is also a local solution of the equality constrained problem

$$\min_{\mathbf{x}\in\overline{\Omega}_E} f(\mathbf{x}), \quad \overline{\Omega}_E = \{\mathbf{x} \in \mathbb{R}^n : [\mathsf{B}\mathbf{x}]_{\mathcal{A}(\overline{\mathbf{x}})} = \mathbf{c}_{\mathcal{A}(\overline{\mathbf{x}})}\}. \tag{2.51}$$

Thus (2.50) is a more difficult problem than the equality constrained problem (2.18) as its solution necessarily enhances the identification of $\mathcal{A}(\overline{\mathbf{x}})$.

2.5.1 Polyhedral Sets

To understand the conditions of solvability of the inequality constrained problem (2.50), it is useful to get some insight into the geometry of polyhedral sets. We shall need a few new concepts.

A set $\mathcal{C} \subseteq \mathbb{R}^n$ is a (convex) *cone* (with its vertex at the origin) if $\mathbf{x}+\mathbf{y} \in \mathcal{C}$ and $\alpha\mathbf{x} \in \mathcal{C}$ for all $\alpha \ge 0$, $\mathbf{x} \in \mathcal{C}$, and $\mathbf{y} \in \mathcal{C}$. We are interested in *polyhedral cones* which are defined by

$$\mathcal{C} = \{\mathbf{x} \in \mathbb{R}^n : \mathsf{B}\mathbf{x} \le \mathbf{o}\},$$

where $\mathsf{B} \in \mathbb{R}^{m\times n}$ is a given matrix. The *Minkowski–Weyl Theorem* (see, e.g., [12]) says that polyhedral cones are *finitely generated,* i.e., there is a matrix C such that

$$\mathcal{C} = \{\mathbf{x} \in \mathbb{R}^n : \mathbf{x} = \mathsf{C}\mathbf{y}, \ \mathbf{y} \ge \mathbf{o}\}.$$

A polyhedral cone is a closed convex set.

A polyhedral set Ω can be represented as the sum of the convex hull of a finite set of points and a polyhedral cone whose elements are the recession directions of Ω. Let us formulate this nontrivial statement more formally.

Proposition 2.13. *A set $\Omega \subseteq \mathbb{R}^n$ is polyhedral if and only if there is a nonempty set of n-vectors $\{\mathbf{x}_1, \dots, \mathbf{x}_k\}$ and a polyhedral cone $\mathcal{C} \subseteq \mathbb{R}^n$ such that*

$$\Omega = \mathcal{C} + \mathrm{Conv}\{\mathbf{x}_1, \dots, \mathbf{x}_k\}.$$

Proof. See, e.g., [12, Proposition B.17]. □

2.5.2 Farkas's Lemma

The main tool for the transformation of the geometrical conditions of optimality for the inequality constrained problems to their algebraic form with Lagrange multipliers is the following lemma by Farkas.

Lemma 2.14. *Let* $\mathsf{B} \in \mathbb{R}^{m\times n}$ *and* $\mathbf{h} \in \mathbb{R}^n$*. Then exactly one of the following problems has a solution:*

$$\begin{array}{lllllll} (I) & \textit{Find} & \mathbf{d} \in \mathbb{R}^n & \textit{such that} & \mathsf{B}\mathbf{d} \le \mathbf{o} & \textit{and} & \mathbf{h}^T\mathbf{d} > 0. \\ (II) & \textit{Find} & \mathbf{y} \in \mathbb{R}^m & \textit{such that} & \mathsf{B}^T\mathbf{y} = \mathbf{h} & \textit{and} & \mathbf{y} \ge \mathbf{o}. \end{array}$$

Proof. Suppose first that (II) has a solution, so that there is $\mathbf{y} \ge \mathbf{o}$ such that $\mathsf{B}^T\mathbf{y} = \mathbf{h}$. Let $\mathbf{d} \in \mathbb{R}^n$ be such that $\mathsf{B}\mathbf{d} \le \mathbf{o}$. Then

$$\mathbf{h}^T\mathbf{d} = \left(\mathsf{B}^T\mathbf{y}\right)^T\mathbf{d} = \mathbf{y}^T\mathsf{B}\mathbf{d} \le 0,$$

so that the problem (I) has no solution.

Now suppose that the problem (II) has no solution and denote

$$\Omega = \{\mathbf{x} \in \mathbb{R}^n : \ \mathbf{x} = \mathsf{B}^T\mathbf{y}, \ \mathbf{y} \ge \mathbf{o}\},$$

so that our assumption amounts to $\mathbf{h} \notin \Omega$. Denoting by $\widehat{\mathbf{h}} \in \Omega$ the projection of $\mathbf{h}$ to Ω and $\mathbf{d} = \mathbf{h} - \widehat{\mathbf{h}}$, we get by (2.16) that for any $\mathbf{x} \in \Omega$

$$\mathbf{d}^T(\mathbf{x} - \widehat{\mathbf{h}}) = (\mathbf{h} - \widehat{\mathbf{h}})^T(\mathbf{x} - \widehat{\mathbf{h}}) \le 0,$$

or alternatively

$$\mathbf{d}^T\mathbf{x} \le \mathbf{d}^T\widehat{\mathbf{h}} = \alpha.$$

Observing that $\mathbf{o} \in \Omega$, we get $\alpha \ge 0$. Moreover, substituting $\mathbf{x} = \mathsf{B}^T\mathbf{y}$, we get for any $\mathbf{y} \ge \mathbf{o}$

$$\mathbf{y}^T\mathsf{B}\mathbf{d} = \mathbf{d}^T\mathsf{B}^T\mathbf{y} \le \alpha.$$

Since the components of $\mathbf{y}$ can be arbitrarily large, $\mathsf{B}\mathbf{d} \le \mathbf{o}$. Thus $\mathbf{d}$ satisfies the first inequality of (I). To check the second one, recall that by our assumption $\mathbf{h} \notin \Omega$. It follows that $\mathbf{d} \ne \mathbf{o}$, so that $\mathbf{d}^T(\mathbf{h} - \widehat{\mathbf{h}}) = \|\mathbf{d}\|^2 > 0$ and

$$\mathbf{d}^T\mathbf{h} > \mathbf{d}^T\widehat{\mathbf{h}} = \alpha \ge 0.$$

Thus $\mathbf{d}$ is a solution of (I). □

Farkas's lemma is used in the proof of the KKT conditions for inequality constrained QP problems in a similar way as the statement that $\mathrm{Im}\mathsf{B}^T$ is the orthogonal complement of $\mathrm{Ker}\mathsf{B}$ in the analysis of equality constrained problems. A geometric illustration of Farkas's lemma is in Fig. 2.10.

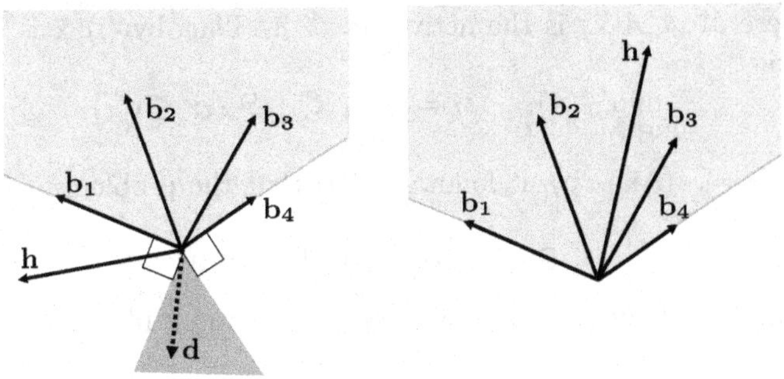

Fig. 2.10. Farkas's lemma: solution of (I) (left) and (II) (right)

2.5.3 Necessary Optimality Conditions for Local Solutions

The structure of inequality constrained QP problems (2.50) is more complicated than that of the equality constrained ones (2.18). We shall start our exposition with the following necessary optimality conditions.

Proposition 2.15. *Let the inequality constrained problem (2.50) be defined by a symmetric matrix* $\mathsf{A} \in \mathbb{R}^{n\times n}$*, the constraint matrix* $\mathsf{B} \in \mathbb{R}^{m\times n}$ *whose column rank is less than* n*, and the vectors* $\mathbf{b}, \mathbf{c}$*. Let* $\mathcal{C}$ *denote the cone of directions of the feasible set* Ω_I*. Then the following statements hold:*
(i) If $\overline{\mathbf{x}} \in \Omega_I$ *is a local solution of (2.50), then*

$$(\mathsf{A}\overline{\mathbf{x}} - \mathbf{b})^T \mathbf{d} \geq 0 \tag{2.52}$$

for any feasible direction $\mathbf{d}$ *of* Ω_I *at* $\overline{\mathbf{x}}$*.*
(ii) If $\overline{\mathbf{x}} \in \Omega_I$ *is a local solution of (2.50), then there is* $\overline{\boldsymbol{\lambda}} \in \mathbb{R}^m$ *such that*

$$\overline{\boldsymbol{\lambda}} \geq \mathbf{o}, \quad \mathsf{A}\overline{\mathbf{x}} - \mathbf{b} + \mathsf{B}^T\overline{\boldsymbol{\lambda}} = \mathbf{o}, \quad \text{and} \quad \overline{\boldsymbol{\lambda}}^T(\mathsf{B}\overline{\mathbf{x}} - \mathbf{c}) = 0. \tag{2.53}$$

Proof. (i) Let $\overline{\mathbf{x}}$ be a local solution of the inequality constrained problem (2.50) and let $\mathbf{d}$ denote a feasible direction of Ω_I at $\overline{\mathbf{x}}$, so that the right-hand side of

$$f(\overline{\mathbf{x}} + \alpha\mathbf{d}) - f(\overline{\mathbf{x}}) = \alpha(\mathsf{A}\overline{\mathbf{x}} - \mathbf{b})^T\mathbf{d} + \frac{\alpha^2}{2}\mathbf{d}^T\mathsf{A}\mathbf{d} \tag{2.54}$$

is nonnegative for all sufficiently small $\alpha > 0$. To prove (2.52), it is enough to take $\alpha > 0$ so small that the nonnegativity of the right-hand side of (2.54) implies that

$$\alpha(\mathsf{A}\overline{\mathbf{x}} - \mathbf{b})^T\mathbf{d} \geq 0.$$

(ii) First observe that if $\overline{\mathbf{x}}$ is a local solution of (2.50), then $\mathbf{d}$ is a feasible direction of Ω_I at $\overline{\mathbf{x}}$ if and only if $\mathbf{d}$ is a feasible direction of

$$\overline{\Omega}_I = \{\mathbf{x} \in \mathbb{R}^n : \mathsf{B}_{\mathcal{A}*}\mathbf{x} \leq \mathbf{c}_{\mathcal{A}}\}$$

at $\overline{\mathbf{x}}$, where $\mathcal{A} = \mathcal{A}(\overline{\mathbf{x}})$ is the active set of $\overline{\mathbf{x}}$. Thus by (i) $\overline{\mathbf{x}}$ is also a local solution of

$$\min_{\mathbf{x} \in \overline{\Omega}_I} f(\mathbf{x}), \quad \overline{\Omega}_I = \{\mathbf{x} \in \mathbb{R}^n : \mathsf{B}_{\mathcal{A}*}\mathbf{x} \le \mathbf{c}_{\mathcal{A}}\}.$$

Denoting $\mathbf{h} = -(\mathsf{A}\overline{\mathbf{x}} - \mathbf{b})$, it follows by (i) that the problem to find $\mathbf{d} \in \mathbb{R}^n$ such that

$$\mathsf{B}_{\mathcal{A}*}\mathbf{d} \le \mathbf{o} \quad \text{and} \quad \mathbf{h}^T\mathbf{d} > 0$$

has no solution. Thus we can apply Farkas's lemma 2.14 to get $\mathbf{y} \in \mathbb{R}^m$ such that

$$(\mathsf{B}_{\overline{\mathcal{A}}*})^T\mathbf{y} = \mathbf{h}_{\mathcal{A}} \quad \text{and} \quad \mathbf{y} \ge \mathbf{o}.$$

Denoting by $\overline{\boldsymbol{\lambda}} \in \mathbb{R}^m$ the vector obtained by padding $\mathbf{y}$ with zeros, so that $[\overline{\boldsymbol{\lambda}}]_i = 0$ for $i \notin \mathcal{A}$ and $\overline{\boldsymbol{\lambda}}_{\mathcal{A}} = \mathbf{y}$, it is easy to check that $\overline{\boldsymbol{\lambda}}$ satisfies (2.53). □

The conditions (2.53) are called the *KKT conditions for inequality constraints.* The last of these conditions, the equation $\overline{\boldsymbol{\lambda}}^T(\mathsf{B}\overline{\mathbf{x}} - \mathbf{c}) = 0$, is called the *condition of complementarity.* Notice that (ii) can be proved without any reference to Farkas's lemma as any solution $\overline{\mathbf{x}}$ of (2.50) solves (2.51), so that by Proposition 2.8(ii) there is $\mathbf{y}$ such that

$$\mathsf{A}\overline{\mathbf{x}} - \mathbf{b} + \mathsf{B}^T_{\mathcal{A}(\overline{\mathbf{x}})}\mathbf{y} = \mathbf{c}_{\mathcal{A}(\overline{\mathbf{x}})},$$

and $\mathbf{y} \ge \mathbf{o}$ by the arguments based on the discussion of sensitivity of the minimum in Sect. 2.4.5.

2.5.4 Existence and Uniqueness

In our discussion of the existence and uniqueness results for the inequality constrained QP problem (2.50), we restrict our attention to the following results that are useful in our applications.

Proposition 2.16. *Let the inequality constrained problem (2.50) be defined by a symmetric matrix* $\mathsf{A} \in \mathbb{R}^{n\times n}$*, a constraint matrix* $\mathsf{B} \in \mathbb{R}^{m\times n}$*, and vectors* $\mathbf{b}, \mathbf{c}$*. Let* C *denote the cone of recession directions of the nonempty feasible set* Ω_I*. Then the following statements hold:*
(i) If problem (2.50) has a solution, then $\mathbf{d}^T\mathsf{A}\mathbf{d} \ge 0$ *for* $\mathbf{d} \in C$ *and*

$$\mathbf{d}^T\mathbf{b} \le 0 \quad for \quad \mathbf{d} \in C \cap \mathrm{Ker}\mathsf{A}. \tag{2.55}$$

(ii) If (2.55) holds and f *is convex, then problem (2.50) has a solution.*
(iii) If f *is convex and* $(\overline{\mathbf{x}}, \overline{\boldsymbol{\lambda}})$ *and* $(\overline{\mathbf{y}}, \overline{\boldsymbol{\mu}})$ *are KKT couples for (2.50), then*

$$\overline{\mathbf{x}} - \overline{\mathbf{y}} \in \mathrm{Ker}\mathsf{A} \quad and \quad \overline{\boldsymbol{\lambda}} - \overline{\boldsymbol{\mu}} \in \mathrm{Ker}\mathsf{B}^T. \tag{2.56}$$

(iv) If A *is positive definite, then the inequality constrained minimization problem (2.50) has the unique solution.*

Proof. (i) Let $\overline{\mathbf{x}}$ be a global solution of the inequality constrained minimization problem (2.50), and recall that

$$f(\overline{\mathbf{x}}+\alpha\mathbf{d}) - f(\overline{\mathbf{x}}) = \alpha(\mathsf{A}\overline{\mathbf{x}} - \mathbf{b})^T\mathbf{d} + \frac{\alpha^2}{2}\mathbf{d}^T\mathsf{A}\mathbf{d} \tag{2.57}$$

for any $\mathbf{d} \in \mathbb{R}^n$ and $\alpha \in \mathbb{R}$. Taking $\mathbf{d} \in \mathcal{C}$ arbitrary but fixed and α sufficiently large, we get that the nonnegativity of the right-hand side requires $\mathbf{d}^T\mathsf{A}\mathbf{d} \geq 0$. Moreover, if $\mathbf{d} \in \mathcal{C} \cap \mathrm{Ker}\mathsf{A}$, then (2.57) reduces to

$$f(\overline{\mathbf{x}}+\alpha\mathbf{d}) - f(\overline{\mathbf{x}}) = -\alpha\mathbf{b}^T\mathbf{d},$$

which is nonnegative for any $\alpha \geq 0$ if and only if $\mathbf{b}^T\mathbf{d} \leq 0$.
(ii) Let us now assume that (2.55) is satisfied and observe that if $\mathbf{c} = \mathbf{o}$, then Ω_I is a cone, so that a solution is known to exist even in infinite dimension (see Zeidler [184, pp. 553–556]).

If $\mathbf{c}$ is arbitrary, then by Proposition 2.13 there are $\mathbf{x}_1, \ldots, \mathbf{x}_k \in \Omega_I$ such that

$$\Omega_I = \mathcal{C} + \mathrm{conv}\{\mathbf{x}_1, \ldots, \mathbf{x}_k\}, \quad \mathcal{C} = \{\mathbf{x} : \mathsf{B}\mathbf{x} \leq \mathbf{o}\}.$$

Observing that $\mathbf{d} \in \mathcal{C}$ if and only if $2\mathbf{d} \in \mathcal{C}$, we get that $\mathbf{x} \in \Omega_I$ if and only if

$$\mathbf{x} = 2\mathbf{d} + \mathbf{y}, \quad \mathbf{d} \in \mathcal{C}, \quad \mathbf{y} \in \mathrm{Conv}\{\mathbf{x}_1, \ldots, \mathbf{x}_k\}.$$

Thus

$$\begin{aligned} f(\mathbf{x}) = f(2\mathbf{d}+\mathbf{y}) &= \mathbf{d}^T\mathsf{A}\mathbf{d} - 2\mathbf{b}^T\mathbf{d} + \mathbf{d}^T\mathsf{A}\mathbf{d} + 2\mathbf{d}^T\mathsf{A}\mathbf{y} + \frac{1}{2}\mathbf{y}^T\mathsf{A}\mathbf{y} - \mathbf{b}^T\mathbf{y} \\ &\geq 2f(\mathbf{d}) + (\mathbf{d}^T\mathsf{A}\mathbf{d} + 2\mathbf{d}^T\mathsf{A}\mathbf{y}) - \mathbf{b}^T\mathbf{y}. \end{aligned} \tag{2.58}$$

We have already seen that f is bounded from below on $\mathcal{C}$. Moreover, using the Euclidean norm, we get

$$-\mathbf{b}^T\mathbf{y} \geq -\|\mathbf{b}\| \max\{\|\mathbf{x}_1\|, \ldots, \|\mathbf{x}_k\|\}$$

and by (2.10)

$$\mathbf{d}^T\mathsf{A}\mathbf{d} + 2\mathbf{d}^T\mathsf{A}\mathbf{y} \geq -(\mathsf{A}\mathbf{y})^T\mathsf{A}^\dagger\mathsf{A}\mathbf{y} = -\mathbf{y}^T\mathsf{A}\mathbf{y} \geq -\|\mathsf{A}\| \max\{\|\mathbf{x}_1\|^2, \ldots, \|\mathbf{x}_k\|^2\},$$

where $\mathsf{A}^\dagger$ denotes the Moore–Penrose generalized inverse to A. Thus f is bounded from below on Ω_I and we can use the Frank–Wolfe theorem (see Proposition 2.5(iii)) to finish the proof of (ii).
(iii) The first inclusion of (2.56) holds by Proposition 2.4(ii) for solutions of any convex problem. The inclusion for multipliers then follows by the KKT condition (2.53).
(iv) If A is positive definite, then f is strictly convex by Proposition 2.2, so that by Proposition 2.5 there is a unique minimizer of f subject to $\mathbf{x} \in \Omega_I$. □

2.5.5 Optimality Conditions for Convex Problems

When the cost function is convex, then the necessary conditions of Sect. 2.5.3 are also sufficient.

Proposition 2.17. *Let f be a convex quadratic function defined by (2.4) with a positive semidefinite Hessian matrix* A. *Then the following statements hold:*
(i) A vector $\overline{\mathbf{x}} \in \Omega_I$ *is a solution of (2.50) if and only if*

$$(\mathsf{A}\overline{\mathbf{x}} - \mathbf{b})^T\mathbf{d} \geq 0 \tag{2.59}$$

for any feasible direction $\mathbf{d}$ *of* Ω_I *at* $\overline{\mathbf{x}}$.
(ii) A vector $\overline{\mathbf{x}} \in \Omega_I$ *is a solution of (2.50) if and only if there is* $\overline{\boldsymbol{\lambda}} \in \mathbb{R}^m$ *such that*

$$\overline{\boldsymbol{\lambda}} \geq \mathbf{o}, \quad \mathsf{A}\overline{\mathbf{x}} - \mathbf{b} + \mathsf{B}^T\overline{\boldsymbol{\lambda}} = \mathbf{o}, \quad \textit{and} \quad \overline{\boldsymbol{\lambda}}^T(\mathsf{B}\overline{\mathbf{x}} - \mathbf{c}) = 0. \tag{2.60}$$

Proof. Since f is convex, it follows by Proposition 2.4(i) that each local minimizer is a global minimizer. Moreover, by Proposition 2.15, each minimizer satisfies (2.59) and (2.60). Thus it is enough to prove that the convexity of f, the feasibility condition, and (2.59) or (2.60) are sufficient for $\overline{\mathbf{x}}$ to be a solution of the inequality constrained minimization problem (2.50).
(i) Let us assume that $\overline{\mathbf{x}} \in \Omega_I$ satisfies (2.59) and $\mathbf{x} \in \Omega_I$. Since Ω_I is convex, it follows that $\mathbf{d} = \mathbf{x} - \overline{\mathbf{x}}$ is a feasible direction of Ω_I at $\overline{\mathbf{x}}$, so that, using Taylor's expansion and the assumptions, we have

$$f(\mathbf{x}) - f(\overline{\mathbf{x}}) = (\mathsf{A}\overline{\mathbf{x}} - \mathbf{b})^T\mathbf{d} + \frac{1}{2}\mathbf{d}^T\mathsf{A}\mathbf{d} \geq 0.$$

(ii) Let us assume that $\overline{\mathbf{x}} \in \Omega_I$ satisfies (2.60) and $\mathsf{B}\mathbf{x} - \mathbf{c} \leq \mathbf{o}$. Then

$$L_0(\overline{\mathbf{x}}, \overline{\boldsymbol{\lambda}}) = f(\overline{\mathbf{x}}) + \overline{\boldsymbol{\lambda}}^T(\mathsf{B}\overline{\mathbf{x}} - \mathbf{c}) = f(\overline{\mathbf{x}}),$$
$$L_0(\mathbf{x}, \overline{\boldsymbol{\lambda}}) = f(\mathbf{x}) + \overline{\boldsymbol{\lambda}}^T(\mathsf{B}\mathbf{x} - \mathbf{c}) \leq f(\mathbf{x}),$$

and

$$\begin{aligned} f(\mathbf{x}) - f(\overline{\mathbf{x}}) &\geq L_0(\mathbf{x}, \overline{\boldsymbol{\lambda}}) - L_0(\overline{\mathbf{x}}, \overline{\boldsymbol{\lambda}}) \\ &= \nabla_{\mathbf{x}} L_0(\overline{\mathbf{x}}, \overline{\boldsymbol{\lambda}})^T(\mathbf{x} - \overline{\mathbf{x}}) + \frac{1}{2}(\mathbf{x} - \overline{\mathbf{x}})^T\mathsf{A}(\mathbf{x} - \overline{\mathbf{x}}) \\ &= (\mathsf{A}\overline{\mathbf{x}} - \mathbf{b} + \mathsf{B}^T\overline{\boldsymbol{\lambda}})^T(\mathbf{x} - \overline{\mathbf{x}}) + \frac{1}{2}(\mathbf{x} - \overline{\mathbf{x}})^T\mathsf{A}(\mathbf{x} - \overline{\mathbf{x}}) \\ &= \frac{1}{2}(\mathbf{x} - \overline{\mathbf{x}})^T\mathsf{A}(\mathbf{x} - \overline{\mathbf{x}}) \geq 0. \end{aligned}$$

□

2.5.6 Optimality Conditions for Bound Constrained Problems

A special case of problem (2.50) is the bound constrained problem

$$\min_{\mathbf{x}\in\Omega_B} f(\mathbf{x}), \quad \Omega_B = \{\mathbf{x} \in \mathbb{R}^n : \ \mathbf{x} \geq \boldsymbol{\ell}\}, \tag{2.61}$$

where f is a quadratic function defined by (2.4) and $\boldsymbol{\ell} \in \mathbb{R}^n$. The optimality conditions for convex bound constrained problems can be written in the form which is more convenient in some applications.

Proposition 2.18. *Let f be a convex quadratic function defined by (2.4) with a positive semidefinite Hessian* A. *Then $\overline{\mathbf{x}} \in \Omega_B$ solves (2.61) if and only if*

$$\mathsf{A}\overline{\mathbf{x}} - \mathbf{b} \geq \mathbf{o} \quad \textit{and} \quad (\mathsf{A}\overline{\mathbf{x}} - \mathbf{b})^T(\overline{\mathbf{x}} - \boldsymbol{\ell}) = 0. \tag{2.62}$$

Proof. First observe that denoting $\mathsf{B} = -\mathsf{I}_n$, $\mathbf{c} = -\boldsymbol{\ell}$, and

$$\Omega_I = \{\mathbf{x} \in \mathbb{R}^n : \ \mathsf{B}\mathbf{x} \leq \mathbf{c}\},$$

the bound constrained problem (2.61) becomes the standard inequality constrained problem(2.50) with $\Omega_I = \Omega_B$. Using Proposition 2.17, it follows that $\overline{\mathbf{x}} \in \Omega_B$ is the solution of (2.61) if and only if there is $\boldsymbol{\lambda} \in \mathbb{R}^n$ such that

$$\boldsymbol{\lambda} \geq \mathbf{o}, \quad \mathsf{A}\overline{\mathbf{x}} - \mathbf{b} - \mathsf{I}\boldsymbol{\lambda} = \mathbf{o}, \quad \text{and} \quad \boldsymbol{\lambda}^T(\overline{\mathbf{x}} - \boldsymbol{\ell}) = 0. \tag{2.63}$$

We complete the proof by observing that (2.62) can be obtained from (2.63) and vice versa by substituting $\boldsymbol{\lambda} = \mathsf{A}\overline{\mathbf{x}} - \mathbf{b}$. □

In the proof, we have shown that $\boldsymbol{\lambda} = \nabla f(\overline{\mathbf{x}})$ is a vector of Lagrange multipliers for the constraints $-\mathbf{x} \leq -\boldsymbol{\ell}$, or, equivalently, for $\mathbf{x} \geq \boldsymbol{\ell}$. Notice that the conditions (2.62) require that none of the vectors $\mathbf{s}_i$ is a feasible decrease direction of Ω_B at $\overline{\mathbf{x}}$, where $\mathbf{s}_i$ denotes a vector of the standard basis of $\mathbb{R}^n$ formed by the columns of I_n, $i \in \mathcal{A}(\overline{\mathbf{x}})$.

2.5.7 Min-max, Dual, and Saddle Point Problems

As in Sect. 2.4.4, we shall assume that A and B are positive definite and full rank matrices, respectively, postponing the analysis of more general problems to Sects. 2.6.4 and 2.6.5. The assumptions imply that the related KKT system

$$\boldsymbol{\lambda} \geq \mathbf{o}, \tag{2.64}$$

$$\nabla_{\mathbf{x}} L_0(\mathbf{x}, \boldsymbol{\lambda}) = \mathsf{A}\mathbf{x} - \mathbf{b} + \mathsf{B}^T\boldsymbol{\lambda} = \mathbf{o}, \tag{2.65}$$

$$\nabla_{\boldsymbol{\lambda}} L_0(\mathbf{x}, \boldsymbol{\lambda}) = \mathsf{B}\mathbf{x} - \mathbf{c} \leq \mathbf{o}, \tag{2.66}$$

$$\boldsymbol{\lambda}^T(\mathsf{B}\mathbf{x} - \mathbf{c}) = 0 \tag{2.67}$$

has a unique solution $(\widehat{\mathbf{x}}, \widehat{\boldsymbol{\lambda}})$. We shall now associate the KKT system (2.64)–(2.67) with some other extremal problems.

First observe that given $\boldsymbol{\lambda} \in \mathbb{R}^m$, we can evaluate $\mathbf{x}$ from (2.65) to get the minimizer

$$\mathbf{x} = \mathbf{x}(\boldsymbol{\lambda}) = \mathsf{A}^{-1}\mathbf{b} - \mathsf{A}^{-1}\mathsf{B}^T\boldsymbol{\lambda}$$

of $L_0(\mathbf{x}, \boldsymbol{\lambda})$ with respect to $\mathbf{x}$. We can use this observation to express explicitly the *dual function*

$$\begin{aligned} \Theta(\boldsymbol{\lambda}) &= \inf_{\mathbf{x}\in\mathbb{R}^n} L_0(\mathbf{x}, \boldsymbol{\lambda}) = \min_{\mathbf{x}\in\mathbb{R}^n} L_0(\mathbf{x}, \boldsymbol{\lambda}) = L_0(\mathbf{x}(\boldsymbol{\lambda}), \boldsymbol{\lambda}) \\ &= -\frac{1}{2}\boldsymbol{\lambda}^T\mathsf{B}\mathsf{A}^{-1}\mathsf{B}^T\boldsymbol{\lambda} + (\mathsf{B}\mathsf{A}^{-1}\mathbf{b} - \mathbf{c})^T\boldsymbol{\lambda} - \frac{1}{2}\mathbf{b}^T\mathsf{A}^{-1}\mathbf{b} \end{aligned} \tag{2.68}$$

and its gradient

$$\nabla\Theta(\boldsymbol{\lambda}) = -\mathsf{B}\mathsf{A}^{-1}\mathsf{B}^T\boldsymbol{\lambda} + (\mathsf{B}\mathsf{A}^{-1}\mathbf{b} - \mathbf{c}). \tag{2.69}$$

Moreover,

$$\mathsf{B}\mathbf{x}(\boldsymbol{\lambda}) - \mathbf{c} = -\mathsf{B}\mathsf{A}^{-1}\mathsf{B}^T\boldsymbol{\lambda} + (\mathsf{B}\mathsf{A}^{-1}\mathbf{b} - \mathbf{c}) = \nabla\Theta(\boldsymbol{\lambda}),$$

so that (2.66) and (2.67) are equivalent to

$$\nabla\Theta(\boldsymbol{\lambda}) \le \mathbf{o} \quad \text{and} \quad \boldsymbol{\lambda}^T\nabla\Theta(\boldsymbol{\lambda}) = 0,$$

which can be rewritten as

$$-\nabla\Theta(\boldsymbol{\lambda}) \ge \mathbf{o} \quad \text{and} \quad -\boldsymbol{\lambda}^T\nabla\Theta(\boldsymbol{\lambda}) = 0. \tag{2.70}$$

As the Hessian $\mathsf{B}\mathsf{A}^{-1}\mathsf{B}^T$ of $-\Theta$ is SPD, we conclude that (2.70) is equivalent to the condition (2.62) for the minimizer of $-\Theta$ subject to $\boldsymbol{\lambda} \ge \mathbf{o}$, or, equivalently, for the maximizer of Θ subject to $\boldsymbol{\lambda} \ge \mathbf{o}$. Therefore the KKT couple $(\widehat{\mathbf{x}}, \widehat{\boldsymbol{\lambda}})$ for the inequality constrained problem (2.50) solves the *min-max problem*

$$L_0(\widehat{\mathbf{x}}, \widehat{\boldsymbol{\lambda}}) = \max_{\boldsymbol{\lambda}\ge\mathbf{o}} \min_{\mathbf{x}\in\mathbb{R}^n} L_0(\mathbf{x}, \boldsymbol{\lambda}), \tag{2.71}$$

$\widehat{\boldsymbol{\lambda}}$ solves the *dual problem*

$$\Theta(\widehat{\boldsymbol{\lambda}}) = \max_{\boldsymbol{\lambda}\ge\mathbf{o}} \Theta(\boldsymbol{\lambda}), \tag{2.72}$$

and

$$f(\widehat{\mathbf{x}}) = L_0(\widehat{\mathbf{x}}, \widehat{\boldsymbol{\lambda}}) = \Theta(\widehat{\boldsymbol{\lambda}}). \tag{2.73}$$

As in Sect. 2.4.4, there is yet another equivalent problem which is related to the penalty method. Since

$$\sup_{\boldsymbol{\lambda}\ge\mathbf{o}} L_0(\mathbf{x}, \boldsymbol{\lambda}) = \infty \text{ for } \mathbf{x} \notin \Omega_I \quad \text{and} \quad \sup_{\boldsymbol{\lambda}\ge\mathbf{o}} L_0(\mathbf{x}, \boldsymbol{\lambda}) = f(\mathbf{x}) \text{ for } \mathbf{x} \in \Omega_I,$$

it follows that the solution $\widehat{\mathbf{x}}$ of the KKT system (2.64)–(2.67) satisfies

$$f(\widehat{\mathbf{x}}) = \min_{\mathbf{x}\in\mathbb{R}^n} \sup_{\boldsymbol{\lambda}\geq\mathbf{o}} L_0(\mathbf{x}, \boldsymbol{\lambda}).$$

Comparing the latter equality with (2.71) and (2.73), we get the well-known *duality relation*

$$\max_{\boldsymbol{\lambda}\geq\mathbf{o}} \min_{\mathbf{x}\in\mathbb{R}^n} L_0(\mathbf{x}, \boldsymbol{\lambda}) = \min_{\mathbf{x}\in\mathbb{R}^n} \sup_{\boldsymbol{\lambda}\geq\mathbf{o}} L_0(\mathbf{x}, \boldsymbol{\lambda}). \tag{2.74}$$

Using the feasibility of $\widehat{\mathbf{x}}$, equations (2.73), and definition (2.68) of the dual function Θ, we also get that $(\widehat{\mathbf{x}}, \widehat{\boldsymbol{\lambda}})$ solves the *saddle point problem* to find $(\widehat{\mathbf{x}}, \widehat{\boldsymbol{\lambda}})$ so that for any $\mathbf{x} \in \mathbb{R}^n$ and $\boldsymbol{\lambda} \geq \mathbf{o}$

$$L_0(\widehat{\mathbf{x}}, \boldsymbol{\lambda}) \leq L_0(\widehat{\mathbf{x}}, \widehat{\boldsymbol{\lambda}}) \leq L_0(\mathbf{x}, \widehat{\boldsymbol{\lambda}}). \tag{2.75}$$

We have obtained again two *bound constrained* problems which are equivalent to the original equality constrained problem (2.50). The mixed formulation enhances explicitly the Lagrange multipliers and is only bound constrained at the cost of two sets of variables, while the dual formulation enjoys both the bound constraints and typically a small dimension at the cost of dealing with more complex matrices. The dual problem may be also better conditioned.

2.6 Equality and Inequality Constrained Problems

In the previous sections, we have obtained the results concerning optimization problems with either equality or inequality constraints. Here we extend these results to the optimization problems whose variables are subjected to both equality and inequality constraints. More formally, we look for

$$\min_{\mathbf{x}\in\Omega_{IE}} f(\mathbf{x}), \quad \Omega_{IE} = \{\mathbf{x} \in \mathbb{R}^n : \ [\mathsf{B}\mathbf{x}]_{\mathcal{I}} \leq \mathbf{c}_{\mathcal{I}}, \ [\mathsf{B}\mathbf{x}]_{\mathcal{E}} = \mathbf{c}_{\mathcal{E}}\}, \tag{2.76}$$

where f is a quadratic function with the symmetric Hessian $\mathsf{A} \in \mathbb{R}^{n\times n}$ and the linear term defined by $\mathbf{b} \in \mathbb{R}^n$, $\mathsf{B} = [\mathbf{b}_1, \ldots, \mathbf{b}_m]^T \in \mathbb{R}^{m\times n}$ is a matrix with possibly dependent rows, $\mathbf{c} = [c_i] \in \mathbb{R}^m$, and $\mathcal{I}$, $\mathcal{E}$ are disjoint sets of indices which decompose $\{1, \ldots, m\}$. We assume that Ω_{IE} is not empty.

If we describe the conditions that define the feasible set Ω_{IE} in detail, we get

$$\Omega_{IE} = \{\mathbf{x} \in \mathbb{R}^n : \ \mathbf{b}_i^T\mathbf{x} \leq c_i, \ i \in \mathcal{I}, \ \mathbf{b}_i^T\mathbf{x} = c_i, \ i \in \mathcal{E}\},$$

which makes sense even for $\mathcal{I} = \emptyset$ or $\mathcal{E} = \emptyset$; we consider the conditions which involve the empty set as always satisfied. For example, $\mathcal{E} = \emptyset$ gives

$$\Omega_{IE} = \{\mathbf{x} \in \mathbb{R}^n : \ \mathbf{b}_i^T\mathbf{x} \leq c_i, \ i \in \mathcal{I}\},$$

and the kernel of an "empty" matrix is defined by

$$\mathrm{Ker}\mathsf{B}_{\mathcal{E}*} = \{\mathbf{x} \in \mathbb{R}^n : \ \mathbf{b}_i^T\mathbf{x} = 0, \ i \in \mathcal{E}\} = \mathbb{R}^n.$$

2.6.1 Optimality Conditions

First observe that Ω_{IE} is a polyhedral set as any equality constraint $\mathbf{b}_i^T\mathbf{x} = c_i$ can be replaced by the couple of inequalities $\mathbf{b}_i^T\mathbf{x} \le c_i$ and $-\mathbf{b}_i^T\mathbf{x} \le -c_i$. We can thus use our results obtained by the analysis of the inequality constrained problems in Sect. 2.5 to get similar results for general bound and equality constrained QP problem (2.76).

Proposition 2.19. *Let the quadratic function f and the feasible set Ω_{IE} be defined by the matrices A, B, the vectors $\mathbf{b}, \mathbf{c}$, and the index sets $\mathcal{I}, \mathcal{E}$ from the definition of problem (2.76). We assume that A is symmetric and that B is not necessarily a full row rank matrix. Let $\mathcal{C}$ denote the cone of directions of the feasible set Ω_{IE}. Then the following statements hold:*
(i) If $\overline{\mathbf{x}} \in \Omega_{IE}$ is a local solution of (2.76), then

$$\nabla f(\overline{\mathbf{x}}) = (\mathsf{A}\overline{\mathbf{x}} - \mathbf{b})^T\mathbf{d} \ge 0 \tag{2.77}$$

for any feasible direction $\mathbf{d}$ of Ω_{IE} at $\overline{\mathbf{x}}$.
(ii) If $\overline{\mathbf{x}} \in \Omega_{IE}$ is a local solution of (2.76), then there is a vector $\overline{\boldsymbol{\lambda}} \in \mathbb{R}^m$ such that

$$\overline{\boldsymbol{\lambda}}_{\mathcal{I}} \ge \mathbf{o}, \quad \mathsf{A}\overline{\mathbf{x}} - \mathbf{b} + \mathsf{B}^T\overline{\boldsymbol{\lambda}} = \mathbf{o}, \quad \textit{and} \quad \overline{\boldsymbol{\lambda}}_{\mathcal{I}}^T[\mathsf{B}\overline{\mathbf{x}} - \mathbf{c}]_{\mathcal{I}} = 0. \tag{2.78}$$

(iii) If A is positive semidefinite, then $\overline{\mathbf{x}} \in \Omega_{IE}$ is a solution of (2.76) if and only if $\overline{\mathbf{x}}$ satisfies (2.77) or (2.78).

Proof. First observe that if $\mathcal{E} = \emptyset$, then the statements of the above proposition reduce to Propositions 2.15 and 2.17, and if $\mathcal{I} = \emptyset$, then they reduce to Proposition 2.8. Thus we can assume in the rest of the proof that $\mathcal{I} \ne \emptyset$ and $\mathcal{E} \ne \emptyset$.

As mentioned above, (2.76) may be rewritten also as

$$\min_{\mathbf{x}\in\Omega_I} f(\mathbf{x}), \ \Omega_I = \{\mathbf{x} \in \mathbb{R}^n : \ [\mathsf{B}\mathbf{x}]_{\mathcal{I}} \le \mathbf{c}_{\mathcal{I}}, \ [\mathsf{B}\mathbf{x}]_{\mathcal{E}} \le \mathbf{c}_{\mathcal{E}}, -[\mathsf{B}\mathbf{x}]_{\mathcal{E}} \le -\mathbf{c}_{\mathcal{E}}\}, \tag{2.79}$$

where $\Omega_I = \Omega_{IE}$. Thus the statement (i) is a special case of Proposition 2.15.

If $\overline{\mathbf{x}} \in \Omega_{IE}$ is a local solution of (2.76), then, using (ii) of Proposition 2.15, we get that there are nonnegative vectors $\mathbf{u}, \mathbf{v}$, and $\mathbf{w}$ such that

$$\mathsf{A}\overline{\mathbf{x}} - \mathbf{b} + \mathsf{B}_{\mathcal{I}*}^T\mathbf{u} + \mathsf{B}_{\mathcal{E}*}^T\mathbf{v} - \mathsf{B}_{\mathcal{E}*}^T\mathbf{w} = \mathbf{o} \quad \text{and} \quad \overline{\boldsymbol{\lambda}}_{\mathcal{I}}^T[\mathsf{B}\overline{\mathbf{x}} - \mathbf{c}]_{\mathcal{I}} = 0.$$

Defining $\overline{\boldsymbol{\lambda}} \in \mathbb{R}^m$ by

$$\overline{\boldsymbol{\lambda}}_{\mathcal{I}} = \mathbf{u}, \quad \overline{\boldsymbol{\lambda}}_{\mathcal{E}} = \mathbf{v} - \mathbf{w},$$

we get that $\overline{\boldsymbol{\lambda}}$ and $\overline{\mathbf{x}}$ satisfy (2.78), which proves (ii).

In the same way as above, we can use Proposition 2.17 to prove the statement (iii). □

2.6.2 Existence and Uniqueness

As recalled in Sect. 2.6.1, problem (2.76) can be considered as a special case of the inequality constrained problem, so that we can use the results of Sect. 2.5.4. We add here one simple proposition for the case that $\mathsf{A}|\mathrm{Ker}\mathsf{B}_{\mathcal{E}*}$ is positive definite.

Proposition 2.20. *Let the quadratic function f and the feasible set Ω_{IE} be defined by the matrices A, B, the vectors $\mathbf{b}, \mathbf{c}$, and the index sets $\mathcal{I}, \mathcal{E}$ from the definition of problem (2.76). Let $\mathsf{A}|\mathrm{Ker}\mathsf{B}_{\mathcal{E}*}$ be positive definite. Then the equality and inequality constrained minimization problem (2.76) has a unique solution $\widehat{\mathbf{x}}$.*

Proof. Let us consider the penalized function

$$f_2(\mathbf{x}) = f(\mathbf{x}) + (\mathsf{B}_{\mathcal{E}*}\mathbf{x} - \mathbf{c}_{\mathcal{E}})^T(\mathsf{B}_{\mathcal{E}*}\mathbf{x} - \mathbf{c}_{\mathcal{E}}).$$

Using the assumption that $\mathsf{A}|\mathrm{Ker}\mathsf{B}_{\mathcal{E}*}$ is positive definite and Lemma 1.2, we get that the Hessian $\mathsf{A}_2 = \mathsf{A} + 2\mathsf{B}_{\mathcal{E}*}^T\mathsf{B}_{\mathcal{E}*}$ of the function f_2 is also positive definite, so that f_2 is strictly convex by Proposition 2.2. Since

$$\Omega_{IE} \subseteq \Omega_{\mathcal{E}}, \quad \Omega_{\mathcal{E}} = \{\mathbf{x} \in \mathbb{R}^n : \ \mathsf{B}_{\mathcal{E}*}\mathbf{x} = \mathbf{c}_{\mathcal{E}}\},$$

it follows that $f|\Omega_{\mathcal{E}} = f_2|\Omega_{\mathcal{E}}$. Thus $f|\Omega_{IE}$ is strictly convex and the statement then follows by Proposition 2.5(i). □

2.6.3 Partially Bound and Equality Constrained Problems

Here we consider the partially bound and equality constrained problem

$$\min_{\mathbf{x}\in\Omega_{BE}} f(\mathbf{x}), \quad \Omega_{BE} = \{\mathbf{x} \in \mathbb{R}^n : \ \mathbf{x}_I \geq \boldsymbol{\ell}_I, \ \mathsf{B}_E\mathbf{x} = \mathbf{c}_E\}, \tag{2.80}$$

where f is a convex quadratic function with a symmetric positive semidefinite Hessian $\mathsf{A} \in \mathbb{R}^{n\times n}$, $\mathsf{B}_E \in \mathbb{R}^{q\times n}$ is a matrix with possibly dependent rows, $\mathbf{c}_E \in \mathbb{R}^q$, and $\boldsymbol{\ell}_I \in \mathbb{R}^p$ is a vector of bounds on the first p components of $\mathbf{x}$ which form variables $\mathbf{x}_I$. Though the partial constraints can be easily implemented by admitting $\ell_i = -\infty$, here we consider them explicitly to simplify the reference in our applications. To unify the references to the corresponding vectors, we denote by $\mathcal{I} = \{1, \ldots, p\}$ and $\mathcal{R} = \{p+1, \ldots, n\}$ the sets of indices of the bound constrained and remaining entries of $\mathbf{x}$, respectively, so that $\mathbf{x}_{\mathcal{I}} = \mathbf{x}_I$. With

$$\mathsf{B}_I = [-\mathsf{I}_p, \mathsf{O}_{pr}], \ r = n - p, \quad \mathsf{B} = \begin{bmatrix} \mathsf{B}_I \\ \mathsf{B}_E \end{bmatrix}, \quad \mathbf{c} = \begin{bmatrix} -\boldsymbol{\ell}_I \\ \mathbf{c}_E \end{bmatrix},$$

and $\mathcal{E} = \{p+1, \ldots, m\}$, the bound and equality constrained problem (2.80) becomes the standard equality and inequality constrained problem (2.76). Introducing the Lagrange multipliers $\boldsymbol{\lambda} \in \mathbb{R}^m$ and denoting

$$\mathbf{g} = \nabla_{\mathbf{x}} L_0(\mathbf{x}, \boldsymbol{\lambda}_{\mathcal{E}}) = \mathsf{A}\mathbf{x} - \mathbf{b} + \mathsf{B}_E^T \boldsymbol{\lambda}_{\mathcal{E}},$$

we get

$$\nabla_{\mathbf{x}} L_0(\mathbf{x}, \boldsymbol{\lambda}) = \mathsf{A}\mathbf{x} - \mathbf{b} + \mathsf{B}_E^T \boldsymbol{\lambda}_{\mathcal{E}} + \mathsf{B}_I^T \boldsymbol{\lambda}_{\mathcal{I}} = \mathbf{g} + \mathsf{B}_I^T \boldsymbol{\lambda}_{\mathcal{I}} = \mathbf{g} - \begin{bmatrix} \mathsf{I} \\ \mathsf{O} \end{bmatrix} \boldsymbol{\lambda}_{\mathcal{I}},$$

so that by Proposition 2.19 the KKT conditions for problem (2.80) read

$$\mathbf{g}_{\mathcal{I}} \geq \mathbf{o}, \quad \mathbf{g}_{\mathcal{R}} = \mathbf{o}, \quad \text{and} \quad \mathbf{g}_{\mathcal{I}}^T (\mathbf{x}_{\mathcal{I}} - \boldsymbol{\ell}_I) = 0. \tag{2.81}$$

Thus the KKT conditions for bound and equality constrained problems can be conveniently expressed by means of the Lagrangian function for equality constrained problems. The Lagrange multipliers for the inequality constraints may be recovered by $\boldsymbol{\lambda}_{\mathcal{I}} = \mathbf{g}_{\mathcal{I}}$. If $\overline{\mathbf{x}} \in \Omega_{BE}$ and $\mathbf{g} = \nabla_{\mathbf{x}} L_0(\mathbf{x}, \overline{\boldsymbol{\lambda}}_{\mathcal{E}})$ satisfy (2.81), then $(\overline{\mathbf{x}}, \overline{\boldsymbol{\lambda}}_{\mathcal{E}})$ is called the *KKT pair for bound and equality constraints.*

Application of the duality typically results in problem

$$\max_{\mathbf{x} \in \Omega_{BE}} f(\mathbf{x}) \tag{2.82}$$

with $-f$ convex. Since

$$\max_{\mathbf{x} \in \Omega_{BE}} f(\mathbf{x}) = - \min_{\mathbf{x} \in \Omega_{BE}} -f(\mathbf{x}),$$

we get easily that the KKT conditions for (2.82) read

$$\mathbf{g}_{\mathcal{I}} \leq \mathbf{o}, \quad \mathbf{g}_{\mathcal{R}} = \mathbf{o}, \quad \text{and} \quad \mathbf{g}_{\mathcal{I}}^T (\mathbf{x}_{\mathcal{I}} - \boldsymbol{\ell}_I) = 0. \tag{2.83}$$

The analysis presented above covers also the *partially bound constrained problem* to find

$$\min_{\mathbf{x} \in \Omega_B} f(\mathbf{x}), \quad \Omega_B = \{\mathbf{x} \in \mathbb{R}^n : \mathbf{x}_I \geq \boldsymbol{\ell}_I\}. \tag{2.84}$$

Skipping the terms concerning the equality constraints and denoting

$$\mathbf{g} = \nabla f(\mathbf{x}) = \mathsf{A}\mathbf{x} - \mathbf{b},$$

we get the *KKT conditions for the bound constrained problem* (2.84) in the form

$$\mathbf{g}_{\mathcal{I}} \geq \mathbf{o}, \quad \mathbf{g}_{\mathcal{R}} = \mathbf{o}, \quad \text{and} \quad \mathbf{g}_{\mathcal{I}}^T (\mathbf{x}_{\mathcal{I}} - \boldsymbol{\ell}_I) = 0. \tag{2.85}$$

As above, we get that the KKT conditions for partially bound constrained problem

$$\max_{\mathbf{x} \in \Omega_B} f(\mathbf{x}), \quad \Omega_B = \{\mathbf{x} \in \mathbb{R}^n : \mathbf{x}_I \geq \boldsymbol{\ell}_I\} \tag{2.86}$$

with f convex read

$$\mathbf{g}_{\mathcal{I}} \leq \mathbf{o}, \quad \mathbf{g}_{\mathcal{R}} = \mathbf{o}, \quad \text{and} \quad \mathbf{g}_{\mathcal{I}}^T (\mathbf{x}_{\mathcal{I}} - \boldsymbol{\ell}_I) = 0. \tag{2.87}$$

2.6.4 Duality for Dependent Constraints

Combining the arguments of the previous subsections, in particular Sects. 2.4.4 and 2.5.7, it is possible to get the dual and mixed formulations of equality and inequality constrained problems under the conditions that A and B are positive definite and full rank matrices, respectively. Since there are important applications where such assumptions are too restrictive, it is useful to extend the duality theory without these assumptions. We first relax the assumptions on the constraints, postponing the discussion of more general cases to the next section.

Duality Relations

Let us consider problem (2.76) with A positive definite and B with dependent rows. Let us recall that by Proposition 2.2 our assumptions imply that f is strictly convex; therefore we can apply Proposition 2.5 to get that there is a unique solution $\widehat{\mathbf{x}}$ of (2.76).

Let $(\widehat{\mathbf{x}}, \overline{\boldsymbol{\lambda}})$ be a solution of the bound and equality problem (2.76). Then

$$\overline{\boldsymbol{\lambda}}^T(\mathsf{B}\widehat{\mathbf{x}} - \mathbf{c}) = \overline{\boldsymbol{\lambda}}_{\mathcal{E}}^T[\mathsf{B}\widehat{\mathbf{x}} - \mathbf{c}]_{\mathcal{E}} + \overline{\boldsymbol{\lambda}}_{\mathcal{I}}^T[\mathsf{B}\widehat{\mathbf{x}} - \mathbf{c}]_{\mathcal{I}} = 0,$$

so that

$$f(\widehat{\mathbf{x}}) = L_0(\widehat{\mathbf{x}}, \overline{\boldsymbol{\lambda}}). \tag{2.88}$$

Next observe that if $\overline{\boldsymbol{\lambda}}$ is a vector of the Lagrange multipliers of the solution, then $\widehat{\mathbf{x}}$ is fully determined by the second condition of (2.78) which reads

$$\nabla_{\mathbf{x}} L_0(\mathbf{x}, \overline{\boldsymbol{\lambda}}) = \mathbf{o}.$$

Since L_0 is strictly convex, the latter is the gradient condition for the unconstrained minimizer of L_0 with respect to $\mathbf{x}$; therefore

$$L_0(\widehat{\mathbf{x}}, \overline{\boldsymbol{\lambda}}) = \min_{\mathbf{x}\in\mathbb{R}^n} L_0(\mathbf{x}, \overline{\boldsymbol{\lambda}}). \tag{2.89}$$

Recalling the definition of the dual function

$$\Theta(\boldsymbol{\lambda}) = \min_{\mathbf{x}\in\mathbb{R}^n} L_0(\mathbf{x}, \boldsymbol{\lambda})$$

and using (2.88) and (2.89), we get

$$f(\widehat{\mathbf{x}}) = L_0(\widehat{\mathbf{x}}, \overline{\boldsymbol{\lambda}}) = \Theta(\overline{\boldsymbol{\lambda}}). \tag{2.90}$$

Dual and Saddle Point Problems

Let us first present the analysis which enhances our earlier results and admits the constraint matrices with possibly dependent rows.

Proposition 2.21. *Let the quadratic function f and the feasible set $\Omega_{IE} \neq \emptyset$ be defined by the matrices A, B, the vectors $\mathbf{b}, \mathbf{c}$, and the index sets $\mathcal{I}, \mathcal{E}$ from the definition of problem (2.76). We assume that A is SPD, but we admit that B is not necessarily a full row rank matrix, $\mathcal{I} = \emptyset$, or $\mathcal{E} = \emptyset$. Let*

$$\Theta(\boldsymbol{\lambda}) = -\frac{1}{2}\boldsymbol{\lambda}^T \mathsf{B}\mathsf{A}^{-1}\mathsf{B}^T\boldsymbol{\lambda} + \boldsymbol{\lambda}^T(\mathsf{B}\mathsf{A}^{-1}\mathbf{b} - \mathbf{c}) - \frac{1}{2}\mathbf{b}^T\mathsf{A}^{-1}\mathbf{b} \tag{2.91}$$

denote the dual function. Then the following statements are equivalent:
(i) $(\widehat{\mathbf{x}}, \overline{\boldsymbol{\lambda}})$ is a KKT pair for problem (2.76).
(ii) $\widehat{\mathbf{x}}$ is a unique solution of the primal problem (2.76) and $\overline{\boldsymbol{\lambda}}$ is a solution of the dual problem

$$\max_{\boldsymbol{\lambda} \in \Omega_B} \Theta(\boldsymbol{\lambda}), \quad \Omega_B = \{\boldsymbol{\lambda} \in \mathbb{R}^m : \ \boldsymbol{\lambda}_{\mathcal{I}} \geq \mathbf{o}\}. \tag{2.92}$$

(iii) $(\widehat{\mathbf{x}}, \overline{\boldsymbol{\lambda}}) \in \mathbb{R}^n \times \mathbb{R}^m$ with $\overline{\boldsymbol{\lambda}}_{\mathcal{I}} \geq \mathbf{o}$ is a saddle point of L_0 in the sense that for any $\mathbf{x} \in \mathbb{R}^n$ and $\boldsymbol{\lambda} \in \mathbb{R}^m$ such that $\boldsymbol{\lambda}_{\mathcal{I}} \geq \mathbf{o}$,

$$L_0(\widehat{\mathbf{x}}, \boldsymbol{\lambda}) \leq L_0(\widehat{\mathbf{x}}, \overline{\boldsymbol{\lambda}}) \leq L_0(\mathbf{x}, \overline{\boldsymbol{\lambda}}). \tag{2.93}$$

Proof. (i) $\Rightarrow$ (ii). Let $(\widehat{\mathbf{x}}, \overline{\boldsymbol{\lambda}})$ be a KKT pair for (2.76), so that $\widehat{\mathbf{x}}$ is a unique solution of (2.76), $f(\widehat{\mathbf{x}}) = \Theta(\overline{\boldsymbol{\lambda}})$ by (2.90), and $(\widehat{\mathbf{x}}, \overline{\boldsymbol{\lambda}})$ is by Proposition 2.19 a solution of

$$\boldsymbol{\lambda}_{\mathcal{I}} \geq \mathbf{o}, \tag{2.94}$$
$$\nabla_{\mathbf{x}} L_0(\mathbf{x}, \boldsymbol{\lambda}) = \mathsf{A}\mathbf{x} - \mathbf{b} + \mathsf{B}^T\boldsymbol{\lambda} = \mathbf{o}, \tag{2.95}$$
$$[\nabla_{\boldsymbol{\lambda}} L_0(\mathbf{x}, \boldsymbol{\lambda})]_{\mathcal{I}} = [\mathsf{B}\mathbf{x} - \mathbf{c}]_{\mathcal{I}} \leq \mathbf{o}, \tag{2.96}$$
$$[\nabla_{\boldsymbol{\lambda}} L_0(\mathbf{x}, \boldsymbol{\lambda})]_{\mathcal{E}} = [\mathsf{B}\mathbf{x} - \mathbf{c}]_{\mathcal{E}} = \mathbf{o}, \tag{2.97}$$
$$\boldsymbol{\lambda}_{\mathcal{I}}^T[\mathsf{B}\mathbf{x} - \mathbf{c}]_{\mathcal{I}} = 0. \tag{2.98}$$

In particular, since A is positive definite, it follows that we can use (2.95) to get

$$\widehat{\mathbf{x}} = \mathsf{A}^{-1}(\mathbf{b} - \mathsf{B}^T\overline{\boldsymbol{\lambda}}).$$

After substituting into (2.96)–(2.98), we get

$$[\,-\mathsf{B}\mathsf{A}^{-1}\mathsf{B}^T\overline{\boldsymbol{\lambda}} + (\mathsf{B}\mathsf{A}^{-1}\mathbf{b} - \mathbf{c})]_{\mathcal{I}} \leq \mathbf{o}, \tag{2.99}$$
$$[\,-\mathsf{B}\mathsf{A}^{-1}\mathsf{B}^T\overline{\boldsymbol{\lambda}} + (\mathsf{B}\mathsf{A}^{-1}\mathbf{b} - \mathbf{c})]_{\mathcal{E}} = \mathbf{o}, \tag{2.100}$$
$$\overline{\boldsymbol{\lambda}}_{\mathcal{I}}^T\,[\,-\mathsf{B}\mathsf{A}^{-1}\mathsf{B}^T\overline{\boldsymbol{\lambda}} + (\mathsf{B}\mathsf{A}^{-1}\mathbf{b} - \mathbf{c})]_{\mathcal{I}} = 0. \tag{2.101}$$

Denoting $\mathbf{g} = \nabla\Theta(\overline{\boldsymbol{\lambda}})$, we can rewrite the relations (2.99)–(2.101) as

$$\mathbf{g}_{\mathcal{I}} \leq \mathbf{o}, \quad \mathbf{g}_{\mathcal{E}} = \mathbf{o}, \quad \text{and} \quad \overline{\boldsymbol{\lambda}}_{\mathcal{I}}^T\mathbf{g}_{\mathcal{I}} = 0. \tag{2.102}$$

Comparing (2.102) with the KKT conditions (2.87) for the partially bound constrained problem (2.86), we conclude that (2.102) are the KKT conditions for (2.92). Since $\overline{\boldsymbol{\lambda}}_{\mathcal{I}} \geq \mathbf{o}$ by (2.94), we have thus proved that $\overline{\boldsymbol{\lambda}}$ is a feasible vector for problem (2.92) which satisfies the related KKT conditions. Recalling that A is positive definite, so that $\mathsf{B}\mathsf{A}^{-1}\mathsf{B}^T$ is positive semidefinite, we conclude that $\overline{\boldsymbol{\lambda}}$ solves (2.92).

(ii) $\Rightarrow$ (iii). Let $\widehat{\mathbf{x}}$ be a unique solution of the primal problem (2.76) and let $\overline{\boldsymbol{\lambda}}$ be a solution of the dual problem (2.92), so that $f(\widehat{\mathbf{x}}) = \Theta(\overline{\boldsymbol{\lambda}})$ by (2.90). Then $\widehat{\mathbf{x}}$ is feasible, i.e.,

$$[\mathsf{B}\widehat{\mathbf{x}} - \mathbf{c}]_{\mathcal{E}} = \mathbf{o}, \quad [\mathsf{B}\widehat{\mathbf{x}} - \mathbf{c}]_{\mathcal{I}} \leq \mathbf{o},$$

and

$$\sup_{\boldsymbol{\lambda}_{\mathcal{I}} \geq \mathbf{o}} L_0(\widehat{\mathbf{x}}, \boldsymbol{\lambda}) = f(\widehat{\mathbf{x}}) + \sup_{\boldsymbol{\lambda}_{\mathcal{I}} \geq \mathbf{o}} \boldsymbol{\lambda}^T(\mathsf{B}\widehat{\mathbf{x}} - \mathbf{c}) = f(\widehat{\mathbf{x}}).$$

Thus

$$\Theta(\overline{\boldsymbol{\lambda}}) = \min_{\mathbf{x} \in \mathbb{R}^n} L_0(\mathbf{x}, \overline{\boldsymbol{\lambda}}) \leq L_0(\widehat{\mathbf{x}}, \overline{\boldsymbol{\lambda}}) \leq \sup_{\boldsymbol{\lambda}_{\mathcal{I}} \geq \mathbf{o}} L_0(\widehat{\mathbf{x}}, \boldsymbol{\lambda}) = f(\widehat{\mathbf{x}}).$$

Since $f(\widehat{\mathbf{x}}) = \Theta(\overline{\boldsymbol{\lambda}})$, it follows that

$$\sup_{\boldsymbol{\lambda}_{\mathcal{I}} \geq \mathbf{o}} L_0(\widehat{\mathbf{x}}, \boldsymbol{\lambda}) = L_0(\widehat{\mathbf{x}}, \overline{\boldsymbol{\lambda}}) = \min_{\mathbf{x} \in \mathbb{R}^n} L_0(\mathbf{x}, \overline{\boldsymbol{\lambda}})$$

and $(\widehat{\mathbf{x}}, \overline{\boldsymbol{\lambda}})$ is the solution of the saddle point problem (2.93).

(iii) $\Rightarrow$ (i). Let us now assume that there is $(\widehat{\mathbf{x}}, \overline{\boldsymbol{\lambda}})$, $\overline{\boldsymbol{\lambda}}_{\mathcal{I}} \geq \mathbf{o}$, such that (2.93) holds for any $\mathbf{x} \in \mathbb{R}^n$ and $\boldsymbol{\lambda} \in \mathbb{R}^m$, $\boldsymbol{\lambda}_{\mathcal{I}} \geq \mathbf{o}$. To show that $(\widehat{\mathbf{x}}, \overline{\boldsymbol{\lambda}})$ is a KKT pair for (2.76), notice that for any $\mathbf{x} \in \mathbb{R}^n$

$$\sup_{\boldsymbol{\lambda}_{\mathcal{I}} \geq \mathbf{o}} L_0(\mathbf{x}, \boldsymbol{\lambda}) = \infty \text{ for } \mathbf{x} \notin \Omega_{IE} \quad \text{and} \quad \sup_{\boldsymbol{\lambda}_{\mathcal{I}} \geq \mathbf{o}} L_0(\mathbf{x}, \boldsymbol{\lambda}) = f(\mathbf{x}) \text{ for } \mathbf{x} \in \Omega_{IE}.$$

Since by the assumptions

$$L_0(\widehat{\mathbf{x}}, \boldsymbol{\lambda}) \leq L_0(\widehat{\mathbf{x}}, \overline{\boldsymbol{\lambda}})$$

for any $\boldsymbol{\lambda} \in \mathbb{R}^m$, $\boldsymbol{\lambda}_{\mathcal{I}} \geq \mathbf{o}$, it follows that $\widehat{\mathbf{x}} \in \Omega_{IE}$ and $f(\widehat{\mathbf{x}}) = L_0(\widehat{\mathbf{x}}, \overline{\boldsymbol{\lambda}})$. The feasibility of $\widehat{\mathbf{x}}$ and the latter equation imply the complementarity condition

$$\overline{\boldsymbol{\lambda}}_{\mathcal{I}}^T[\mathsf{B}\widehat{\mathbf{x}} - \mathbf{c}]_{\mathcal{I}} = 0.$$

Taking into account the right saddle point inequality in (2.93), we get that

$$L_0(\widehat{\mathbf{x}}, \overline{\boldsymbol{\lambda}}) = \min_{\mathbf{x} \in \mathbb{R}^n} L_0(\mathbf{x}, \overline{\boldsymbol{\lambda}}).$$

Thus $\widehat{\mathbf{x}}$ is the unconstrained minimizer of $L_0(\mathbf{x}, \overline{\boldsymbol{\lambda}})$ with respect to $\mathbf{x}$. Since L_0 is a convex function of $\mathbf{x}$, we conclude that $\nabla_{\mathbf{x}} L_0(\widehat{\mathbf{x}}, \overline{\boldsymbol{\lambda}}) = \mathbf{o}$ and $(\widehat{\mathbf{x}}, \overline{\boldsymbol{\lambda}})$ is a KKT pair for problem (2.76). □

2.6.5 Duality for Semicoercive Problems

Now let us examine what happens when the matrices A and B in the definition of the bound and equality constrained QP problem (2.76) are positive semidefinite and rank deficient, respectively. A special case of (2.76) is a linear programming problem with $\mathsf{A} = \mathsf{O}$.

Constrained Dual Problem

First observe that if A is only positive semidefinite and $\mathbf{b} \neq \mathbf{o}$, then the cost function f need not be bounded from below. Thus $-\infty$ can be in the range of the dual function Θ. We resolve this problem by keeping Θ quadratic at the cost of introducing equality constraints. An alternative development of duality for special semicoercive QP problems can be found in the papers by Dorn [37, 38].

Proposition 2.22. *Let matrices* A, B, *vectors* $\mathbf{b}, \mathbf{c}$, *and index sets* $\mathcal{I}, \mathcal{E}$ *be those from the definition of problem (2.76) with* A *positive semidefinite and* $\Omega_{IE} \neq \emptyset$. *Let* $\mathsf{R} \in \mathbb{R}^{n \times d}$ *be a full rank matrix such that*

$$\mathrm{Im}\mathsf{R} = \mathrm{Ker}\mathsf{A},$$

let A^+ *denote a symmetric positive semidefinite generalized inverse of* A, *and let*

$$\Theta(\boldsymbol{\lambda}) = -\frac{1}{2}\boldsymbol{\lambda}^T\mathsf{B}\mathsf{A}^+\mathsf{B}^T\boldsymbol{\lambda} + \boldsymbol{\lambda}^T(\mathsf{B}\mathsf{A}^+\mathbf{b} - \mathbf{c}) - \frac{1}{2}\mathbf{b}^T\mathsf{A}^+\mathbf{b}. \tag{2.103}$$

Then the following statements hold:
(i) If $(\overline{\mathbf{x}}, \overline{\boldsymbol{\lambda}})$ *is a KKT pair for (2.76), then* $\overline{\boldsymbol{\lambda}}$ *is a solution of*

$$\max_{\boldsymbol{\lambda} \in \Omega_{BE}} \Theta(\boldsymbol{\lambda}), \quad \Omega_{BE} = \{\boldsymbol{\lambda} \in \mathbb{R}^m : \; \boldsymbol{\lambda}_{\mathcal{I}} \geq \mathbf{o}, \; \mathsf{R}^T\mathsf{B}^T\boldsymbol{\lambda} = \mathsf{R}^T\mathbf{b}\}. \tag{2.104}$$

Moreover, there is $\overline{\boldsymbol{\alpha}} \in \mathbb{R}^d$ *such that* $(\overline{\boldsymbol{\lambda}}, \overline{\boldsymbol{\alpha}})$ *is a KKT pair for problem (2.104) and*

$$\overline{\mathbf{x}} = \mathsf{A}^+(\mathbf{b} - \mathsf{B}^T\overline{\boldsymbol{\lambda}}) + \mathsf{R}\overline{\boldsymbol{\alpha}}. \tag{2.105}$$

(ii) If $(\overline{\boldsymbol{\lambda}}, \overline{\boldsymbol{\alpha}})$ *is a KKT pair for problem (2.104), then* $\overline{\mathbf{x}}$ *defined by (2.105) is a solution of the equality and inequality constrained problem (2.76).*
(iii) If $(\overline{\mathbf{x}}, \overline{\boldsymbol{\lambda}})$ *is a KKT pair for problem (2.76), then*

$$f(\overline{\mathbf{x}}) = \Theta(\overline{\boldsymbol{\lambda}}). \tag{2.106}$$

Proof. (i) Assume that $(\overline{\mathbf{x}}, \overline{\boldsymbol{\lambda}})$ is a KKT pair for (2.76), so that $(\overline{\mathbf{x}}, \overline{\boldsymbol{\lambda}})$ is by Proposition 2.19 a solution of

$$\boldsymbol{\lambda}_{\mathcal{I}} \geq \mathbf{o}, \tag{2.107}$$

$$\nabla_{\mathbf{x}} L_0(\mathbf{x}, \boldsymbol{\lambda}) = \mathsf{A}\mathbf{x} - \mathbf{b} + \mathsf{B}^T\boldsymbol{\lambda} = \mathbf{o}, \tag{2.108}$$

$$[\nabla_{\boldsymbol{\lambda}} L_0(\mathbf{x}, \boldsymbol{\lambda})]_{\mathcal{I}} = [\mathsf{B}\mathbf{x} - \mathbf{c}]_{\mathcal{I}} \leq \mathbf{o}, \tag{2.109}$$

$$[\nabla_{\boldsymbol{\lambda}} L_0(\mathbf{x}, \boldsymbol{\lambda})]_{\mathcal{E}} = [\mathsf{B}\mathbf{x} - \mathbf{c}]_{\mathcal{E}} = \mathbf{o}, \tag{2.110}$$

$$\boldsymbol{\lambda}_{\mathcal{I}}^T[\mathsf{B}\mathbf{x} - \mathbf{c}]_{\mathcal{I}} = 0. \tag{2.111}$$

Notice that given a vector $\boldsymbol{\lambda} \in \mathbb{R}^m$, we can express the condition

$$\mathbf{b} - \mathsf{B}^T\boldsymbol{\lambda} \in \mathrm{Im}\mathsf{A},$$

which guarantees solvability of (2.108) with respect to $\mathbf{x}$, conveniently as

$$\mathsf{R}^T(\mathsf{B}^T\boldsymbol{\lambda} - \mathbf{b}) = \mathbf{o}. \tag{2.112}$$

If the latter condition is satisfied, then we can use any symmetric left generalized inverse A^+ to find all the solutions of (2.108) with respect to $\mathbf{x}$ in the form

$$\mathbf{x}(\boldsymbol{\lambda}, \boldsymbol{\alpha}) = \mathsf{A}^+(\mathbf{b} - \mathsf{B}^T\boldsymbol{\lambda}) + \mathsf{R}\boldsymbol{\alpha}, \ \boldsymbol{\alpha} \in \mathbb{R}^d,$$

where d is the dimension of KerA. After substituting for $\mathbf{x}$ into (2.109)–(2.111), we get

$$[\,-\mathsf{B}\mathsf{A}^+\mathsf{B}^T\boldsymbol{\lambda} + (\mathsf{B}\mathsf{A}^+\mathbf{b} - \mathbf{c}) + \mathsf{B}\mathsf{R}\boldsymbol{\alpha}]_{\mathcal{I}} \le \mathbf{o}, \tag{2.113}$$

$$[\,-\mathsf{B}\mathsf{A}^+\mathsf{B}^T\boldsymbol{\lambda} + (\mathsf{B}\mathsf{A}^+\mathbf{b} - \mathbf{c}) + \mathsf{B}\mathsf{R}\boldsymbol{\alpha}]_{\mathcal{E}} = \mathbf{o}, \tag{2.114}$$

$$\boldsymbol{\lambda}_{\mathcal{I}}^T\,[\,-\mathsf{B}\mathsf{A}^+\mathsf{B}^T\boldsymbol{\lambda} + (\mathsf{B}\mathsf{A}^+\mathbf{b} - \mathbf{c}) + \mathsf{B}\mathsf{R}\boldsymbol{\alpha}]_{\mathcal{I}} = 0. \tag{2.115}$$

The formulae in (2.113)–(2.115) look like something that we have already seen. Indeed, introducing the vector of Lagrange multipliers $\boldsymbol{\alpha}$ for (2.112) and denoting

$$\begin{aligned}\Lambda(\boldsymbol{\lambda}, \boldsymbol{\alpha}) &= \Theta(\boldsymbol{\lambda}) + \boldsymbol{\alpha}^T(\mathsf{R}^T\mathsf{B}^T\boldsymbol{\lambda} - \mathsf{R}^T\mathbf{b})\\ &= -\frac{1}{2}\boldsymbol{\lambda}^T\mathsf{B}\mathsf{A}^+\mathsf{B}^T\boldsymbol{\lambda} + \boldsymbol{\lambda}^T(\mathsf{B}\mathsf{A}^+\mathbf{b} - \mathbf{c}) - \frac{1}{2}\mathbf{b}^T\mathsf{A}^+\mathbf{b}\\ &\quad + \boldsymbol{\alpha}^T(\mathsf{R}^T\mathsf{B}^T\boldsymbol{\lambda} - \mathsf{R}^T\mathbf{b}),\end{aligned}$$

$$\mathbf{g} = \nabla_{\boldsymbol{\lambda}}\Lambda(\boldsymbol{\lambda}, \boldsymbol{\alpha}) = -\mathsf{B}\mathsf{A}^+\mathsf{B}^T\boldsymbol{\lambda} + (\mathsf{B}\mathsf{A}^+\mathbf{b} - \mathbf{c}) + \mathsf{B}\mathsf{R}\boldsymbol{\alpha},$$

we can rewrite the relations (2.113)–(2.115) as

$$\mathbf{g}_{\mathcal{I}} \le \mathbf{o}, \quad \mathbf{g}_{\mathcal{E}} = \mathbf{o}, \quad \text{and} \quad \boldsymbol{\lambda}_{\mathcal{I}}^T\mathbf{g}_{\mathcal{I}} = 0. \tag{2.116}$$

Comparing (2.116) with the KKT conditions (2.83) for the bound and equality constrained problem (2.82), we conclude that (2.116) are the KKT conditions for

$$\max \Theta(\boldsymbol{\lambda}) \quad \text{subject to} \quad \mathsf{R}^T\mathsf{B}^T\boldsymbol{\lambda} - \mathsf{R}^T\mathbf{b} = \mathbf{o} \quad \text{and} \quad \boldsymbol{\lambda}_{\mathcal{I}} \ge \mathbf{o}. \tag{2.117}$$

We have thus proved that if $(\overline{\mathbf{x}}, \overline{\boldsymbol{\lambda}})$ solves (2.107)–(2.111), then $\overline{\boldsymbol{\lambda}}$ is a feasible vector for problem (2.117) which satisfies the related KKT conditions. Recalling that A^+ is by the assumption symmetric positive semidefinite, so that $\mathsf{B}\mathsf{A}^+\mathsf{B}^T$ is also positive semidefinite, we conclude that $\overline{\boldsymbol{\lambda}}$ solves (2.104). Moreover, we have shown that any solution $\overline{\mathbf{x}}$ can be obtained in the form (2.105) with a KKT pair $(\overline{\boldsymbol{\lambda}}, \overline{\boldsymbol{\alpha}})$, where $\overline{\boldsymbol{\alpha}}$ is a vector of the Lagrange multipliers for (2.104).

(ii) Let $(\overline{\boldsymbol{\lambda}}, \overline{\boldsymbol{\alpha}})$ be a KKT pair for problem (2.104), so that $(\overline{\boldsymbol{\lambda}}, \overline{\boldsymbol{\alpha}})$ satisfies (2.112)–(2.115) and $\overline{\boldsymbol{\lambda}}_{\mathcal{I}} \geq \mathbf{o}$. If we denote

$$\overline{\mathbf{x}} = \mathsf{A}^{+}(\mathbf{b} - \mathsf{B}^T\overline{\boldsymbol{\lambda}}) + \mathsf{R}\overline{\boldsymbol{\alpha}},$$

we can use (2.113)–((2.115) to verify directly that $\overline{\mathbf{x}}$ is feasible and that $(\overline{\mathbf{x}}, \overline{\boldsymbol{\lambda}})$ satisfies the complementarity conditions, respectively. Finally, using (2.112), we get that there is $\mathbf{y} \in \mathbb{R}^n$ such that

$$\mathbf{b} - \mathsf{B}^T\overline{\boldsymbol{\lambda}} = \mathsf{A}\mathbf{y}.$$

Thus

$$\begin{aligned}\mathsf{A}\overline{\mathbf{x}} - \mathbf{b} + \mathsf{B}^T\overline{\boldsymbol{\lambda}} &= \mathsf{A}\big(\mathsf{A}^{+}(\mathbf{b} - \mathsf{B}^T\overline{\boldsymbol{\lambda}}) + \mathsf{R}\overline{\alpha}\big) - \mathbf{b} + \mathsf{B}^T\overline{\boldsymbol{\lambda}} \\ &= \mathsf{A}\mathsf{A}^{+}\mathsf{A}\mathbf{y} - \mathbf{b} + \mathsf{B}^T\overline{\boldsymbol{\lambda}} = \mathbf{b} - \mathsf{B}^T\overline{\boldsymbol{\lambda}} - \mathbf{b} + \mathsf{B}^T\overline{\boldsymbol{\lambda}} = \mathbf{o},\end{aligned}$$

which proves that $(\overline{\mathbf{x}}, \overline{\boldsymbol{\lambda}})$ is a KKT pair for (2.76).
(iii) Let $(\overline{\mathbf{x}}, \overline{\boldsymbol{\lambda}})$ be a KKT pair for (2.76). Using the feasibility condition (2.109) and the complementarity condition (2.111), we get

$$\overline{\boldsymbol{\lambda}}^T(\mathsf{B}\overline{\mathbf{x}} - \mathbf{c}) = \overline{\boldsymbol{\lambda}}_{\mathcal{E}}^T[\mathsf{B}\overline{\mathbf{x}} - \mathbf{c}]_{\mathcal{E}} + \overline{\boldsymbol{\lambda}}_{\mathcal{I}}^T[\mathsf{B}\overline{\mathbf{x}} - \mathbf{c}]_{\mathcal{I}} = 0.$$

Hence

$$f(\overline{\mathbf{x}}) = f(\overline{\mathbf{x}}) + \overline{\boldsymbol{\lambda}}^T(\mathsf{B}\overline{\mathbf{x}} - \mathbf{c}) = L_0(\overline{\mathbf{x}}, \overline{\boldsymbol{\lambda}}).$$

Next recall that if $(\overline{\mathbf{x}}, \overline{\boldsymbol{\lambda}})$ is a KKT pair, then

$$\nabla_{\mathbf{x}} L_0(\overline{\mathbf{x}}, \overline{\boldsymbol{\lambda}}) = \mathbf{o}.$$

Since L_0 is convex, the latter is the gradient condition for the unconstrained minimizer of L_0 with respect to $\mathbf{x}$; therefore

$$L_0(\overline{\mathbf{x}}, \overline{\boldsymbol{\lambda}}) = \min_{\mathbf{x} \in \mathbb{R}^n} L_0(\mathbf{x}, \overline{\boldsymbol{\lambda}}) = \Theta(\overline{\boldsymbol{\lambda}}).$$

Thus

$$f(\overline{\mathbf{x}}) = L_0(\overline{\mathbf{x}}, \overline{\boldsymbol{\lambda}}) = \Theta(\overline{\boldsymbol{\lambda}}).$$

□

The result which we have just proved is useful for reformulation of the convex quadratic problems with general inequality constraints to the problems with bound and equality constraints. See Chaps. 7 and 8 for examples.

Since the constant term is not essential in our applications and we formulate our algorithms for minimization problems, we shall consider the function

$$\theta(\boldsymbol{\lambda}) = -\Theta(\boldsymbol{\lambda}) - \frac{1}{2}\mathbf{b}^T\mathsf{A}^{+}\mathbf{b} = \frac{1}{2}\boldsymbol{\lambda}^T\mathsf{B}\mathsf{A}^{+}\mathsf{B}^T\boldsymbol{\lambda} - \boldsymbol{\lambda}^T(\mathsf{B}\mathsf{A}^{+}\mathbf{b} - \mathbf{c}), \qquad (2.118)$$

so that

$$\arg\min_{\lambda \in \Omega_{BE}} \theta(\boldsymbol{\lambda}) = \arg\max_{\lambda \in \Omega_{BE}} \Theta(\boldsymbol{\lambda}).$$

Uniqueness of a KKT Pair

We shall complete our exposition of duality by formulating the results concerning the uniqueness of the solution for the *constrained dual problem*

$$\min_{\boldsymbol{\lambda}\in\Omega_{BE}} \theta(\boldsymbol{\lambda}), \quad \Omega_{BE} = \{\boldsymbol{\lambda} \in \mathbb{R}^m : \ \boldsymbol{\lambda}_{\mathcal{I}} \geq \mathbf{o}, \ \mathsf{R}^T\mathsf{B}^T\boldsymbol{\lambda} = \mathsf{R}^T\mathbf{b}\}, \tag{2.119}$$

where θ is defined by (2.118).

Proposition 2.23. *Let the matrices* A, B, *the vectors* $\mathbf{b}, \mathbf{c}$, *and the index sets* $\mathcal{I}, \mathcal{E}$ *be those from the definition of problem (2.76) with* A *positive semidefinite,* $\Omega_{IE} \neq \emptyset$, *and* $\Omega_{BE} \neq \emptyset$. *Let* $\mathsf{R} \in \mathbb{R}^{n\times d}$ *be a full rank matrix such that*

$$\mathrm{Im}\mathsf{R} = \mathrm{Ker}\mathsf{A}.$$

Then the following statements hold:
(i) If B^T *and* BR *are full column rank matrices, then there is a unique solution* $\widehat{\boldsymbol{\lambda}}$ *of problem (2.119).*
(ii) If $\widehat{\boldsymbol{\lambda}}$ *is a unique solution of the constrained dual problem (2.119),*

$$\mathcal{A} = \{i : \ [\boldsymbol{\lambda}]_i > 0\} \cup \mathcal{E},$$

and $\mathsf{B}_{\mathcal{A}*}\mathsf{R}$ *is a full column rank matrix, then there is a unique triple* $(\widehat{\mathbf{x}}, \widehat{\boldsymbol{\lambda}}, \widehat{\boldsymbol{\alpha}})$ *such that* $(\widehat{\mathbf{x}}, \widehat{\boldsymbol{\lambda}})$ *solves the primal problem (2.76) and* $(\widehat{\boldsymbol{\lambda}}, \widehat{\boldsymbol{\alpha}})$ *solves the constrained dual problem (2.119). If* $\widehat{\boldsymbol{\lambda}}$ *is known, then*

$$\widehat{\boldsymbol{\alpha}} = (\mathsf{R}^T\mathsf{B}_{\mathcal{A}*}^T\mathsf{B}_{\mathcal{A}*}\mathsf{R})^{-1}\mathsf{R}^T\mathsf{B}_{\mathcal{A}*}^T\left(\mathsf{B}_{\mathcal{A}*}\mathsf{A}^+\mathsf{B}^T\widehat{\boldsymbol{\lambda}} - (\mathsf{B}_{\mathcal{A}*}\mathsf{A}^+\mathbf{b} - \mathbf{c}_{\mathcal{A}})\right) \tag{2.120}$$

and

$$\widehat{\mathbf{x}} = \mathsf{A}^+(\mathbf{b} - \mathsf{B}^T\widehat{\boldsymbol{\lambda}}) + \mathsf{R}\widehat{\boldsymbol{\alpha}}. \tag{2.121}$$

(iii) If B^T *and* $\mathsf{B}_{\mathcal{E}*}\mathsf{R}$ *are full column rank matrices, then there is a unique triple* $(\widehat{\mathbf{x}}, \widehat{\boldsymbol{\lambda}}, \widehat{\boldsymbol{\alpha}})$ *such that* $(\widehat{\mathbf{x}}, \widehat{\boldsymbol{\lambda}})$ *solves the primal problem (2.76) and* $(\widehat{\boldsymbol{\lambda}}, \widehat{\boldsymbol{\alpha}})$ *solves the constrained dual problem (2.119).*

Proof. (i) Let B^T and BR be full column rank matrices. To show that there is a unique solution of (2.119), we examine the Hessian $\mathsf{BA}^+\mathsf{B}^T$ of θ. Let $\mathsf{R}^T\mathsf{B}^T\boldsymbol{\lambda} = \mathbf{o}$ and $\mathsf{BA}^+\mathsf{B}^T\boldsymbol{\lambda} = \mathbf{o}$. Using the definition of R, it follows that $\mathsf{B}^T\boldsymbol{\lambda} \in \mathrm{Im}\mathsf{A}$. Hence there is $\boldsymbol{\mu} \in \mathbb{R}^n$ such that

$$\mathsf{B}^T\boldsymbol{\lambda} = \mathsf{A}\boldsymbol{\mu}$$

and

$$\boldsymbol{\mu}^T\mathsf{A}\boldsymbol{\mu} = \boldsymbol{\mu}^T\mathsf{AA}^+\mathsf{A}\boldsymbol{\mu} = \boldsymbol{\lambda}^T\mathsf{BA}^+\mathsf{B}^T\boldsymbol{\lambda} = 0.$$

Thus $\boldsymbol{\mu} \in \mathrm{Ker}\mathsf{A}$ and

$$\mathsf{B}^T\boldsymbol{\lambda} = \mathsf{A}\boldsymbol{\mu} = \mathbf{o}.$$

Since we assume that B^T has independent columns, we conclude that $\boldsymbol{\lambda} = \mathbf{o}$. We have thus proved that the restriction of $\mathsf{BA}^+\mathsf{B}^T$ to $\mathrm{Ker}(\mathsf{R}^T\mathsf{B}^T)$ is positive definite, so that $\theta|\mathrm{Ker}\mathsf{R}^T\mathsf{B}^T$ is by Proposition 2.5 strictly convex, and by Corollary 2.3 it is strictly convex on

$$\mathcal{U} = \{\boldsymbol{\lambda} \in \mathbb{R}^m : \mathsf{R}^T\mathsf{B}^T\boldsymbol{\lambda} = \mathsf{R}^T\mathbf{b}\}.$$

Since $\Omega_{BE} \neq \emptyset$ and $\Omega_{BE} \subseteq \mathcal{U}$, we have that θ is strictly convex on Ω_{BE}, and it follows by Proposition 2.4 that there is a unique solution $\widehat{\boldsymbol{\lambda}}$ of (2.119).
(ii) Let $\widehat{\boldsymbol{\lambda}}$ be a unique solution of problem (2.119). Since the solution satisfies the related KKT conditions, it follows that there is $\widehat{\boldsymbol{\alpha}}$ such that

$$\mathsf{B}_{\mathcal{A}*}\mathsf{A}^+\mathsf{B}^T\widehat{\boldsymbol{\lambda}} - (\mathsf{B}_{\mathcal{A}*}\mathsf{A}^+\mathbf{b} - \mathbf{c}_{\mathcal{A}}) - \mathsf{B}_{\mathcal{A}*}\mathsf{R}\widehat{\boldsymbol{\alpha}} = \mathbf{o}.$$

After multiplying on the left by $\mathsf{R}^T\mathsf{B}^T_{\mathcal{A}*}$ and simple manipulations, we get (2.120). The inverse exists and the solution $\widehat{\boldsymbol{\alpha}}$ is unique due to the uniqueness of $\widehat{\boldsymbol{\lambda}}$ and the assumption on the full column rank of $\mathsf{B}_{\mathcal{A}*}\mathsf{R}$.
(iii) If B^T and $\mathsf{B}_{\mathcal{E}*}\mathsf{R}$ are full column rank matrices, then BR is also a full column rank matrix. Hence there is a unique solution $\widehat{\boldsymbol{\lambda}}$ of problem (2.119) by (i). Since $\mathcal{E} \subseteq \mathcal{A}$ and $\mathsf{B}_{\mathcal{E}*}\mathsf{R}$ has independent columns, it follows that $\mathsf{B}_{\mathcal{A}*}\mathsf{R}$ has also independent columns. Thus we can use (ii) to finish the proof. □

The reconstruction formula (2.120) can be modified in order to work whenever the dual problem has a solution $\overline{\boldsymbol{\lambda}}$. The resulting formula obtained by analysis of the related KKT conditions then reads

$$\overline{\boldsymbol{\alpha}} = (\mathsf{R}^T\mathsf{B}^T_{\mathcal{A}*}\mathsf{B}_{\mathcal{A}*}\mathsf{R})^+\mathsf{R}^T\mathsf{B}^T_{\mathcal{A}*}\left(\mathsf{B}_{\mathcal{A}*}\mathsf{A}^+\mathsf{B}^T\overline{\boldsymbol{\lambda}} - (\mathsf{B}_{\mathcal{A}*}\mathsf{A}^+\mathbf{b} - \mathbf{c}_{\mathcal{A}})\right). \qquad (2.122)$$

The duality theory can be illustrated on a problem to find the displacement $\mathbf{x}$ of an elastic body under traction $\mathbf{b}$. After the finite element discretization, we get a convex QP problem. We assume that the body is fixed on a part of the boundary in normal direction, so that the vector of nodal displacements satisfies $\mathsf{B}_{\mathcal{E}*}\mathbf{x} = \mathbf{c}_{\mathcal{E}}$ as in Fig. 2.12. Moreover, the body may not be allowed to penetrate an obstacle, so that $\mathsf{B}_{\mathcal{I}*}\mathbf{x} \leq \mathbf{c}_{\mathcal{I}}$ as in Fig. 2.11.

The displacement $\overline{\mathbf{x}}$ of the body in equilibrium is a minimizer of the convex energy function f. The Hessian A of f is positive semidefinite if the constraints admit rigid body motions. The Lagrange multipliers solve the dual problem. The condition $\mathsf{R}^T\mathbf{b} = \mathsf{R}^T\mathsf{B}^T\widehat{\boldsymbol{\lambda}}$ requires that the resulting forces are balanced in the directions of the rigid body motions and $\widehat{\boldsymbol{\lambda}}_{\mathcal{I}} \geq \mathbf{o}$ guarantees that the body is not glued to the obstacle. If the reaction forces $\mathsf{B}^T\widehat{\boldsymbol{\lambda}}$ determine the components of $\widehat{\boldsymbol{\lambda}}$, then $\widehat{\boldsymbol{\lambda}}$ is uniquely determined by the conditions of equilibrium. Notice that $\mathsf{B}^T\widehat{\boldsymbol{\lambda}}$ is always uniquely determined by the conditions of equilibrium. If no rigid body motion is possible due to the active constraints $\mathsf{B}_{\mathcal{A}*}\mathbf{x} = \mathbf{c}_{\mathcal{A}}$ as in Fig. 2.11, then the displacement $\mathbf{x}$ is uniquely determined. If this is not the case, then the displacement is determined up to some rigid body motion as in Fig. 2.12.

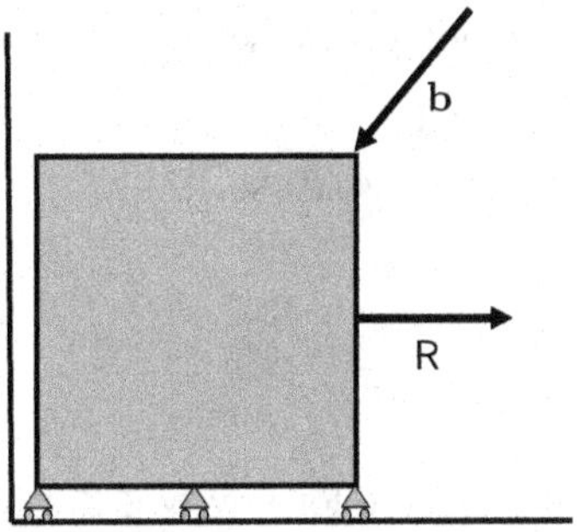

Fig. 2.11. Unique displacement

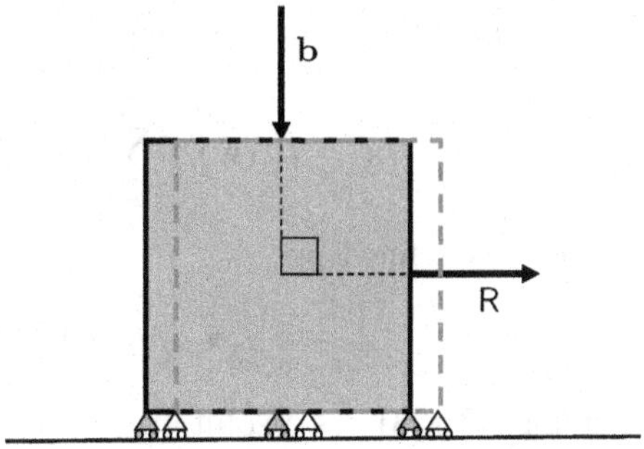

Fig. 2.12. Nonunique displacement

2.7 Linear Programming

We shall finish our review of optimization theory by recalling some results on minimization of linear functions subject to linear constraints. More formally, we shall look for the solution of *linear programming problem* to find

$$\min_{\mathbf{x}\in\Omega_{IE}} \ell(\mathbf{x}), \quad \Omega_{IE} = \{\mathbf{x} \in \mathbb{R}^n : [\mathsf{C}\mathbf{x}]_{\mathcal{I}} \geq \mathbf{d}_{\mathcal{I}}, [\mathsf{C}\mathbf{x}]_{\mathcal{E}} = \mathbf{d}_{\mathcal{E}}\}, \tag{2.123}$$

where $\ell(\mathbf{x}) = \mathbf{f}^T\mathbf{x}$ is a linear function defined by $\mathbf{f} \in \mathbb{R}^n$, $\mathsf{C} \in \mathbb{R}^{m\times n}$ is a matrix with possibly dependent rows, $\mathbf{d} \in \mathbb{R}^m$, and $\mathcal{I}$, $\mathcal{E}$ are disjoint sets of indices which decompose $\{1,\ldots,m\}$. As in the definition of (2.76), we admit $\mathcal{I} = \emptyset$ or $\mathcal{E} = \emptyset$ and assume that Ω_{IE} is not empty. Let us mention that (2.123) may be easily reduced to the standard form (2.80) with $\mathsf{C}_{\mathcal{I}*} = \mathsf{I}_n$ and $\mathbf{d}_{\mathcal{I}} = \mathbf{o}$. Here we restrict our attention to the very basic results on linear programming that we shall use in what follows. More information on linear programming may be found, e.g., in the books by Gass [98], Chvátal [22], Bertsekas and Tsitsiklis [13], Nocedal and Wright [155], or Vanderbei [177].

2.7.1 Solvability and Localization of Solutions

Let us recall that by Proposition 2.13 there exists a nonempty finite set of n-vectors $\{\mathbf{x}_1,\ldots,\mathbf{x}_k\}$ and a polyhedral cone $\mathcal{C} \subset \mathbb{R}^n$ such that

$$\Omega_{BI} = \mathcal{C} + \mathrm{Conv}\{\mathbf{x}_1,\ldots,\mathbf{x}_k\};$$

hence any $\mathbf{x} \in \Omega_{EI}$ can be written in the form

$$\mathbf{x} = \alpha\mathbf{w} + \sum_{i=1}^{k}\alpha_i\mathbf{x}_i, \ \mathbf{w}\in\mathcal{C},\ \alpha\geq 0, \sum_{i=1}^{k}\alpha_i = 1,\ 0\leq\alpha_i\leq 1,\ i=1,\ldots,k.$$

The vectors $\{\mathbf{x}_1,\ldots,\mathbf{x}_k\}$ can be chosen in such a way that none of the vectors $\mathbf{x}_i$ can be expressed as a convex combination of the remaining ones; the vectors

$\{\mathbf{x}_1, \ldots, \mathbf{x}_k\}$ are then called the *vertices* of the polyhedral set Ω_{BI}. It follows that

$$\ell(\mathbf{x}) = \alpha\ell(\mathbf{w}) + \sum_{i=1}^{k} \alpha_i \ell(\mathbf{x}_i) \geq \alpha\ell(\mathbf{w}) + \alpha_m \ell(\mathbf{x}_m),$$

where m is defined by

$$\ell(\mathbf{x}_m) \leq \ell(\mathbf{x}_i), \quad i = 1, \ldots, k;$$

therefore (2.123) has a solution if and only if $\ell(\mathbf{w}) \geq 0$ for any $\mathbf{w} \in \mathcal{C}$, i.e., if ℓ is bounded from below on $\mathcal{C}$. Moreover, if a solution exists, then it is achieved in at least one vertex $\mathbf{x}_i$.

2.7.2 Duality in Linear Programming

Noticing that problem (2.123) can be considered as a special case of the bound and equality constrained problem (2.76) with $\mathsf{A} = \mathsf{O}$, $\mathbf{b} = -\mathbf{f}$, $\mathsf{B} = -\mathsf{C}$, and $\mathbf{c} = -\mathbf{d}$, we can get the following proposition on duality in linear programming as a special case of Proposition 2.22.

Proposition 2.24. *Let the matrix* C, *vectors* $\mathbf{f}, \mathbf{d}$, *and index sets* $\mathcal{I}, \mathcal{E}$ *be those of the definition of problem (2.123) with* $\Omega_{IE} \neq \emptyset$. *Let let*

$$\zeta(\boldsymbol{\lambda}) = \boldsymbol{\lambda}^T \mathbf{d}. \tag{2.124}$$

Then the following statements hold:
(i) Problem (2.123) has a solution if and only if $\Omega_{IE} \neq \emptyset$ *and* ℓ *is bounded from below.*
(ii) The dual problem

$$\max_{\boldsymbol{\lambda} \in \Omega_{BE}} \zeta(\boldsymbol{\lambda}), \quad \Omega_{BE} = \{\boldsymbol{\lambda} \in \mathbb{R}^m : \; \boldsymbol{\lambda}_{\mathcal{I}} \geq \mathbf{o}, \; \mathsf{C}^T \boldsymbol{\lambda} = \mathbf{f}\} \tag{2.125}$$

has a solution if and only if $\Omega_{BE} \neq \emptyset$ *and* ζ *is bounded from above.*
(iii) Problem (2.123) has a solution if and only if the dual problem (2.125) has a solution.
(iv) A vector $\overline{\mathbf{x}}$ *is a solution of (2.123) if and only if* $\overline{\boldsymbol{\lambda}}$ *is a solution of (2.125). Moreover,*

$$\ell(\overline{\mathbf{x}}) = \zeta(\overline{\boldsymbol{\lambda}}). \tag{2.126}$$

Proof. The statements (i) and (ii) can be considered as special cases of Frank–Wolfe theorem [93]. See also Eaves [79] or Blum and Oettli [15]. To finish the proof, consider problem (2.123) as a special case of (2.76) with $\mathsf{A} = \mathsf{O}$, $\mathbf{b} = -\mathbf{f}$, $\mathsf{B} = -\mathsf{C}$, and $\mathbf{c} = -\mathbf{d}$. Choose $\mathsf{R} = \mathsf{I}_n$ and $\mathsf{O}_{nn}^+ = \mathsf{O}_{nn}$. The statements (iii) and (iv) then become special cases of Proposition 2.22. □

Part II

Algorithms

3

Conjugate Gradients for Unconstrained Minimization

We shall begin our development of scalable algorithms by description of the *conjugate gradient method* for the solution of

$$\min_{\mathbf{x}\in\mathbb{R}^n} f(\mathbf{x}), \tag{3.1}$$

where $f(\mathbf{x}) = \frac{1}{2}\mathbf{x}^T\mathsf{A}\mathbf{x} - \mathbf{x}^T\mathbf{b}$, $\mathbf{b}$ is a given column n-vector, and A is an $n \times n$ symmetric positive definite or positive semidefinite matrix. We are interested especially in problems with n large and A sparse and reasonably conditioned. We have already seen in Sect. 2.2.2 that (3.1) is equivalent to the solution of a system of linear equations $\mathsf{A}\mathbf{x} = \mathbf{b}$, but our main goal here is not to solve large systems of linear equations, but rather to describe our basic tool for dealing with the auxiliary linear systems that are generated by algorithms for the solution of constrained quadratic programming problems.

We shall use the conjugate gradient (CG) method as an *iterative method* which generates improving approximations to the solution at each step. The cost of one step of the CG method is typically dominated by the cost of the multiplication of a vector by the matrix A, which is proportional to the number of nonzero entries of A. The memory requirements are also proportional to the number of nonzero entries of A.

To develop optimal algorithms for more general quadratic programming problems, it is important that the rate of convergence of the conjugate gradient method depends on the distribution of the spectrum of A. In particular, given a positive interval $[a_{\min}, a_{\max}]$ with the spectrum of A, it is possible to give a bound in terms of $a_{\max}/a_{\min}$ on a number of the conjugate gradient iterations that are necessary to solve problem (3.1) to a given relative precision. It is also important that the number of steps that are necessary to obtain an approximate solution of a given problem is typically proportional to the logarithm of prescribed precision, so that the algorithm can return a low-precision solution at a reduced time.

Zdeněk Dostál, *Optimal Quadratic Programming Algorithms*,
Springer Optimization and Its Applications, DOI 10.1007/978-0-387-84806-8_3,

Overview of Algorithms

The first algorithm of this chapter is the *method of conjugate directions* defined by the two simple formulae (3.6). The algorithm assumes that we are given an A-orthogonal basis of $\mathbb{R}^n$, leaving open the problem how to get it.

The *conjugate gradient algorithm*, Algorithm 3.1, the main hero of this chapter, combines the conjugate gradient direction method with a clever construction of conjugate directions. It is the best method as it exploits effectively all the information gathered during the solution in order to maximize the decrease of the cost function. The CG method can be considered both as a direct method and an iterative method.

A step of the *restarted conjugate gradient method* described in Sect. 3.4 comprises a fixed number of the conjugate gradient steps. Such algorithm is more robust, but usually less efficient. If the chain of the CG iterations reduces to just one iteration, we get the *gradient method*, known also as the *method of the steepest descent.* It is the most robust and most simple variant of the restarted CG method. See Algorithm 3.2 for a more formal description.

If we are able to find an easily invertible approximation of the Hessian, we can use it to improve the performance of the CG method in the *preconditioned conjugate gradient method* described in Sect. 3.6 as Algorithm 3.3. The construction of preconditioners is problem dependent. The *preconditioning by a conjugate projector* described in Sect. 3.7 as Algorithm 3.4 is useful in the minimization problems arising from the discretization of elliptic partial differential equations and variational inequalities.

3.1 Conjugate Directions and Minimization

The conjugate gradient method, an ingenious and powerful engine of our algorithms, is based on simple observations. In this section we examine the first one, namely, that it is possible to reduce the solution of (3.1) to the solution of a sequence of one-dimensional problems.

Let $\mathsf{A} \in \mathbb{R}^{n\times n}$ be an SPD matrix and let us assume that there are nonzero n-vectors $\mathbf{p}^1, \dots, \mathbf{p}^n$ such that

$$(\mathbf{p}^i, \mathbf{p}^j)_{\mathsf{A}} = (\mathbf{p}^i)^T \mathsf{A} \mathbf{p}^j = 0 \quad \text{for} \quad i \neq j.$$

We call such vectors A-*conjugate* or briefly *conjugate.* Specializing the arguments of Sect. 1.7, we get that $\mathbf{p}^1, \dots, \mathbf{p}^n$ are independent. Thus $\mathbf{p}^1, \dots, \mathbf{p}^n$ form the basis of $\mathbb{R}^n$ and any $\mathbf{x} \in \mathbb{R}^n$ can be written in the form

$$\mathbf{x} = \xi_1 \mathbf{p}^1 + \dots + \xi_n \mathbf{p}^n.$$

Substituting into f and using the conjugacy results in

$$\begin{aligned} f(\mathbf{x}) &= \left(\frac{1}{2}\xi_1^2 (\mathbf{p}^1)^T \mathsf{A} \mathbf{p}^1 - \xi_1 \mathbf{b}^T \mathbf{p}^1\right) + \dots + \left(\frac{1}{2}\xi_n^2 (\mathbf{p}^n)^T \mathsf{A} \mathbf{p}^n - \xi_n \mathbf{b}^T \mathbf{p}^n\right) \\ &= f(\xi_1 \mathbf{p}^1) + \dots + f(\xi_n \mathbf{p}^n). \end{aligned}$$

Thus

$$f(\widehat{\mathbf{x}}) = \min_{\mathbf{x}\in\mathbb{R}^n} f(\mathbf{x}) = \min_{\xi_1\in\mathbb{R}} f(\xi_1\mathbf{p}^1) + \cdots + \min_{\xi_n\in\mathbb{R}} f(\xi_n\mathbf{p}^n).$$

We have thus managed to decompose the original problem (3.1) into n one-dimensional problems. Since

$$\left.\frac{\mathrm{d}f\left(\xi\mathbf{p}^i\right)}{\mathrm{d}\xi}\right|_{\xi_i} = \xi_i(\mathbf{p}^i)^T\mathsf{A}\mathbf{p}^i - \mathbf{b}^T\mathbf{p}^i = 0,$$

the solution $\widehat{\mathbf{x}}$ of (3.1) is given by

$$\widehat{\mathbf{x}} = \xi_1\mathbf{p}^1 + \cdots + \xi_n\mathbf{p}^n, \quad \xi_i = \mathbf{b}^T\mathbf{p}^i/(\mathbf{p}^i)^T\mathsf{A}\mathbf{p}^i,\ i = 1,\ldots,n. \tag{3.2}$$

If the dimension of problem (3.1) is large, the task to evaluate $\widehat{\mathbf{x}}$ may be too ambitious. In this case it may be useful to modify the procedure that we have just described so that it can be used to find an approximation $\widetilde{\mathbf{x}}$ to the solution $\widehat{\mathbf{x}}$ for (3.1) by means of some initial guess $\mathbf{x}^0$ and a few vectors $\mathbf{p}^1,\ldots,\mathbf{p}^k,\ k \ll n$. A natural choice for the approximation $\widetilde{\mathbf{x}}$ is the minimizer $\mathbf{x}^k$ of f in $\mathcal{S}^k = \mathbf{x}^0 + \mathrm{Span}\{\mathbf{p}^1,\ldots,\mathbf{p}^k\}$. To find it, notice that any $\mathbf{x}\in\mathcal{S}^k$ can be written in the form

$$\mathbf{x} = \mathbf{x}^0 + \xi_1\mathbf{p}^1 + \cdots + \xi_k\mathbf{p}^k,$$

so, after substituting into f and using that $\mathbf{p}^1,\ldots,\mathbf{p}^k$ are conjugate, we get

$$\begin{aligned} f(\mathbf{x}) = f(\mathbf{x}^0) &+ \left(\frac{1}{2}\xi_1^2(\mathbf{p}^1)^T\mathsf{A}\mathbf{p}^1 + \xi_1\left(\mathsf{A}\mathbf{x}^0 - \mathbf{b}\right)^T\mathbf{p}^1\right) + \ldots \\ &+ \left(\frac{1}{2}\xi_k^2(\mathbf{p}^k)^T\mathsf{A}\mathbf{p}^k + \xi_k\left(\mathsf{A}\mathbf{x}^0 - \mathbf{b}\right)^T\mathbf{p}^k\right). \end{aligned}$$

Denoting $\mathbf{g}^0 = \mathbf{g}(\mathbf{x}^0) = \nabla f(\mathbf{x}^0) = \mathsf{A}\mathbf{x}^0 - \mathbf{b}$ and

$$f_0(\mathbf{x}) = \frac{1}{2}\mathbf{x}^T\mathsf{A}\mathbf{x} + \mathbf{x}^T\mathbf{g}^0,$$

we have

$$f(\mathbf{x}) = f(\mathbf{x}^0) + f_0(\xi_1\mathbf{p}^1) + \cdots + f_0(\xi_k\mathbf{p}^k)$$

and

$$f(\mathbf{x}^k) = \min_{\mathbf{x}\in\mathcal{S}^k} f(\mathbf{x}) = f(\mathbf{x}^0) + \min_{\xi_1\in\mathbb{R}} f_0(\xi_1\mathbf{p}^1) + \cdots + \min_{\xi_k\in\mathbb{R}} f_0(\xi_k\mathbf{p}^k). \tag{3.3}$$

We have thus again reduced our problem to the solution of a sequence of simple one-dimensional problems. The approximation $\mathbf{x}^k$ is given by

$$\mathbf{x}^k = \mathbf{x}^0 + \xi_1\mathbf{p}^1 + \cdots + \xi_k\mathbf{p}^k, \quad \xi_i = -(\mathbf{g}^0)^T\mathbf{p}^i/(\mathbf{p}^i)^T\mathsf{A}\mathbf{p}^i,\ i = 1,\ldots,k, \tag{3.4}$$

as

$$\left.\frac{\mathrm{d}f\left(\xi\mathbf{p}^i\right)}{\mathrm{d}\xi}\right|_{\xi_i} = \xi_i(\mathbf{p}^i)^T\mathsf{A}\mathbf{p}^i + (\mathbf{g}^0)^T\mathbf{p}^i = 0.$$

Since by (3.3) for $k \geq 1$

$$f(\mathbf{x}^k) = \min_{\mathbf{x}\in\mathcal{S}^k} f(\mathbf{x}) = f(\mathbf{x}^{k-1}) + \min_{\xi\in\mathbb{R}} f_0(\xi\mathbf{p}^k), \tag{3.5}$$

we can generate the approximations $\mathbf{x}^k$ iteratively. The *conjugate direction method* starts from an arbitrary initial guess $\mathbf{x}^0$. If $\mathbf{x}^{k-1}$ is given, then $\mathbf{x}^k$ is generated by the formula

$$\mathbf{x}^k = \mathbf{x}^{k-1} - \alpha_k\mathbf{p}^k, \quad \alpha_k = (\mathbf{g}^0)^T\mathbf{p}^i/(\mathbf{p}^i)^T\mathsf{A}\mathbf{p}^i. \tag{3.6}$$

Thus $f(\mathbf{x}^{k-1} + \xi\mathbf{p}^k)$ achieves its minimum at $\xi = -\alpha_k$ and the procedure guarantees that the *successive iterates* $\mathbf{x}^k$ *minimize* f *over a progressively expanding manifold* $\mathcal{S}^k$ *that eventually includes the global minimum of* f.

The coefficients α_k can be evaluated by alternative formulae. For example, using Corollary 2.9 and the definition of $\mathcal{S}^k$, we get

$$(\mathbf{g}^k)^T\mathbf{p}^i = 0, \quad i = 1,\dots,k. \tag{3.7}$$

Since for $i \geq 1$

$$\begin{aligned}\mathbf{g}^i &= \mathsf{A}\mathbf{x}^i - \mathbf{b} = \mathsf{A}\left(\mathbf{x}^{i-1} - \alpha_i\mathbf{p}^i\right) - \mathbf{b} = \left(\mathsf{A}\mathbf{x}^{i-1} - \mathbf{b}\right) - \alpha_i\mathsf{A}\mathbf{p}^i\\ &= \mathbf{g}^{i-1} - \alpha_i\mathsf{A}\mathbf{p}^i,\end{aligned}$$

we get for $k \geq 1$ and $i = 1,\dots,k-1$, by using the conjugacy, that

$$(\mathbf{g}^i)^T\mathbf{p}^k = (\mathbf{g}^{i-1})^T\mathbf{p}^k - \alpha_i(\mathbf{p}^i)^T\mathsf{A}\mathbf{p}^k = (\mathbf{g}^{i-1})^T\mathbf{p}^k.$$

Thus

$$(\mathbf{g}^0)^T\mathbf{p}^k = (\mathbf{g}^1)^T\mathbf{p}^k = \dots = (\mathbf{g}^{k-1})^T\mathbf{p}^k$$

and

$$\alpha_k = \frac{(\mathbf{g}^0)^T\mathbf{p}^k}{(\mathbf{p}^k)^T\mathsf{A}\mathbf{p}^k} = \dots = \frac{(\mathbf{g}^{k-1})^T\mathbf{p}^k}{(\mathbf{p}^k)^T\mathsf{A}\mathbf{p}^k}. \tag{3.8}$$

Combining the latter formula with the Taylor expansion, we get

$$f\left(\mathbf{x}^k\right) = f\left(\mathbf{x}^{k-1}\right) - \frac{1}{2}\frac{\left((\mathbf{g}^{k-1})^T\mathbf{p}^k\right)^2}{(\mathbf{p}^k)^T\mathsf{A}\mathbf{p}^k}. \tag{3.9}$$

So far, we have not discussed how to get the vectors $\mathbf{p}^1,\dots,\mathbf{p}^n$. Are we able to generate them efficiently? Positive answer in the next section is a key to the success of the conjugate gradient method.

3.2 Generating Conjugate Directions and Krylov Spaces

Let us now recall how to generate conjugate directions with the *Gramm–Schmidt procedure.* Let us first suppose that $\mathbf{p}^1, \ldots, \mathbf{p}^k$ are nonzero conjugate directions, $1 \le k < n$, and let us examine how to use $\mathbf{h}^k \notin \text{Span}\{\mathbf{p}^1, \ldots, \mathbf{p}^k\}$ to generate a new member $\mathbf{p}^{k+1}$ in the form

$$\mathbf{p}^{k+1} = \mathbf{h}^k + \beta_{k1}\mathbf{p}^1 + \cdots + \beta_{kk}\mathbf{p}^k. \tag{3.10}$$

Since $\mathbf{p}^{k+1}$ should be conjugate to $\mathbf{p}^1, \ldots, \mathbf{p}^k$, we get

$$\begin{aligned} 0 = (\mathbf{p}^i)^T \mathsf{A}\mathbf{p}^{k+1} &= (\mathbf{p}^i)^T \mathsf{A}\mathbf{h}^k + \beta_{k1}(\mathbf{p}^i)^T \mathsf{A}\mathbf{p}^1 + \cdots + \beta_{kk}(\mathbf{p}^i)^T \mathsf{A}\mathbf{p}^k \\ &= (\mathbf{p}^i)^T \mathsf{A}\mathbf{h}^k + \beta_{ki}(\mathbf{p}^i)^T \mathsf{A}\mathbf{p}^i, \quad i = 1, \ldots, k. \end{aligned}$$

Thus

$$\beta_{ki} = -\frac{(\mathbf{p}^i)^T \mathsf{A}\mathbf{h}^k}{(\mathbf{p}^i)^T \mathsf{A}\mathbf{p}^i}, \quad i = 1, \ldots, k. \tag{3.11}$$

Obviously

$$\text{Span}\{\mathbf{p}^1, \ldots, \mathbf{p}^{k+1}\} = \text{Span}\{\mathbf{p}^1, \ldots, \mathbf{p}^k, \mathbf{h}^k\}.$$

Therefore, given any independent vectors $\mathbf{h}^0, \ldots, \mathbf{h}^{k-1}$, we can start from $\mathbf{p}^1 = \mathbf{h}^0$ and use (3.10) and (3.11) to construct a set of mutually A-conjugate directions $\mathbf{p}^1, \ldots, \mathbf{p}^k$ such that

$$\text{Span}\{\mathbf{h}^0, \ldots, \mathbf{h}^{i-1}\} = \text{Span}\{\mathbf{p}^1, \ldots, \mathbf{p}^i\}, \ i = 1, \ldots, k.$$

For $\mathbf{h}^0, \ldots, \mathbf{h}^{k-1}$ arbitrary, the construction is increasingly expensive as it requires both the storage for the vectors $\mathbf{p}^1, \ldots, \mathbf{p}^k$ and heavy calculations including evaluation of $k(k+1)/2$ scalar products. However, it turns out that we can adapt the procedure so that it generates very efficiently the conjugate basis of the *Krylov spaces*

$$\mathcal{K}^k = \mathcal{K}^k(\mathsf{A}, \mathbf{g}^0) = \text{Span}\{\mathbf{g}^0, \mathsf{A}\mathbf{g}^0, \ldots, \mathsf{A}^{k-1}\mathbf{g}^0\}, \quad k = 1, \ldots, n,$$

with $\mathbf{g}^0 = \mathsf{A}\mathbf{x}^0 - \mathbf{b}$ defined by a suitable initial vector $\mathbf{x}^0$ and $\mathcal{K}^0 = \{\mathbf{o}\}$. The powerful method is again based on a few simple observations.

First assume that $\mathbf{p}^1, \ldots, \mathbf{p}^i$ form a conjugate basis of $\mathcal{K}^i$, $i = 1, \ldots, k$, and observe that if $\mathbf{x}^k$ denotes the minimizer of f on $\mathbf{x}^0 + \mathcal{K}^k$, then by Corollary 2.9 the gradient $\mathbf{g}^k = \nabla f(\mathbf{x}^k)$ is orthogonal to the Krylov space $\mathcal{K}^k$, that is,

$$(\mathbf{g}^k)^T \mathbf{x} = 0 \ \text{ for any } \ \mathbf{x} \in \mathcal{K}^k.$$

In particular, if $\mathbf{g}^k \ne \mathbf{o}$, then

$$\mathbf{g}^k \notin \mathcal{K}^k.$$

Since $\mathbf{g}^k \in \mathcal{K}^{k+1}$, we can use (3.10) with $\mathbf{h}^k = \mathbf{g}^k$ to expand any conjugate basis of $\mathcal{K}^k$ to the conjugate basis of $\mathcal{K}^{k+1}$. Obviously

$$\mathcal{K}^k(\mathsf{A}, \mathbf{g}^0) = \mathrm{Span}\{\mathbf{g}^0, \ldots, \mathbf{g}^{k-1}\}.$$

Next observe that for any $\mathbf{x} \in \mathcal{K}^{k-1}$ and $k \geq 1$

$$\mathsf{A}\mathbf{x} \in \mathcal{K}^k,$$

or briefly $\mathsf{A}\mathcal{K}^{k-1} \subseteq \mathcal{K}^k$. Since $\mathbf{p}^i \in \mathcal{K}^i \subseteq \mathcal{K}^{k-1}$, $i = 1, \ldots, k-1$, we have

$$(\mathsf{A}\mathbf{p}^i)^T \mathbf{g}^k = (\mathbf{p}^i)^T \mathsf{A}\mathbf{g}^k = 0, \; i = 1, \ldots, k-1.$$

It follows that

$$\beta_{ki} = -\frac{(\mathbf{p}^i)^T \mathsf{A}\mathbf{g}^k}{(\mathbf{p}^i)^T \mathsf{A}\mathbf{p}^i} = 0, \; i = 1, \ldots, k-1.$$

Summing up, if we have a set of such conjugate vectors $\mathbf{p}^1, \ldots, \mathbf{p}^k$ that

$$\mathrm{Span}\{\mathbf{p}^1, \ldots, \mathbf{p}^i\} = \mathcal{K}^i, \; i = 1, \ldots k,$$

then the formula (3.10) applied to $\mathbf{p}^1, \ldots, \mathbf{p}^k$ and $\mathbf{h}^k = \mathbf{g}^k$ simplifies to

$$\mathbf{p}^{k+1} = \mathbf{g}^k + \beta_k \mathbf{p}^k \tag{3.12}$$

with

$$\beta_k = \beta_{kk} = -\frac{(\mathbf{p}^k)^T \mathsf{A}\mathbf{g}^k}{(\mathbf{p}^k)^T \mathsf{A}\mathbf{p}^k}. \tag{3.13}$$

Finally, observe that the orthogonality of $\mathbf{g}^k$ to $\mathrm{Span}\{\mathbf{p}^1, \ldots, \mathbf{p}^k\}$ and (3.12) imply that

$$\|\mathbf{p}^{k+1}\| \geq \|\mathbf{g}^k\|. \tag{3.14}$$

In particular, if $\mathbf{g}^{k-1} \neq \mathbf{o}$, then $\mathbf{p}^k \neq \mathbf{o}$, so the formula (3.13) is well defined provided $\mathbf{g}^{k-1} \neq \mathbf{o}$.

3.3 Conjugate Gradient Method

In the previous two sections, we have found that the conjugate directions can be used to reduce the minimization of any convex quadratic function to the solution of a sequence of one-dimensional problems, and that the conjugate directions can be generated very efficiently. The famous *conjugate gradient (CG) method* just puts these two observations together.

The algorithm starts from an initial guess $\mathbf{x}^0$, $\mathbf{g}^0 = \mathsf{A}\mathbf{x}^0 - \mathbf{b}$, and $\mathbf{p}^1 = \mathbf{g}^0$. If $\mathbf{x}^{k-1}$ and $\mathbf{g}^{k-1}$ are given, $k \geq 1$, it first checks if $\mathbf{x}^{k-1}$ is the solution. If not, then the algorithm generates

$$\mathbf{x}^k = \mathbf{x}^{k-1} - \alpha_k \mathbf{p}^k \;\; \text{with} \;\; \alpha_k = (\mathbf{g}^{k-1})^T \mathbf{p}^k / (\mathbf{p}^k)^T \mathsf{A}\mathbf{p}^k$$

and

$$\begin{aligned} \mathbf{g}^k &= \mathsf{A}\mathbf{x}^k - \mathbf{b} = \mathsf{A}\left(\mathbf{x}^{k-1} - \alpha_k \mathbf{p}^k\right) - \mathbf{b} = \left(\mathsf{A}\mathbf{x}^{k-1} - \mathbf{b}\right) - \alpha_k \mathsf{A}\mathbf{p}^k \\ &= \mathbf{g}^{k-1} - \alpha_k \mathsf{A}\mathbf{p}^k. \end{aligned} \tag{3.15}$$

Finally the new conjugate direction $\mathbf{p}^{k+1}$ is generated by (3.12) and (3.13).

The decision if $\mathbf{x}^{k-1}$ is an acceptable solution is typically based on the value of $\|\mathbf{g}^{k-1}\|$, so the norm of the gradient must be evaluated at each step. It turns out that the norm can also be used to replace the scalar products involving the gradient in the definition of α_k and β_k. To find the formulae, let us replace k in (3.12) by $k-1$ and multiply the resulting identity by $(\mathbf{g}^{k-1})^T$. Using the orthogonality, we get

$$(\mathbf{g}^{k-1})^T \mathbf{p}^k = \|\mathbf{g}^{k-1}\|^2 + \beta_{k-1}(\mathbf{g}^{k-1})^T \mathbf{p}^{k-1} = \|\mathbf{g}^{k-1}\|^2, \tag{3.16}$$

so by (3.8)

$$\alpha_k = \frac{\|\mathbf{g}^{k-1}\|^2}{(\mathbf{p}^k)^T \mathsf{A}\mathbf{p}^k}. \tag{3.17}$$

To find an alternative formula for β_k, notice that $\alpha_k > 0$ for $\mathbf{g}^{k-1} \neq \mathbf{o}$ and that by (3.15)

$$\mathsf{A}\mathbf{p}^k = \frac{1}{\alpha_k}(\mathbf{g}^{k-1} - \mathbf{g}^k),$$

so that

$$\alpha_k (\mathbf{g}^k)^T \mathsf{A}\mathbf{p}^k = (\mathbf{g}^k)^T(\mathbf{g}^{k-1} - \mathbf{g}^k) = -\|\mathbf{g}^k\|^2$$

and

$$\beta_k = -\frac{(\mathbf{p}^k)^T \mathsf{A}\mathbf{g}^k}{(\mathbf{p}^k)^T \mathsf{A}\mathbf{p}^k} = \frac{\|\mathbf{g}^k\|^2}{\alpha_k (\mathbf{p}^k)^T \mathsf{A}\mathbf{p}^k} = \frac{\|\mathbf{g}^k\|^2}{\|\mathbf{g}^{k-1}\|^2}. \tag{3.18}$$

The complete CG method is presented as Algorithm 3.1.

Algorithm 3.1. Conjugate gradient method (CG).

Given a symmetric positive definite matrix $\mathsf{A} \in \mathbb{R}^{n\times n}$ *and* $\mathbf{b} \in \mathbb{R}^n$.

Step 0. {*Initialization.*}
Choose $\mathbf{x}^0 \in \mathbb{R}^n$, *set* $\mathbf{g}^0 = \mathsf{A}\mathbf{x}^0 - \mathbf{b}$, $\mathbf{p}^1 = \mathbf{g}^0$, $k = 1$

Step 1. {*Conjugate gradient loop.* }
while $\|\mathbf{g}^{k-1}\| > 0$
$\alpha_k = \|\mathbf{g}^{k-1}\|^2/(\mathbf{p}^k)^T \mathsf{A}\mathbf{p}^k$
$\mathbf{x}^k = \mathbf{x}^{k-1} - \alpha_k \mathbf{p}^k$
$\mathbf{g}^k = \mathbf{g}^{k-1} - \alpha_k \mathsf{A}\mathbf{p}^k$
$\beta_k = \|\mathbf{g}^k\|^2/\|\mathbf{g}^{k-1}\|^2 = -(\mathsf{A}\mathbf{p}^k)^T \mathbf{g}^k / \left((\mathbf{p}^k)^T \mathsf{A}\mathbf{p}^k\right)$
$\mathbf{p}^{k+1} = \mathbf{g}^k + \beta_k \mathbf{p}^k$
$k = k + 1$
end while

Step 2. {*Return the solution.*}
$\widehat{\mathbf{x}} = \mathbf{x}^k$

Each step of the CG method can be implemented with just one matrix–vector multiplication. This multiplication by the Hessian matrix A typically dominates the cost of the step. Only one generation of vectors $\mathbf{x}^k, \mathbf{p}^k$, and $\mathbf{g}^k$ is typically stored, so the memory requirements are modest.

Let us recall that the algorithm finds at each step the minimizer $\mathbf{x}^k$ of f on $\mathbf{x}^0 + \mathcal{K}^k = \mathbf{x}^0 + \mathcal{K}^k(\mathsf{A}, \mathbf{g}^0)$ and expands the conjugate basis of $\mathcal{K}^k$ to that of $\mathcal{K}^{k+1}$ provided $\mathbf{g}^k \neq \mathbf{o}$. Since the dimension of $\mathcal{K}^k$ is less than or equal to k, it follows that for some $k \leq n$

$$\mathcal{K}^k = \mathcal{K}^{k+1}.$$

Since $\mathbf{g}^k \in \mathcal{K}^{k+1}$ and $\mathbf{g}^k$ is orthogonal to $\mathcal{K}^k$, Algorithm 3.1 implemented in the exact arithmetics finds the solution $\widehat{\mathbf{x}}$ of (3.1) in at most n steps. We can sum up the most important properties of Algorithm 3.1 into the following theorem.

Theorem 3.1. *Let $\{\mathbf{x}^k\}$ be generated by Algorithm 3.1 to find the solution $\widehat{\mathbf{x}}$ of (3.1) starting from $\mathbf{x}^0 \in \mathbb{R}^n$. Then the algorithm is well defined and there is $k \leq n$ such that $\mathbf{x}^k = \widehat{\mathbf{x}}$. Moreover, the following statements hold for $i = 1, \ldots, k$:*

(i) $f(\mathbf{x}^i) = \min\{f(\mathbf{x}) : \mathbf{x} \in \mathbf{x}^0 + \mathcal{K}^i(\mathsf{A}, \mathbf{g}^0)\}$.
(ii) $\|\mathbf{p}^{i+1}\| \geq \|\mathbf{g}^i\|$.
(iii) $(\mathbf{g}^i)^T\mathbf{g}^j = 0$ *for* $i \neq j$.
(iv) $(\mathbf{p}^i)^T\mathsf{A}\mathbf{p}^j = 0$ *for* $i \neq j$.
(v) $\mathcal{K}^i(\mathsf{A}, \mathbf{g}^0) = \mathrm{Span}\{\mathbf{g}^0, \ldots, \mathbf{g}^{i-1}\} = \mathrm{Span}\{\mathbf{p}^1, \ldots, \mathbf{p}^i\}$.

It is usually sufficient to find $\mathbf{x}^k$ such that $\|\mathbf{g}^k\|$ is small. For example, given a small $\varepsilon > 0$, we can consider $\mathbf{g}^k$ small if

$$\|\mathbf{g}^k\| \leq \varepsilon\|\mathbf{b}\|.$$

Then $\widetilde{\mathbf{x}} = \mathbf{x}^k$ is an approximate solution which satisfies

$$\|\mathsf{A}(\widetilde{\mathbf{x}} - \widehat{\mathbf{x}})\| \leq \varepsilon\|\mathbf{b}\|, \quad \|\widetilde{\mathbf{x}} - \widehat{\mathbf{x}}\| \leq \varepsilon\lambda_{\min}(\mathsf{A})^{-1},$$

where $\lambda_{\min}(\mathsf{A})$ denotes the least eigenvalue of A. It is easy to check that the approximate solution $\widetilde{\mathbf{x}}$ solves the perturbed problem

$$\min_{\mathbf{x}\in\mathbb{R}^n} \widetilde{f}(\mathbf{x}) = \frac{1}{2}\mathbf{x}^T\mathsf{A}\mathbf{x} - \widetilde{\mathbf{b}}^T\mathbf{x}, \quad \widetilde{\mathbf{b}} = \mathbf{b} + \mathbf{g}^k.$$

What is "small" depends on the problem solved. To keep our exposition general, we shall often not specify the test in what follows. Of course $\mathbf{g}^k = \mathbf{o}$ is always considered small.

3.4 Restarted CG and the Gradient Method

Given an approximation $\mathbf{x}^0$ of the solution $\widehat{\mathbf{x}}$, we can use k conjugate gradient iterations to find an improved approximation $\mathbf{x}^k$. Repeating the procedure with $\mathbf{x}^0 = \mathbf{x}^k$, we get the *restarted conjugate gradient method.*

A special case with $k = 1$ and $\mathbf{p}^1 = \nabla f(\mathbf{x}^0)$ is of independent interest. Given $\mathbf{x}^k$, the *gradient method* (also called the *steepest descent method*) generates $\mathbf{x}^{k+1}$ by

$$\mathbf{x}^{k+1} = \arg\min_{\alpha \in \mathbb{R}} f(\mathbf{x}^k - \alpha \mathbf{g}^k), \quad \mathbf{g}^k = \nabla f(\mathbf{x}^k).$$

The name "steepest descent" is derived from observation that the linear model of f at $\mathbf{x}$ achieves its minimum on the set of all unit vectors

$$\mathcal{U} = \{\mathbf{d} \in \mathbb{R}^n, \|\mathbf{d}\| = 1\}$$

at $\widehat{\mathbf{d}} = -\|\nabla f(\mathbf{x})\|^{-1}\nabla f(\mathbf{x})$. Indeed, for any $\mathbf{d} \in \mathcal{U}$

$$\nabla f(\mathbf{x})^T\mathbf{d} \geq -\|\nabla f(\mathbf{x})\|\|\mathbf{d}\| = -\|\nabla f(\mathbf{x})\| = \nabla f(\mathbf{x})^T\widehat{\mathbf{d}}.$$

The complete steepest descent method reads as follows:

Algorithm 3.2. Gradient (steepest descent) method.

Given a symmetric positive definite matrix $\mathsf{A} \in \mathbb{R}^{n\times n}$ *and* $\mathbf{b} \in \mathbb{R}^n$.

Step 0. {*Initialization.*}
  Choose $\mathbf{x}^0 \in \mathbb{R}^n$, *set* $\mathbf{g}^0 = \mathsf{A}\mathbf{x}^0 - \mathbf{b}$, $k = 0$

Step 1. {*Steepest descent loop.* }
  while $\|\mathbf{g}^k\|$ *is not small*
    $\alpha_k = \|\mathbf{g}^k\|^2/(\mathbf{g}^k)^T\mathsf{A}\mathbf{g}^k$
    $\mathbf{x}^{k+1} = \mathbf{x}^k - \alpha_k\mathbf{g}^k$
    $\mathbf{g}^{k+1} = \mathbf{g}^k - \alpha_k\mathsf{A}\mathbf{g}^k$
    $k = k + 1$
  end while

Step 2. {*Return a (possibly approximate) solution.*}
    $\widetilde{\mathbf{x}} = \mathbf{x}^k$

The gradient method is known to converge, but its convergence is for ill-conditioned problems considerably slower than that of the conjugate gradient method, as we shall see in the next section. The slow convergence is illustrated in Fig. 3.1.

In spite of its slow convergence, the gradient method is useful as it is easy to implement and uses a robust decrease direction. It is illustrated in Fig. 3.2 that even if $\partial\mathbf{g}$ is a relatively large perturbation of the gradient $\mathbf{g}$, the vector $-\mathbf{g} - \partial\mathbf{g}$ is still a decrease direction, while a small perturbation $\partial\mathbf{p}$ of the CG direction $\mathbf{p}$ can cause that $-\mathbf{p} - \partial\mathbf{p}$ is not a decrease direction.

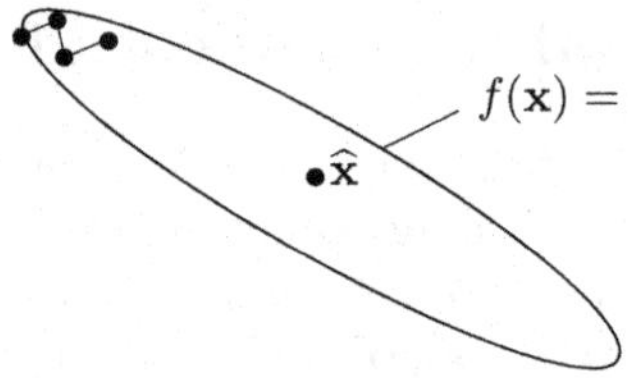

Fig. 3.1. Slow convergence of the steepest descent method

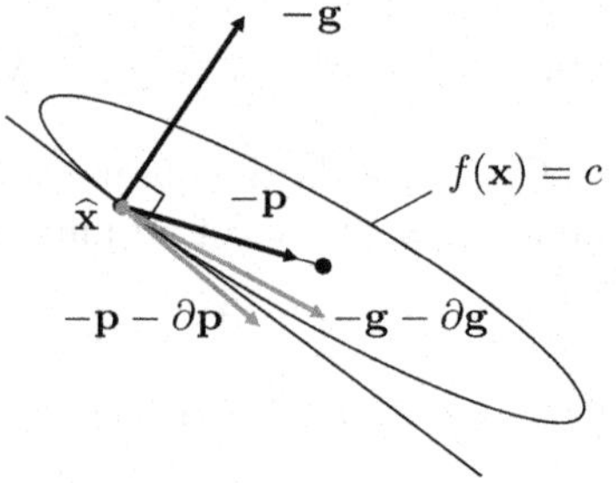

Fig. 3.2. Robustness of the gradient and CG decrease directions $\mathbf{g}$ and $\mathbf{p}$

3.5 Rate of Convergence and Optimality

Although the conjugate gradient method finds by Theorem 3.1 the exact solution $\widehat{\mathbf{x}}$ of (3.1) in a number of steps which does not exceed the dimension of the problem, it turns out that it can often produce a sufficiently accurate approximation $\widetilde{\mathbf{x}}$ of $\widehat{\mathbf{x}}$ in a much smaller number of steps than required for exact termination. This observation suggests that the conjugate gradient method may also be considered as an iterative method. In this section we present the results which substantiate this claim and help us to identify the favorable cases.

3.5.1 Min-max Estimate

Let us denote the *solution error* as

$$\mathbf{e} = \mathbf{e}(\mathbf{x}) = \mathbf{x} - \widehat{\mathbf{x}}$$

and observe that

$$\mathbf{g}(\widehat{\mathbf{x}}) = \mathsf{A}\widehat{\mathbf{x}} - \mathbf{b} = \mathbf{o}.$$

It follows that

$$\mathbf{g}^k = \mathsf{A}\mathbf{x}^k - \mathbf{b} = \mathsf{A}\mathbf{x}^k - \mathsf{A}\widehat{\mathbf{x}} = \mathsf{A}(\mathbf{x}^k - \widehat{\mathbf{x}}) = \mathsf{A}\mathbf{e}^k,$$

so in particular

$$\mathcal{K}^k(\mathsf{A}, \mathbf{g}^0) = \mathrm{Span}\{\mathbf{g}^0, \mathsf{A}\mathbf{g}^0, \dots, \mathsf{A}^{k-1}\mathbf{g}^0\} = \mathrm{Span}\{\mathsf{A}\mathbf{e}^0, \dots, \mathsf{A}^k\mathbf{e}^0\}.$$

We start our analysis of the solution error by using the Taylor expansion (2.5) to obtain the identity

$$\begin{aligned} f(\mathbf{x}) - f(\widehat{\mathbf{x}}) &= f(\widehat{\mathbf{x}} + (\mathbf{x} - \widehat{\mathbf{x}})) - f(\widehat{\mathbf{x}}) \\ &= f(\widehat{\mathbf{x}}) + \mathbf{g}(\widehat{\mathbf{x}})^T(\mathbf{x} - \widehat{\mathbf{x}}) + \frac{1}{2}\|\mathbf{x} - \widehat{\mathbf{x}}\|_{\mathsf{A}}^2 - f(\widehat{\mathbf{x}}) \\ &= \frac{1}{2}\|\mathbf{x} - \widehat{\mathbf{x}}\|_{\mathsf{A}}^2 = \frac{1}{2}\|\mathbf{e}\|_{\mathsf{A}}^2. \end{aligned}$$

Combining the latter identity with Theorem 3.1, we get

$$\begin{aligned} \|\mathbf{e}^k\|_{\mathsf{A}}^2 &= 2\left(f(\mathbf{x}^k) - f(\widehat{\mathbf{x}})\right) = \min_{\mathbf{x} \in \mathbf{x}^0 + \mathcal{K}^k(\mathsf{A},\mathbf{g}^0)} 2\left(f(\mathbf{x}) - f(\widehat{\mathbf{x}})\right) \\ &= \min_{\mathbf{x} \in \mathbf{x}^0 + \mathcal{K}^k(\mathsf{A},\mathbf{g}^0)} \|\mathbf{x} - \widehat{\mathbf{x}}\|_{\mathsf{A}}^2 = \min_{\mathbf{x} \in \mathbf{x}^0 + \mathcal{K}^k(\mathsf{A},\mathbf{g}^0)} \|\mathbf{e}(\mathbf{x})\|_{\mathsf{A}}^2. \end{aligned}$$

Since any $\mathbf{x} \in \mathbf{x}^0 + \mathcal{K}^k(\mathsf{A}, \mathbf{g}^0)$ may be written in the form

$$\mathbf{x} = \mathbf{x}^0 + \xi_1 \mathbf{g}^0 + \xi_2 \mathsf{A}\mathbf{g}^0 + \cdots + \xi_k \mathsf{A}^{k-1}\mathbf{g}^0 = \mathbf{x}^0 + \xi_1 \mathsf{A}\mathbf{e}^0 + \cdots + \xi_k \mathsf{A}^k \mathbf{e}^0,$$

it follows that

$$\mathbf{x} - \widehat{\mathbf{x}} = \mathbf{e}^0 + \xi_1 \mathsf{A}\mathbf{e}^0 + \cdots + \xi_k \mathsf{A}^k \mathbf{e}^0 = p(\mathsf{A})\mathbf{e}^0,$$

where p denotes the polynomial defined for any $x \in \mathbb{R}$ by

$$p(x) = 1 + \xi_1 x + \xi_2 x^2 + \cdots + \xi_k x^k.$$

Thus denoting by $\mathcal{P}^k$ the set of all kth degree polynomials p which satisfy $p(0) = 1$, we have

$$\|\mathbf{e}^k\|_{\mathsf{A}}^2 = \min_{\mathbf{x} \in \mathbf{x}^0 + \mathcal{K}^k(\mathsf{A},\mathbf{g}^0)} \|\mathbf{e}(\mathbf{x})\|_{\mathsf{A}}^2 = \min_{p \in \mathcal{P}^k} \|p(\mathsf{A})\mathbf{e}^0\|_{\mathsf{A}}^2. \tag{3.19}$$

We shall now derive a bound on the expression on the right-hand side of (3.19) that depends on the spectrum of A, but is independent of the direction of the initial error $\mathbf{e}^0$. Let a spectral decomposition of A be written as $\mathsf{A} = \mathsf{U}\mathsf{D}\mathsf{U}^T$, where U is an orthogonal matrix and $\mathsf{D} = \mathrm{diag}(\lambda_1, \ldots, \lambda_n)$ is a diagonal matrix defined by the eigenvalues of A. Since A is assumed to be positive definite, the square root of A is well defined by

$$\mathsf{A}^{\frac{1}{2}} = \mathsf{U}\mathsf{D}^{\frac{1}{2}}\mathsf{U}^T.$$

Using $p(\mathsf{A}) = \mathsf{U}p(\mathsf{D})\mathsf{U}^T$, it is also easy to check that

$$\mathsf{A}^{\frac{1}{2}} p(\mathsf{A}) = p(\mathsf{A})\mathsf{A}^{\frac{1}{2}}.$$

Moreover, for any vector $\mathbf{v} \in \mathbb{R}^n$

$$\|\mathbf{v}\|_{\mathsf{A}}^2 = \mathbf{v}^T \mathsf{A}\mathbf{v} = \mathbf{v}^T \mathsf{A}^{\frac{1}{2}} \mathsf{A}^{\frac{1}{2}} \mathbf{v} = (\mathsf{A}^{\frac{1}{2}}\mathbf{v})^T \mathsf{A}^{\frac{1}{2}} \mathbf{v} = \|\mathsf{A}^{\frac{1}{2}}\mathbf{v}\|^2.$$

Using the latter identities, (3.19), and the properties of norms, we get

$$\begin{aligned}\|\mathbf{e}^k\|_A^2 &= \min_{p\in\mathcal{P}^k} \|p(\mathsf{A})\mathbf{e}^0\|_{\mathsf{A}}^2 = \min_{p\in\mathcal{P}^k} \|\mathsf{A}^{\frac{1}{2}}p(\mathsf{A})\mathbf{e}^0\|^2 = \min_{p\in\mathcal{P}^k} \|p(\mathsf{A})\mathsf{A}^{\frac{1}{2}}\mathbf{e}^0\|^2 \\ &\le \min_{p\in\mathcal{P}^k} \|p(\mathsf{A})\|^2\|\mathsf{A}^{\frac{1}{2}}\mathbf{e}^0\|^2 = \min_{p\in\mathcal{P}^k} \|p(\mathsf{D})\|^2\|\mathbf{e}^0\|_{\mathsf{A}}^2.\end{aligned}$$

Since

$$\|p(\mathsf{D})\| = \max_{i\in\{1,\dots,n\}} |p(\lambda_i)|,$$

we can write

$$\|\mathbf{e}^k\|_{\mathsf{A}} \le \min_{p\in\mathcal{P}^k} \max_{i\in\{1,\dots,n\}} |p(\lambda_i)|\ \|\mathbf{e}^0\|_{\mathsf{A}}. \tag{3.20}$$

3.5.2 Estimate in the Condition Number

The estimate (3.20) reduces the analysis of convergence of the CG method to the analysis of approximation of the zero function on the spectrum of A by a kth degree polynomial with the value one at origin. This result helps us to identify the favorable cases when the conjugate gradient method is effective. For example, if the spectrum of A is clustered around a single point ξ, then the minimization by the CG should be very effective because $|(1 - x/\xi)^k|$ is small near ξ. We shall use (3.20) to get a "global" estimate of the rate of convergence of the CG method in terms of the condition number of A.

Theorem 3.2. *Let $\{\mathbf{x}^k\}$ be generated by Algorithm 3.1 to find the solution $\widehat{\mathbf{x}}$ of (3.1) starting from $\mathbf{x}^0 \in \mathbb{R}^n$. Then the error*

$$\mathbf{e}^k = \mathbf{x}^k - \widehat{\mathbf{x}}$$

satisfies

$$\|\mathbf{e}^k\|_{\mathsf{A}} \le 2\left(\frac{\sqrt{\kappa(\mathsf{A})}-1}{\sqrt{\kappa(\mathsf{A})}+1}\right)^k \|\mathbf{e}^0\|_{\mathsf{A}}, \tag{3.21}$$

where $\kappa(\mathsf{A})$ denotes the spectral condition number of A.

Proof. First notice that if $\mathcal{P}^k$ is the set of all kth degree polynomials p such that $p(0) = 1$, then for any $t \in \mathcal{P}^k$

$$\min_{p\in\mathcal{P}^k} \max_{\lambda\in[\lambda_{\min},\lambda_{\max}]} |p(\lambda)| \le \max_{\lambda\in[\lambda_{\min},\lambda_{\max}]} |t(\lambda)|. \tag{3.22}$$

A natural choice for t is the kth (weighted and shifted) Chebyshev polynomial on the interval $[\lambda_{\min}, \lambda_{\max}]$

$$t_k(\lambda) = T_k\left(\frac{2\lambda - \lambda_{\max} - \lambda_{\min}}{\lambda_{\max} - \lambda_{\min}}\right) / T_k\left(-\frac{\lambda_{\max} + \lambda_{\min}}{\lambda_{\max} - \lambda_{\min}}\right),$$

where $T_k(x)$ is the Chebyshev polynomial of the first kind on the interval $[-1,1]$ given by

$$T_k(x) = \frac{1}{2}\left(x+\sqrt{x^2-1}\right)^k + \frac{1}{2}\left(x-\sqrt{x^2-1}\right)^k.$$

This t_k is known to minimize the right-hand side of (3.22) (see, e.g., [172]). Obviously $t_k \in \mathcal{P}^k$, so that we can use its well-known properties to get

$$\max_{\lambda\in[\lambda_{\min},\lambda_{\max}]} |t_k(\lambda)| = 1/T_k\left(\frac{\lambda_{\max}+\lambda_{\min}}{\lambda_{\max}-\lambda_{\min}}\right).$$

Simple manipulations then show that

$$T_k\left(\frac{\lambda_{\max}+\lambda_{\min}}{\lambda_{\max}-\lambda_{\min}}\right) = \frac{1}{2}\left(\frac{\sqrt{\kappa(\mathsf{A})}+1}{\sqrt{\kappa(\mathsf{A})}-1}\right)^k + \frac{1}{2}\left(\frac{\sqrt{\kappa(\mathsf{A})}-1}{\sqrt{\kappa(\mathsf{A})}+1}\right)^k.$$

Thus for any $\lambda \in [\lambda_{\min}, \lambda_{\max}]$

$$|p_k(\lambda)| \le 2\left(\frac{\sqrt{\kappa(\mathsf{A})}-1}{\sqrt{\kappa(\mathsf{A})}+1}\right)^k.$$

Substituting this bound into (3.20) then gives the required result. □

The estimate (3.21) can be improved for some special distributions of the eigenvalues. For example, if the spectrum of A is in a positive interval $[a_{\min}, a_{\max}]$ except for m isolated eigenvalues $\lambda_1, \ldots, \lambda_m$, then we can use special polynomials $p \in \mathcal{P}^{k+m}$ of the form

$$p(\lambda) = \left(1-\frac{\lambda}{\lambda_1}\right)\ldots\left(1-\frac{\lambda}{\lambda_m}\right)q(\lambda),\ q\in\mathcal{P}^k$$

to get the estimate

$$\|\mathbf{e}^{k+m}\|_{\mathsf{A}} \le 2\left(\frac{\sqrt{\widetilde{\kappa}}-1}{\sqrt{\widetilde{\kappa}}+1}\right)^k \|\mathbf{e}^0\|_{\mathsf{A}}, \tag{3.23}$$

where $\widetilde{\kappa} = a_{\max}/a_{\min}$.

If the spectrum of A is distributed in two positive intervals $[a_{\min}, a_{\max}]$ and $[a_{\min}+d, a_{\max}+d]$, $d>0$, then

$$\|\mathbf{e}^{k}\|_{\mathsf{A}} \le 2\left(\frac{\sqrt{\overline{\kappa}}-1}{\sqrt{\overline{\kappa}}+1}\right)^k \|\mathbf{e}^0\|_{\mathsf{A}}, \tag{3.24}$$

where $\overline{\kappa} = 4a_{\max}/a_{\min}$ approximates the *effective condition number* of a matrix A with the spectrum in $[a_{\min}, a_{\max}] \cup [a_{\min}+d, a_{\max}+d]$. An interesting feature of the estimates (3.23) and (3.24) is that the *upper bound is independent of the values of some eigenvalues or d.* The proofs of the above and some other interesting estimates can be found in papers by Axelsson [3] and Axelsson and Lindskøg [5].

3.5.3 Convergence Rate of the Gradient Method

Observing that the step of the gradient method defined by Algorithm 3.2 is just the first step of the CG algorithm, we can use the results of Sect. 3.5.1 to find the rate of convergence of the gradient method. The estimate is formulated in the following proposition.

Proposition 3.3. *Let $\{\mathbf{x}^k\}$ be generated by Algorithm 3.2 to find the solution $\widehat{\mathbf{x}}$ of (3.1) starting from $\mathbf{x}^0 \in \mathbb{R}^n$. Then the error*

$$\mathbf{e}^k = \mathbf{x}^k - \widehat{\mathbf{x}}$$

satisfies

$$\|\mathbf{e}^k\|_{\mathsf{A}} \le \left(\frac{\kappa(\mathsf{A}) - 1}{\kappa(\mathsf{A}) + 1}\right)^k \|\mathbf{e}^0\|_{\mathsf{A}}, \tag{3.25}$$

where $\kappa(\mathsf{A})$ denotes the spectral condition number of A.

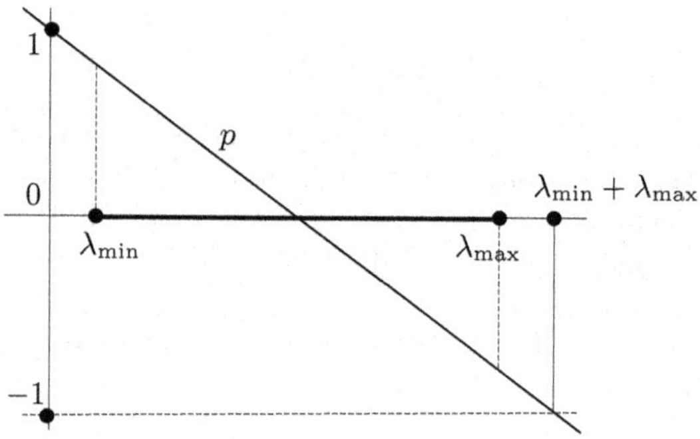

Fig. 3.3. The best approximation of zero on $\sigma(\mathsf{A})$ by linear polynomial with $p(0) = 1$

Proof. Let $\mathbf{x}^{k+1}$ be generated by the gradient method from $\mathbf{x}^k \in \mathbb{R}^n$ and let $\mathcal{P}^1$ denote the set of all linear polynomials p such that $p(0) = 1$. Then the energy norm $\|\mathbf{e}^k\|_{\mathsf{A}}$ of the error

$$\mathbf{e}^k = \mathbf{x}^k - \widehat{\mathbf{x}}$$

is by (3.20) reduced by a factor which can be estimated from

$$\|\mathbf{e}^{k+1}\|_{\mathsf{A}} \le \min_{p \in \mathcal{P}^1} \max_{i \in \{1,\ldots,n\}} |p(\lambda_i)| \, \|\mathbf{e}^k\|_{\mathsf{A}} = \min_{\xi_1 \in \mathbb{R}} \max_{i = \{1,\ldots,n\}} |\xi_1 \lambda_i + 1| \, \|\mathbf{e}^k\|_{\mathsf{A}}.$$

Using elementary properties of linear functions or Fig. 3.3, we get that the minimizer $\overline{\xi}_1$ satisfies

$$\overline{\xi}_1\lambda_{\min} + 1 = -(\overline{\xi}_1\lambda_{\max} + 1).$$

It follows that

$$\overline{\xi}_1 = -2/(\lambda_{\min} + \lambda_{\max})$$

and

$$\|\mathbf{e}^{k+1}\|_{\mathsf{A}} \le \left(\frac{-2\lambda_{\min}}{\lambda_{\min} + \lambda_{\max}} + 1\right)\|\mathbf{e}^k\|_{\mathsf{A}} = \frac{\lambda_{\max} - \lambda_{\min}}{\lambda_{\max} + \lambda_{\min}}\|\mathbf{e}^k\|_{\mathsf{A}}. \tag{3.26}$$

The estimate (3.25) can be obtained from (3.26) by simple manipulations. □

Notice that the estimate (3.21) for the first step of the conjugate gradient method may give worse bound than the estimate (3.25) for the gradient method, but for large k, the estimate (3.21) for the kth step of the conjugate gradient method is much better than the estimate (3.25) for the kth step of the gradient method. The reason is that (3.21) captures the global performance of the CG method, in particular its capability to exploit the information from the previous steps, while (3.25) is based on analysis of just one step, in agreement with the one-step information used by the gradient method.

3.5.4 Optimality

Theorem 3.2 implies an easy optimality result concerning the number of iterations of the CG algorithm. To formulate it, let $\mathcal{T}$ denote any set of indices and assume that for any $t \in \mathcal{T}$ there is defined the problem

$$\text{minimize } f_t(\mathbf{x})$$

with $f_t(\mathbf{x}) = \frac{1}{2}\mathbf{x}^T\mathsf{A}_t\mathbf{x} - \mathbf{b}_t^T\mathbf{x}$, $\mathsf{A}_t \in \mathbb{R}^{n_t \times n_t}$ symmetric positive definite, and $\mathbf{b}_t, \mathbf{x} \in \mathbb{R}^{n_t}$. Moreover, assume that the eigenvalues of any A_t are in the interval $[a_{\min}, a_{\max}]$, $0 < a_{\min} \le a_{\max}$. Then the number of the CG iterations that are necessary to reduce the error by a given factor ε is uniformly bounded. It easily follows that *the CG algorithm starting from* $\mathbf{x}_t^0 = \mathbf{o}$ *finds* $\mathbf{x}_t^k$ *such that*

$$\|\mathsf{A}_t\mathbf{x}_t^k - \mathbf{b}_t\| \le \epsilon\|\mathbf{b}_t\|$$

at $O(1)$ *iterations.* It follows that if the matrices A_t have $O(n_t)$ elements, then we can get approximate solutions at the optimal $O(n_t)$ arithmetic operations.

3.6 Preconditioned Conjugate Gradients

The analysis of the previous section shows that the rate of convergence of the conjugate gradient algorithm depends on the distribution of the eigenvalues of the Hessian A of f. In particular, we argued that CG converges very rapidly if the eigenvalues of A are clustered around one point, i.e., if the condition number $\kappa(\mathsf{A})$ is close to one. We shall now show that we can reduce our minimization problem to this favorable case if we have a symmetric positive definite matrix M such that $\mathsf{M}^{-1}\mathbf{x}$ can be easily evaluated for any $\mathbf{x}$ and M approximates A in the sense that $\mathsf{M}^{-1}\mathsf{A}$ is close to the identity.

First assume that M is available in the form

$$\mathsf{M} = \widetilde{\mathsf{L}}\widetilde{\mathsf{L}}^T,$$

so that $\mathsf{M}^{-1}\mathsf{A}$ is similar to $\widetilde{\mathsf{L}}^{-1}\mathsf{A}\widetilde{\mathsf{L}}^{-T}$ and the latter matrix is close to the identity. Then

$$f(\mathbf{x}) = \frac{1}{2}(\widetilde{\mathsf{L}}^T\mathbf{x})^T(\widetilde{\mathsf{L}}^{-1}\mathsf{A}\widetilde{\mathsf{L}}^{-T})(\widetilde{\mathsf{L}}^T\mathbf{x}) - (\widetilde{\mathsf{L}}^{-1}\mathbf{b})^T(\widetilde{\mathsf{L}}^T\mathbf{x})$$

and we can replace our original problem (3.1) by the *preconditioned problem* to find

$$\min_{\mathbf{y}\in\mathbb{R}^n} \bar{f}(\mathbf{y}), \tag{3.27}$$

where we substituted $\mathbf{y} = \widetilde{\mathsf{L}}^T\mathbf{x}$ and set

$$\bar{f}(\mathbf{y}) = \frac{1}{2}\mathbf{y}^T(\widetilde{\mathsf{L}}^{-1}\mathsf{A}\widetilde{\mathsf{L}}^{-T})\mathbf{y} - (\widetilde{\mathsf{L}}^{-1}\mathbf{b})^T\mathbf{y}.$$

The solution $\widehat{\mathbf{y}}$ of the preconditioned problem (3.27) is related to the solution $\widehat{\mathbf{x}}$ of the original problem by

$$\widehat{\mathbf{x}} = \widetilde{\mathsf{L}}^{-T}\widehat{\mathbf{y}}.$$

If the CG algorithm is applied directly to the preconditioned problem (3.27) with a given $\mathbf{y}^0$, then the algorithm is initialized by

$$\mathbf{y}^0 = \widetilde{\mathsf{L}}^T\mathbf{x}^0, \quad \bar{\mathbf{g}}^0 = \widetilde{\mathsf{L}}^{-1}\mathsf{A}\widetilde{\mathsf{L}}^{-T}\mathbf{y}^0 - \widetilde{\mathsf{L}}^{-1}\mathbf{b} = \widetilde{\mathsf{L}}^{-1}\mathbf{g}^0, \quad \text{and } \bar{\mathbf{p}}^1 = \bar{\mathbf{g}}^0;$$

the iterates are defined by

$$\begin{aligned}
\bar{\alpha}_k &= \|\bar{\mathbf{g}}^{k-1}\|^2/(\bar{\mathbf{p}}^k)^T\widetilde{\mathsf{L}}^{-1}\mathsf{A}\widetilde{\mathsf{L}}^{-T}\bar{\mathbf{p}}^k,\\
\mathbf{y}^k &= \mathbf{y}^{k-1} - \bar{\alpha}_k\bar{\mathbf{p}}^k,\\
\bar{\mathbf{g}}^k &= \bar{\mathbf{g}}^{k-1} - \bar{\alpha}_k\widetilde{\mathsf{L}}^{-1}\mathsf{A}\widetilde{\mathsf{L}}^{-T}\bar{\mathbf{p}}^k,\\
\bar{\beta}_k &= \|\bar{\mathbf{g}}^k\|^2/\|\bar{\mathbf{g}}^{k-1}\|^2,\\
\bar{\mathbf{p}}^{k+1} &= \bar{\mathbf{g}}^k + \bar{\beta}_k\bar{\mathbf{p}}^k.
\end{aligned}$$

Substituting

$$\mathbf{y}^k = \widetilde{\mathsf{L}}^T\mathbf{x}^k, \quad \bar{\mathbf{g}}^k = \widetilde{\mathsf{L}}^{-1}\mathbf{g}^k, \quad \text{and} \quad \bar{\mathbf{p}}^k = \widetilde{\mathsf{L}}^T\mathbf{p}^k,$$

and denoting

$$\mathbf{z}^k = \widetilde{\mathsf{L}}^{-T}\widetilde{\mathsf{L}}^{-1}\mathbf{g}^k = \mathsf{M}^{-1}\mathbf{g}^k,$$

we obtain the *preconditioned conjugate gradient algorithm* (PCG) in the original variables.

Algorithm 3.3. Preconditioned conjugate gradient method (PCG).

Given a symmetric positive definite matrix $\mathsf{A} \in \mathbb{R}^{n\times n}$, *its symmetric positive definite approximation* $\mathsf{M} \in \mathbb{R}^{n\times n}$, *and* $\mathbf{b} \in \mathbb{R}^n$.

Step 0. {*Initialization.*}
Choose $\mathbf{x}^0 \in \mathbb{R}^n$, *set* $\mathbf{g}^0 = \mathsf{A}\mathbf{x}^0 - \mathbf{b}$, $\mathbf{z}^0 = \mathsf{M}^{-1}\mathbf{g}^0$, $\mathbf{p}^1 = \mathbf{z}^0$, $k = 1$

Step 1. {*Conjugate gradient loop.*}
while $\|\mathbf{g}^{k-1}\|$ *is not small*
$\alpha_k = (\mathbf{z}^{k-1})^T\mathbf{g}^{k-1}/(\mathbf{p}^k)^T\mathsf{A}\mathbf{p}^k$
$\mathbf{x}^k = \mathbf{x}^{k-1} - \alpha_k\mathbf{p}^k$
$\mathbf{g}^k = \mathbf{g}^{k-1} - \alpha_k\mathsf{A}\mathbf{p}^k$
$\mathbf{z}^k = \mathsf{M}^{-1}\mathbf{g}^k$
$\beta_k = (\mathbf{z}^k)^T\mathbf{g}^k/(\mathbf{z}^{k-1})^T\mathbf{g}^{k-1}$
$\mathbf{p}^{k+1} = \mathbf{z}^k + \beta_k\mathbf{p}^k$
$k = k + 1$
end while

Step 2. {*Return a (possibly approximate) solution.*}
$\widetilde{\mathbf{x}} = \mathbf{x}^k$

Notice that the PCG algorithm does not exploit explicitly the Cholesky factorization of the preconditioner M. The *pseudoresiduals* $\mathbf{z}^k$ are typically obtained by solving $\mathsf{M}\mathbf{z}^k = \mathbf{g}^k$. If M is a good approximation of A, then $\mathbf{z}^k$ is close to the error vector $\mathbf{e}^k$. The rate of convergence of the PCG algorithm depends on the condition number of the Hessian of the transformed function $\bar{f}$, i.e., on $\kappa(\mathsf{M}^{-1}\mathsf{A}) = \kappa(\widetilde{\mathsf{L}}^{-1}\mathsf{A}\widetilde{\mathsf{L}}^{-T})$. Thus the efficiency of the preconditioned conjugate gradient method depends critically on the choice of a preconditioner, which should balance the cost of its application with the preconditioning effect. We refer interested readers to specialized books like Saad [163] or Axelsson [4] for more information. Since the choice of the preconditioner is problem dependent, we limit our attention here to the brief discussion of a few general strategies.

The most simple preconditioners may be defined by means of the decomposition

$$\mathsf{A} = \mathsf{D} + \mathsf{E} + \mathsf{E}^T,$$

where D is the diagonal of A and E is its strict lower part with the entries $[\mathsf{E}]_{ij} = [\mathsf{A}]_{ij}$ for $i > j$ and $[\mathsf{E}]_{ij} = 0$ otherwise.

The *Jacobi preconditioner* $\mathsf{M}_J = \mathsf{D}$ is the easiest one to implement, but its efficiency is very limited. Better approximation of A can be achieved by choosing the *block diagonal Jacobi preconditioner*

$$\mathsf{M}_{BJ} = \begin{bmatrix} \mathsf{A}_{11} & \mathsf{O} & \dots & \mathsf{O} \\ \mathsf{O} & \mathsf{A}_{22} & \dots & \mathsf{O} \\ . & . & \dots & . \\ \mathsf{O} & \mathsf{O} & \dots & \mathsf{A}_{kk} \end{bmatrix}, \tag{3.28}$$

where A_{ii} are diagonal blocks of A (see, e.g., Greenbaum [106, Sect. 10.5]). The *pseudoresiduals* $\mathbf{z}^k$ are typically obtained by solving $\mathsf{A}_{ii}\mathbf{z}_i^k = \mathbf{g}_i^k$.

Good results may be often achieved with the *symmetric Gauss-Seidel* preconditioner

$$\mathsf{M}_{SGS} = (\mathsf{D} + \mathsf{E})\mathsf{D}^{-1}(\mathsf{D} + \mathsf{E}^T).$$

Notice that

$$\widetilde{\mathsf{L}} = (\mathsf{D} + \mathsf{E})\mathsf{D}^{-\frac{1}{2}}$$

is a regular lower triangular matrix, so we have the triangular factorization

$$\mathsf{M}_{SGS} = \widetilde{\mathsf{L}}\widetilde{\mathsf{L}}^T$$

for free.

More generally, the factorized preconditioners can be produced by *incomplete Cholesky (IC)* which neglects some fill-in elements in the factor L. When the elements of L are neglected because they are smaller than a certain threshold, the factorization is called "IC-by-value", and when they are omitted because they do not belong to a certain sparsity pattern, we have "IC-by-position". See for example Axelsson [4] or Saad [163]. The drawback of this method is that it can fail on the generation of diagonal entries.

3.7 Preconditioning by Conjugate Projector

So far we have assumed that the preconditioners to a symmetric positive definite matrix A are nonsingular matrices that approximate A. In this section we describe an alternative strategy which is useful when we are able to find the minimizer $\mathbf{x}^0$ of f over a subspace $\mathcal{U}$ of $\mathbb{R}^n$. We shall show that in this case we can get the preconditioning effect by reducing the conjugate gradient iterations to the conjugate complement of $\mathcal{U}$.

3.7.1 Conjugate Projectors

Our main tools will be the projectors with conjugate range and kernel. We shall use the basic relations introduced in Sect. 1.3 and some observations that we review in this subsection.

Let $\mathsf{A} \in \mathbb{R}^{n\times n}$ be a symmetric positive definite matrix. A projector P is an *A-conjugate projector* or briefly a *conjugate projector* if $\mathrm{Im}\mathsf{P}$ is A-conjugate to $\mathrm{Ker}\mathsf{P}$, or equivalently

$$\mathsf{P}^T\mathsf{A}(\mathsf{I} - \mathsf{P}) = \mathsf{P}^T\mathsf{A} - \mathsf{P}^T\mathsf{A}\mathsf{P} = \mathsf{O}.$$

It follows that $\mathsf{Q} = \mathsf{I} - \mathsf{P}$ is also a conjugate projector,

$$\mathsf{P}^T\mathsf{A} = \mathsf{A}\mathsf{P} = \mathsf{P}^T\mathsf{A}\mathsf{P}, \quad \text{and} \quad \mathsf{Q}^T\mathsf{A} = \mathsf{A}\mathsf{Q} = \mathsf{Q}^T\mathsf{A}\mathsf{Q}. \tag{3.29}$$

Let us denote $\mathcal{V} = \text{ImQ}$. If $\mathbf{x} \in \mathsf{A}\mathcal{V}$, then $\mathbf{y} = \mathsf{Q}\mathbf{x}$ satisfies $\mathbf{y} \in \mathcal{V}$ and

$$\mathsf{Q}^T\mathsf{A}\mathsf{Q}\mathbf{x} = \mathsf{A}\mathsf{Q}\mathbf{x} = \mathsf{A}\mathbf{y},$$

so that

$$\mathsf{Q}^T\mathsf{A}\mathsf{Q}(\mathsf{A}\mathcal{V}) \subseteq \mathsf{A}\mathcal{V}. \tag{3.30}$$

Thus $\mathsf{A}\mathcal{V}$ is an invariant subspace of $\mathsf{Q}^T\mathsf{A}\mathsf{Q}$.

The following lemma shows that the mapping which assigns to each $\mathbf{x} \in \mathsf{A}\mathcal{V}$ the vector $\mathsf{Q}\mathbf{x} \in \mathcal{V}$ is expansive as in Fig. 3.4.

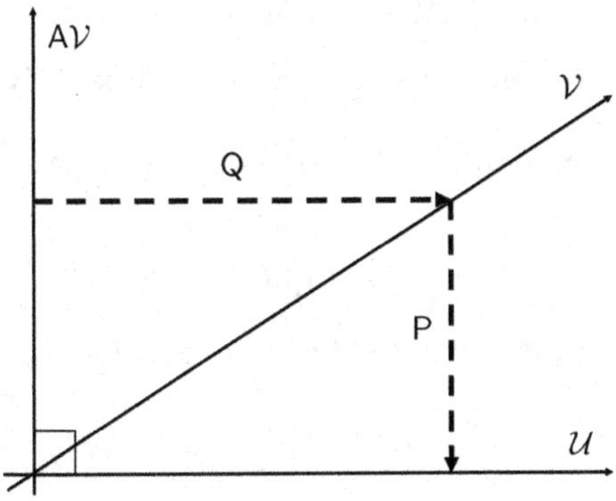

Fig. 3.4. Geometric illustration of Lemma 3.4

Lemma 3.4. *Let* Q *denote a conjugate projector on* $\mathcal{V}$ *and* $\mathbf{x} \in \mathsf{A}\mathcal{V}$. *Then*

$$\|\mathsf{Q}\mathbf{x}\| \geq \|\mathbf{x}\|.$$

Proof. For any $\mathbf{x} \in \mathsf{A}\mathcal{V}$, there is $\mathbf{y} \in \mathbb{R}^n$ such that $\mathbf{x} = \mathsf{A}\mathsf{Q}\mathbf{y}$. It follows that

$$\mathsf{Q}^T\mathbf{x} = \mathsf{Q}^T\mathsf{A}\mathsf{Q}\mathbf{y} = \mathsf{A}\mathsf{Q}\mathbf{y} = \mathbf{x}.$$

Since $\mathbf{x}^T\mathsf{Q}\mathbf{x} = \mathbf{x}^T\mathsf{Q}^T\mathbf{x} = \|\mathbf{x}\|^2$, we have

$$\|\mathsf{Q}\mathbf{x}\|^2 = \mathbf{x}^T\mathsf{Q}^T\mathsf{Q}\mathbf{x} = \mathbf{x}^T\left((\mathsf{Q}^T - \mathsf{I}) + \mathsf{I}\right)\left((\mathsf{Q} - \mathsf{I}) + \mathsf{I}\right)\mathbf{x} = \|(\mathsf{Q} - \mathsf{I})\mathbf{x}\|^2 + \|\mathbf{x}\|^2.$$

□

3.7.2 Minimization in Subspace

Let us assume that $\mathcal{U}$ is the subspace spanned by the columns of a full column rank matrix $\mathsf{U} \in \mathbb{R}^{n \times n}$ and notice that $\mathsf{U}^T\mathsf{A}\mathsf{U}$ is regular. Indeed, if $\mathsf{U}^T\mathsf{A}\mathsf{U}\mathbf{x} = \mathbf{o}$, then

$$\|\mathsf{U}\mathbf{x}\|_{\mathsf{A}}^2 = \mathbf{x}^T(\mathsf{U}^T\mathsf{A}\mathsf{U}\mathbf{x}) = 0,$$

so $\mathbf{x} = \mathbf{o}$ by the assumptions that U is the full column rank matrix and A is SPD. Thus we can define

$$\mathsf{P} = \mathsf{U}(\mathsf{U}^T\mathsf{A}\mathsf{U})^{-1}\mathsf{U}^T\mathsf{A}.$$

It is easy to check directly that P is a conjugate projector onto $\mathcal{U}$ as

$$\mathsf{P}^2 = \mathsf{U}(\mathsf{U}^T\mathsf{A}\mathsf{U})^{-1}\mathsf{U}^T\mathsf{A}\mathsf{U}(\mathsf{U}^T\mathsf{A}\mathsf{U})^{-1}\mathsf{U}^T\mathsf{A} = \mathsf{P}$$

and

$$\mathsf{P}^T\mathsf{A}(\mathsf{I} - \mathsf{P}) = \mathsf{A}\mathsf{U}(\mathsf{U}^T\mathsf{A}\mathsf{U})^{-1}\mathsf{U}^T\mathsf{A}(\mathsf{I} - \mathsf{U}(\mathsf{U}^T\mathsf{A}\mathsf{U})^{-1}\mathsf{U}^T\mathsf{A}) = \mathsf{O}.$$

Since any vector $\mathbf{x} \in \mathcal{U}$ can be written in the form $\mathbf{x} = \mathsf{U}\mathbf{y}$, $\mathbf{y} \in \mathbb{R}^m$, and

$$\mathsf{P}\mathbf{x} = \mathsf{U}(\mathsf{U}^T\mathsf{A}\mathsf{U})^{-1}\mathsf{U}^T\mathsf{A}\mathsf{U}\mathbf{y} = \mathsf{U}\mathbf{y} = \mathbf{x},$$

it follows that

$$\mathcal{U} = \mathrm{Im}\mathsf{P}.$$

The conjugate projector P onto $\mathcal{U}$ can be used for the solution of

$$\min_{\mathbf{x}\in\mathcal{U}} f(\mathbf{x}) = \min_{\mathbf{y}\in R^m} f(\mathsf{U}\mathbf{y}) = \min_{\mathbf{y}\in R^m} \frac{1}{2}\mathbf{y}^T\mathsf{U}^T\mathsf{A}\mathsf{U}\mathbf{y} - \mathbf{b}^T\mathsf{U}\mathbf{y}.$$

Using the gradient argument of Proposition 2.1, we get that the minimizer $\mathbf{y}^0$ of the latter problem satisfies

$$\mathsf{U}^T\mathsf{A}\mathsf{U}\mathbf{y}^0 = \mathsf{U}^T\mathbf{b}, \tag{3.31}$$

so that the minimizer $\mathbf{x}^0$ of f over $\mathcal{U}$ satisfies

$$\mathbf{x}^0 = \mathsf{U}\mathbf{y}^0 = \mathsf{U}(\mathsf{U}^T\mathsf{A}\mathsf{U})^{-1}\mathsf{U}^T\mathbf{b} = \mathsf{P}\mathsf{A}^{-1}\mathbf{b}. \tag{3.32}$$

Our assumption concerning the ability to find the minimum of f over $\mathcal{U}$ effectively amounts to the assumption that we are able to solve (3.31). Notice that we can evaluate the product $\mathsf{P}\mathsf{A}^{-1}\mathbf{b}$ without solving any system of linear equations with the matrix A.

3.7.3 Conjugate Gradients in Conjugate Complement

In the next step we shall use the conjugate projectors P and $\mathsf{Q} = \mathsf{I} - \mathsf{P}$ to decompose our minimization problem (3.1) into the minimization on $\mathcal{U}$ and the minimization on $\mathcal{V} = \mathrm{Im}\mathsf{Q}$. We shall use the conjugate gradient method to solve the latter problem.

Two observations are needed to exploit the special structure of our problem. First, using Lemma 3.4, $\dim\mathcal{V} = \dim\mathsf{A}\mathcal{V}$, and (1.2), we get that the mapping which assigns to each $\mathbf{x} \in \mathsf{A}\mathcal{V}$ a vector $\mathsf{Q}\mathbf{x} \in \mathcal{V}$ is an isomorphism, so that

$$\mathsf{Q}(\mathsf{A}\mathcal{V}) = \mathcal{V}.$$

Second, using (3.29) and (3.32), we get

$$\mathbf{g}^0 = \mathsf{A}\mathbf{x}^0 - \mathbf{b} = \mathsf{APA}^{-1}\mathbf{b} - \mathbf{b} = \mathsf{P}^T\mathbf{b} - \mathbf{b} = -\mathsf{Q}^T\mathbf{b}. \tag{3.33}$$

Using both observations, we get

$$\begin{aligned}\min_{\mathbf{x}\in\mathbb{R}^n} f(\mathbf{x}) &= \min_{\mathbf{y}\in\mathcal{U},\mathbf{z}\in\mathcal{V}} f(\mathbf{y}+\mathbf{z}) = \min_{\mathbf{y}\in\mathcal{U}} f(\mathbf{y}) + \min_{\mathbf{z}\in\mathcal{V}} f(\mathbf{z})\\ &= f(\mathbf{x}^0) + \min_{\mathbf{z}\in\mathcal{V}} f(\mathbf{z}) = f(\mathbf{x}^0) + \min_{\mathbf{x}\in\mathsf{A}\mathcal{V}} \frac{1}{2}\mathbf{x}^T\mathsf{Q}^T\mathsf{AQ}\mathbf{x} - \mathbf{b}^T\mathsf{Q}\mathbf{x}\\ &= f(\mathbf{x}^0) + \min_{\mathbf{x}\in\mathsf{A}\mathcal{V}} \frac{1}{2}\mathbf{x}^T\mathsf{Q}^T\mathsf{AQ}\mathbf{x} + \left(\mathbf{g}^0\right)^T\mathbf{x},\end{aligned} \tag{3.34}$$

where $\mathbf{x}^0$ is determined by (3.32).

It remains to solve the minimization problem (3.34). First observe that using Lemma 3.4, we get that $\mathsf{Q}^T\mathsf{AQ}|\mathsf{A}\mathcal{V}$ is positive definite. Since by (3.33) $\mathbf{g}^0 \in \mathrm{Im}\mathsf{Q}^T$,

$$\mathrm{Im}\mathsf{Q}^T = \mathrm{Im}(\mathsf{Q}^T\mathsf{A}) = \mathrm{Im}(\mathsf{AQ}) = \mathsf{A}\mathcal{V}, \tag{3.35}$$

and $\mathsf{A}\mathcal{V}$ is an invariant subspace of $\mathsf{Q}^T\mathsf{AQ}$, we can use the procedure described in Sect. 3.2 to generate $\mathsf{Q}^T\mathsf{AQ}$-conjugate vectors $\mathbf{p}^1, \ldots, \mathbf{p}^k$ of

$$\mathcal{K}^k = \mathcal{K}^k(\mathsf{Q}^T\mathsf{AQ}, \mathbf{g}^0) = \mathrm{Span}\{\mathbf{g}^0, \mathsf{Q}^T\mathsf{AQ}\mathbf{g}^0, \ldots, (\mathsf{Q}^T\mathsf{AQ})^{k-1}\mathbf{g}^0\}.$$

It simply follows that

$$\mathbf{q}^1 = \mathsf{Q}\mathbf{p}^1, \mathbf{q}^2 = \mathsf{Q}\mathbf{p}^2, \ldots$$

are A-conjugate vectors of $\mathcal{V}$. Using (3.14), $\mathbf{p}^i \in \mathsf{A}\mathcal{V}$, and Lemma 3.4, it is easy to see that

$$\|\mathbf{q}^k\| \geq \|\mathbf{p}^k\| \geq \|\mathbf{g}^{k-1}\|,$$

so that we can generate a new conjugate direction $\mathbf{q}^k$ whenever $\mathbf{g}^{k-1} \neq \mathbf{o}$. We can sum up the most important properties of the CG algorithm with the preconditioning by the conjugate projector into the following theorem.

Theorem 3.5. *Let $\mathbf{x}^k$ be generated by Algorithm 3.4 to find the solution $\widehat{\mathbf{x}}$ of (3.1) with a full column rank matrix $\mathsf{U} \in \mathbb{R}^{n\times m}$. Then the algorithm is well defined and there is $k \leq n - m$ such that $\mathbf{x}^k = \widehat{\mathbf{x}}$. Moreover, the following statements hold for $i = 1, \ldots, k$:*

(i) $f(\mathbf{x}^i) = \min\{f(\mathbf{x}) : \ \mathbf{x} \in \mathcal{U} + \mathsf{Q}\mathcal{K}^i(\mathsf{Q}^T\mathsf{AQ}, \mathbf{g}^0)\}$.
(ii) $\|\mathbf{q}^i\| \geq \|\mathbf{g}^{i-1}\|$.
(iii) $(\mathbf{q}^i)^T\mathsf{A}\mathbf{q}^j = 0$ *for* $i > j$.
(iv) $(\mathbf{q}^i)^T\mathsf{A}\mathbf{x} = 0$ *for* $\mathbf{x} \in \mathcal{U}$.

The complete conjugate gradient algorithm with the preconditioning by the conjugate projector reads as follows:

Algorithm 3.4. Conjugate gradients with projector preconditioning (CGPP).

Given a symmetric positive definite matrix $\mathsf{A} \in \mathbb{R}^{n\times n}$, *a full column rank matrix* $\mathsf{U} \in \mathbb{R}^{n\times m}$, *and* $\mathbf{b} \in \mathbb{R}^n$.

Step 0. {*Initialization.*}
$\mathsf{P} = \mathsf{U}(\mathsf{U}^T\mathsf{A}\mathsf{U})^{-1}\mathsf{U}^T\mathsf{A},\ \mathsf{Q} = \mathsf{I} - \mathsf{P}$
$\mathbf{x}^0 = \mathsf{P}\mathsf{A}^{-1}\mathbf{b} = \mathsf{U}(\mathsf{U}^T\mathsf{A}\mathsf{U})^{-1}\mathsf{U}^T\mathbf{b}$
$k = 1,\ \mathbf{g}^0 = \mathsf{A}\mathbf{x}^0 - \mathbf{b},\ \mathbf{q}^1 = \mathsf{Q}\mathbf{g}^0$

Step 1. {*Conjugate gradient loop.* }
while $\|\mathbf{g}^{k-1}\| > 0$
$\alpha_k = (\mathbf{g}^{k-1})^T\mathbf{q}^k/(\mathbf{q}^k)^T\mathsf{A}\mathbf{q}^k$
$\mathbf{x}^k = \mathbf{x}^{k-1} - \alpha_k\mathbf{q}^k$
$\mathbf{g}^k = \mathbf{g}^{k-1} - \alpha_k\mathsf{A}\mathbf{q}^k$
$\beta_k = (\mathbf{g}^k)^T\mathsf{A}\mathbf{q}^k/(\mathbf{q}^k)^T\mathsf{A}\mathbf{q}^k$
$\mathbf{q}^{k+1} = \mathsf{Q}\mathbf{g}^k + \beta_k\mathbf{q}^k$
$k = k + 1$
end while

Step 2. {*Return a (possibly approximate) solution.*}
$\widetilde{\mathbf{x}} = \mathbf{x}^k$

3.7.4 Preconditioning Effect

As we have seen in the previous section, the iterations of Algorithm 3.4 may be considered as the conjugate gradient iterations for the minimization of

$$f_{0,\mathsf{Q}}(\mathbf{x}) = \frac{1}{2}\mathbf{x}^T\mathsf{Q}^T\mathsf{A}\mathsf{Q}\mathbf{x} + (\mathbf{g}^0)^T\mathbf{x}$$

that generate the iterations

$$\mathbf{x}^k \in \mathcal{K}^k(\mathsf{Q}^T\mathsf{A}\mathsf{Q}, \mathbf{g}^0) \subseteq \mathsf{A}\mathcal{V}.$$

Thus only the positive definite restriction $\mathsf{Q}^T\mathsf{A}\mathsf{Q}|\mathsf{A}\mathcal{V}$ of $\mathsf{Q}^T\mathsf{A}\mathsf{Q}$ to $\mathsf{A}\mathcal{V}$ takes part in the process of solution, and the rate of convergence may be estimated by the spectral condition number $\kappa(\mathsf{Q}^T\mathsf{A}\mathsf{Q}|\mathsf{A}\mathcal{V})$ of $\mathsf{Q}^T\mathsf{A}\mathsf{Q}|\mathsf{A}\mathcal{V}$.

It is rather easy to see that

$$\kappa(\mathsf{Q}^T\mathsf{A}\mathsf{Q}|\mathsf{A}\mathcal{V}) \le \kappa(\mathsf{A}).$$

Indeed, denoting by $\lambda_1 \ge \cdots \ge \lambda_n$ the eigenvalues of A, we can observe that if $\mathbf{x} \in \mathsf{A}\mathcal{V}$ and $\|\mathbf{x}\| = 1$, then by Lemma 3.4

$$\mathbf{x}^T\mathsf{Q}^T\mathsf{A}\mathsf{Q}\mathbf{x} \ge (\mathsf{Q}\mathbf{x})^T\mathsf{A}(\mathsf{Q}\mathbf{x})/\|\mathsf{Q}\mathbf{x}\|^2 \ge \lambda_n$$

and

$$\mathbf{x}^T\mathsf{Q}^T\mathsf{A}\mathsf{Q}\mathbf{x} \le \mathbf{x}^T\mathsf{Q}^T\mathsf{A}\mathsf{Q}\mathbf{x} + \mathbf{x}^T\mathsf{P}^T\mathsf{A}\mathsf{P}\mathbf{x} = \mathbf{x}^T\mathsf{A}\mathbf{x} \le \lambda_1. \tag{3.36}$$

To see the preconditioning effect of the algorithm in more detail, let us denote by $\mathcal{E}$ the m-dimensional subspace spanned by the eigenvectors corresponding to the m smallest eigenvalues $\lambda_{n-m+1} \geq \cdots \geq \lambda_n$ of A, and let $\mathsf{R}_{\mathsf{A}\mathcal{U}}$ and $\mathsf{R}_{\mathcal{E}}$ denote the orthogonal projectors on $\mathsf{A}\mathcal{U}$ and $\mathcal{E}$, respectively. Let

$$\gamma = \|\mathsf{R}_{\mathsf{A}\mathcal{U}} - \mathsf{R}_{\mathcal{E}}\|$$

denote the *gap* between $\mathsf{A}\mathcal{U}$ and $\mathcal{E}$. It can be evaluated provided we have matrices U and E whose columns form the orthonormal bases of $\mathsf{A}\mathcal{U}$ and $\mathcal{E}$, respectively. It is known [170] that if σ is the least singular value of $\mathsf{U}^T\mathsf{E}$, then

$$\gamma = \sqrt{1-\sigma^2} \leq 1.$$

Theorem 3.6. *Let $\mathcal{U}, \mathcal{V}, \mathsf{Q}$ be those of Algorithm 3.4, let $\lambda_1 \geq \cdots \geq \lambda_n$ denote the eigenvalues of A, and let $\overline{\lambda}_{\min}$ denote the least nonzero eigenvalue of $\mathsf{Q}^T\mathsf{A}\mathsf{Q}$. Then*

$$\lambda_n \leq \sqrt{(1-\gamma^2)\lambda_{n-m}^2 + \gamma^2\lambda_n^2} \leq \overline{\lambda}_{\min} \tag{3.37}$$

and

$$\kappa(\mathsf{Q}^T\mathsf{A}\mathsf{Q}|\mathsf{A}\mathcal{V}) \leq \frac{\lambda_1}{\sqrt{(1-\gamma^2)\lambda_{n-m}^2 + \gamma^2\lambda_n^2}}.$$

Proof. Let $\mathbf{x} \in \mathsf{A}\mathcal{V}$, $\|\mathbf{x}\| = 1$, so that $\|\mathsf{Q}\mathbf{x}\| \geq 1$ by Lemma 3.4. Observing that $\mathrm{Im}\mathsf{R}_{\mathcal{E}}$ and $\mathrm{Im}(\mathsf{I}-\mathsf{R}_{\mathcal{E}})$ are orthogonal invariant subspaces of A, we get that

$$\begin{aligned}
\|\mathsf{A}\mathsf{Q}\mathbf{x}\|^2 &= \|\mathsf{A}(\mathsf{I}-\mathsf{R}_{\mathcal{E}})\mathsf{Q}\mathbf{x}\|^2 + \|\mathsf{A}\mathsf{R}_{\mathcal{E}}\mathsf{Q}\mathbf{x}\|^2 \\
&\geq \lambda_{n-m}^2\|(\mathsf{I}-\mathsf{R}_{\mathcal{E}})\mathsf{Q}\mathbf{x}\|^2 + \lambda_n^2\|\mathsf{R}_{\mathcal{E}}\mathsf{Q}\mathbf{x}\|^2 \\
&\geq \left(\lambda_{n-m}^2\|(\mathsf{I}-\mathsf{R}_{\mathcal{E}})\mathsf{Q}\mathbf{x}\|^2 + \lambda_n^2\|\mathsf{R}_{\mathcal{E}}\mathsf{Q}\mathbf{x}\|^2\right)/\|\mathsf{Q}\mathbf{x}\|^2 \\
&\geq \lambda_{n-m}^2(1-\xi^2) + \lambda_n^2\xi^2,
\end{aligned} \tag{3.38}$$

where $\xi = \|\mathsf{Q}\mathbf{x}\|^{-1}\|\mathsf{R}_{\mathcal{E}}\mathsf{Q}\mathbf{x}\|$. We have used that

$$\|(\mathsf{I}-\mathsf{R}_{\mathcal{E}})\mathsf{Q}\mathbf{x}\|^2 + \|\mathsf{R}_{\mathcal{E}}\mathsf{Q}\mathbf{x}\|^2 = \|\mathsf{Q}\mathbf{x}\|^2.$$

Since $\mathrm{Im}\mathsf{Q} = \mathcal{V}$, it follows by the definition of $\mathsf{R}_{\mathsf{A}\mathcal{U}}$ that $\mathsf{R}_{\mathsf{A}\mathcal{U}}\mathsf{Q} = \mathsf{O}$ and

$$\xi = \|\mathsf{Q}\mathbf{x}\|^{-1}\|(\mathsf{R}_{\mathcal{E}} - \mathsf{R}_{\mathsf{A}\mathcal{U}})\mathsf{Q}\mathbf{x}\| \leq \gamma.$$

As the last expression in (3.38) is a decreasing function of ξ for $\xi \geq 0$, it follows that

$$\|\mathsf{Q}^T\mathsf{A}\mathsf{Q}\mathbf{x}\|^2 = \|\mathsf{A}\mathsf{Q}\mathbf{x}\|^2 \geq \lambda_{n-m}^2(1-\gamma^2) + \lambda_n^2\gamma^2.$$

The rest is an easy consequence of (3.36). □

The above theorem suggests that the preconditioning by the conjugate projector is efficient when $\mathcal{U}$ approximates the subspace spanned by the eigenvectors which correspond to the smallest eigenvalues of A. If $\mathsf{U}^T\mathsf{E}$ is nonsingular and $\lambda_n < \lambda_{n-m}$, then $\gamma < 1$ and

$$\kappa(\mathsf{Q}^T\mathsf{A}\mathsf{Q}|\mathsf{A}\mathcal{V}) < \kappa(\mathsf{A}).$$

If the minimization problem arises from the discretization of elliptic partial differential equations, than U can be obtained by aggregation. It turns out that even the subspace with a very small dimension can considerably improve the rate of convergence. See Sect. 3.10.1 for a numerical example.

3.8 Conjugate Gradients for More General Problems

Let A be only positive semidefinite, so that the cost function f is convex but not strictly convex, and let the unconstrained minimization problem (3.1) be well posed, i.e., $\mathbf{b} \in \mathrm{Im}\mathsf{A}$ by Proposition 2.1.

If we start the conjugate gradient algorithm from arbitrary $\mathbf{x}^0 \in \mathbb{R}^n$, then the gradient $\mathbf{g}^0$ and the Krylov space $\mathcal{K}^n(\mathsf{A}, \mathbf{g}^0)$ satisfy

$$\mathbf{g}^0 \in \mathrm{Im}\mathsf{A} \quad \text{and} \quad \mathcal{K}^n(\mathsf{A}, \mathbf{g}^0) \subseteq \mathrm{Im}\mathsf{A}.$$

Since the CG method picks the conjugate directions from $\mathcal{K}^n(\mathsf{A}, \mathbf{g}^0)$, it follows that the method works only on the range of A. Thus the algorithm generates the iterates $\mathbf{x}^k$ which converge to a solution $\overline{\mathbf{x}}$ with the rate of convergence which can be described by the distribution of the spectrum of the restriction $\mathsf{A}|\mathcal{K}^n(\mathsf{A}, \mathbf{g}^0)$. Observing that the least eigenvalue of $\mathsf{A}|\mathcal{K}^n(\mathsf{A}, \mathbf{g}^0)$ is bounded from below by the least nonzero eigenvalue $\overline{\lambda}_{\min}$ of A, we get the error estimate

$$\|\mathbf{e}^k\|_{\mathsf{A}} \le 2\left(\frac{\sqrt{\overline{\kappa}(\mathsf{A})}-1}{\sqrt{\overline{\kappa}(\mathsf{A})}+1}\right)^k \|\mathbf{e}^0\|_{\mathsf{A}}, \tag{3.39}$$

where $\overline{\kappa}(\mathsf{A})$ denotes the *regular spectral condition number* of A defined by

$$\overline{\kappa}(\mathsf{A}) = \kappa(\mathsf{A}|\mathrm{Im}\mathsf{A}) = \lambda_{\max}/\overline{\lambda}_{\min}.$$

Let P and $\mathsf{Q} = \mathsf{I} - \mathsf{P}$ denote the orthogonal projectors on $\mathrm{Im}\mathsf{A}$ and $\mathrm{Ker}\mathsf{A}$, respectively, so that

$$\overline{\mathbf{x}} = \mathsf{P}\overline{\mathbf{x}} + \mathsf{Q}\overline{\mathbf{x}}.$$

Since the reduction $\mathsf{A}|\mathrm{Im}\mathsf{A}$ is nonsingular, it follows that there is a unique solution $\overline{\mathbf{x}}_{\mathrm{LS}} \in \mathrm{Im}\mathsf{A}$ of (3.1), and by Proposition 2.1 any solution $\overline{\mathbf{x}}$ satisfies

$$\overline{\mathbf{x}} = \widehat{\mathbf{x}}_{\mathrm{LS}} + \mathbf{d}, \quad \mathbf{d} \in \mathrm{Ker}\mathsf{A}.$$

Thus if $\overline{\mathbf{x}}$ is a solution of (3.1), then $\mathsf{P}\overline{\mathbf{x}} = \overline{\mathbf{x}}_{\mathrm{LS}}$ and

$$\mathsf{Q}\overline{\mathbf{x}} = \mathsf{Q}\mathbf{x}^0.$$

It follows that $\overline{\mathbf{x}}_{\mathrm{LS}}$ is the least square solution of $\mathsf{A}\mathbf{x} = \mathbf{b}$, and to get it by the conjugate gradient algorithm, it is enough to take $\mathbf{x}^0 \in \mathrm{Im}\mathsf{A}$.

If A is indefinite, then, using the arguments of Sect. 3.2, it is easy to check that the conjugate gradient method still generates conjugate directions, but it fails when $(\mathbf{p}^k)^T\mathsf{A}\mathbf{p}^k = 0$. The latter case may happen with $\mathbf{p}^k \notin \mathrm{Ker}\mathsf{A}$, as in

$$[1,\ 1]\begin{bmatrix} 1 & 0 \\ 0 & -1 \end{bmatrix}\begin{bmatrix} 1 \\ 1 \end{bmatrix} = 0.$$

It follows that there is no guarantee that the CG Algorithm 3.1 is able, at least without modifications, to find a stationary point of f.

3.9 Convergence in Presence of Rounding Errors

The elegant mathematical theory presented above assumes implementation of the conjugate gradient algorithm in exact arithmetic and captures well the performance of only a limited number of conjugate gradient iterations in computer arithmetics. Since we are going to use the conjugate gradient method mainly for a low-precision approximation of well-conditioned auxiliary problems, we shall base our exposition on this theory in what follows. However, it is still useful to be aware of possible effects of rounding errors that accompany any computer implementation of the conjugate gradient algorithm.

It has been known since the introduction of the CG method and the Lanczos method [140], which generates the same iterates, that, when used in finite precision arithmetic, the vectors generated by these algorithms can seriously violate their theoretical properties. In particular, it has been observed that the evaluated gradients can lose their orthogonality after as small a number of iterations as twenty, and that nearly dependent conjugate directions can be generated. In spite of these effects, it has been observed that the conjugate gradient method still converges in finite precision arithmetic, but that the convergence is delayed [105, 107].

Undesirable effects of the rounding errors can be reduced by reorthogonalization. A simple analysis reveals that the full reorthogonalization of the gradients is costly and requires large memory. A key to an efficient implementation of the reorthogonalization is based on observation that accumulation of the rounding errors has a regular pattern, namely, that large perturbations of the generated vectors belong to the space generated by the eigenvectors of A which can be approximated well by the vectors from the current Krylov space. This has led to the efficient implementation of the conjugate gradient method based on the *selective orthogonalization* proposed by Parlett and Scott [158]. More details and information about the effects of rounding errors and implementation of the conjugate gradient method in finite arithmetic can be found in the comprehensive review paper by Meurant and Strakoš [152].

3.10 Numerical Experiments

Here we illustrate the performance of the CG algorithm and the effect of preconditioning on the solution of an ill-conditioned benchmark and a class of well-conditioned problems. The latter was proposed to resemble the class of problems arising from application of the multigrid or domain decomposition methods to the elliptic partial differential equations. The cost functions $f_{\mathrm{L},h}$ and $f_{\mathrm{LW},h}$ introduced here are used in Sects. 4.8, 5.11, and 6.12 as benchmarks for the solution of constrained problems, so that we can assess additional complexity arising from implementation of various constraints and better understand our algorithms. Moreover, using the same cost functions in our benchmarks considerably simplifies their implementation.

3.10.1 Basic CG and Preconditioning

Let $\Omega = (0,1) \times (0,1)$ denote an open domain with the boundary Γ and its part $\Gamma_u = \{0\} \times [0,1]$. Let $H^1(\Omega)$ denote the Sobolev space of the first order in the space $L^2(\Omega)$ of functions on Ω whose squares are integrable in the Lebesgue sense, let

$$\mathcal{V} = \{u \in H^1(\Omega) : \ u = 0 \ \text{ on } \ \Gamma_u\},$$

and let us define for any $u \in H^1(\Omega)$

$$f_{\mathrm{L}}(u) = \frac{1}{2}\int_\Omega \|\nabla u(x)\|^2 \mathrm{d}\Omega + \int_\Omega u \mathrm{d}\Omega.$$

Thus we can define the continuous problem to find

$$\min_{u \in \mathcal{V}} f_{\mathrm{L}}(u). \tag{3.40}$$

Our ill-conditioned benchmark was obtained from (3.40) by the finite element discretization using a regular grid defined by the discretization parameter h and linear elements. The Dirichlet conditions were enhanced into the Hessian $\mathsf{A}_{\mathrm{L},h}$ of the discretized cost function $f_{\mathrm{L},h}$, so that $\mathsf{A}_{\mathrm{L},h} \in \mathbb{R}^{n \times n}$ is positive definite, $n = p(p-1)$, and $p = 1/h + 1$. Moreover, $\mathsf{A}_{\mathrm{L},h}$ is known to be ill-conditioned with the condition number $\kappa(\mathsf{A}_{\mathrm{L},h}) \approx h^{-2}$. The computations were carried out with $h = 1/32$, which corresponds to $n = 1056$ unknowns.

We used the benchmark to compare the performance of CG, CG with SSOR preconditioning, and CG with preconditioning by the conjugate projector. To define the conjugate projector, we decomposed the domain into 4×4 squares with typically 8×8 variables which were aggregated by means of the matrix U with 16 columns.

The graph of the norm of the gradient (vertical axis) against the number of iterations for the basic CG algorithm (CG), the CG algorithm with SSOR preconditioning (CG–SSOR), and the CG algorithm with preconditioning by the conjugate projector (CG–CP) is in Fig. 3.5. We can see that though the performance of the CG algorithm is poor if the Hessian of the cost function is ill-conditioned, it can be considerably improved by preconditioning.

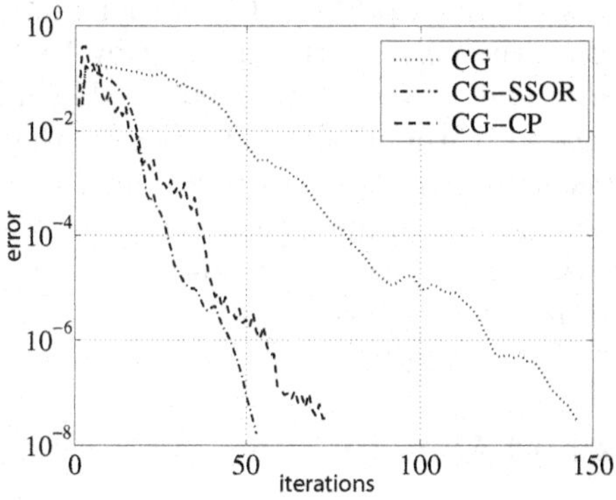

Fig. 3.5. Convergence of CG, CG–SSOR, and CG–CP algorithms

3.10.2 Numerical Demonstration of Optimality

To illustrate the concept of optimality, let us consider the class of problems to minimize

$$f_{\mathrm{LW},h}(\mathbf{x}) = \frac{1}{2}\mathbf{x}^T \mathsf{A}_{\mathrm{LW},h}\mathbf{x} - \mathbf{b}_{\mathrm{LW},h}^T\mathbf{x},$$

where

$$\mathsf{A}_{\mathrm{LW},h} = \mathsf{A}_{\mathrm{L},h} + 2\mathsf{I}, \quad [\mathbf{b}_{\mathrm{LW},h}]_i = -1, \quad i = 1, \ldots, n, \quad n = 1/h + 1.$$

The class of problems can be given a mechanical interpretation associated to the expanding spring systems on Winkler's foundation. Using Gershgorin's theorem, it can be proved that the spectrum of the Hessian $\mathsf{A}_{\mathrm{LW},h}$ of $f_{\mathrm{LW},h}$ is located in the interval $[2, 10]$, so that $\kappa(\mathsf{A}_{\mathrm{LW},h}) \le 5$.

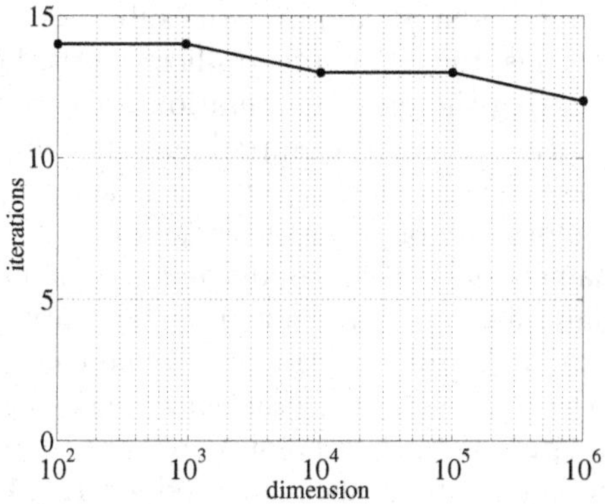

Fig. 3.6. Optimality of CG for a class of well-conditioned problems

In Fig. 3.6, we can see the numbers of CG iterations k_n (vertical axis) that were necessary to reduce the norm of the gradient by 10^{-6} for the problems with the dimension n ranging from 100 to 1000000. The dimension n on the horizontal axis is in the logarithmic scale. We can see that k_n varies mildly with varying n, in agreement with the theory developed in Sect. 3.5. Moreover, since the cost of the matrix–vector multiplications is in our case proportional to the dimension n of the matrix $\mathsf{A}_{\mathrm{LW},h}$, it follows that the cost of the solution is also proportional to n.

3.11 Comments and Conclusions

The development of the conjugate gradient method was preceded by the method of conjugate directions [92]. If the conjugate directions are generated by means of a suitable matrix decomposition, the method can be considered as a variant of the direct methods of Sect. 1.5 (see, e.g., [169]).

Since its introduction in the early 1950s by Hestenes and Stiefel [117], a lot of research related to the development of the CG method has been carried out, so that there are many references concerning this subject. We refer an interested reader to the textbooks and research monographs by Saad [163], van der Vorst [178], Greenbaum [106], Hackbusch [110], Chen [21], and Axelsson [4] for more information. A comprehensive account of development of the CG method up to 1989 may be found in the paper by Golub and O'Leary [102]. Most of the research is concentrated on the development and analysis of preconditioners.

Preconditioning by conjugate projector presented in Sect. 3.7 was introduced by Dostál [39]. The same preconditioning with different analysis was presented independently by Marchuk and Kuznetsov [150] as the *conjugate gradients in subspace* or the *generalized conjugate gradient method* and by Nicolaides [154] as the *deflation method.*

Finding at each step the minimum over the subspace generated by all the previous search directions, the conjugate gradient method exploits all the information gathered during the previous iterations. To use this feature in the algorithms for the solution of constrained problems, it is important to generate long uninterrupted sequences of the conjugate gradient iterations. This strategy also supports exploitation of yet another unique feature of the conjugate gradient method, namely, its self-preconditioning capabilities that were described by van der Sluis and van der Vorst [168]. The latter property can also be described in terms of the preconditioning by the conjugate projector. Indeed, if Q_k denotes the conjugate projector onto the conjugate complement $\mathcal{V}$ of $\mathcal{U} = \mathrm{Span}\{\mathbf{p}_1, \ldots, \mathbf{p}_k\}$, then it is possible to give the bound on the rate of convergence of the conjugate gradient method starting from $\mathbf{x}_{k+1}$ in terms of the regular condition number $\overline{\kappa}_k = \overline{\kappa}(\mathsf{Q}_k^T \mathsf{A} \mathsf{Q}_k | \mathcal{V})$ of $\mathsf{Q}_k^T \mathsf{A} \mathsf{Q}_k | \mathcal{V}$ and observe that $\overline{\kappa}_k$ decreases with the increasing k.

For the solution of large problems, the basic CG algorithm is most successful when it is combined with the preconditioning which exploits additional information about A, often obtained by tracing its generation. Thus the *multigrid* (see, e.g., Hackbusch [109] or Trottenberg et al. [176]) or *FETI* (see, e.g., Farhat, Mandel, and Roux [85], or Toselli and Widlund [175])-based preconditioners for the solution of problems arising from the discretization of elliptic partial differential equations exploit the information about the original continuous problems so efficiently that the discretized problems can be solved at a cost proportional to the number of unknowns. It follows that the conjugate gradient method should outperform direct solvers at least for some large problems. Special preconditioners for singular or nearly singular systems arising in optimization were proposed, e.g., by Hager [114].

4

Equality Constrained Minimization

We shall now be interested in the development of efficient algorithms for

$$\min_{\mathbf{x} \in \Omega_E} f(\mathbf{x}), \tag{4.1}$$

where $f(\mathbf{x}) = \frac{1}{2}\mathbf{x}^T \mathsf{A}\mathbf{x} - \mathbf{x}^T \mathbf{b}$, $\mathbf{b}$ is a given column n-vector, A is an $n \times n$ symmetric positive definite matrix, $\Omega_E = \{\mathbf{x} \in \mathbb{R}^n : \mathsf{B}\mathbf{x} = \mathbf{c}\}$, $\mathsf{B} \in \mathbb{R}^{m \times n}$, and $\mathbf{c} \in \mathrm{Im}\mathsf{B}$. We assume that $\mathsf{B} \neq \mathsf{O}$ is not a full column rank matrix, so that $\mathrm{Ker}\mathsf{B} \neq \{\mathbf{o}\}$, but we allow dependent rows of B. Using a simple observation of Sect. 4.6.7, we can extend our results to the solution of problems with A positive definite on $\mathrm{Ker}\mathsf{B}$.

There are several reasons why we consider the constraint matrix B with dependent rows. First, for large problems, it may be expensive to identify the dependent rows, as this can often be done only by an expensive rank revealing decomposition. Second, the removal of the dependent constraints may complicate the precision control of the removed equations when we accept approximate solutions. For example, if we carry out the minimization subject to $x_1 = x_2 = x_3$, but control only that $|x_1 - x_2| \leq \varepsilon$ and $|x_2 - x_3| \leq \varepsilon$, then it can easily happen that $|x_1 - x_3| > \varepsilon$. Finally, the whole concept of the dependence assumes that all computations are carried out in exact arithmetics, so that it is better to avoid such assumption whenever we assume our algorithms to be implemented in computer arithmetics.

Here we are interested in large sparse problems with a well-conditioned A, and in algorithms that can be used also for the solution of equality and inequality constrained problems. Our choice is the class of inexact augmented Lagrangian algorithms which enforce the feasibility condition by the Lagrange multipliers generated in the outer loop, while unconstrained minimization is carried out by the conjugate gradient algorithm in the inner loop. A new feature of our approach is that the algorithm is viewed as a repeated implementation of the penalty method. We combine this approach with an adaptive precision control of the inner loop to get the convergence results which are independent of the representation of Ω_E.

Zdeněk Dostál, *Optimal Quadratic Programming Algorithms*,
Springer Optimization and Its Applications, DOI 10.1007/978-0-387-84806-8_4,

Overview of Algorithms

If we add the penalization function, which is zero on the feasible domain and which achieves large values outside the feasible region, to the original cost function, we can approximate a solution of the original equality constrained problem by the solution of the unconstrained minimization problem with the modified (penalized) cost function. The resulting *penalty method* presented in Sect. 4.2 is probably the most simple way to reduce the equality constrained problem to the unconstrained one. If the penalized problem is solved by an iterative method, the Hessian of the penalized problem can be preconditioned by a special *preconditioner* of Sect. 4.2.6 which preserves the gap in the spectrum.

A prototype of the method studied in this chapter is the *exact augmented Lagrangian method* and its specialization called the *Uzawa algorithm.* See Algorithm 4.2 for their formal description. These methods reduce the original bound constrained problem to a sequence of the unconstrained, optionally moderately penalized problems that are solved exactly, typically by the direct methods of Sect. 1.5.

The auxiliary problems of the augmented Lagrangian method need not be solved exactly. An extreme case is Algorithm 4.1, known as the *Arrows–Hurwitz algorithm*, which carries out only one gradient iteration with the fixed steplength to approximate the solution of the auxiliary problem.

The *asymptotically exact augmented Lagrangian method*, which is described in Sect. 4.4 as Algorithm 4.3, controls the precision of the solution of the auxiliary unconstrained problems by a forcing sequence decreasing to zero. The forcing sequence should be defined by the user.

The precision of the solution of the auxiliary unconstrained problems can also be controlled by the current feasibility error. To achieve convergence, the *adaptive augmented Lagrangian method* modifies also the regularization parameter by means of the forcing sequence generated in the process of solution. The method is described in Sect. 4.5 as Algorithm 4.4.

The most sophisticated method presented in this chapter is the *semimonotonic augmented Lagrangian method for equality constraints* referred to as SMALE. The algorithm is described in Sect. 4.6 as Algorithm 4.5. Similarly to the adaptive augmented Lagrangian method, SMALE controls the precision of the solution of the auxiliary unconstrained problems by the feasibility error, but the penalty parameter is adapted in order to guarantee a sufficient increase of the augmented Lagrangians. The unique theoretical results concerning SMALE include a small explicit bound on the penalty parameter which guarantees that the number of iterations that are necessary to find an approximate solution can be bounded by a number independent of the constraints. The preconditioning preserving the bound on the rate of convergence of Sect. 4.2.6 can be applied also to SMALE.

4.1 Review of Alternative Methods

Before we embark on the study of inexact augmented Lagrangians, let us briefly review alternative methods for the solution of the equality constrained problem (4.1).

Using Proposition 2.8, it follows that (4.1) is equivalent to the solution of the *saddle point system of linear equations*

$$\begin{bmatrix} \mathsf{A} & \mathsf{B}^T \\ \mathsf{B} & \mathsf{O} \end{bmatrix} \begin{bmatrix} \mathbf{x} \\ \boldsymbol{\lambda} \end{bmatrix} = \begin{bmatrix} \mathbf{b} \\ \mathbf{c} \end{bmatrix}. \tag{4.2}$$

If B is a full row rank matrix, we can solve (4.2) effectively by the *Gauss elimination* with a suitable pivoting strategy, or by a *symmetric factorization* which takes into account that (4.2) is indefinite. Alternatively, we can also use *MINRES*, a Krylov space method which generates the iterates minimizing the Euclidean norm of the residual in the Krylov space. The performance of MINRES depends on the distribution of the spectrum of the KKT system (4.2) similarly as the performance of the CG method. A recent comprehensive review of the methods for the solution of saddle point systems with many references is in Benzi, Golub, and Liesen [10]; see also Elman, Sylvester, and Wathen [81].

We can also reduce (4.2) to a symmetric positive definite case. If B is a full row rank matrix, and if we are able to evaluate the action of A^{-1} effectively, we can multiply the first block row in (4.2) by $\mathsf{B}\mathsf{A}^{-1}$, subtract the second row from the result, and change the signs to obtain the symmetric positive definite *Schur complement system*

$$\mathsf{B}\mathsf{A}^{-1}\mathsf{B}^T\boldsymbol{\lambda} = \mathsf{B}\mathsf{A}^{-1}\mathbf{b} - \mathbf{c}, \tag{4.3}$$

which can be solved by the methods described in Chap. 3. The method is also known as the *range-space method.* Let us point out that if we solve (4.3) by the CG method, then it is not necessary to evaluate $\mathsf{B}\mathsf{A}^{-1}\mathsf{B}^T$ explicitly. Using the CG method and the left generalized inverse of Sect. 1.4, the method can be extended to A positive semidefinite and B with dependent rows. We shall see that the range-space method is closely related to the Uzawa-type methods that we shall study later in this chapter.

Alternatively, we can use the *null-space* method, provided we have a basis Z of $\mathrm{Ker}\mathsf{B}$ and a feasible $\mathbf{x}_0$,

$$\mathsf{B}\mathbf{x}_0 = \mathbf{c}.$$

Observing that $\Omega_E = \{\mathbf{x}_0 + \mathsf{Z}\mathbf{y} :\ \mathbf{y} \in \mathbb{R}^d\}$, we can substitute into (4.1) to get

$$\min_{\mathbf{x}\in\Omega_E} f(\mathbf{x}) = \min_{\mathbf{y}\in\mathbb{R}^d} f(\mathbf{x}_0 + \mathsf{Z}\mathbf{y}) = \frac{1}{2}\mathbf{y}^T\mathsf{Z}^T\mathsf{A}\mathsf{Z}\mathbf{y} - (\mathbf{b} - \mathsf{A}\mathbf{x}_0)^T\mathsf{Z}\mathbf{y} + \frac{1}{2}\mathbf{x}_0^T\mathsf{A}\mathbf{x}_0,$$

so that we can evaluate $\mathbf{y}$ by solving the gradient equation

$$\mathsf{Z}^T\mathsf{A}\mathsf{Z}\mathbf{y} = \mathsf{Z}^T(\mathbf{b} - \mathsf{A}\mathbf{x}_0).$$

If the resulting system is solved by the CG method, then the method can be directly applied to the problems with A positive semidefinite and B with dependent rows.

Results concerning application of domain decomposition methods can be found in the monograph by Toselli and Widlund [175].

A class of algorithms which is important for our exposition is based on the mixed formulation

$$L_0(\widehat{\mathbf{x}}, \widehat{\boldsymbol{\lambda}}) = \max_{\boldsymbol{\lambda} \in \mathbb{R}^m} \min_{\mathbf{x} \in \mathbb{R}^n} L_0(\mathbf{x}, \boldsymbol{\lambda})$$

for the problems with full row rank B. As an example let us recall the *Arrow–Hurwitz algorithm* 4.1, which exploits the first-order approximation of L_0 given by

$$L_0(\mathbf{x} + \alpha \mathbf{d}, \boldsymbol{\lambda} + r\boldsymbol{\delta}) \approx L_0(\mathbf{x}, \boldsymbol{\lambda}) + \alpha \nabla_{\mathbf{x}} L_0(\mathbf{x}, \boldsymbol{\lambda})\mathbf{d} + r\nabla_{\boldsymbol{\lambda}} L_0(\mathbf{x}, \boldsymbol{\lambda})\boldsymbol{\delta}$$

to improve the approximations of the solution $\widehat{\mathbf{x}}$ by taking small steps in the direction opposite to the gradient

$$\nabla_{\mathbf{x}} L_0(\mathbf{x}, \boldsymbol{\lambda}) = \mathsf{A}\mathbf{x} - \mathbf{b} + \mathsf{B}^T \boldsymbol{\lambda},$$

and to improve the approximations of the Lagrange multipliers $\widehat{\boldsymbol{\lambda}}$ by taking small steps in the direction

$$\nabla_{\boldsymbol{\lambda}} L_0(\mathbf{x}, \boldsymbol{\lambda}) = \mathsf{B}\mathbf{x} - \mathbf{c}.$$

Algorithm 4.1. Arrow–Hurwitz algorithm.

Given a symmetric positive definite matrix $\mathsf{A} \in \mathbb{R}^{n \times n}$, *a matrix* $\mathsf{B} \in \mathbb{R}^{m \times n}$ *with the nonempty kernel,* $\mathbf{b} \in \mathbb{R}^n$, *and* $\mathbf{c} \in \mathrm{Im}\mathsf{B}$.

Step 0. {*Initialization.*}
Choose $\boldsymbol{\lambda}^0 \in \mathbb{R}^m$, $\mathbf{x}^{-1} \in \mathbb{R}^n$, $\alpha > 0$, $r > 0$

for *k=0,1,2,...*
Step 1. {*Reducing the value of* L_0 *in* $\mathbf{x}$ *direction.*}
$\mathbf{x}^k = \mathbf{x}^{k-1} - \alpha \nabla_{\mathbf{x}} L_0(\mathbf{x}^{k-1}, \boldsymbol{\lambda}^k) = \mathbf{x}^{k-1} - \alpha(\mathsf{A}\mathbf{x} - \mathbf{b} + \mathsf{B}^T \boldsymbol{\lambda})$

Step 2. {*Increasing the value of* L_0 *in* $\boldsymbol{\lambda}$ *direction.*}
$\boldsymbol{\lambda}^{k+1} = \boldsymbol{\lambda}^k + r\nabla_{\boldsymbol{\lambda}} L_0(\mathbf{x}^k, \boldsymbol{\lambda}^k) = \boldsymbol{\lambda}^k + r(\mathsf{B}\mathbf{x}^k - \mathbf{c})$
end for

The Arrow–Hurwitz algorithm is known to converge for sufficiently small steplengths α and r. Even though its convergence is known to be slow, the algorithm has found its applications due to the low cost of the iterations and minimal memory requirements. It can be considered as an extreme case of the inexact Uzawa-type algorithms, the main topic of this chapter.

4.2 Penalty Method

Probably the most simple way to reduce the equality constrained quadratic programming problem (4.1) to the unconstrained one is to enhance the constraints into the objective function by adding a suitable term which penalizes the violation of the constraints. In this section we consider the *quadratic penalty method* which approximates the solution $\widehat{\mathbf{x}}$ of (4.1) by the solution $\widehat{\mathbf{x}}_\varrho$ of

$$\min_{\mathbf{x}\in\mathbb{R}^n} f_\varrho(\mathbf{x}), \quad f_\varrho(\mathbf{x}) = f(\mathbf{x}) + \frac{\varrho}{2}\|\mathsf{B}\mathbf{x}-\mathbf{c}\|^2, \tag{4.4}$$

where $\varrho \geq 0$ is the *penalty parameter* and $\|\mathsf{B}\mathbf{x}-\mathbf{c}\|^2$ is the *penalty function.*

Intuitively, if the penalty parameter ϱ is large, then the solution $\widehat{\mathbf{x}}_\varrho$ of (4.4) can hardly be far from the solution of (4.1). Indeed, if ϱ were infinite, then the minimizer of f_ϱ would solve the equality constrained problem (4.1). Thus it is natural to expect that if ϱ is sufficiently large, then the penalty approximation $\widehat{\mathbf{x}}_\varrho$ is a suitable approximation to the solution $\widehat{\mathbf{x}}$ of (4.1). The effect of the penalty term is illustrated in Fig. 4.1. Notice that the penalty approximation is typically near the feasible set, but does not belong to it. That is why our penalty method is also called the *exterior penalty method.*

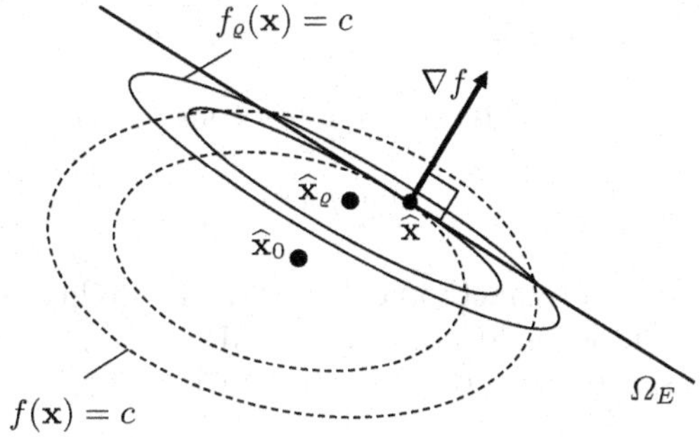

Fig. 4.1. The effect of the quadratic penalty

In the following sections, we shall often use the more general *augmented Lagrangian* penalty function $L : \mathbb{R}^{n+m+1} \to \mathbb{R}$ which is defined by

$$L(\mathbf{x}, \boldsymbol{\lambda}, \varrho) = f(\mathbf{x}) + (\mathsf{B}\mathbf{x}-\mathbf{c})^T\boldsymbol{\lambda} + \frac{\varrho}{2}\|\mathsf{B}\mathbf{x}-\mathbf{c}\|^2 = L_0(\mathbf{x}, \lambda) + \frac{\varrho}{2}\|\mathsf{B}\mathbf{x}-\mathbf{c}\|^2, \tag{4.5}$$

where $\boldsymbol{\lambda} \in \mathbb{R}^m$ is arbitrary and $L_0(\mathbf{x}, \boldsymbol{\lambda}) = L(\mathbf{x}, \boldsymbol{\lambda}, 0)$ is the Lagrangian function (2.20). Notice that $f_\varrho(\mathbf{x}) = L(\mathbf{x}, \mathbf{o}, \varrho)$. Since $\varrho\|\mathsf{B}\mathbf{x}-\mathbf{c}\|^2$ and $(\mathsf{B}\mathbf{x}-\mathbf{c})^T\boldsymbol{\lambda}$ vanish when $\mathsf{B}\mathbf{x} = \mathbf{c}$, it follows that

$$f(\mathbf{x}) = f_\varrho(\mathbf{x}) = L(\mathbf{x}, \boldsymbol{\lambda}, \varrho)$$

for any $\mathbf{x} \in \Omega_E$, $\boldsymbol{\lambda} \in \mathbb{R}^m$, and $\varrho \geq 0$.

4.2.1 Minimization of Augmented Lagrangian

Let us start with the modified problem

$$\min_{\mathbf{x}\in\Omega_E} L(\mathbf{x},\boldsymbol{\lambda},\varrho). \tag{4.6}$$

Since the gradient of the augmented Lagrangian is given by

$$\nabla_{\mathbf{x}} L(\mathbf{x},\boldsymbol{\lambda},\varrho) = \mathsf{A}\mathbf{x} - \mathbf{b} + \mathsf{B}^T\left(\boldsymbol{\lambda} + \varrho(\mathsf{B}\mathbf{x} - \mathbf{c})\right), \tag{4.7}$$

it follows that the KKT system for (4.6) reads

$$\begin{bmatrix} \mathsf{A}_\varrho & \mathsf{B}^T \\ \mathsf{B} & \mathsf{O} \end{bmatrix} \begin{bmatrix} \mathbf{x} \\ \boldsymbol{\lambda} \end{bmatrix} = \begin{bmatrix} \mathbf{b} + \varrho\mathsf{B}^T\mathbf{c} \\ \mathbf{c} \end{bmatrix}, \tag{4.8}$$

where $\mathsf{A}_\varrho = \mathsf{A} + \varrho\mathsf{B}^T\mathsf{B}$. Eliminating $\mathbf{x}$, we get that any multiplier $\overline{\boldsymbol{\lambda}}$ satisfies

$$\mathsf{B}\mathsf{A}_\varrho^{-1}\mathsf{B}^T\overline{\boldsymbol{\lambda}} = \mathsf{B}\mathsf{A}_\varrho^{-1}(\mathbf{b} + \varrho\mathsf{B}^T\mathbf{c}) - \mathbf{c}. \tag{4.9}$$

Moreover, if we substitute $\mathsf{B}\mathbf{x} = \mathbf{c}$ into the first block equation, we get that (4.8) is equivalent to the KKT system (2.31), so the saddle points of L_0 are exactly the saddle points of L. This result is not surprising as

$$L(\mathbf{x},\boldsymbol{\lambda},\varrho) = f_\varrho(\mathbf{x}) + (\mathsf{B}\mathbf{x} - \mathbf{c})^T\boldsymbol{\lambda}$$

is the Lagrangian for the penalized equality constrained problem

$$\min_{\mathbf{x}\in\Omega_E} f_\varrho(\mathbf{x}).$$

To see how the penalty method enforces the feasibility, let us assume that $\boldsymbol{\lambda} \in \mathbb{R}^m$ is fixed, and let us denote by $\widehat{\mathbf{x}}_0$ and $\widehat{\mathbf{x}}_\varrho$ the minimizers of $L_0(\mathbf{x},\boldsymbol{\lambda})$ and $L(\mathbf{x},\boldsymbol{\lambda},\varrho)$, respectively. Then the solution $\widehat{\mathbf{x}}$ satisfies

$$L_0(\widehat{\mathbf{x}}_\varrho,\boldsymbol{\lambda}) + \frac{\varrho}{2}\|\mathsf{B}\widehat{\mathbf{x}}_\varrho - \mathbf{c}\|^2 = L(\widehat{\mathbf{x}}_\varrho,\boldsymbol{\lambda},\varrho) \le L(\widehat{\mathbf{x}},\boldsymbol{\lambda},\varrho) = f(\widehat{\mathbf{x}}),$$

so that, using $L_0(\widehat{\mathbf{x}}_0,\boldsymbol{\lambda}) \le L_0(\widehat{\mathbf{x}}_\varrho,\boldsymbol{\lambda})$, we get

$$\|\mathsf{B}\widehat{\mathbf{x}}_\varrho - \mathbf{c}\|^2 \le \frac{2}{\varrho}\left(f(\widehat{\mathbf{x}}) - L_0(\widehat{\mathbf{x}}_\varrho,\boldsymbol{\lambda})\right) \le \frac{2}{\varrho}\left(f(\widehat{\mathbf{x}}) - L_0(\widehat{\mathbf{x}}_0,\boldsymbol{\lambda})\right).$$

It follows that the feasibility error $\|\mathsf{B}\widehat{\mathbf{x}}_\varrho - \mathbf{c}\|$, which corresponds to the second block equation of the KKT system (2.31), can be made arbitrarily small. We shall give stronger or easier computable bounds later in this section.

To see how $\widehat{\mathbf{x}}_\varrho$ satisfies the first block equation of the KKT conditions (4.2), let us recall that the gradient of the augmented Lagrangian is given by (4.7) and denote

$$\widetilde{\boldsymbol{\lambda}} = \boldsymbol{\lambda} + \varrho(\mathsf{B}\widehat{\mathbf{x}}_\varrho - \mathbf{c}). \tag{4.10}$$

Then

$$\mathsf{A}\widehat{\mathbf{x}}_\varrho - \mathbf{b} + \mathsf{B}^T\widetilde{\boldsymbol{\lambda}} = \nabla_{\mathbf{x}} L_0(\widehat{\mathbf{x}}_\varrho, \widetilde{\boldsymbol{\lambda}}) = \nabla_{\mathbf{x}} L(\widehat{\mathbf{x}}_\varrho, \boldsymbol{\lambda}, \varrho) = \mathbf{o},$$

so that $(\widehat{\mathbf{x}}_\varrho, \widetilde{\boldsymbol{\lambda}})$ satisfies the first block equation of the KKT conditions exactly. Moreover, if $\boldsymbol{\lambda}$ is considered as an approximation of a vector of Lagrange multipliers of the solution of (4.1), then our observations indicate that $\widetilde{\boldsymbol{\lambda}}$ is a better approximation. Using Proposition 2.12, we conclude that $(\widehat{\mathbf{x}}_\varrho, \widetilde{\boldsymbol{\lambda}})$ can approximate the KKT pair of (4.1) with arbitrarily small error.

4.2.2 An Optimal Feasibility Error Estimate for Homogeneous Constraints

Let us first examine the feasibility error of an approximate solution $\mathbf{x}$ of the problem

$$\min_{\mathbf{x}\in\mathbb{R}^n} f_\varrho(\mathbf{x}), \quad f_\varrho(\mathbf{x}) = f(\mathbf{x}) + \frac{\varrho}{2}\|\mathsf{B}\mathbf{x}\|^2, \tag{4.11}$$

where f and B are from the definition of problem (4.1) and $\varrho > 0$. We assume that $\mathbf{x}$ satisfies

$$\|\nabla f_\varrho(\mathbf{x})\| \le \varepsilon\|\mathbf{b}\|, \tag{4.12}$$

where $\varepsilon > 0$ is a small number.

Notice that our $\mathbf{x}$ can be considered as an approximation to the solution $\widehat{\mathbf{x}}$ of the equality constrained problem (4.1) in case that the equality constraints are homogeneous, i.e., $\mathbf{c} = \mathbf{o}$. To check that $\mathbf{x}$ satisfies approximately the first part of the KKT conditions (4.2), observe that

$$\nabla f_\varrho(\mathbf{x}) = (\mathsf{A} + \varrho\mathsf{B}^T\mathsf{B})\mathbf{x} - \mathbf{b}.$$

After denoting $\boldsymbol{\lambda} = \varrho\mathsf{B}\mathbf{x}$ and $\mathbf{g} = \nabla f_\varrho(\mathbf{x})$, we get

$$\mathsf{A}\mathbf{x} - \mathbf{b} + \mathsf{B}^T\boldsymbol{\lambda} = \mathbf{g}, \tag{4.13}$$

which can be considered as an approximation of the first block equation of the KKT conditions (4.2).

The feasibility error is considered in the next theorem.

Theorem 4.1. *Let* A, B, *and* $\mathbf{b}$ *be those of the definition of problem (4.1) with* B *not necessarily a full rank matrix, let* $\lambda_{\min} = \lambda_{\min}(\mathsf{A}) > 0$ *denote the smallest eigenvalue of* A, *and let* $\varepsilon \ge 0$ *and* $\varrho > 0$.

If $\mathbf{x}$ *is an approximate solution of (4.11) such that*

$$\|\nabla f_\varrho(\mathbf{x})\| \le \varepsilon\|\mathbf{b}\|,$$

then

$$\|\mathsf{B}\mathbf{x}\| \le \frac{1+\varepsilon}{\sqrt{\lambda_{\min}\varrho}}\|\mathbf{b}\|. \tag{4.14}$$

Proof. Let us denote

$$\mathsf{A}_\varrho = \mathsf{A} + \varrho \mathsf{B}^T \mathsf{B}$$

and notice that for any $\mathbf{x}, \mathbf{d} \in \mathbb{R}^n$

$$f_\varrho(\mathbf{x}+\mathbf{d}) = f_\varrho(\mathbf{x}) + \mathbf{g}^T\mathbf{d} + \frac{1}{2}\mathbf{d}^T\mathsf{A}_\varrho\mathbf{d} \geq f_\varrho(\mathbf{x}) - \|\mathbf{g}\|\|\mathbf{d}\| + \frac{1}{2}\lambda_{\min}\|\mathbf{d}\|^2$$

$$\geq \min_{\xi\in\mathbb{R}}\left(f_\varrho(\mathbf{x}) - \|\mathbf{g}\|\xi + \frac{1}{2}\lambda_{\min}\xi^2\right) \geq f_\varrho(\mathbf{x}) - \frac{1}{2\lambda_{\min}}\|\mathbf{g}\|^2,$$

where $\mathbf{g} = \nabla f_\varrho(\mathbf{x})$. Recalling that by (2.11)

$$\min_{\mathbf{d}\in\mathbb{R}^n} f_\varrho(\mathbf{x}+\mathbf{d}) = \min_{\mathbf{y}\in\mathbb{R}^n} f_\varrho(\mathbf{y}) = -\frac{1}{2}\mathbf{b}^T\mathsf{A}_\varrho^{-1}\mathbf{b},$$

we get

$$0 \geq -\frac{1}{2}\mathbf{b}^T\mathsf{A}_\varrho^{-1}\mathbf{b} \geq f(\mathbf{x}) + \frac{\varrho}{2}\|\mathsf{B}\mathbf{x}\|^2 - \frac{1}{2\lambda_{\min}}\|\mathbf{g}\|^2.$$

Let us now assume that $\mathbf{x}$ satisfies $\|\mathbf{g}\| \leq \varepsilon\|\mathbf{b}\|$. After substituting into the last inequality and using (2.11), (1.24), and the properties of the Euclidean norm, we get

$$0 \geq f(\mathbf{x}) + \frac{\varrho}{2}\|\mathsf{B}\mathbf{x}\|^2 - \frac{1}{2\lambda_{\min}}\|\mathbf{g}\|^2 \geq \min_{\mathbf{y}\in\mathbb{R}^n} f(\mathbf{y}) + \frac{\varrho}{2}\|\mathsf{B}\mathbf{x}\|^2 - \frac{\varepsilon^2}{2\lambda_{\min}}\|\mathbf{b}\|^2$$

$$= -\frac{1}{2}\mathbf{b}^T\mathsf{A}^{-1}\mathbf{b} + \frac{\varrho}{2}\|\mathsf{B}\mathbf{x}\|^2 - \frac{\varepsilon^2}{2\lambda_{\min}}\|\mathbf{b}\|^2$$

$$\geq -\frac{1}{2\lambda_{\min}}\|\mathbf{b}\|^2 + \frac{\varrho}{2}\|\mathsf{B}\mathbf{x}\|^2 - \frac{\varepsilon^2}{2\lambda_{\min}}\|\mathbf{b}\|^2 \geq \frac{\varrho}{2}\|\mathsf{B}\mathbf{x}\|^2 - \frac{1+\varepsilon^2}{2\lambda_{\min}}\|\mathbf{b}\|^2.$$

Equation (4.14) can be obtained by simple manipulations with application of $1+\varepsilon^2 \leq (1+\varepsilon)^2$. □

An interesting feature of Theorem 4.1 is that *the estimate is independent of the constraint matrix* B. In particular, the estimate (4.14) is valid even if B has dependent rows. The assumption that the constraints are homogeneous was used to get that the unconstrained minimum of f_ϱ is not positive.

Theorem 4.1 implies a simple optimality result concerning the approximation by the penalty method. To formulate it, let $\mathcal{T}$ denote any set of indices and assume that for any $t \in \mathcal{T}$, there is defined a problem

$$\min_{\mathbf{x}\in\Omega_E^t} f_t(\mathbf{x}), \tag{4.15}$$

where $f_t(\mathbf{x}) = \frac{1}{2}\mathbf{x}^T\mathsf{A}_t\mathbf{x} - \mathbf{b}_t^T\mathbf{x}$, $\mathsf{A}_t \in \mathbb{R}^{n_t\times n_t}$ is SPD with the eigenvalues in the interval $[a_{\min}, a_{\max}]$, $0 < a_{\min} < a_{\max}$, $\mathbf{b}_t, \mathbf{x} \in \mathbb{R}^{n_t}$, $\mathsf{B}_t \in \mathbb{R}^{m_t\times n_t}$, and $\Omega_E^t = \{\mathbf{x}\in\mathbb{R}^{n_t} : \mathsf{B}_t\mathbf{x} = \mathbf{o}\}$.

Corollary 4.2. *For each* $\varepsilon > 0$, *there is* $\varrho > 0$ *such that if approximate solutions* $\mathbf{x}_{t,\varrho}$ *of* (4.15) *satisfy*

$$\nabla f_{t,\varrho}(\mathbf{x}_{t,\varrho}) \le \varepsilon \|\mathbf{b}_t\|, \quad t \in \mathcal{T},$$

then

$$\|\mathsf{B}_t \mathbf{x}_{t,\varrho}\| \le \varepsilon \|\mathbf{b}_t\|, \quad t \in \mathcal{T}.$$

Proof. Notice that by Theorem 4.1

$$\|\mathsf{B}_t \widehat{\mathbf{x}}_{t,\varrho}\| \le \frac{1}{\sqrt{a_{\min}\varrho}} \|\mathbf{b}_t\|$$

for any $\varrho > 0$. It is enough to set $\varrho = 1/(a_{\min}\varepsilon^2)$. □

We conclude that the *prescribed bound on the relative feasibility error for all problems (4.15) can be achieved with one value of the penalty parameter* $\varrho_t = \varrho$. Numerical experiments which illustrate the optimal feasibility estimates in the framework of FETI methods can be found in Dostál and Horák [65, 66].

4.2.3 Approximation Error and Convergence

Using the feasibility estimate (4.14) of the previous subsection and an error bound on the violation of the first block equation of the KKT conditions (2.46), we can bound the approximation error of the penalty method for homogeneous constraints.

Theorem 4.3. *Let problem (4.1) be defined by* $\mathsf{A}, \mathsf{B}, \mathbf{b}$, *and* $\mathbf{c} = \mathbf{o}$, *with* $\mathsf{B} \neq \mathsf{O}$ *not necessarily a full rank matrix, let* $(\widehat{\mathbf{x}}, \boldsymbol{\lambda}_{\mathrm{LS}})$ *denote the least square KKT pair for (4.1) with* $\mathbf{c} = \mathbf{o}$, *let* $\lambda_{\min}$ *denote the least eigenvalue of* A, *let* $\overline{\sigma}_{\min}$ *denote the least nonzero singular value of* B, *let* $\varepsilon > 0$, $\varrho > 0$, *and let*

$$\boldsymbol{\lambda} = \varrho \mathsf{B}\mathbf{x}. \tag{4.16}$$

If $\mathbf{x}$ *is such that*

$$\|\nabla f_\varrho(\mathbf{x})\| \le \varepsilon \|\mathbf{b}\|,$$

then

$$\|\mathbf{x} - \widehat{\mathbf{x}}\| \le \varepsilon \frac{\kappa(\mathsf{A}) + 1}{\lambda_{\min}} \|\mathbf{b}\| + \frac{1+\varepsilon}{\sqrt{\varrho}} \frac{\kappa(\mathsf{A})}{\overline{\sigma}_{\min}\sqrt{\lambda_{\min}}} \|\mathbf{b}\| \tag{4.17}$$

and

$$\|\boldsymbol{\lambda} - \boldsymbol{\lambda}_{\mathrm{LS}}\| \le \frac{1}{\overline{\sigma}_{\min}} \left(\varepsilon \kappa(\mathsf{A}) \|\mathbf{b}\| + \frac{1+\varepsilon}{\sqrt{\varrho}} \frac{\|\mathsf{A}\|}{\overline{\sigma}_{\min}\sqrt{\lambda_{\min}}} \|\mathbf{b}\| \right). \tag{4.18}$$

Proof. Let us denote $\mathbf{g} = \nabla f_\varrho(\mathbf{x})$ and $\mathbf{e} = \mathsf{B}\mathbf{x}$, so that

$$\mathsf{A}\mathbf{x} + \mathsf{B}^T\boldsymbol{\lambda} = \mathbf{b} + \mathbf{g} \quad \text{and} \quad \mathsf{B}\mathbf{x} = \mathbf{e}, \tag{4.19}$$

and notice that by the assumptions $\boldsymbol{\lambda} \in \mathrm{Im}\mathsf{B}$. Assuming that

$$\|\mathbf{g}\| = \|\nabla f_\varrho(\mathbf{x})\| \le \varepsilon \|\mathbf{b}\|,$$

it follows by Theorem 4.1 that

$$\|\mathsf{B}\mathbf{x}\| \le \frac{1+\varepsilon}{\sqrt{\lambda_{\min}\varrho}}\|\mathbf{b}\|.$$

Substituting into the estimates (2.48) and (2.49) of Proposition 2.12, we get (4.17) and (4.18). □

Notice that the error bounds (4.17) and (4.18) depend on the representation of Ω_E, namely, on the constraint matrix B.

The performance of the penalty method can also be described in terms of convergence. Let $\varepsilon_k > 0$ denote a sequence converging to zero, let $\varrho_k > 0$ denote an increasing unbounded sequence, let $\mathbf{g}^k = \nabla f_{\varrho_k}(\mathbf{x}^k)$, and let $\mathbf{x}^k$ satisfy

$$\|\mathbf{g}^k\| = \|\nabla f_{\varrho_k}(\mathbf{x}^k)\| \le \varepsilon_k \|\mathbf{b}\|.$$

Let us denote

$$\boldsymbol{\lambda}^k = \varrho_k \mathsf{B}\mathbf{x}^k.$$

Then by (4.17) there is a constant C_1 dependent on A and C_2 dependent on A, B such that

$$\|\mathbf{x}^k - \widehat{\mathbf{x}}\| \le \varepsilon_k C_1 \|\mathbf{b}\| + \frac{1+\varepsilon_k}{\sqrt{\varrho_k}} C_2 \|\mathbf{b}\|,$$

and by (4.18) there are constants C_3 and C_4 dependent on A, B such that

$$\|\boldsymbol{\lambda}^k - \boldsymbol{\lambda}_{\mathrm{LS}}\| \le \varepsilon_k C_3 \|\mathbf{b}\| + \frac{1+\varepsilon_k}{\sqrt{\varrho_k}} C_4 \|\mathbf{b}\|. \tag{4.20}$$

It follows that $\boldsymbol{\lambda}^k$ converges to $\boldsymbol{\lambda}_{\mathrm{LS}}$ and $\mathbf{x}^k$ converges to $\widehat{\mathbf{x}}$.

4.2.4 Improved Feasibility Error Estimate

We shall now give a feasibility error estimate for the penalty approximation of (4.1) which is valid for nonhomogeneous constraints with $\mathbf{c} \ne \mathbf{o}$. Our new bound on the error is proportional to ϱ^{-1}, but *dependent on* B and $\mathbf{c}$.

Theorem 4.4. *Let $\mathsf{A}, \mathsf{B}, \mathbf{b}$, and $\mathbf{c}$ be those of the definition of problem (4.1) with $\mathsf{B} \ne \mathsf{O}$ not necessarily a full rank matrix, let $\overline{\beta}_{\min} > 0$ denote the smallest nonzero eigenvalue of $\mathsf{B}^T\mathsf{A}^{-1}\mathsf{B}$, let ε denote a given positive number, and let $\varrho > 0$.*

If $\mathbf{x}$ *is such that*

$$\|\nabla f_\varrho(\mathbf{x})\| \le \varepsilon\|\mathbf{b}\|,$$

then the feasibility error satisfies

$$\|\mathsf{B}\mathbf{x} - \mathbf{c}\| \le \left(1 + \overline{\beta}_{\min}\varrho\right)^{-1} \left((1+\varepsilon)\|\mathsf{B}\mathsf{A}^{-1}\|\|\mathbf{b}\| + \|\mathbf{c}\|\right). \tag{4.21}$$

Proof. Let us recall that for any vector $\mathbf{x}$

$$\nabla f_\varrho(\mathbf{x}) = (\mathsf{A} + \varrho\mathsf{B}^T\mathsf{B})\mathbf{x} - \mathbf{b} - \varrho\mathsf{B}^T\mathbf{c},$$

so that, after denoting $\mathbf{g} = \nabla f_\varrho(\mathbf{x})$ and $\mathsf{A}_\varrho = \mathsf{A} + \varrho\mathsf{B}^T\mathsf{B}$,

$$\mathbf{x} = \mathsf{A}_\varrho^{-1}(\mathbf{g} + \mathbf{b} + \varrho\mathsf{B}^T\mathbf{c}).$$

It follows that

$$\mathsf{B}\mathbf{x} = \mathsf{B}\mathsf{A}_\varrho^{-1}(\mathbf{g} + \mathbf{b}) + \varrho\mathsf{B}\mathsf{A}_\varrho^{-1}\mathsf{B}^T\mathbf{c}.$$

Using equation (1.41) of Lemma 1.4 and simple manipulations, we get

$$\begin{aligned}
\mathsf{B}\mathbf{x} - \mathbf{c} &= \mathsf{B}\mathsf{A}_\varrho^{-1}(\mathbf{g} + \mathbf{b}) + \varrho(\mathsf{I} + \varrho\mathsf{B}\mathsf{A}^{-1}\mathsf{B}^T)^{-1}\mathsf{B}\mathsf{A}^{-1}\mathsf{B}^T\mathbf{c} - \mathbf{c} \\
&= \mathsf{B}\mathsf{A}_\varrho^{-1}(\mathbf{g} + \mathbf{b}) + (\mathsf{I} + \varrho\mathsf{B}\mathsf{A}^{-1}\mathsf{B}^T)^{-1}\left((\mathsf{I} + \varrho\mathsf{B}\mathsf{A}^{-1}\mathsf{B}^T) - \mathsf{I}\right)\mathbf{c} - \mathbf{c} \\
&= \mathsf{B}\mathsf{A}_\varrho^{-1}(\mathbf{g} + \mathbf{b}) - (\mathsf{I} + \varrho\mathsf{B}\mathsf{A}^{-1}\mathsf{B}^T)^{-1}\mathbf{c}.
\end{aligned}$$

To finish the proof, use the assumptions that $\mathbf{c} \in \mathrm{Im}\mathsf{B}$ and $\|\mathbf{g}\| \le \varepsilon\|\mathbf{b}\|$, Lemma 1.6, and the properties of norms. □

Numerical experiments which illustrate (4.21) can be found in Dostál and Horák [65, 66].

4.2.5 Improved Approximation Error Estimate

Using the improved feasibility estimate (4.21) of the previous section, we can improve the bounds on the approximation error of the penalty method given in Sect. 4.2.3.

Theorem 4.5. *Let* $\mathsf{A}, \mathsf{B}, \mathbf{b}$, *and* $\mathbf{c}$ *be those of the definition of problem (4.1) with* B *not necessarily a full rank matrix, let* $\lambda_{\min}$ *denote the least eigenvalue of* A, *let* $\overline{\sigma}_{\min}$ *denote the least nonzero singular value of* B, *let* $(\widehat{\mathbf{x}}, \boldsymbol{\lambda}_{\mathrm{LS}})$ *denote the least square KKT pair for (4.1), let* $\overline{\beta}_{\min} > 0$ *denote the least nonzero eigenvalue of the matrix* $\mathsf{B}\mathsf{A}^{-1}\mathsf{B}^T$, *let* $\varepsilon > 0$, $\varrho > 0$, *and*

$$\boldsymbol{\lambda} = \varrho(\mathsf{B}\mathbf{x} - \mathbf{c}). \tag{4.22}$$

If $\mathbf{x}$ *is such that*

$$\|\nabla f_\varrho(\mathbf{x})\| \le \varepsilon\|\mathbf{b}\|,$$

then

$$\|\boldsymbol{\lambda} - \boldsymbol{\lambda}_{\mathrm{LS}}\| \le \varepsilon \frac{\kappa(\mathsf{A})\|\mathbf{b}\|}{\overline{\sigma}_{\min}} + \frac{\|\mathsf{A}\|\left((1+\varepsilon)\|\mathsf{BA}^{-1}\|\|\mathbf{b}\| + \|\mathbf{c}\|\right)}{\overline{\sigma}^2_{\min}(1+\varrho\overline{\beta}_{\min})} \tag{4.23}$$

and

$$\|\mathbf{x} - \widehat{\mathbf{x}}\| \le \varepsilon \frac{\kappa(\mathsf{A})+1}{\lambda_{\min}}\|\mathbf{b}\| + \frac{\kappa(\mathsf{A})\left((1+\varepsilon)\|\mathsf{BA}^{-1}\|\|\mathbf{b}\| + \|\mathbf{c}\|\right)}{\overline{\sigma}_{\min}(1+\varrho\overline{\beta}_{\min})}. \tag{4.24}$$

Proof. Let us denote $\mathbf{g} = \nabla f_\varrho(\mathbf{x})$ and $\mathbf{e} = \mathsf{B}\mathbf{x} - \mathbf{c}$, so that

$$\mathsf{A}\mathbf{x} + \mathsf{B}^T\boldsymbol{\lambda} = \mathbf{b} + \mathbf{g} \quad \text{and} \quad \mathsf{B}\mathbf{x} = \mathbf{c} + \mathbf{e}.$$

If

$$\|\mathbf{g}\| = \|\nabla f_\varrho(\mathbf{x})\| \le \varepsilon\|\mathbf{b}\|,$$

then by Theorem 4.4

$$\|\mathsf{B}\mathbf{x} - \mathbf{c}\| \le \frac{1}{1+\varrho\overline{\beta}_{\min}}\left((1+\varepsilon)\|\mathsf{BA}^{-1}\|\|\mathbf{b}\| + \|\mathbf{c}\|\right).$$

Substituting into the estimates (2.47) and (2.48) of Proposition 2.12, we get

$$\|\mathsf{B}^T(\boldsymbol{\lambda} - \boldsymbol{\lambda}_{\mathrm{LS}})\| \le \varepsilon\kappa(\mathsf{A})\|\mathbf{b}\| + \frac{\|\mathsf{A}\|\left((1+\varepsilon)\|\mathsf{BA}^{-1}\|\|\mathbf{b}\| + \|\mathbf{c}\|\right)}{\overline{\sigma}_{\min}(1+\varrho\overline{\beta}_{\min})} \tag{4.25}$$

and (4.24). To finish the proof, notice that $\boldsymbol{\lambda} - \boldsymbol{\lambda}_{\mathrm{LS}} \in \mathrm{Im}\mathsf{B}$, so that by (1.34)

$$\overline{\sigma}_{\min}\|\boldsymbol{\lambda} - \boldsymbol{\lambda}_{\mathrm{LS}}\| \le \|\mathsf{B}^T(\boldsymbol{\lambda} - \boldsymbol{\lambda}_{\mathrm{LS}})\|,$$

apply the latter estimate to the left-hand side of (4.25), and divide the resulting chain of inequalities by $\overline{\sigma}_{\min}$. □

We can also get the improved rates of convergence compared with those of Sect. 4.2.3. Let $\varepsilon_k \ge 0$ denote again a sequence converging to zero, let $\varrho_k > 0$ denote an increasing unbounded sequence, let $\mathbf{x}^k$ satisfy

$$\|\mathbf{g}^k\| = \|\nabla f_{\varrho_k}(\mathbf{x}^k)\| \le \varepsilon_k\|\mathbf{b}\|,$$

and let us denote

$$\boldsymbol{\lambda}^k = \varrho_k(\mathsf{B}\mathbf{x}^k - \mathbf{c}).$$

Then by (4.23) there are constants C_1, C_2, and C_3 dependent on A, B such that

$$\|\boldsymbol{\lambda}^k - \boldsymbol{\lambda}_{\mathrm{LS}}\| \le \varepsilon_k C_1\|\mathbf{b}\| + \frac{1+\varepsilon_k}{\varrho_k}C_2\|\mathbf{b}\| + \frac{C_3}{\varrho_k}\|\mathbf{c}\|,$$

and by (4.24) there is a constant C_4 dependent on A and constants C_5, C_6 dependent on A, B such that

$$\|\mathbf{x}^k - \widehat{\mathbf{x}}\| \le \varepsilon_k C_4\|\mathbf{b}\| + \frac{1+\varepsilon_k}{\varrho_k}C_5\|\mathbf{b}\| + \frac{C_6}{\varrho_k}\|\mathbf{c}\|.$$

Thus $\boldsymbol{\lambda}^k$ converges to $\boldsymbol{\lambda}_{\mathrm{LS}}$ and $\mathbf{x}^k$ converges to $\widehat{\mathbf{x}}$.

4.2.6 Preconditioning Preserving Gap in the Spectrum

We have seen that the penalty method reduces the solution of the equality constrained minimization problem (4.1) to the unconstrained penalized problem (4.4). The resulting problem may be solved either by a suitable direct method such as the Cholesky decomposition, or by an iterative method such as the conjugate gradient method. If the penalty parameter ϱ is large, then the Hessian matrix

$$\mathsf{A}_\varrho = \mathsf{A} + \varrho \mathsf{B}^T \mathsf{B}$$

of the cost function f_ϱ of the penalized problem (4.4) is obviously ill-conditioned. Thus the estimates based on the condition number do not guarantee fast convergence of the conjugate gradient method, and a natural idea is to reduce the condition number of A_ϱ by a suitable preconditioning. This is indeed possible as has been shown, e.g., by Hager [111, 113].

Here we consider an alternative approach which exploits the fast convergence of the conjugate gradient method for the problems with a gap in the spectrum. The method is based on two results: the bounds on the rate of convergence independent of ϱ given by (3.23) and (3.24) and Lemma 1.7 on the distribution of the spectrum of A_ϱ. The method presented here is applicable for large ϱ provided we have an effective preconditioner M for A that can be used by the preconditioned conjugate gradient algorithm of Sect. 3.6. To simplify our exposition, we assume that $\mathsf{M} = \mathsf{L}\mathsf{L}^T$, where L is a sparse lower triangular matrix.

To express briefly the effects of the preconditioning strategies presented in this section, let $k(\mathsf{W}, \varepsilon)$ denote the number of the conjugate gradient iterations that are necessary to reduce the residual of any system with the symmetric positive definite matrix W by ε, and let

$$\overline{k}(\mathsf{W}, \varepsilon) = \mathrm{int}(\frac{1}{2}\sqrt{\kappa(\mathsf{W})}\ln(2/\varepsilon) + 1) \tag{4.26}$$

denote the upper bound on $k(\mathsf{W}, \varepsilon)$ which may be easily obtained from (3.23).

Let us first assume that the rank m of the constraint matrix $\mathsf{B} \in \mathbb{R}^{p \times n}$ in the original problem (4.1) is small. Then it is possible to use L to redistribute the spectrum of the penalized matrix A_ϱ directly. In this case (1.51) and the estimate (3.23) of the rate of the conjugate gradient method for the case that the Hessian of the cost function has m isolated eigenvalues give the bound

$$k(L^{-1}\mathsf{A}_\varrho \mathsf{L}^{-T}, \varepsilon) \leq \overline{k}(\mathsf{L}^{-1}\mathsf{A}\mathsf{L}^{-T}, \varepsilon) + m. \tag{4.27}$$

Such preconditioning can be implemented even without the factorization of the preconditioner $\mathsf{M} = \mathsf{L}\mathsf{L}^T$ as in Algorithm 3.3, provided we can solve efficiently the linear systems with the matrix M.

If m dominates in the expression on the right-hand side of (4.27), then the bound (4.27) can be improved at the cost of increased computational complexity. In particular, this may be useful when we have several problems (4.4) with the same matrix B. Noticing that for any nonsingular matrix Q

$$\Omega_E = \{\mathbf{x} \in \mathbb{R}^n : \ \mathsf{QB}\mathbf{x} = \mathsf{Q}\mathbf{c}\},$$

choosing the matrix Q in such a way that the rows of QBL^{-T} are orthonormal, and denoting $\overline{\mathsf{B}} = \mathsf{QB}$, we can observe that minimizer of the penalized function with the Hessian

$$\overline{\mathsf{A}}_\varrho = \mathsf{A} + \varrho \overline{\mathsf{B}}^T \overline{\mathsf{B}}$$

also approximates the solution of (4.1), but the spectrum $\sigma(\mathsf{L}^{-1}\overline{\mathsf{A}}_\varrho\mathsf{L}^{-T})$ of the preconditioned Hessian $\mathsf{L}^{-1}\overline{\mathsf{A}}_\varrho\mathsf{L}^{-T}$ satisfies by (1.50) and (1.51)

$$\sigma(\mathsf{L}^{-1}\overline{\mathsf{A}}_\varrho\mathsf{L}^{-T}) \subseteq [a_{\min}, a_{\max}] \cup [a_{\min} + \varrho, a_{\max} + \varrho],$$

where $a_{\min} = \lambda_1(\mathsf{L}^{-1}\mathsf{A}\mathsf{L}^{-T})$ and $a_{\max} = \lambda_n(\mathsf{L}^{-1}\mathsf{A}\mathsf{L}^{-T})$. Since the spectrum is located in two intervals of the same length, we can use (3.24) to get the bound

$$k(\mathsf{L}^{-1}\overline{\mathsf{A}}_\varrho\mathsf{L}^{-T}, \varepsilon) \le \min\{\overline{k}(\mathsf{L}^{-1}\mathsf{A}\mathsf{L}^{-T}, \varepsilon) + m, 2\overline{k}(\mathsf{L}^{-1}\mathsf{A}\mathsf{L}^{-T}, \varepsilon)\}, \tag{4.28}$$

which is optimal with respect to both ϱ and m. Results of some numerical experiments with this strategy can be found in [44].

Observe that $\mathsf{Q}^T\mathsf{Q}$ represents a scalar product on $\mathbb{R}^m$. The method can be efficient also in the case that the rows of QBL^{-T} are orthonormal only approximately [144]. If $\mathsf{A} = \mathsf{LL}^T$ and QBL^{-T} are orthonormal, then $\sigma(\mathsf{L}^{-1}\overline{\mathsf{A}}_\varrho\mathsf{L}^{-T}) = \{1, \varrho\}$ and the CG algorithm finds the solution in just two steps.

4.3 Exact Augmented Lagrangian Method

Because of its simplicity and intuitive appeal, the penalty method is often used in computations. However, a good approximation of the solution may require a very large penalty parameter, which can cause serious problems in computer implementation. The remedy can be based on the observation that having a solution $\mathbf{x}_\varrho$ of the penalized problem (4.4), we can modify the linear term of f in such a way that the unconstrained minimum of the modified cost function $\overline{f}$ *without the penalization term* is achieved again at $\mathbf{x}_\varrho$. Then we can hopefully find a better approximation by adding the penalization term to the modified cost function $\overline{f}$, possibly with the same value of the penalty parameter, and look for the minimizer of $\overline{f}_\varrho$ as in Fig. 4.2. The result is the well-known classical *augmented Lagrangian algorithm*, also named the *method of multipliers*, which was proposed by Hestenes [116] and Powell [160].

In this section, we present as the augmented Lagrangian algorithm a little more general algorithm; its special cases are the classical Uzawa algorithm [1] and the original algorithm by Hestenes and Powell. We review and slightly extend the well-known arguments presented, e.g., in the classical monographs by Bertsekas [11] and Glowinski and Le Tallec [100].

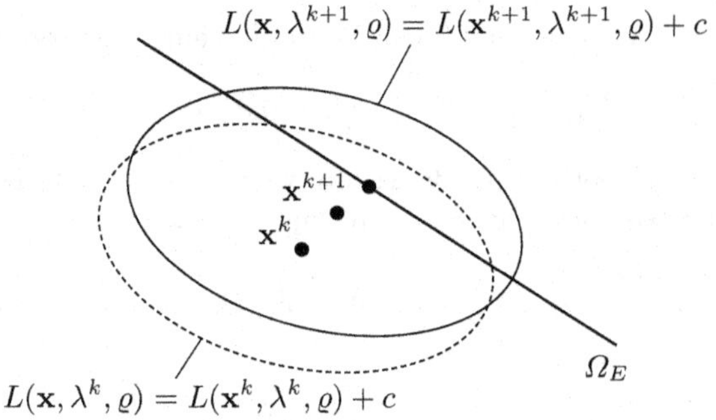

Fig. 4.2. Augmented Lagrangian iteration

4.3.1 Algorithm

The augmented Lagrangian algorithm is based, similarly as the Arrow–Hurwitz algorithm 4.1, on the mixed formulation (2.38) of the equality constrained problem (4.1). However, the augmented Lagrangian algorithm differs from the Arrow–Hurwitz algorithm applied to the penalized problem (4.4) in Step 1, where the former algorithm assigns $\mathbf{x}^k$ the minimizer of $L(\mathbf{x}, \lambda^k, \varrho_k)$ with respect to $\mathbf{x} \in \mathbb{R}^n$. Both algorithms use the same update rule for the Lagrange multipliers in Step 2. Here we present a variant of the augmented Lagrangian algorithm whose special cases are the original *Uzawa algorithm* [1], which corresponds to $\varrho_k = 0$, $k = 0, 1, \ldots$, and the original *method of multipliers*, which corresponds to $r_k = \varrho_k$. Our augmented Lagrangian algorithm reads as follows.

Algorithm 4.2. Exact augmented Lagrangian algorithm.

Given a symmetric positive definite matrix $\mathsf{A} \in \mathbb{R}^{n \times n}$, $\mathsf{B} \in \mathbb{R}^{m \times n}$, $\mathbf{b} \in \mathbb{R}^n$, *and* $\mathbf{c} \in \mathrm{Im}\mathsf{B}$.
Step 0. {*Initialization.*}
Choose $\boldsymbol{\lambda}^0 \in \mathbb{R}^m$, $r > 0$, $r_k \geq r$, $\varrho_k \geq 0$
for k=0,1,2,...
Step 1. {*Minimization with respect to* $\mathbf{x}$.}
$\mathbf{x}^k = \arg\ \min\{L(\mathbf{x}, \boldsymbol{\lambda}^k, \varrho_k) :\ \mathbf{x} \in \mathbb{R}^n\}$
Step 2. {*Updating the Lagrange multipliers.*}
$\boldsymbol{\lambda}^{k+1} = \boldsymbol{\lambda}^k + r_k(\mathsf{B}\mathbf{x}^k - \mathbf{c})$
end for

Since $\mathbf{x}^k$ is the unconstrained minimizer of the Lagrangian L with respect to its first variable, it follows that

$$\nabla_{\mathbf{x}} L(\mathbf{x}^k, \boldsymbol{\lambda}^k, \varrho_k) = (\mathsf{A} + \varrho_k \mathsf{B}^T\mathsf{B})\mathbf{x}^k - \mathbf{b} - \varrho_k \mathsf{B}^T\mathbf{c} + \mathsf{B}^T\boldsymbol{\lambda}^k = \mathbf{o},$$

so that Step 1 of Algorithm 4.2 can be implemented by solving the system

$$(\mathsf{A} + \varrho_k \mathsf{B}^T\mathsf{B})\mathbf{x}^k = \mathbf{b} + \varrho_k \mathsf{B}^T\mathbf{c} - \mathsf{B}^T\boldsymbol{\lambda}^k. \tag{4.29}$$

To understand better the algorithm, we shall examine its alternative formulation which we obtain after eliminating $\mathbf{x}^k$ or $\boldsymbol{\lambda}^k$ from Algorithm 4.2. Thus denoting for any $\varrho \in \mathbb{R}$

$$\mathsf{A}_\varrho = \mathsf{A} + \varrho\mathsf{B}^T\mathsf{B},$$

we can use (4.29) to get

$$\mathbf{x}^k = \mathsf{A}_{\varrho_k}^{-1}(\mathbf{b} + \varrho_k \mathsf{B}^T\mathbf{c} - \mathsf{B}^T\boldsymbol{\lambda}^k).$$

After substituting for $\mathbf{x}^k$ into Step 2 of Algorithm 4.2 and simple manipulations, we can rewrite our augmented Lagrangian algorithm as

$$\text{Choose } \boldsymbol{\lambda}^0 \in \mathbb{R}^m, \tag{4.30}$$
$$\boldsymbol{\lambda}^{k+1} = \boldsymbol{\lambda}^k - r_k\left(\mathsf{B}\mathsf{A}_{\varrho_k}^{-1}\mathsf{B}^T\boldsymbol{\lambda}^k - \mathsf{B}\mathsf{A}_{\varrho_k}^{-1}(\mathbf{b} + \varrho_k\mathsf{B}^T\mathbf{c}) + \mathbf{c}\right). \tag{4.31}$$

To understand the formula (4.31), notice that

$$f_\varrho(\mathbf{x}) = \frac{1}{2}\mathbf{x}^T\mathsf{A}_\varrho\mathbf{x} - (\mathbf{b} + \varrho\mathsf{B}^T\mathbf{c})^T\mathbf{x} + \frac{\varrho}{2}\|\mathbf{c}\|^2.$$

Using the formula (2.36) for the dual function Θ for problem (4.1), we can check that the explicit expression for the dual function Θ_ϱ for the minimum of $f_\varrho(\mathbf{x})$ subject to $\mathbf{x} \in \Omega_E$ reads

$$\begin{aligned}\Theta_\varrho(\boldsymbol{\lambda}) = & -\frac{1}{2}\boldsymbol{\lambda}^T\mathsf{B}\mathsf{A}_\varrho^{-1}\mathsf{B}^T\boldsymbol{\lambda} + \left(\mathsf{B}\mathsf{A}_\varrho^{-1}(\mathbf{b} + \varrho\mathsf{B}^T\mathbf{c}) - \mathbf{c}\right)\boldsymbol{\lambda}\\ & -\frac{1}{2}(\mathbf{b} + \varrho\mathsf{B}^T\mathbf{c})^T\mathsf{A}^{-1}(\mathbf{b} + \varrho\mathsf{B}^T) + \frac{\varrho}{2}\|\mathbf{c}\|^2.\end{aligned}$$

It follows that

$$\nabla\Theta_{\varrho_k}(\boldsymbol{\lambda}^k) = -\mathsf{B}\mathsf{A}_{\varrho_k}^{-1}\mathsf{B}^T\boldsymbol{\lambda}^k + \mathsf{B}\mathsf{A}_{\varrho_k}^{-1}(\mathbf{b} + \varrho_k\mathsf{B}^T\mathbf{c}) - \mathbf{c}. \tag{4.32}$$

Comparing the latter formula with (4.31), we conclude that

$$\boldsymbol{\lambda}^{k+1} = \boldsymbol{\lambda}^k + r_k\nabla\Theta_{\varrho_k}(\boldsymbol{\lambda}^k).$$

Thus the augmented Lagrangian algorithm may be interpreted as the gradient method for maximization of the dual function Θ_ϱ for the penalized problem (4.4) with the steplength r_k.

Alternatively, we can eliminate $\boldsymbol{\lambda}^k$ from Algorithm 4.2 to get

$$\mathbf{x}^0 = \mathsf{A}_{\varrho_0}^{-1}(\mathbf{b} + \varrho_k\mathsf{B}^T\mathbf{c} - \mathsf{B}^T\boldsymbol{\lambda}^0), \tag{4.33}$$
$$\mathbf{x}^{k+1} = \mathbf{x}^k - r_k\mathsf{A}_{\varrho_k}^{-1}\mathsf{B}^T(\mathsf{B}\mathbf{x}^k - \mathbf{c}). \tag{4.34}$$

4.3.2 Convergence of Lagrange Multipliers

Let us first recall that, by Proposition 2.10(iii) and the discussion at the end of Sect. 2.4.2, any Lagrange multiplier $\overline{\boldsymbol{\lambda}}$ of the equality constrained problem (4.1) can be expressed as

$$\overline{\boldsymbol{\lambda}} = \boldsymbol{\lambda}_{\mathrm{LS}} + \boldsymbol{\delta}, \qquad \boldsymbol{\lambda}_{\mathrm{LS}} \in \mathrm{Im}\mathsf{B}, \qquad \boldsymbol{\delta} \in \mathrm{Ker}\mathsf{B}^T,$$

where $\boldsymbol{\lambda}_{\mathrm{LS}}$ is the Lagrange multiplier with the minimal Euclidean norm. If we denote by P and $\mathsf{Q} = \mathsf{I} - \mathsf{P}$ the orthogonal projectors on $\mathrm{Im}\mathsf{B}$ and $\mathrm{Ker}\mathsf{B}^T$, respectively, then the components of $\overline{\boldsymbol{\lambda}}$ are given by

$$\boldsymbol{\lambda}_{\mathrm{LS}} = \mathsf{P}\overline{\boldsymbol{\lambda}}, \qquad \boldsymbol{\nu} = \mathsf{Q}\overline{\boldsymbol{\lambda}}.$$

To simplify the notations, we shall assume that $\varrho_k = \varrho$ and $r_k = r$.

To study the convergence of $\boldsymbol{\lambda}^k$ generated by Algorithm 4.2, let $\boldsymbol{\lambda}^0 \in \mathbb{R}^m$, let us denote

$$\overline{\boldsymbol{\lambda}} = \boldsymbol{\lambda}_{\mathrm{LS}} + \mathsf{Q}\boldsymbol{\lambda}^0,$$

and observe that

$$\begin{aligned}
\boldsymbol{\lambda}^0 - \overline{\boldsymbol{\lambda}} &= \mathsf{P}\boldsymbol{\lambda}^0 + \mathsf{Q}\boldsymbol{\lambda}^0 - \boldsymbol{\lambda}_{\mathrm{LS}} - \mathsf{Q}\boldsymbol{\lambda}^0 = \mathsf{P}(\boldsymbol{\lambda}^0 - \boldsymbol{\lambda}_{\mathrm{LS}}) \in \mathrm{Im}\mathsf{B},\\
\boldsymbol{\lambda}^{k+1} - \overline{\boldsymbol{\lambda}} &= (\boldsymbol{\lambda}^k - \overline{\boldsymbol{\lambda}}) - r(\mathsf{B}\mathsf{A}_\varrho^{-1}\mathsf{B}^T\boldsymbol{\lambda}^k - \mathsf{B}\mathsf{A}_\varrho^{-1}(\mathbf{b} + \varrho\mathsf{B}^T\mathbf{c}) + \mathbf{c})\\
&= (\boldsymbol{\lambda}^k - \overline{\boldsymbol{\lambda}}) - r\mathsf{B}\mathsf{A}_\varrho^{-1}\mathsf{B}^T(\boldsymbol{\lambda}^k - \overline{\boldsymbol{\lambda}}),
\end{aligned}$$

where we used $\mathsf{P}\boldsymbol{\lambda}_{\mathrm{LS}} = \boldsymbol{\lambda}_{\mathrm{LS}}$ and (4.9). It follows that

$$\boldsymbol{\lambda}^{k+1} - \overline{\boldsymbol{\lambda}} = (\mathsf{I} - r\mathsf{B}\mathsf{A}_\varrho^{-1}\mathsf{B}^T)(\boldsymbol{\lambda}^k - \overline{\boldsymbol{\lambda}}) \tag{4.35}$$

and

$$\boldsymbol{\lambda}^{k+1} - \overline{\boldsymbol{\lambda}} \in \mathrm{Im}\mathsf{B}, \quad k = 0, 1, \ldots.$$

Therefore the analysis of convergence of $\boldsymbol{\lambda}^k$ reduces to the analysis of the spectrum of the restriction of the iteration matrix $\mathsf{I} - r\mathsf{B}\mathsf{A}_\varrho^{-1}\mathsf{B}^T$ to its invariant subspace $\mathrm{Im}\mathsf{B}$.

Using (1.26) and Lemma 1.5, we get that the eigenvalues μ_i of the iteration matrix are related to the eigenvalues $\overline{\beta}_i$ of $\mathsf{B}\mathsf{A}^{-1}\mathsf{B}^T|\mathrm{Im}\mathsf{B}$ by

$$\mu_i = 1 - \frac{r\overline{\beta}_i}{1 + \varrho\overline{\beta}_i} = \frac{1 + (\varrho - r)\overline{\beta}_i}{1 + \varrho\overline{\beta}_i},$$

so that

$$\|(\mathsf{I} - r\mathsf{B}\mathsf{A}_\varrho^{-1}\mathsf{B}^T)|\mathrm{Im}\mathsf{B}\| = \max_{\substack{i \in \{1,\ldots,m\}\\ \beta_i > 0}} \frac{|1 + (\varrho - r)\beta_i|}{1 + \varrho\beta_i}.$$

Denoting

$$R(\varrho, r) = \max_{\substack{i \in \{1,\ldots,m\}\\ \beta_i > 0}} \frac{|1 + (\varrho - r)\beta_i|}{1 + \varrho\beta_i} \tag{4.36}$$

and using that the norm is submultiplicative, we get for $R(\varrho, r) < 1$ the linear rate of convergence

$$\|\boldsymbol{\lambda}^{k+1} - \overline{\boldsymbol{\lambda}}\| \leq R(\varrho, r)\|\boldsymbol{\lambda}^k - \overline{\boldsymbol{\lambda}}\|. \tag{4.37}$$

We have thus reduced the study of convergence of Algorithm 4.2 to the analysis of $R(\varrho, r)$. We shall formulate the result on the convergence of the Lagrange multipliers in the following theorem.

Theorem 4.6. *Let $\boldsymbol{\lambda}^k$, $k = 0, 1, \ldots$, denote the sequence of vectors generated by Algorithm 4.2 for problem (4.1) with a given ϱ_k and r_k starting from a given vector $\boldsymbol{\lambda}^0 \in \mathbb{R}^m$. Let $\boldsymbol{\lambda}_{\mathrm{LS}}$ denote the least square Lagrange multiplier, let P denote the orthogonal projector on* $\mathrm{Im}\mathsf{B}$, *let $\beta_{\max}$ denote the largest eigenvalue of $\mathsf{B}\mathsf{A}^{-1}\mathsf{B}^T$, and denote*

$$\overline{\boldsymbol{\lambda}} = \boldsymbol{\lambda}_{\mathrm{LS}} + (\mathsf{I} - \mathsf{P})\boldsymbol{\lambda}^0.$$

If there are $\varepsilon > 0$ and $M > 0$ such that

$$\varepsilon \leq r_k \leq \frac{2}{\beta_{\max}} + 2\varrho_k - \varepsilon \leq M, \tag{4.38}$$

then $\boldsymbol{\lambda}^k$ converge to $\overline{\boldsymbol{\lambda}}$ and the rate of convergence is at least linear, i.e., there is $R < 1$ such that

$$\|\boldsymbol{\lambda}^{k+1} - \overline{\boldsymbol{\lambda}}\| \leq R\|\boldsymbol{\lambda}^k - \overline{\boldsymbol{\lambda}}\|.$$

Proof. Elementary, but a bit laborious analysis of $R(\varrho_k, r_k)$, where R is defined by (4.36), reveals that if ϱ_k, r_k satisfy (4.38), then

$$\sup_{k=0,1,\ldots} R(\varrho_k, r_k) = R < 1.$$

To finish the proof, it is enough to substitute this result into (4.37). □

Using different arguments, it is possible to prove convergence of Algorithm 4.2 under more relaxed conditions. For example, Glowinski and Le Tallec [100] give the condition

$$0 < \varepsilon \leq r_k \leq 2\varrho_k.$$

4.3.3 Effect of the Steplength

Let us now examine possible options of the steplength $r = r(\varrho)$ as a function of ϱ, including their effect on $R(\varrho, r)$. We shall denote by $\overline{\beta}_{\min}$ the smallest nonzero eigenvalues of $\mathsf{B}\mathsf{A}^{-1}\mathsf{B}^T$, i.e., the smallest eigenvalue of $\mathsf{B}\mathsf{A}^{-1}\mathsf{B}^T|\mathrm{Im}\mathsf{B}$. Our examples are from Glowinski and Le Tallec [100].

Optimal choice of r with $\varrho = 0$.

In this case, which corresponds to the original Uzawa algorithm,

$$R(\varrho, r) = R(0, r) = \max_{\substack{i\in\{1,\dots,m\}\\ \beta_i>0}} |1 - r\beta_i|, \tag{4.39}$$

so that the best choice of r is given by

$$\begin{aligned} R(0, r_{\text{opt}}) &= \min_r \max_{\substack{i\in\{1,\dots,m\}\\ \beta_i>0}} |1 - r\beta_i| = \min_r \max_{\substack{i\in\{1,\dots,m\}\\ \beta_i>0}} \{1 - r\beta_i, r\beta_i - 1\} \\ &= \min_r \max\{1 - r\overline{\beta}_{\min}, r\beta_{\max} - 1\}. \end{aligned}$$

A simple analysis reveals that r_{opt} satisfies

$$1 - r\overline{\beta}_{\min} = r\beta_{\max} - 1.$$

Solving the last equation with respect to r, we get

$$r_{\text{opt}} = \frac{2}{\overline{\beta}_{\min} + \beta_{\max}}$$

and

$$R(0, r_{\text{opt}}) = 1 - r_{\text{opt}}\overline{\beta}_{\min} = \frac{\beta_{\max} - \overline{\beta}_{\min}}{\beta_{\max} + \overline{\beta}_{\min}} = \frac{\beta_{\max}/\overline{\beta}_{\min} - 1}{\beta_{\max}/\overline{\beta}_{\min} + 1}.$$

This is of course the optimal rate of convergence of the gradient method of Sect. 3.4 applied to the dual function. Inspection of (4.39) reveals that $R(0, r) < 1$ requires that $1 - r\beta_{\max} > -1$, i.e.,

$$r < 2/\beta_{\max},$$

so that r_{opt} is typically near the bound which guarantees the convergence.

Choice $r = \varrho$.

In this case, which is natural from the point of view of our analysis of the penalty method,

$$R(\varrho, \varrho) = \max_{\substack{i\in\{1,\dots,m\}\\ \beta_i>0}} \frac{1}{1 + \varrho\beta_i} = \frac{1}{1 + \varrho\overline{\beta}_{\min}}. \tag{4.40}$$

An interesting feature of this choice is that

$$\lim_{\varrho\to\infty} R(\varrho, \varrho) = 0,$$

so that by increasing ϱ, it is possible to achieve arbitrary preconditioning effect.

Choice $r = (1+\delta)\varrho$.

Let us now consider the choice $r = (1+\delta)\varrho$ with $\delta > -1$. In this case

$$R\big(\varrho, (1+\delta)\varrho\big) = \max_{\substack{i\in\{1,\dots,m\}\\ \beta_i>0}} \frac{|1-\delta\varrho\beta_i|}{1+\varrho\beta_i},$$

so that $r>0$ and $R\big(\varrho,(1+\delta)\varrho\big) < 1$ if and only if $-1 < \delta < 1 + 2/(\varrho\beta_{\max})$. Moreover

$$\lim_{\varrho\to\infty} R(\varrho, (1+\delta)\varrho) = |\delta|.$$

It follows that the preconditioning effect which can be achieved by increasing the penalty parameter is limited when $\delta \neq 0$.

Optimal steplength for a given ϱ.

If ϱ is given, then the optimal steplength $r_{\mathrm{opt}}(\varrho)$ is given by

$$r_{\mathrm{opt}} = \arg\min_{r\geq 0}\left(\max_{\substack{i\in\{1,\dots,m\}\\ \beta_i>0}} \frac{|1+(\varrho-r)\beta_i|}{1+\varrho\beta_i}\right).$$

To find it, let us denote

$$\varphi_i(r) = \frac{1+(\varrho-r)\beta_i}{1+\varrho\beta_i} = \frac{(1+\varrho\beta_i)-r\beta_i}{1+\varrho\beta_i},$$

and observe that if $\beta_i>0$, then $\varphi_i(r)$ is decreasing. Since $\varphi_i(0)=1$, it follows that for $r\geq 0$

$$\max_{\substack{i\in\{1,\dots,m\}\\ \beta_i>0}} \varphi_i(r) = \max_{\substack{i\in\{1,\dots,m\}\\ \beta_i>0}} \frac{1+(\varrho-r)\beta_i}{1+\varrho\beta_i} = \frac{1+(\varrho-r)\overline{\beta}_{\min}}{1+\varrho\overline{\beta}_{\min}}.$$

Similarly $-\varphi_i(0) = -1$, and if $\beta_i>0$, then $-\varphi_i(r)$ is increasing. Therefore

$$\max_{\substack{i\in\{1,\dots,m\}\\ \beta_i>0}} -\varphi_i(r) = \max_{\substack{i\in\{1,\dots,m\}\\ \beta_i>0}} -\frac{1+(\varrho-r)\beta_i}{1+\varrho\beta_i} = -\frac{1+(\varrho-r)\overline{\beta}_{\max}}{1+\varrho\overline{\beta}_{\max}}$$

for nonnegative r. Since the both maxima are nonnegative on the positive interval $[\varrho+1/\beta_{\max}, \varrho+1/\beta_{\min}]$, it follows that the optimal choice $r_{\mathrm{opt}}(\varrho)$ is a nonnegative solution of

$$\frac{1+(\varrho-r)\overline{\beta}_{\min}}{1+\varrho\overline{\beta}_{\min}} = -\frac{1+(\varrho-r)\beta_{\max}}{1+\varrho\beta_{\max}}.$$

Carrying out the computations, we get that

$$r_{\mathrm{opt}}(\varrho) = \varrho + \frac{2+\varrho(\overline{\beta}_{\min}+\beta_{\max})}{2\varrho\overline{\beta}_{\min}\beta_{\max} + \overline{\beta}_{\min} + \beta_{\max}}.$$

We conclude that the optimal steplength $r_{\mathrm{opt}}(\varrho)$ based on the estimate (4.37) is *longer* than the penalization parameter ϱ, but $r_{\mathrm{opt}}(\varrho)$ approaches ϱ for large values of ϱ as

$$\lim_{\varrho\to\infty} r_{\mathrm{opt}}(\varrho)/\varrho = 1.$$

This is in agreement with the above discussion and our analysis of the penalty method in Sect. 4.2.1, which suggests that a suitable steplength for large ϱ is given by $r = \varrho$.

Given $\mathbf{x}^k$ which minimizes $L(\mathbf{x}, \boldsymbol{\lambda}^k, \varrho)$ with respect to $\mathbf{x} \in \mathbb{R}^n$, then $\mathbf{x}^k$ satisfies

$$\mathsf{A}\mathbf{x}^k + \mathsf{B}^T\left(\boldsymbol{\lambda}^k + \varrho(\mathsf{B}\mathbf{x}^k - \mathbf{c})\right) - \mathbf{b} = \mathbf{o}.$$

Thus the choice $r = \varrho$ results in

$$\nabla L_0(\mathbf{x}^k, \boldsymbol{\lambda}^{k+1}) = \mathbf{o},$$

so that it is optimal also in the sense that

$$\varrho = \arg\min_{r\geq 0} \|\mathsf{A}\mathbf{x}^k + \mathsf{B}^T\left(\boldsymbol{\lambda}^k + r(\mathsf{B}\mathbf{x}^k - \mathbf{c})\right) - \mathbf{b}\|.$$

Maximizing the Gradient Ascent.

In Sect. 4.3.1, we have shown that Algorithm 4.2 may be interpreted as a gradient algorithm applied to the maximization of the dual function Θ_ϱ. Thus it seems natural to define r_k by maximizing the quadratic function

$$\phi(r) = \Theta_{\varrho_k}(\boldsymbol{\lambda}^k + r\mathbf{g}^k),$$

where $\mathbf{g}^k = \nabla\Theta_{\varrho_k}(\boldsymbol{\lambda}^k)$, with respect to r. Denoting $\mathsf{A}_{\varrho_k} = \mathsf{A} + \varrho_k\mathsf{B}^T\mathsf{B}$,

$$\mathsf{F}_{\varrho_k} = \mathsf{B}\mathsf{A}_{\varrho_k}^{-1}\mathsf{B}^T, \qquad \text{and} \qquad \mathbf{d} = \mathbf{d}_{\varrho_k} = \mathsf{B}\mathsf{A}_{\varrho_k}^{-1}(\mathbf{b} + \varrho_k\mathsf{B}^T\mathbf{c}) - \mathbf{c},$$

we can write

$$\phi(r) = -\frac{1}{2}(\boldsymbol{\lambda}^k + r\mathbf{g}^k)^T\mathsf{F}_{\varrho_k}(\boldsymbol{\lambda}^k + r\mathbf{g}^k) + (\boldsymbol{\lambda}^k + r\mathbf{g}^k)^T\mathbf{d},$$

so that the maximizer satisfies

$$\frac{d}{dr}\phi(r) = -r(\mathbf{g}^k)^T\mathsf{F}_{\varrho_k}\mathbf{g}^k - (\mathbf{g}^k)^T(\mathsf{F}_{\varrho_k}\boldsymbol{\lambda}^k - \mathbf{d}) = -r(\mathbf{g}^k)^T\mathsf{F}_{\varrho_k}\mathbf{g}^k + (\mathbf{g}^k)^T\mathbf{g}^k = 0.$$

Thus we can use the steepest ascent formula

$$r_k = \frac{\|\mathbf{g}^k\|^2}{(\mathbf{g}^k)^T\mathsf{F}_{\varrho_k}\mathbf{g}^k}, \tag{4.41}$$

which may be applied to obtain the largest increase of Θ_{ϱ_k} in step k. For large ϱ_k, we get by (1.47) and (1.48) that r_k is close to ϱ_k in agreement with the optimal choice of the steplength based on the estimate (4.37). Notice that the steplength based on (4.41) depends on the current iteration.

4.3.4 Convergence of the Feasibility Error

To estimate the feasibility error $\|\mathsf{B}\mathbf{x}^k - \mathbf{c}\|$, let us multiply equation (4.34) by B and then subtract $\mathbf{c}$ from both sides of the result to get

$$\mathsf{B}\mathbf{x}^{k+1} - \mathbf{c} = \mathsf{B}\mathbf{x}^k - \mathbf{c} - r_k \mathsf{B}\mathsf{A}_{\varrho_k}^{-1}\mathsf{B}^T(\mathsf{B}\mathbf{x}^k - \mathbf{c}),$$

where $\mathsf{A}_{\varrho_k} = \mathsf{A} + \varrho_k \mathsf{B}^T\mathsf{B}$. It follows that

$$\|\mathsf{B}\mathbf{x}^{k+1} - \mathbf{c}\| \le \|(\mathsf{I} - r_k \mathsf{B}\mathsf{A}_{\varrho_k}^{-1}\mathsf{B}^T)|\mathrm{Im}\mathsf{B}\|\|\mathsf{B}\mathbf{x}^k - \mathbf{c}\|, \tag{4.42}$$

so that, under the assumptions of Theorem 4.6, we can use the same arguments to prove the linear convergence of the feasibility error. We can thus state the following theorem.

Theorem 4.7. *Let $\mathbf{x}^k$, $k = 0, 1, \dots$, be generated by Algorithm 4.2 for problem (4.1) with given ϱ_k, r_k, and $\boldsymbol{\lambda}^0 \in \mathbb{R}^m$. Let $\beta_{\max}$ denote the largest eigenvalue of $\mathsf{B}\mathsf{A}^{-1}\mathsf{B}^T$.*

If there are $\varepsilon > 0$ and $M > 0$ such that

$$\varepsilon \le r_k \le \frac{2}{\beta_{\max}} + 2\varrho_k - \varepsilon \le M, \tag{4.43}$$

then the feasibility error $\|\mathsf{B}\mathbf{x}^k - \mathbf{c}\|$ converges to zero and the rate of convergence is at least linear, i.e., there is $R < 1$ such that

$$\|\mathsf{B}\mathbf{x}^{k+1} - \mathbf{c}\| \le R\|\mathsf{B}\mathbf{x}^k - \mathbf{c}\|. \tag{4.44}$$

We have thus obtained exactly the same rate of convergence of the feasibility error as that for the Lagrange multipliers.

4.3.5 Convergence of Primal Variables

Having the proofs of convergence of the Lagrange multipliers and of the feasibility error, we may use Proposition 2.12 to prove the convergence of the primal variables.

Theorem 4.8. *Let $\mathbf{x}^k$, $k = 0, 1, \dots$, be generated by Algorithm 4.2 for problem (4.1) with given ϱ_k, r_k, and $\boldsymbol{\lambda}^0 \in \mathbb{R}^m$, let $\widehat{\mathbf{x}}$ denote the unique solution of (4.1), let $\overline{\sigma}_{\min}$ denote the least nonzero singular value of B, and let $\beta_{\max}$ denote the largest eigenvalue of $\mathsf{B}\mathsf{A}^{-1}\mathsf{B}^T$.*

If there are $\varepsilon > 0$ and $M > 0$ such that

$$\varepsilon \le r_k \le \frac{2}{\beta_{\max}} + 2\varrho_k - \varepsilon \le M, \tag{4.45}$$

then $\|\mathbf{x}^k - \widehat{\mathbf{x}}\|$ converges to zero and the rate of convergence is at least R-linear, i.e., there is $R < 1$ such that

$$\|\mathbf{x}^k - \widehat{\mathbf{x}}\| \le R^k \frac{\kappa(\mathsf{A})\|B\|}{\overline{\sigma}_{\min}}\|\mathbf{x}^0 - \widehat{\mathbf{x}}\|. \tag{4.46}$$

Proof. First recall that by the assumptions and (4.44),

$$\|\mathsf{B}\mathbf{x}^k - \mathbf{c}\| \le R^k \|\mathsf{B}\mathbf{x}^0 - \mathbf{c}\| = R^k \|\mathsf{B}(\mathbf{x}^0 - \widehat{\mathbf{x}})\| \le R^k \|\mathsf{B}\|\|\mathbf{x}^0 - \widehat{\mathbf{x}}\|,$$

where $R < 1$. Using (2.48), we get that

$$\|\mathbf{x}^k - \widehat{\mathbf{x}}\| \le \frac{\kappa(\mathsf{A})}{\overline{\sigma}_{\min}} \|\mathsf{B}\mathbf{x}^k - \mathbf{c}\| \le R^k \frac{\kappa(\mathsf{A})\|\mathsf{B}\|}{\overline{\sigma}_{\min}} \|\mathbf{x}^0 - \widehat{\mathbf{x}}\|.$$

We have used the fact that $\nabla_{\mathbf{x}} L(\mathbf{x}^k, \boldsymbol{\lambda}^k, \varrho) = \mathbf{o}$. □

4.3.6 Implementation

Since it is possible to approximate the solution of (4.1) with a single step of the penalty method, the above discussion suggests to take $r_k = \varrho_k = \varrho$ as large as possible.

The auxiliary problems in Step 1 can be effectively solved by the Cholesky factorization

$$\mathsf{L}\mathsf{L}^T = \mathsf{A}_\varrho,$$

which should be carried out after each update of ϱ, and the multiplication of a vector $\boldsymbol{\lambda}$ by $\mathsf{B}\mathsf{A}_\varrho^{-1}\mathsf{B}^T$ should be carried out as

$$\mathsf{B}\mathsf{A}_\varrho^{-1}\mathsf{B}^T\boldsymbol{\lambda} = \mathsf{B}\big(\mathsf{L}^{-T}\left(\mathsf{L}^{-1}(\mathsf{B}^T\boldsymbol{\lambda})\right)\big).$$

The multiplication by the inverse factors should be implemented as the solution of the related triangular systems. Since the sensitivity to round-off errors is greater when ϱ is large, the algorithm should be implemented in double precision.

Application of iterative solvers can hardly be efficient for implementation of Step 1 of Algorithm 4.2, where an exact solution is required, but it can be very efficient for the implementation of inexact augmented Lagrangian algorithms discussed in the rest of this chapter.

On Application of the Conjugate Gradient Method

Since the augmented Lagrangian algorithm maximizes the dual function, we can alternatively forget it and apply the CG algorithm of Sect. 3.3 to maximize the dual function Θ. This strategy may be very efficient as indicated by the success of the FETI methods introduced by Farhat and Roux [86, 87]. The large penalty parameters result in efficient preconditioning of the Hessian of Θ (1.48), so that, due to the optimal properties of the conjugate gradient method, the latter is a natural choice provided we can solve exactly the auxiliary linear problems. The picture changes when inexact solutions of the auxiliary problems are considered, as a perturbed conjugate gradient need not be even a decrease direction as indicated in Fig. 3.2. Thus it is mainly the *capability to accept the inexact solutions and treat separately the constraints and minimization* that makes the augmented Lagrangian algorithm an attractive alternative for the solution of equality constrained QP problems.

4.4 Asymptotically Exact Augmented Lagrangian Method

The augmented Lagrangian method considered in the previous section assumed that the minimization in Step 1 is carried out exactly. Since such iterations are expensive, there is a good chance to reduce the cost of the outer iterates without a large increase of the number of iterations due to the approximate minimization, especially when we recall that the gradient is a robust ascent direction.

In this section we carry out the analysis of convergence of the augmented Lagrangian algorithm when the precisions of the solutions of the auxiliary problems in Step 1 are determined by the bounds on the norm of the gradient. We assume that the bounds are prescribed by the *forcing sequence* $\{\varepsilon_k\}$, where $\varepsilon^k > 0$ and $\lim_{k\to\infty} \varepsilon_k = 0$. The latter condition implies that the stopping criterion becomes more stringent with the increasing index of the outer iterations so that the minimization is asymptotically exact. Taking into account the discussion of Sect. 4.3.3, we consider the steplength $r_k = \varrho_k$.

4.4.1 Algorithm

The augmented Lagrangian algorithm with asymptotically exact solution of auxiliary unconstrained problems differs from the exact algorithm only in Step 1. We restrict our attention to the inexact version of the original augmented Lagrangian method which reads as follows.

Algorithm 4.3. Asymptotically Exact Augmented Lagrangians.

Given a symmetric positive definite matrix $\mathsf{A} \in \mathbb{R}^{n\times n}$, $\mathsf{B} \in \mathbb{R}^{m\times n}$, $\mathbf{b} \in \mathbb{R}^n$, *and* $\mathbf{c} \in \mathrm{Im}\mathsf{B}$.

Step 0. {*Initialization.*}
Choose $\varepsilon_i > 0$ *so that* $\lim_{i\to\infty} \varepsilon_i = 0$, $\boldsymbol{\lambda}^0 \in \mathbb{R}^m$, $\varrho_i \ge \varrho > 0$
for *k=0,1,2,...*

Step 1. {*Minimization with respect to* $\mathbf{x}$.}
Choose $\mathbf{x}^k \in \mathbb{R}^n$ *so that* $\|\nabla_{\mathbf{x}} L(\mathbf{x}^k, \boldsymbol{\lambda}^k, \varrho_k)\| \le \varepsilon_k$

Step 2. {*Updating the Lagrange multipliers.*}
$\boldsymbol{\lambda}^{k+1} = \boldsymbol{\lambda}^k + \varrho_k(\mathsf{B}\mathbf{x}^k - \mathbf{c})$
end for

We assume that the inexact solution of the auxiliary problems in Step 1 of Algorithm 4.3 is implemented by a suitable iterative method such as the conjugate gradient method introduced in Chap. 3. Thus the algorithm solves approximately the auxiliary unconstrained problems in the inner loop while it generates the approximations of the Lagrange multipliers in the outer loop. Let us recall that effective application of the conjugate gradient method assumes that the iterations are carried out with the matrix which has a favorable distribution of the spectrum. This can be achieved by a problem-dependent preconditioning discussed in Sect. 3.6 in combination with the gap-preserving strategy of Sect. 4.7.

4.4.2 Auxiliary Estimates

Our analysis of the augmented Lagrangian algorithm is based on the following lemma.

Lemma 4.9. *Let* $\mathsf{A}, \mathsf{B}, \mathbf{b}$, *and* $\mathbf{c}$ *be those of the definition of problem (4.1) with* $\mathsf{B} \neq \mathsf{O}$ *not necessarily a full rank matrix. For any vectors* $\mathbf{x} \in \mathbb{R}^n$ *and* $\boldsymbol{\lambda} \in \mathbb{R}^m$, *let us denote*

$$\widetilde{\boldsymbol{\lambda}} = \boldsymbol{\lambda} + \varrho(\mathsf{B}\mathbf{x} - \mathbf{c}),$$
$$\mathbf{g} = \nabla_{\mathbf{x}} L(\mathbf{x}, \boldsymbol{\lambda}, \varrho) = \mathsf{A}_\varrho \mathbf{x} - \mathbf{b} + \mathsf{B}^T \boldsymbol{\lambda} - \varrho \mathsf{B}^T \mathbf{c}.$$

Let $\boldsymbol{\lambda}_{\mathrm{LS}}$ *denote the vector of the least square Lagrange multipliers for problem (4.1), and let* $\overline{\beta}_{\min}$ *denote the least nonzero eigenvalue of* $\mathsf{B}\mathsf{A}^{-1}\mathsf{B}$.

Then for any $\boldsymbol{\lambda} \in \mathrm{Im}\mathsf{B}$

$$\|\widetilde{\boldsymbol{\lambda}} - \boldsymbol{\lambda}_{\mathrm{LS}}\| \leq \frac{\|\mathsf{B}\mathsf{A}^{-1}\|}{\overline{\beta}_{\min} + \varrho^{-1}} \|\mathbf{g}\| + \frac{\varrho^{-1}}{\overline{\beta}_{\min} + \varrho^{-1}} \|\boldsymbol{\lambda} - \boldsymbol{\lambda}_{\mathrm{LS}}\|. \tag{4.47}$$

Proof. The definitions of $\widetilde{\boldsymbol{\lambda}}$ and $\mathbf{g}$ imply that

$$\begin{aligned} \mathsf{A}\mathbf{x} + \mathsf{B}^T \widetilde{\boldsymbol{\lambda}} &= \mathbf{b} + \mathbf{g}, \\ \mathsf{B}\mathbf{x} - \varrho^{-1} \widetilde{\boldsymbol{\lambda}} &= -\varrho^{-1}(\boldsymbol{\lambda} - \boldsymbol{\lambda}_{\mathrm{LS}}) - \varrho^{-1}\boldsymbol{\lambda}_{\mathrm{LS}} + \mathbf{c}, \end{aligned} \tag{4.48}$$

and the solution $\widehat{\mathbf{x}}$ and $\boldsymbol{\lambda}_{\mathrm{LS}}$ satisfy by the assumptions

$$\begin{aligned} \mathsf{A}\widehat{\mathbf{x}} + \mathsf{B}^T \boldsymbol{\lambda}_{\mathrm{LS}} &= \mathbf{b}, \\ \mathsf{B}\widehat{\mathbf{x}} - \varrho^{-1}\boldsymbol{\lambda}_{\mathrm{LS}} &= -\varrho^{-1}\boldsymbol{\lambda}_{\mathrm{LS}} + \mathbf{c}. \end{aligned} \tag{4.49}$$

Subtracting (4.49) from (4.48) and switching to the matrix notation, we get

$$\begin{bmatrix} \mathsf{A} & \mathsf{B}^T \\ \mathsf{B} & -\varrho^{-1}\mathsf{I} \end{bmatrix} \begin{bmatrix} \mathbf{x} - \widehat{\mathbf{x}} \\ \widetilde{\boldsymbol{\lambda}} - \boldsymbol{\lambda}_{\mathrm{LS}} \end{bmatrix} = \begin{bmatrix} \mathbf{g} \\ -\varrho^{-1}(\boldsymbol{\lambda} - \boldsymbol{\lambda}_{\mathrm{LS}}) \end{bmatrix}. \tag{4.50}$$

After multiplying the first block row of (4.50) by $-\mathsf{B}\mathsf{A}^{-1}$, adding the result to the second block row, and simple manipulations, we get

$$\widetilde{\boldsymbol{\lambda}} - \boldsymbol{\lambda}_{\mathrm{LS}} = \mathsf{S}_\varrho^{-1}\mathsf{B}\mathsf{A}^{-1}\mathbf{g} + \varrho^{-1}\mathsf{S}_\varrho^{-1}(\boldsymbol{\lambda} - \boldsymbol{\lambda}_{\mathrm{LS}}), \tag{4.51}$$

where $\mathsf{S}_\varrho = \mathsf{B}\mathsf{A}^{-1}\mathsf{B}^T + \varrho^{-1}\mathsf{I}$.

Noticing that $\boldsymbol{\lambda}_{\mathrm{LS}} - \boldsymbol{\lambda} \in \mathrm{Im}\mathsf{B}$ and taking norms, we get

$$\|\widetilde{\boldsymbol{\lambda}} - \boldsymbol{\lambda}\| \leq \|\mathsf{S}_\varrho^{-1}|\mathrm{Im}\mathsf{B}\| \left(\|\mathsf{B}\mathsf{A}^{-1}\|\|\mathbf{g}\| + \varrho^{-1}\|\boldsymbol{\lambda}_{\mathrm{LS}} - \boldsymbol{\lambda}\|\right). \tag{4.52}$$

To estimate the first factor on the right-hand side, notice that by (1.43) our task reduces to finding an upper bound for

$$\|(\varrho^{-1}\mathsf{I} + \mathsf{B}\mathsf{A}^{-1}\mathsf{B}^T)^{-1}|\mathrm{Im}\mathsf{B}\mathsf{A}^{-1}\mathsf{B}^{\mathrm{T}}\|.$$

Since $\mathrm{Im}\mathsf{B}^{\mathrm{T}}\mathsf{A}^{-1}\mathsf{B}$ is an invariant subspace of any matrix function of $\mathsf{B}^T\mathsf{A}^{-1}\mathsf{B}$, and the eigenvectors of $\mathsf{B}\mathsf{A}^{-1}\mathsf{B}^T|\mathrm{Im}\mathsf{B}\mathsf{A}^{-1}\mathsf{B}^{\mathrm{T}}$ are just the eigenvectors of $\mathsf{B}\mathsf{A}^{-1}\mathsf{B}^T$ which correspond to the nonzero eigenvalues, it follows by (1.24) and (1.26) that

$$\|\mathsf{S}_\varrho^{-1}|\mathrm{Im}\mathsf{B}\| = \|(\varrho^{-1}\mathsf{I} + \mathsf{B}\mathsf{A}^{-1}\mathsf{B}^{\mathrm{T}})^{-1}|\mathrm{Im}\mathsf{B}\| = 1/(\varrho^{-1} + \overline{\beta}_{\min}).$$

After substituting into (4.52), we get (4.47). □

To simplify applications of Lemma 4.9 for $\boldsymbol{\lambda}^0 \notin \mathrm{Im}\mathsf{B}$, let us formulate the following easy lemma.

Lemma 4.10. *Let* $\boldsymbol{\lambda}^0 \in \mathbb{R}^m$, *let*

$$\boldsymbol{\lambda}^{k+1} = \boldsymbol{\lambda}^k + \mathbf{u}^k, \quad \mathbf{u}^k \in \mathrm{Im}\mathsf{B}, \quad k = 0, 1, \ldots,$$

let $\boldsymbol{\lambda}_{\mathrm{LS}}$ *denote the vector of the least square Lagrange multipliers for problem (4.1), so that* $\boldsymbol{\lambda}_{\mathrm{LS}} \in \mathrm{Im}\mathsf{B}$, *let* P *denote the orthogonal projector onto* $\mathrm{Im}\mathsf{B}$, *and let*

$$\overline{\boldsymbol{\lambda}} = \boldsymbol{\lambda}_{\mathrm{LS}} + (\mathsf{I} - \mathsf{P})\boldsymbol{\lambda}^0. \tag{4.53}$$

Then

$$\boldsymbol{\lambda}^k - \overline{\boldsymbol{\lambda}} = \mathsf{P}\boldsymbol{\lambda}^k - \boldsymbol{\lambda}_{\mathrm{LS}}, \quad k = 0, 1, \ldots. \tag{4.54}$$

Proof. Since for $k = 0, 1, \ldots$

$$(\mathsf{I} - \mathsf{P})\boldsymbol{\lambda}^{k+1} = (\mathsf{I} - \mathsf{P})(\boldsymbol{\lambda}^k + \mathbf{u}^k) = (\mathsf{I} - \mathsf{P})\boldsymbol{\lambda}^k = \cdots = (\mathsf{I} - \mathsf{P})\boldsymbol{\lambda}^0,$$

we have

$$\boldsymbol{\lambda}^k - \overline{\boldsymbol{\lambda}} = \mathsf{P}\boldsymbol{\lambda}^k + (\mathsf{I} - \mathsf{P})\boldsymbol{\lambda}^k - \boldsymbol{\lambda}_{\mathrm{LS}} - (\mathsf{I} - \mathsf{P})\boldsymbol{\lambda}^0 = \mathsf{P}\boldsymbol{\lambda}^k - \boldsymbol{\lambda}_{\mathrm{LS}}.$$

□

4.4.3 Convergence Analysis

Now we are ready to use Lemma 4.9 in the proof of convergence of Algorithm 4.3.

Theorem 4.11. *Let $\mathbf{x}^k$, $\boldsymbol{\lambda}^k$, $k = 0, 1, \ldots$, be generated by Algorithm 4.3 for the solution of (4.1) with given $\boldsymbol{\lambda}^0 \in \mathbb{R}^m$, $\varrho > 0$, $\varrho_k \geq \varrho$, and $\varepsilon_k > 0$ such that $\lim_{k\to\infty} \varepsilon_k = 0$. Let $(\widehat{\mathbf{x}}, \boldsymbol{\lambda}_{\mathrm{LS}})$ denote the least square KKT pair for (4.1). Let P denote the orthogonal projector on $\mathrm{Im}\mathsf{B}$, let $\overline{\beta}_{\min}$ denote the least nonzero eigenvalue of $\mathsf{BA}^{-1}\mathsf{B}^T$, let $\lambda_{\min}$ denote the least eigenvalue of A, and denote*

$$\overline{\boldsymbol{\lambda}} = \mathsf{P}\boldsymbol{\lambda}_{\mathrm{LS}} + (\mathsf{I}-\mathsf{P})\boldsymbol{\lambda}^0 = \boldsymbol{\lambda}_{\mathrm{LS}} + (\mathsf{I}-\mathsf{P})\boldsymbol{\lambda}^0.$$

Then

$$\lim_{i\to\infty} \mathbf{x}^k = \widehat{\mathbf{x}}, \qquad \lim_{i\to\infty} \boldsymbol{\lambda}^k = \overline{\boldsymbol{\lambda}},$$

and for any positive integers $k, s,\ k+s = i$,

$$\|\boldsymbol{\lambda}^{k+s} - \overline{\boldsymbol{\lambda}}\| \leq C\overline{\varepsilon}_k \frac{1}{1-\nu} + \nu^s C\overline{\varepsilon}_0 \frac{1}{1-\nu} + \nu^{k+s}\|\boldsymbol{\lambda}^0 - \overline{\boldsymbol{\lambda}}\|, \tag{4.55}$$

$$\|\mathbf{x}^i - \widehat{\mathbf{x}}\| \leq \lambda_{\min}^{-1}\|\mathsf{B}\| \left(\|\boldsymbol{\lambda}^i - \overline{\boldsymbol{\lambda}}\| + \varepsilon_i\right), \tag{4.56}$$

where $\overline{\varepsilon}_k = \max\{\varepsilon_k, \varepsilon_{k+1}, \ldots\}$,

$$C = \frac{\|\mathsf{BA}^{-1}\|}{\overline{\beta}_{\min} + \varrho^{-1}}, \quad \textit{and} \quad \nu = \frac{\varrho^{-1}}{\overline{\beta}_{\min} + \varrho^{-1}} < 1. \tag{4.57}$$

Proof. First notice that by the assumptions

$$(\mathsf{A} + \varrho_k \mathsf{B}^T\mathsf{B})\mathbf{x}^k = \mathbf{b} + \varrho_k \mathsf{B}^T\mathbf{c} - \mathsf{B}^T\boldsymbol{\lambda}^k + \mathbf{g}^k, \quad \|\mathbf{g}^k\| \leq \varepsilon_k, \tag{4.58}$$

where $\mathbf{g}^k = \nabla_{\mathbf{x}} L(\mathbf{x}^k, \boldsymbol{\lambda}^k, \varrho_k)$, and observe that the update rule in Step 2 of Algorithm 4.3 and Lemma 4.10 with $\mathbf{u}^k = \varrho_k(\mathsf{B}\mathbf{x}^k - \mathbf{c})$ imply that

$$\boldsymbol{\lambda}^k - \overline{\boldsymbol{\lambda}} = \mathsf{P}\boldsymbol{\lambda}^k - \boldsymbol{\lambda}_{\mathrm{LS}}, \ k = 0, 1, \ldots.$$

Since

$$\mathsf{P}\boldsymbol{\lambda}^{k+1} = \mathsf{P}\boldsymbol{\lambda}^k + \varrho_k(\mathsf{B}\mathbf{x}^k - \mathbf{c})$$

and $\mathsf{P}\boldsymbol{\lambda}^k \in \mathrm{Im}\mathsf{B}$, we can apply Lemma 4.9 with

$$\boldsymbol{\lambda} = \mathsf{P}\boldsymbol{\lambda}^k \quad \text{and} \quad \widetilde{\boldsymbol{\lambda}} = \mathsf{P}\boldsymbol{\lambda}^k + \varrho_k(\mathsf{B}\mathbf{x}^k - \mathbf{c}) = \mathsf{P}\boldsymbol{\lambda}^{k+1}$$

and use the assumptions to get

$$\begin{aligned}\|\boldsymbol{\lambda}^{k+1} - \overline{\boldsymbol{\lambda}}\| = \|\mathsf{P}\boldsymbol{\lambda}^{k+1} - \boldsymbol{\lambda}_{\mathrm{LS}}\| &\leq \frac{\|\mathsf{BA}^{-1}\|}{\overline{\beta}_{\min} + \varrho_k^{-1}}\|\mathbf{g}^k\| + \frac{\varrho_k^{-1}}{\overline{\beta}_{\min} + \varrho_k^{-1}}\|\mathsf{P}\boldsymbol{\lambda}^k - \boldsymbol{\lambda}_{\mathrm{LS}}\| \\ &\leq C\varepsilon_k + \nu\|\boldsymbol{\lambda}^k - \overline{\boldsymbol{\lambda}}\|,\end{aligned}$$

where C and ν are defined by (4.57).

It follows that for any positive integer s and $k = 0, 1, \ldots$, we have

$$\begin{aligned}\|\boldsymbol{\lambda}^{k+s}-\overline{\boldsymbol{\lambda}}\| &\le C\varepsilon_{k+s-1}+\nu\|\boldsymbol{\lambda}^{k+s-1}-\overline{\boldsymbol{\lambda}}\|\\ &\le C(\varepsilon_{k+s-1}+\nu\varepsilon_{k+s-2}+\cdots+\nu^{s-1}\varepsilon_k)+\nu^s\|\boldsymbol{\lambda}^k-\overline{\boldsymbol{\lambda}}\|\\ &\le C\overline{\varepsilon}_k(1+\nu+\nu^2+\cdots+\nu^{s-1})+\nu^s\|\boldsymbol{\lambda}^k-\overline{\boldsymbol{\lambda}}\|\\ &\le C\overline{\varepsilon}_k\frac{1}{1-\nu}+\nu^s\|\boldsymbol{\lambda}^k-\overline{\boldsymbol{\lambda}}\|.\end{aligned}$$

To prove (4.55), it is enough to use the above inequalities to bound the last term by

$$\|\boldsymbol{\lambda}^k-\overline{\boldsymbol{\lambda}}\|=\|\boldsymbol{\lambda}^{0+k}-\overline{\boldsymbol{\lambda}}\|\le C\overline{\varepsilon}_0\frac{1}{1-\nu}+\nu^k\|\boldsymbol{\lambda}^0-\overline{\boldsymbol{\lambda}}\|.$$

Observing that any large integer may be expressed as a sum of two large integers, and that $\overline{\varepsilon}_k$ converges to zero, we conclude that $\boldsymbol{\lambda}^k$ converges to $\overline{\boldsymbol{\lambda}}$.

To prove the convergence of the primal variables, denote $\mathsf{A}_{\varrho_k}=\mathsf{A}+\varrho_k\mathsf{B}^T\mathsf{B}$ and observe that

$$\mathsf{A}_{\varrho_k}\widehat{\mathbf{x}}=\mathbf{b}+\varrho_k\mathsf{B}^T\mathbf{c}-\mathsf{B}^T\overline{\boldsymbol{\lambda}}.$$

After subtracting the last equation from (4.58) and simple manipulations, we get

$$\mathbf{x}^k-\widehat{\mathbf{x}}=\mathsf{A}_{\varrho_k}^{-1}\mathsf{B}^T\left(\overline{\boldsymbol{\lambda}}-\boldsymbol{\lambda}^k+\mathbf{g}^k\right).$$

Taking norms, using the properties of norms, the assumptions, and

$$\|\mathsf{A}_{\varrho_k}^{-1}\|\le\lambda_{\min}^{-1},$$

we get (4.56). It follows by assumptions that $\mathbf{x}^k$ converges to $\widehat{\mathbf{x}}$. □

The analysis of the asymptotically exact augmented Lagrangian algorithm for more general equality constrained problems may be found in Chap. 2 of Bertsekas [11].

4.5 Adaptive Augmented Lagrangian Method

The analysis of the previous section reveals that it is possible to use an inexact solution of the auxiliary problems in Step 1 of the augmented Lagrangian algorithm. However, the terms related to the precision in the inequalities (4.55) and (4.56) indicate that the convergence can be considerably slowed down if the precision control is relaxed. The price paid for the inexact minimization is an additional term in the estimate of the rate of convergence.

Here we present a different approach which arises from the intuitive argument that the precision of the solution $\mathbf{x}^k$ of the auxiliary problems should be related to the feasibility of $\mathbf{x}^k$, i.e., $\|\mathsf{B}\mathbf{x}^k-\mathbf{c}\|$, since it does not seem reasonable to solve the auxiliary problems to high precision at the early stage of computations with $\boldsymbol{\lambda}^k$ far from the Lagrange multiplier of the solution. Our approach is based on the observation of Sect. 4.3 that the rate of convergence of the augmented Lagrangian algorithm with the steplength $r_k=\varrho_k$ can be controlled by the penalty parameter (4.40).

4.5.1 Algorithm

The new features of the algorithm that we present here are the precision control in Step 1 and the update rule for the penalty parameter.

Algorithm 4.4. Augmented Lagrangians with Adaptive Precision Control.

Given a symmetric positive definite matrix $\mathsf{A} \in \mathbb{R}^{n\times n}$, $\mathsf{B} \in \mathbb{R}^{m\times n}$, $\mathbf{b} \in \mathbb{R}^n$, *and* $\mathbf{c} \in \mathrm{Im}\mathsf{B}$.

Step 0. {*Initialization.*}
Choose $\eta_0 > 0,\ 0 < \alpha < 1,\ \beta > 1,\ M > 0,\ \varrho_0 > 0,\ \boldsymbol{\lambda}^0 \in \mathbb{R}^m$

for *k=0,1,2,...*

Step 1. {*Approximate minimization with respect to* $\mathbf{x}$.}
Choose $\mathbf{x}^k \in \mathbb{R}^n$ *so that*

$$\|\nabla_{\mathbf{x}} L(\mathbf{x}^k, \boldsymbol{\lambda}^k, \varrho_k)\| \le M\|\mathsf{B}\mathbf{x}^k - \mathbf{c}\| \tag{4.59}$$

Step 2. {*Updating the Lagrange multipliers.*}
$\boldsymbol{\lambda}^{k+1} = \boldsymbol{\lambda}^k + \varrho_k(\mathsf{B}\mathbf{x}^k - \mathbf{c})$

Step 3. {*Updating* $\varrho_k,\ \eta_k$.}
if $\|\mathsf{B}\mathbf{x}^k - \mathbf{c}\| \le \eta_k$

$$\varrho_{k+1} = \varrho_k,\ \ \eta_{k+1} = \alpha\eta_k \tag{4.60}$$

else

$$\varrho_{k+1} = \beta\varrho_k,\ \ \eta_{k+1} = \eta_k \tag{4.61}$$

end *if*
end *for*

The next lemma shows that Algorithm 4.4 is well defined, that is, any convergent algorithm for the solution of the auxiliary problems required in Step 1 generates either $\mathbf{x}^k$ that satisfies (4.59) in a finite number of steps, or a sequence of approximations that converge to the solution of (4.1). Thus there is no hidden enforcement of exact solution in (4.59) and consequently typically inexact solutions of the auxiliary unconstrained problems are obtained in Step 1.

Lemma 4.12. *Let* $M > 0$, $\boldsymbol{\lambda} \in \mathbb{R}^m$, *and* $\varrho \ge 0$ *be given and let* $\{\mathbf{y}^k\}$ *denote any sequence that converges to the unique solution* $\widehat{\mathbf{y}}$ *of the problem*

$$\min_{\mathbf{y}\in\mathbb{R}^n} L(\mathbf{y}, \boldsymbol{\lambda}, \varrho). \tag{4.62}$$

Then $\{\mathbf{y}^k\}$ *either converges to the solution* $\widehat{\mathbf{x}}$ *of problem (4.1), or there is an index* k *such that*

$$\|\nabla L(\mathbf{y}^k, \boldsymbol{\lambda}, \varrho)\| \le M\|\mathsf{B}\mathbf{y}^k - \mathbf{c}\|. \tag{4.63}$$

Proof. First observe that $\nabla L(\mathbf{y}^k, \boldsymbol{\lambda}, \varrho)$ converges to zero by the assumptions. Thus if (4.63) does not hold for any k, then we must have $\mathsf{B}\widehat{\mathbf{y}} = \mathbf{c}$. In this case, since $\widehat{\mathbf{y}}$ is the solution of (4.62), it follows that

$$\mathsf{A}\widehat{\mathbf{y}} - \mathbf{b} + \mathsf{B}^T\boldsymbol{\lambda} + \varrho\mathsf{B}^T(\mathsf{B}\widehat{\mathbf{y}} - \mathbf{c}) = \mathbf{o}. \tag{4.64}$$

After substituting $\mathsf{B}\widehat{\mathbf{y}} = \mathbf{c}$ into (4.64), we get

$$\mathsf{B}\widehat{\mathbf{y}} - \mathbf{b} + \mathsf{B}^T\boldsymbol{\lambda} = \mathbf{o}. \tag{4.65}$$

However, since (4.65) and $\mathsf{B}\widehat{\mathbf{y}} = \mathbf{c}$ are sufficient conditions for $\widehat{\mathbf{y}}$ to be the unique solution of (4.1), we have $\widehat{\mathbf{y}} = \widehat{\mathbf{x}}$. □

4.5.2 Convergence of Lagrange Multipliers for Large ϱ

The convergence analysis of Algorithm 4.4 is based on the following lemma.

Lemma 4.13. *Let* $\mathsf{A}, \mathsf{B}, \mathbf{b}$, *and* $\mathbf{c}$ *be those of the definition of problem (4.1) with* $\mathsf{B} \neq \mathsf{O}$ *not necessarily a full rank matrix, let* $M > 0$, *and let*

$$\overline{\varrho} = \|\mathsf{B}\mathsf{A}^{-1}\|M/\overline{\beta}_{\min}, \tag{4.66}$$

where $\overline{\beta}_{\min}$ *denotes the least nonzero eigenvalue of* $\mathsf{B}\mathsf{A}^{-1}\mathsf{B}^T$. *Let* $\boldsymbol{\lambda}_{\mathrm{LS}}$ *denote the vector of the least square Lagrange multipliers for problem (4.1), let* P *denote the orthogonal projector onto* $\mathrm{Im}\mathsf{B}$, *and let for any* $\boldsymbol{\lambda} \in \mathbb{R}^m$

$$\overline{\boldsymbol{\lambda}} = \boldsymbol{\lambda}_{\mathrm{LS}} + (\mathsf{I} - \mathsf{P})\boldsymbol{\lambda} \quad \text{and} \quad \widetilde{\boldsymbol{\lambda}} = \boldsymbol{\lambda} + \varrho(\mathsf{B}\mathbf{x} - \mathbf{c}). \tag{4.67}$$

If $\varrho \geq 2\overline{\varrho}$, $\mathbf{x} \in \mathbb{R}^n, \boldsymbol{\lambda} \in \mathbb{R}^m$, *and*

$$\|\nabla_{\mathbf{x}} L(\mathbf{x}, \boldsymbol{\lambda}, \varrho)\| \leq M\|\mathsf{B}\mathbf{x} - \mathbf{c}\|, \tag{4.68}$$

then

$$\|\widetilde{\boldsymbol{\lambda}} - \overline{\boldsymbol{\lambda}}\| \leq \frac{2}{\varrho}(\overline{\varrho} + \overline{\beta}_{\min}^{-1})\|\boldsymbol{\lambda} - \overline{\boldsymbol{\lambda}}\|. \tag{4.69}$$

Proof. Let us first denote $\boldsymbol{\mu} = \mathsf{P}\boldsymbol{\lambda}$ and $\widetilde{\boldsymbol{\mu}} = \mathsf{P}\widetilde{\boldsymbol{\lambda}}$, so that $\boldsymbol{\mu} \in \mathrm{Im}\mathsf{B}$ and $\widetilde{\boldsymbol{\mu}} \in \mathrm{Im}\mathsf{B}$, and observe that by the definition of $\overline{\boldsymbol{\lambda}}$

$$\boldsymbol{\lambda} - \overline{\boldsymbol{\lambda}} = \boldsymbol{\lambda} - (\boldsymbol{\lambda}_{\mathrm{LS}} + (\mathsf{I} - \mathsf{P})\boldsymbol{\lambda}) = \boldsymbol{\mu} - \boldsymbol{\lambda}_{\mathrm{LS}}, \tag{4.70}$$

$$\widetilde{\boldsymbol{\lambda}} - \overline{\boldsymbol{\lambda}} = \boldsymbol{\lambda} + \varrho(\mathsf{B}\mathbf{x} - \mathbf{c}) - \left(\boldsymbol{\lambda}_{\mathrm{LS}} + (\mathsf{I} - \mathsf{P})\boldsymbol{\lambda}\right) = \widetilde{\boldsymbol{\mu}} - \boldsymbol{\lambda}_{\mathrm{LS}}. \tag{4.71}$$

Since $\mathsf{P}\mathsf{B} = \mathsf{B}$, we have

$$\mathsf{B}^T\boldsymbol{\lambda} = (\mathsf{P}\mathsf{B})^T\boldsymbol{\lambda} = \mathsf{B}^T\mathsf{P}\boldsymbol{\lambda} = \mathsf{B}^T\boldsymbol{\mu},$$
$$\nabla_{\mathbf{x}} L(\mathbf{x}, \boldsymbol{\lambda}, \varrho) = \mathsf{A}_\varrho\mathbf{x} - \mathbf{b} + \mathsf{B}^T\boldsymbol{\lambda} - \varrho\mathsf{B}^T\mathbf{c} = \mathsf{A}_\varrho\mathbf{x} - \mathbf{b} + \mathsf{B}^T\boldsymbol{\mu} - \varrho\mathsf{B}^T\mathbf{c},$$

where $\mathsf{A}_\varrho = \mathsf{A} + \varrho \mathsf{B}^T\mathsf{B}$. Thus the assumption (4.68) is equivalent to

$$\|\nabla_{\mathbf{x}} L(\mathbf{x}, \boldsymbol{\mu}, \varrho)\| = \|\nabla_{\mathbf{x}} L(\mathbf{x}, \boldsymbol{\lambda}, \varrho)\| \leq M\|\mathsf{B}\mathbf{x} - \mathbf{c}\|. \tag{4.72}$$

Finally, notice that by the definition of $\widetilde{\boldsymbol{\lambda}}$ in (4.67), we have

$$\mathsf{B}\mathbf{x} - \mathbf{c} = \varrho^{-1}(\widetilde{\boldsymbol{\lambda}} - \boldsymbol{\lambda}). \tag{4.73}$$

Let us now denote

$$\mathbf{g} = \nabla_{\mathbf{x}} L(\mathbf{x}, \boldsymbol{\mu}, \varrho)$$

and assume that $\mathbf{x}$, $\boldsymbol{\lambda}$, and ϱ satisfy the assumptions including (4.68), so that by (4.72)

$$\|\mathbf{g}\| \leq M\|\mathsf{B}\mathbf{x} - \mathbf{c}\|. \tag{4.74}$$

Using (4.71), Lemma 4.9, (4.70), (4.74), (4.73), and notation (4.66), we get

$$\begin{aligned}
\|\widetilde{\boldsymbol{\lambda}} - \overline{\boldsymbol{\lambda}}\| &= \|\widetilde{\boldsymbol{\mu}} - \boldsymbol{\lambda}_{\mathrm{LS}}\| \\
&\leq \frac{\|\mathsf{B}\mathsf{A}^{-1}\|}{\overline{\beta}_{\min} + \varrho^{-1}}\|\mathbf{g}\| + \frac{\varrho^{-1}}{\overline{\beta}_{\min} + \varrho^{-1}}\|\boldsymbol{\mu} - \boldsymbol{\lambda}_{\mathrm{LS}}\| \\
&\leq \frac{\|\mathsf{B}\mathsf{A}^{-1}\|}{\overline{\beta}_{\min} + \varrho^{-1}} M\|\mathsf{B}\mathbf{x} - \mathbf{c}\| + \frac{\varrho^{-1}}{\overline{\beta}_{\min} + \varrho^{-1}}\|\boldsymbol{\lambda} - \overline{\boldsymbol{\lambda}}\| \\
&= \frac{\|\mathsf{B}\mathsf{A}^{-1}\|}{\overline{\beta}_{\min} + \varrho^{-1}} \frac{M}{\varrho}\|\widetilde{\boldsymbol{\lambda}} - \boldsymbol{\lambda}\| + \frac{\varrho^{-1}}{\overline{\beta}_{\min} + \varrho^{-1}}\|\boldsymbol{\lambda} - \overline{\boldsymbol{\lambda}}\| \\
&\leq \frac{\overline{\varrho}}{\varrho}\left(\|\widetilde{\boldsymbol{\lambda}} - \overline{\boldsymbol{\lambda}}\| + \|\boldsymbol{\lambda} - \overline{\boldsymbol{\lambda}}\|\right) + \frac{1}{\overline{\beta}_{\min}\varrho}\|\boldsymbol{\lambda} - \overline{\boldsymbol{\lambda}}\|.
\end{aligned}$$

Thus, since $\varrho \geq 2\overline{\varrho}$, it follows that

$$\|\widetilde{\boldsymbol{\lambda}} - \overline{\boldsymbol{\lambda}}\| \leq \left(\frac{\overline{\varrho}}{\varrho} + \frac{1}{\overline{\beta}_{\min}\varrho}\right)\|\boldsymbol{\lambda} - \overline{\boldsymbol{\lambda}}\| / \left(1 - \frac{\overline{\varrho}}{\varrho}\right) \leq \frac{2}{\varrho}(\overline{\varrho} + \overline{\beta}_{\min}^{-1})\|\boldsymbol{\lambda} - \overline{\boldsymbol{\lambda}}\|.$$

□

Lemma 4.13 suggests that the Lagrange multipliers generated by Algorithm 4.4 converge to the solution $\overline{\boldsymbol{\lambda}}$ linearly when the penalty parameter is sufficiently large. We shall formulate this result explicitly.

Corollary 4.14. *Let $\{\boldsymbol{\lambda}^k\}$,$\{\mathbf{x}^k\}$, and $\{\varrho_k\}$ be generated by Algorithm 4.4 for problem (4.1) with the initialization defined in Step 0. Using the notation of Lemma 4.13, let for any index $k \geq 0$*

$$\varrho_k \geq 2\alpha_0^{-1}(\overline{\varrho} + \overline{\beta}_{\min}^{-1}), \tag{4.75}$$

where $\alpha_0 < 1$ is a positive constant.

Then

$$\|\boldsymbol{\lambda}^{k+1} - \overline{\boldsymbol{\lambda}}\| \leq \alpha_0\|\boldsymbol{\lambda}^k - \overline{\boldsymbol{\lambda}}\|. \tag{4.76}$$

Proof. Let k satisfy (4.75). Comparing (4.59) with (4.68), we can check that all the assumptions of Lemma 4.13 are satisfied for $\mathbf{x} = \mathbf{x}^k$, $\boldsymbol{\lambda} = \boldsymbol{\lambda}^k$, and $\varrho = \varrho_k$. Substituting into (4.69) and using $\boldsymbol{\lambda}^{k+1} = \widetilde{\boldsymbol{\lambda}}$, we get

$$\|\boldsymbol{\lambda}^{k+1} - \overline{\boldsymbol{\lambda}}\| \le 2\varrho_k{}^{-1}(\overline{\varrho} + \overline{\beta}_{\min}^{-1})\|\boldsymbol{\lambda}^k - \overline{\boldsymbol{\lambda}}\| \le \alpha_0\|\boldsymbol{\lambda}^k - \overline{\boldsymbol{\lambda}}\|.$$

□

Notice that in (4.76), there is no term which accounts for inexact solutions of auxiliary problems. This compares favorably with (4.55).

4.5.3 R-Linear Convergence for Any Initialization of ϱ

The following lemma gives us a simple key to the proof of R-Linear convergence of Algorithm 4.4 for any initial regularization parameter $\varrho_0 \ge 0$.

Lemma 4.15. *Let $\{\boldsymbol{\lambda}^k\}$, $\{\mathbf{x}^k\}$, and $\{\varrho_k\}$ be generated by Algorithm 4.4 for problem (4.1) with the assumptions of Lemma 4.13 and the initialization defined in Step 0.*

Then ϱ_k is bounded and there is a constant C such that

$$\|\mathsf{B}\mathbf{x}^k - \mathbf{c}\| \le C\alpha^k, \tag{4.77}$$

where $\alpha < 1$ is a positive constant defined in Step 0 of Algorithm 4.4.

Proof. Using the notation of Lemma 4.13, let us first assume that for any index k, $\varrho_k < 2(\overline{\varrho} + \overline{\beta}_{\min}^{-1})/\alpha$, so that there is k_0 such that for $k \ge k_0$ the values of ϱ_k and η_k are updated by the rule (4.60) in Step 3 of Algorithm 4.4. It follows that for any $k \ge k_0$,

$$\|\mathsf{B}\mathbf{x}^k - \mathbf{c}\| \le \eta_k = \alpha^{k-k_0}\eta_{k_0} = C\alpha^k,$$

where $\alpha < 1$ is defined in Step 0 of Algorithm 4.4.

If there is k_0 such that $\varrho_{k_0} \ge 2(\overline{\varrho} + \overline{\beta}_{\min}^{-1})/\alpha$, then, since $\{\varrho_k\}$ is nondecreasing, we can use Corollary 4.14 to get that for $k > k_0$

$$\|\boldsymbol{\lambda}^k - \overline{\boldsymbol{\lambda}}\| \le \alpha^{k-k_0}\|\boldsymbol{\lambda}^{k_0} - \overline{\boldsymbol{\lambda}}\|. \tag{4.78}$$

Using the update rule of Step 2 of Algorithm 4.4, we get

$$\|\mathsf{B}\mathbf{x}^k - \mathbf{c}\| = \varrho_k^{-1}\|\boldsymbol{\lambda}^{k+1} - \boldsymbol{\lambda}^k\| \le \varrho_k^{-1}(\|\boldsymbol{\lambda}^{k+1} - \overline{\boldsymbol{\lambda}}\| + \|\boldsymbol{\lambda}^k - \overline{\boldsymbol{\lambda}}\|).$$

Combining the last inequality with (4.78), we get

$$\|\mathsf{B}\mathbf{x}^k - \mathbf{c}\| \le \varrho_k^{-1}(\alpha^{k-k_0+1} + \alpha^{k-k_0})\|\boldsymbol{\lambda}^{k_0} - \overline{\boldsymbol{\lambda}}\| \le 2\alpha^{k-k_0}\varrho_{k_0}^{-1}\|\boldsymbol{\lambda}^{k_0} - \overline{\boldsymbol{\lambda}}\| = C\alpha^k.$$

This proves (4.77).

To prove that $\{\varrho_k\}$ is bounded, notice that we can express each $k \geq 0$ as a sum $k = k_1 + k_2$, where $\eta_k = \alpha^{k_1}\eta_0$ and $\varrho_k = \beta^{k_2}\varrho_0$. Hence given k, k_1 and k_2 denote the numbers of preceding steps that invoked the updates (4.60) and (4.61), respectively. Moreover, $\varrho_{k+1} = \beta\varrho_k > \varrho_k$ if and only if $\|\mathsf{B}\mathbf{x}^k - \mathbf{c}\| > \eta^k$, and for such k

$$\alpha^{k_1}\eta_0 = \eta_k < \|\mathsf{B}\mathbf{x}^k - \mathbf{c}\| \leq C\alpha^k = C\alpha^{k_1+k_2}.$$

Since $\alpha < 1$, it follows that k_2 is finite and ϱ_k is bounded. □

Using that ϱ_k is uniformly bounded, it is now easy to show that $\{\boldsymbol{\lambda}^k\}$ and $\{\mathbf{x}^k\}$ converge R-linearly.

Corollary 4.16. *Let $\{\boldsymbol{\lambda}^k\}$,$\{\mathbf{x}^k\}$, and $\{\varrho_k\}$ be generated by Algorithm 4.4 for problem (4.1) with the initialization defined in Step 0. Then there are constants C_1 and C_2 such that*

$$\|\mathbf{x}^k - \widehat{\mathbf{x}}\| \leq C_1\alpha^k \quad \textit{and} \quad \|\boldsymbol{\lambda}^k - \overline{\boldsymbol{\lambda}}\| \leq C_2\alpha^k, \tag{4.79}$$

where $\overline{\boldsymbol{\lambda}}$ is defined by (4.67), $\widehat{\mathbf{x}}$ is a unique solution of (4.1), and $\alpha < 1$ is a parameter of Algorithm 4.4.

Proof. Observe that Lemma 4.15 and the condition (4.59) in the definition of Step 1 imply that there is a constant C such that

$$\|\mathsf{B}\mathbf{x}^k - \mathbf{c}\| \leq C\alpha^k \quad \text{and} \quad \|\mathbf{g}^k\| \leq C\alpha^k.$$

To finish the proof, it is enough to use Proposition 2.12 and simple manipulations. □

4.6 Semimonotonic Augmented Lagrangians (SMALE)

In the previous section, we have shown that Algorithm 4.4 always achieves the R-linear rate of convergence given by the constant α which controls the decrease of the feasibility error. This looks like not a bad result, its only drawback being that such a rate of convergence is achieved *only with the penalty parameter ϱ_k which exceeds a threshold which depends on the constraint matrix* B. Is it possible to propose an inexact algorithm with any reasonable kind of convergence *independent of the constraint matrix* B? A key to getting a positive answer is to return to the augmented Lagrangian algorithm viewed as an alternative implementation of the penalty method with the adaptive precision control used by Algorithm 4.4. We shall also see that the convergence can be achieved with a rather small threshold on the penalty parameter independent of the singular values of the constraint matrix B.

4.6.1 SMALE Algorithm

The algorithm presented here is based on the observation that, having for a sufficiently large ϱ an *approximate* minimizer $\mathbf{x}_\varrho$ of the augmented Lagrangian $L(\mathbf{x}, \boldsymbol{\lambda}, \varrho)$ with respect to $\mathbf{x}$, we can modify $\boldsymbol{\lambda}$ in such a way that $\mathbf{x}_\varrho$ is also an *approximate* unconstrained minimizer of $L(\mathbf{x}, \widetilde{\boldsymbol{\lambda}}, 0)$. Thus we can hopefully find a better approximation by minimizing $L(\mathbf{x}, \widetilde{\boldsymbol{\lambda}}, \varrho)$. Since the better penalty approximation results in an increased value of the Lagrangian, it is natural to increase the penalty parameter until increasing values of the Lagrangian are generated. We shall show that the threshold value for the penalty parameter is rather small and independent of the constraint matrix B. The algorithm that we consider here reads as follows.

Algorithm 4.5. Semimonotonic augmented Lagrangians (SMALE).

Given a symmetric positive definite matrix $\mathsf{A} \in \mathbb{R}^{n\times n}$, $\mathsf{B} \in \mathbb{R}^{m\times n}$, $\mathbf{b} \in \mathbb{R}^n$, *and* $\mathbf{c} \in \mathrm{Im}\mathsf{B}$.

Step 0. {*Initialization.*}
  Choose $\eta > 0,\ \beta > 1,\ M > 0,\ \varrho_0 > 0,\ \boldsymbol{\lambda}^0 \in \mathbb{R}^m$

  for *k=0,1,2,...*

Step 1. {*Inner iteration with adaptive precision control.*}
  Find $\mathbf{x}^k$ *such that*

$$\|\mathbf{g}(\mathbf{x}^k, \boldsymbol{\lambda}^k, \varrho_k)\| \le \min\{M\|\mathsf{B}\mathbf{x}^k - \mathbf{c}\|, \eta\}. \tag{4.80}$$

Step 2. {*Updating the Lagrange multipliers.*}

$$\boldsymbol{\lambda}^{k+1} = \boldsymbol{\lambda}^k + \varrho_k(\mathsf{B}\mathbf{x}^k - \mathbf{c}) \tag{4.81}$$

Step 3. {*Update* ϱ *provided the increase of the Lagrangian is not sufficient.*}
  if $k > 0$ *and*

$$L(\mathbf{x}^k, \boldsymbol{\lambda}^k, \varrho_k) < L(\mathbf{x}^{k-1}, \boldsymbol{\lambda}^{k-1}, \varrho_{k-1}) + \frac{\varrho_k}{2}\|\mathsf{B}\mathbf{x}^k - \mathbf{c}\|^2 \tag{4.82}$$

  $\varrho_{k+1} = \beta\varrho_k$
  else
  $\varrho_{k+1} = \varrho_k.$
  end if
  end for

In Step 1 we can use any convergent algorithm for the minimization of the strictly convex quadratic function such as the preconditioned conjugate gradient method of Sect. 3.3. Let us point out that Algorithm 4.5 differs from Algorithm 4.4 by the condition (4.82) on the update of the penalization parameter in Step 3.

To see that Algorithm 4.5 is well defined, let $\{\mathbf{y}^k\}$ be a sequence generated by any convergent algorithm for the solution of the auxiliary problem

$$\text{minimize } \{L(\mathbf{y}, \boldsymbol{\lambda}, \varrho) : \mathbf{y} \in \mathbb{R}^n\}.$$

Then there is an integer k_0 such that for $k \geq k_0$

$$\|\mathbf{g}(\mathbf{y}^k, \boldsymbol{\lambda}, \varrho)\| \leq \eta$$

and we can use Lemma 4.12 to show that either $\{\mathbf{y}^k\}$ converges to the solution $\widehat{\mathbf{x}}$ of (4.1) or there is k such that (4.80) holds. Thus there is no hidden enforcement of the exact solution in (4.80) and consequently typically inexact solutions of the auxiliary unconstrained problems are obtained in Step 1.

4.6.2 Relations for Augmented Lagrangians

In this section we establish the basic inequalities that relate the bound on the norm of the gradient $\mathbf{g}$ of the augmented Lagrangian L to the values of the augmented Lagrangian L. These inequalities are the key ingredients in the proof of convergence of Algorithm 4.5.

Lemma 4.17. *Let* A, B, $\mathbf{b}$, *and* $\mathbf{c}$ *be those of problem (4.1),* $\mathbf{x} \in \mathbb{R}^n$, $\boldsymbol{\lambda} \in \mathbb{R}^m$, $\varrho > 0$, $\eta > 0$, *and* $M > 0$. *Let* $\lambda_{\min}$ *denote the least eigenvalue of* A *and* $\widetilde{\boldsymbol{\lambda}} = \boldsymbol{\lambda} + \varrho(\mathsf{B}\mathbf{x} - \mathbf{c})$.
(i) If

$$\|\mathbf{g}(\mathbf{x}, \boldsymbol{\lambda}, \varrho)\| \leq M\|\mathsf{B}\mathbf{x} - \mathbf{c}\|, \tag{4.83}$$

then for any $\mathbf{y} \in \mathbb{R}^n$

$$L(\mathbf{y}, \widetilde{\boldsymbol{\lambda}}, \varrho) \geq L(\mathbf{x}, \boldsymbol{\lambda}, \varrho) + \frac{1}{2}\left(\varrho - \frac{M^2}{\lambda_{\min}}\right)\|\mathsf{B}\mathbf{x} - \mathbf{c}\|^2 + \frac{\varrho}{2}\|\mathsf{B}\mathbf{y} - \mathbf{c}\|^2. \tag{4.84}$$

(ii) If

$$\|\mathbf{g}(\mathbf{x}, \boldsymbol{\lambda}, \varrho)\| \leq \eta, \tag{4.85}$$

then for any $\mathbf{y} \in \mathbb{R}^n$

$$L(\mathbf{y}, \widetilde{\boldsymbol{\lambda}}, \varrho) \geq L(\mathbf{x}, \boldsymbol{\lambda}, \varrho) + \frac{\varrho}{2}\|\mathsf{B}\mathbf{x} - \mathbf{c}\|^2 + \frac{\varrho}{2}\|\mathsf{B}\mathbf{y} - \mathbf{c}\|^2 - \frac{\eta^2}{2\lambda_{\min}}. \tag{4.86}$$

(iii) If (4.85) holds and $\mathbf{z}_0$ *is any vector such that* $\mathsf{B}\mathbf{z}_0 = \mathbf{c}$, *then*

$$L(\mathbf{x}, \boldsymbol{\lambda}, \varrho) \leq f(\mathbf{z}_0) + \frac{\eta^2}{2\lambda_{\min}}. \tag{4.87}$$

Proof. Let us denote $\boldsymbol{\delta} = \mathbf{y} - \mathbf{x}$, $\mathsf{A}_\varrho = \mathsf{A} + \varrho\mathsf{B}^T\mathsf{B}$, $\mathbf{g} = \mathbf{g}(\mathbf{x}, \boldsymbol{\lambda}, \varrho)$, and $\widetilde{\mathbf{g}} = \mathbf{g}(\mathbf{x}, \widetilde{\boldsymbol{\lambda}}, \varrho)$. Using

$$L(\mathbf{x}, \widetilde{\boldsymbol{\lambda}}, \varrho) = L(\mathbf{x}, \boldsymbol{\lambda}, \varrho) + \varrho\|\mathsf{B}\mathbf{x} - \mathbf{c}\|^2 \text{ and } \mathbf{g}(\mathbf{x}, \widetilde{\boldsymbol{\lambda}}, \varrho) = \mathbf{g}(\mathbf{x}, \boldsymbol{\lambda}, \varrho) + \varrho\mathsf{B}^T(\mathsf{B}\mathbf{x} - \mathbf{c}),$$

we get

$$\begin{aligned}
L(\mathbf{y},\widetilde{\boldsymbol{\lambda}},\varrho) &= L(\mathbf{x},\widetilde{\boldsymbol{\lambda}},\varrho) + \boldsymbol{\delta}^T\widetilde{\mathbf{g}} + \frac{1}{2}\boldsymbol{\delta}^T\mathsf{A}_\varrho\boldsymbol{\delta}\\
&= L(\mathbf{x},\boldsymbol{\lambda},\varrho) + \boldsymbol{\delta}^T\mathbf{g} + \frac{1}{2}\boldsymbol{\delta}^T\mathsf{A}_\varrho\boldsymbol{\delta} + \varrho\boldsymbol{\delta}^T\mathsf{B}^T(\mathsf{B}\mathbf{x}-\mathbf{c}) + \varrho\|\mathsf{B}\mathbf{x}-\mathbf{c}\|^2\\
&\ge L(\mathbf{x},\boldsymbol{\lambda},\varrho) + \boldsymbol{\delta}^T\mathbf{g} + \frac{\lambda_{\min}}{2}\|\boldsymbol{\delta}\|^2 + \varrho\boldsymbol{\delta}^T\mathsf{B}^T(\mathsf{B}\mathbf{x}-\mathbf{c}) + \frac{\varrho}{2}\|\mathsf{B}\boldsymbol{\delta}\|^2\\
&\quad +\varrho\|\mathsf{B}\mathbf{x}-\mathbf{c}\|^2.
\end{aligned}$$

Noticing that

$$\frac{\varrho}{2}\|\mathsf{B}\mathbf{y}-\mathbf{c}\|^2 = \frac{\varrho}{2}\|\mathsf{B}\boldsymbol{\delta}+(\mathsf{B}\mathbf{x}-\mathbf{c})\|^2 = \varrho\boldsymbol{\delta}^T\mathsf{B}^T(\mathsf{B}\mathbf{x}-\mathbf{c}) + \frac{\varrho}{2}\|\mathsf{B}\boldsymbol{\delta}\|^2 + \frac{\varrho}{2}\|\mathsf{B}\mathbf{x}-\mathbf{c}\|^2,$$

we get

$$L(\mathbf{y},\widetilde{\boldsymbol{\lambda}},\varrho) \ge L(\mathbf{x},\boldsymbol{\lambda},\varrho) + \boldsymbol{\delta}^T\mathbf{g} + \frac{\lambda_{\min}}{2}\|\boldsymbol{\delta}\|^2 + \frac{\varrho}{2}\|\mathsf{B}\mathbf{x}-\mathbf{c}\|^2 + \frac{\varrho}{2}\|\mathsf{B}\mathbf{y}-\mathbf{c}\|^2. \tag{4.88}$$

Assuming (4.83) and using simple manipulations, we get

$$\begin{aligned}
L(\mathbf{y},\widetilde{\boldsymbol{\lambda}},\varrho) &\ge L(\mathbf{x},\boldsymbol{\lambda},\varrho) - M\|\boldsymbol{\delta}\|\|\mathsf{B}\mathbf{x}-\mathbf{c}\| + \frac{\lambda_{\min}}{2}\|\boldsymbol{\delta}\|^2\\
&\qquad +\frac{\varrho}{2}\|\mathsf{B}\mathbf{x}-\mathbf{c}\|^2 + \frac{\varrho}{2}\|\mathsf{B}\mathbf{y}-\mathbf{c}\|^2\\
&= L(\mathbf{x},\boldsymbol{\lambda},\varrho) + \left(\frac{\lambda_{\min}}{2}\|\boldsymbol{\delta}\|^2 - M\|\boldsymbol{\delta}\|\|\mathsf{B}\mathbf{x}-\mathbf{c}\| + \frac{M^2\|\mathsf{B}\mathbf{x}-\mathbf{c}\|^2}{2\lambda_{\min}}\right)\\
&\quad -\frac{M^2\|\mathsf{B}\mathbf{x}-\mathbf{c}\|^2}{2\lambda_{\min}} + \frac{\varrho}{2}\|\mathsf{B}\mathbf{x}-\mathbf{c}\|^2 + \frac{\varrho}{2}\|\mathsf{B}\mathbf{y}-\mathbf{c}\|^2\\
&\ge L(\mathbf{x},\boldsymbol{\lambda},\varrho) + \frac{1}{2}\left(\varrho - \frac{M^2}{\lambda_{\min}}\right)\|\mathsf{B}\mathbf{x}-\mathbf{c}\|^2 + \frac{\varrho}{2}\|\mathsf{B}\mathbf{y}-\mathbf{c}\|^2,
\end{aligned}$$

which proves (i).

If we assume that (4.85) holds, then by (4.88) and similar manipulations as above

$$\begin{aligned}
L(\mathbf{y},\widetilde{\boldsymbol{\lambda}},\varrho) &\ge L(\mathbf{x},\boldsymbol{\lambda},\varrho) - \|\boldsymbol{\delta}\|\eta + \frac{\lambda_{\min}}{2}\|\boldsymbol{\delta}\|^2 + \frac{\varrho}{2}\|\mathsf{B}\mathbf{x}-\mathbf{c}\|^2 + \frac{\varrho}{2}\|\mathsf{B}\mathbf{y}-\mathbf{c}\|^2\\
&\ge L(\mathbf{x},\boldsymbol{\lambda},\varrho) + \frac{\varrho}{2}\|\mathsf{B}\mathbf{x}-\mathbf{c}\|^2 + \frac{\varrho}{2}\|\mathsf{B}\mathbf{y}-\mathbf{c}\|^2 - \frac{\eta^2}{2\lambda_{\min}},
\end{aligned}$$

which proves (ii).

Finally, let $\widehat{\mathbf{y}}$ denote the solution of the auxiliary problem

$$\text{minimize}\ \ L(\mathbf{y},\boldsymbol{\lambda},\varrho)\ \ \text{s.t.}\ \ \mathbf{y}\in\mathbb{R}^n, \tag{4.89}$$

$\mathsf{B}\mathbf{z}_0 = \mathbf{c}$, and $\widehat{\boldsymbol{\delta}} = \widehat{\mathbf{y}} - \mathbf{x}$. If (4.85) holds, then

$$0 \geq L(\widehat{\mathbf{y}}, \boldsymbol{\lambda}, \varrho) - L(\mathbf{x}, \boldsymbol{\lambda}, \varrho) = \widehat{\boldsymbol{\delta}}^T \mathbf{g} + \frac{1}{2}\widehat{\boldsymbol{\delta}}^T \mathsf{A}_\varrho \widehat{\boldsymbol{\delta}} \geq -\|\widehat{\boldsymbol{\delta}}\|\eta + \frac{1}{2}\lambda_{\min}\|\widehat{\boldsymbol{\delta}}\|^2 \geq -\frac{\eta^2}{2\lambda_{\min}}.$$

Since $L(\widehat{\mathbf{y}}, \boldsymbol{\lambda}, \varrho) \leq L(\mathbf{z}_0, \boldsymbol{\lambda}, \varrho) = f(\mathbf{z}_0)$, we conclude that

$$L(\mathbf{x}, \boldsymbol{\lambda}, \varrho) \leq L(\mathbf{x}, \boldsymbol{\lambda}, \varrho) - L(\widehat{\mathbf{y}}, \boldsymbol{\lambda}, \varrho) + f(\mathbf{z}_0) \leq f(\mathbf{z}_0) + \frac{\eta^2}{2\lambda_{\min}}.$$

□

4.6.3 Convergence and Monotonicity

The analysis of SMALE is based on the following lemma.

Lemma 4.18. *Let $\{\mathbf{x}^k\}$, $\{\boldsymbol{\lambda}^k\}$, and $\{\varrho_k\}$ be generated by Algorithm 4.5 for the solution of problem (4.1) with $\eta > 0$, $\beta > 1$, $M > 0$, $\varrho_0 > 0$, and $\boldsymbol{\lambda}^0 \in \mathbb{R}^m$. Let $\lambda_{\min}$ denote the least eigenvalue of the Hessian A of f.*
(i) If $k \geq 0$ and

$$\varrho_k \geq M^2/\lambda_{\min}, \tag{4.90}$$

then

$$L(\mathbf{x}^{k+1}, \boldsymbol{\lambda}^{k+1}, \varrho_{k+1}) \geq L(\mathbf{x}^k, \boldsymbol{\lambda}^k, \varrho_k) + \frac{\varrho_{k+1}}{2}\|\mathsf{B}\mathbf{x}^{k+1} - \mathbf{c}\|^2. \tag{4.91}$$

(ii) For any $k \geq 0$

$$\begin{aligned} L(\mathbf{x}^{k+1}, \boldsymbol{\lambda}^{k+1}, \varrho_{k+1}) \geq L(\mathbf{x}^k, \boldsymbol{\lambda}^k, \varrho_k) &+ \frac{\varrho_k}{2}\|\mathsf{B}\mathbf{x}^k - \mathbf{c}\|^2 \\ &+ \frac{\varrho_{k+1}}{2}\|\mathsf{B}\mathbf{x}^{k+1} - \mathbf{c}\|^2 - \frac{\eta^2}{2\lambda_{\min}}. \end{aligned} \tag{4.92}$$

(iii) For any $k \geq 0$ and $\mathbf{z}_0$ such that $\mathsf{B}\mathbf{z}_0 = \mathbf{c}$

$$L(\mathbf{x}^k, \boldsymbol{\lambda}^k, \varrho_k) \leq f(\mathbf{z}_0) + \frac{\eta^2}{2\lambda_{\min}}. \tag{4.93}$$

Proof. In Lemma 4.17, let us substitute $\mathbf{x} = \mathbf{x}^k$, $\boldsymbol{\lambda} = \boldsymbol{\lambda}^k$, $\varrho = \varrho_k$, and $\mathbf{y} = \mathbf{x}^{k+1}$, so that inequality (4.83) holds by (4.80), and by (4.81) $\widetilde{\boldsymbol{\lambda}} = \boldsymbol{\lambda}^{k+1}$.
If (4.90) holds, we get by (4.84)

$$L(\mathbf{x}^{k+1}, \boldsymbol{\lambda}^{k+1}, \varrho_k) \geq L(\mathbf{x}^k, \boldsymbol{\lambda}^k, \varrho_k) + \frac{\varrho_k}{2}\|\mathsf{B}\mathbf{x}^{k+1} - \mathbf{c}\|^2. \tag{4.94}$$

To prove (4.91), it is enough to add

$$\frac{\varrho_{k+1} - \varrho_k}{2}\|\mathsf{B}\mathbf{x}^{k+1} - \mathbf{c}\|^2 \tag{4.95}$$

to both sides of (4.94) and to notice that

$$L(\mathbf{x}^{k+1}, \boldsymbol{\lambda}^{k+1}, \varrho_{k+1}) = L(\mathbf{x}^{k+1}, \boldsymbol{\lambda}^{k+1}, \varrho_k) + \frac{\varrho_{k+1} - \varrho_k}{2} \|\mathsf{B}\mathbf{x}^{k+1} - \mathbf{c}\|^2. \quad (4.96)$$

If we notice that by the definition of Step 1 of Algorithm 4.5

$$\|\mathbf{g}(\mathbf{x}^k, \boldsymbol{\lambda}^k, \varrho_k)\| \le \eta,$$

we can apply the same substitution as above to Lemma 4.17(ii) to get

$$\begin{aligned} L(\mathbf{x}^{k+1}, \boldsymbol{\lambda}^{k+1}, \varrho_k) \ge \quad & L(\mathbf{x}^k, \boldsymbol{\lambda}^k, \varrho_k) \\ & + \frac{\varrho_k}{2} \|\mathsf{B}\mathbf{x}^k - \mathbf{c}\|^2 + \frac{\varrho_k}{2} \|\mathsf{B}\mathbf{x}^{k+1} - \mathbf{c}\|^2 - \frac{\eta^2}{2\lambda_{\min}}. \quad (4.97) \end{aligned}$$

After adding the nonnegative expression (4.95) to both sides of (4.97) and using (4.96), we get (4.92). Similarly, inequality (4.93) results from application of the substitution to Lemma 4.17(iii). □

Theorem 4.19. *Let $\{\mathbf{x}^k\}$, $\{\boldsymbol{\lambda}^k\}$, and $\{\varrho_k\}$ be generated by Algorithm 4.5 for the solution of problem (4.1) with $\eta > 0$, $\beta > 1$, $M > 0$, $\varrho_0 > 0$, and $\boldsymbol{\lambda}^0 \in \mathbb{R}^m$. Let $\lambda_{\min}$ denote the least eigenvalue of the Hessian A of f and let $s \ge 0$ denote the smallest integer such that*

$$\beta^s \varrho_0 \ge M^2/\lambda_{\min}.$$

(i) The sequence $\{\varrho_k\}$ is bounded and

$$\varrho_k \le \beta^s \varrho_0. \quad (4.98)$$

(ii) If $\mathbf{z}_0$ denotes any vector such that $\mathsf{B}\mathbf{z}_0 = \mathbf{c}$, then

$$\sum_{k=1}^{\infty} \frac{\varrho_k}{2} \|\mathsf{B}\mathbf{x}^k - \mathbf{c}\|^2 \le f(\mathbf{z}_0) - L(\mathbf{x}^0, \boldsymbol{\lambda}^0, \varrho_0) + (1+s)\frac{\eta^2}{2\lambda_{\min}}. \quad (4.99)$$

(iii) The sequence $\{\mathbf{x}^k\}$ converges to the solution $\widehat{\mathbf{x}}$ of (4.1).
(iv) The sequence $\{\boldsymbol{\lambda}^k\}$ converges to the vector

$$\overline{\boldsymbol{\lambda}} = \boldsymbol{\lambda}_{\mathrm{LS}} + (\mathsf{I} - \mathsf{P})\boldsymbol{\lambda}^0,$$

where P is the orthogonal projector onto $\mathrm{Im}\mathsf{B}$, *and $\boldsymbol{\lambda}_{\mathrm{LS}}$ is the least square Lagrange multiplier of (4.1).*

Proof. Let $s \ge 0$ denote the smallest integer such that $\beta^s \varrho_0 \ge M^2/\lambda_{\min}$ and let $\mathcal{I}$ denote the set of all indices k_i such that $\{\varrho_{k_i} > \varrho_{k_{i-1}}\}$. Using Lemma 4.18(i), $\varrho_{k_i} = \beta \varrho_{k_{i-1}} = \beta^i \varrho_0$ for $k_i \in \mathcal{I}$, and $\beta^s \varrho_0 \ge M^2/\lambda_{\min}$, we conclude that there is no k such that $\varrho_k > \beta^s \varrho_0$. Thus $\mathcal{I}$ has at most s elements and (4.98) holds.

By the definition of Step 3, for $k > 0$ either $k+1 \notin \mathcal{I}$ and

$$\frac{\varrho_k}{2}\|\mathsf{B}\mathbf{x}^k - \mathbf{c}\|^2 \le L(\mathbf{x}^k, \boldsymbol{\lambda}^k, \varrho_k) - L(\mathbf{x}^{k-1}, \boldsymbol{\lambda}^{k-1}, \varrho_{k-1}),$$

or $k+1 \in \mathcal{I}$ and by (4.92)

$$\begin{aligned}\frac{\varrho_k}{2}\|\mathsf{B}\mathbf{x}^k - \mathbf{c}\|^2 &\le \frac{\varrho_{k-1}}{2}\|\mathsf{B}\mathbf{x}^{k-1} - \mathbf{c}\|^2 + \frac{\varrho_k}{2}\|\mathsf{B}\mathbf{x}^k - \mathbf{c}\|^2 \\ &\le L(\mathbf{x}^k, \boldsymbol{\lambda}^k, \varrho_k) - L(\mathbf{x}^{k-1}, \boldsymbol{\lambda}^{k-1}, \varrho_{k-1}) + \frac{\eta^2}{2\lambda_{\min}}.\end{aligned}$$

Summing up the appropriate cases of the last two inequalities for $k = 1, \dots, n$ and taking into account that $\mathcal{I}$ has at most s elements, we get

$$\sum_{k=1}^{n} \frac{\varrho_k}{2}\|\mathsf{B}\mathbf{x}^k - \mathbf{c}\|^2 \le L(\mathbf{x}^n, \boldsymbol{\lambda}^n, \varrho_n) - L(\mathbf{x}^0, \boldsymbol{\lambda}^0, \varrho_0) + s\frac{\eta^2}{2\lambda_{\min}}. \tag{4.100}$$

To get (4.99), it is enough to replace $L(\mathbf{x}^n, \boldsymbol{\lambda}^n, \varrho_n)$ by the upper bound (4.93).

To prove (iii) and (iv), let us denote

$$\mathbf{g}^k = \mathbf{g}(\mathbf{x}^k, \boldsymbol{\lambda}^k, \varrho_k) = \mathsf{A}_{\varrho_k}\mathbf{x}^k + \mathsf{B}^T\boldsymbol{\lambda}^k - \mathbf{b} - \varrho_k\mathsf{B}^T\mathbf{c}, \quad \mathsf{A}_{\varrho_k} = \mathsf{A} + \varrho_k\mathsf{B}^T\mathsf{B},$$

and let us assume that B is a full row rank matrix. Since the unique KKT pair $(\widehat{\mathbf{x}}, \widehat{\boldsymbol{\lambda}})$ is fully determined by

$$\begin{aligned}\mathsf{A}\widehat{\mathbf{x}} + \mathsf{B}^T\widehat{\boldsymbol{\lambda}} &= \mathbf{b}, \\ \mathsf{B}\widehat{\mathbf{x}} \qquad &= \mathbf{c},\end{aligned}$$

we can rewrite $\mathbf{g}^k$ as

$$\mathbf{g}^k = \mathsf{A}_{\varrho_k}(\mathbf{x}^k - \widehat{\mathbf{x}}) + \mathsf{B}^T(\boldsymbol{\lambda}^k - \widehat{\boldsymbol{\lambda}}). \tag{4.101}$$

The last equation together with

$$\mathsf{B}(\mathbf{x}^k - \widehat{\mathbf{x}}) = \mathsf{B}\mathbf{x}^k - \mathbf{c} \tag{4.102}$$

may be written in the matrix form as

$$\begin{pmatrix} \mathsf{A}_{\varrho_k} & \mathsf{B}^T \\ \mathsf{B} & 0 \end{pmatrix} \begin{pmatrix} \mathbf{x}^k - \widehat{\mathbf{x}} \\ \boldsymbol{\lambda}^k - \widehat{\boldsymbol{\lambda}} \end{pmatrix} = \begin{pmatrix} \mathbf{g}^k \\ \mathsf{B}\mathbf{x}^k - \mathbf{c} \end{pmatrix}. \tag{4.103}$$

Since $\|\mathsf{B}\mathbf{x}^k - \mathbf{c}\|$ converges to zero due to (4.99), $\|\mathbf{g}^k\| \le M\|\mathsf{B}\mathbf{x}^k - \mathbf{c}\|$, and the matrix of the system (4.103) is regular, we conclude, using Proposition 2.12, that $\mathbf{x}^k$ converges to $\widehat{\mathbf{x}}$ and $\boldsymbol{\lambda}^k$ converges to $\widehat{\boldsymbol{\lambda}}$. Since B is a full rank matrix, it follows that $\widehat{\boldsymbol{\lambda}} = \boldsymbol{\lambda}_{\mathrm{LS}} = \overline{\boldsymbol{\lambda}}$.

If B is not a full rank matrix, then the augmented matrix on the left-hand side of (4.103) is singular, but the solution $\widehat{\mathbf{x}}$ is still uniquely determined, as $\mathrm{Ker}\mathsf{A} \cap \mathrm{Ker}\mathsf{B} \subseteq \mathrm{Ker}\mathsf{A} = \{\mathbf{o}\}$ by the assumptions. Since any KKT pair $(\widehat{\mathbf{x}}, \overline{\boldsymbol{\lambda}})$ satisfies

$$\begin{pmatrix} \mathsf{A}_{\varrho_k} & \mathsf{B}^T \\ \mathsf{B} & 0 \end{pmatrix} \begin{pmatrix} \mathbf{x}^k - \widehat{\mathbf{x}} \\ \boldsymbol{\lambda}^k - \overline{\boldsymbol{\lambda}} \end{pmatrix} = \begin{pmatrix} \mathbf{g}^k \\ \mathsf{B}\mathbf{x}^k - \mathbf{c} \end{pmatrix}, \tag{4.104}$$

we can use the same arguments as above and Proposition 2.12 to find out again that $\mathbf{x}^k$ converges to $\widehat{\mathbf{x}}$, but now we shall get only that $\mathsf{B}^T\boldsymbol{\lambda}^k$ converges to $\mathsf{B}^T\overline{\boldsymbol{\lambda}}$. However, using Lemma 4.10, we get

$$\boldsymbol{\lambda}^k - \overline{\boldsymbol{\lambda}} = \mathsf{P}\boldsymbol{\lambda}^k - \boldsymbol{\lambda}_{\mathrm{LS}},$$

so that in particular $\boldsymbol{\lambda}^k - \overline{\boldsymbol{\lambda}} \in \mathrm{Im}\mathsf{B}$. It follows by (1.34) that

$$\|\mathsf{B}^T(\boldsymbol{\lambda}^k - \overline{\boldsymbol{\lambda}})\| \geq \overline{\sigma}_{\min}\|\boldsymbol{\lambda}^k - \overline{\boldsymbol{\lambda}}\|.$$

Since the right-hand side converges to zero, we conclude that $\boldsymbol{\lambda}^k$ converges to $\overline{\boldsymbol{\lambda}}$, which completes the proof of (iii) and (iv). □

4.6.4 Linear Convergence for Large ϱ_0

Using the estimates of the previous section, we can prove that Algorithm 4.5 converges to the solution $\overline{\boldsymbol{\lambda}}$ linearly provided ϱ_0 is sufficiently large. We shall formulate this result explicitly.

Proposition 4.20. *Let $\{\boldsymbol{\lambda}^k\}$,$\{\mathbf{x}^k\}$, and $\{\varrho_k\}$ be generated by Algorithm 4.5 for problem (4.1) with the initialization defined in Step 0 and*

$$\varrho_0 \geq 2\alpha^{-1}(\overline{\varrho} + \overline{\beta}_{\min}^{-1}), \tag{4.105}$$

where we use the notation of Lemma 4.13 and α is an arbitrary constant such that $0 < \alpha < 1$.

(i) For any index $k \geq 0$

$$\|\boldsymbol{\lambda}^{k+1} - \overline{\boldsymbol{\lambda}}\| \leq \alpha\|\boldsymbol{\lambda}^k - \overline{\boldsymbol{\lambda}}\|. \tag{4.106}$$

(ii) There is a constant C_1 such that for any index $k \geq 0$

$$\|\mathsf{B}\mathbf{x}^k - \mathbf{c}\| \leq C_1\alpha^k. \tag{4.107}$$

(iii) There is a constant C_2 such that for any index $k \geq 0$

$$\|\mathbf{x}^k - \widehat{\mathbf{x}}\| \leq C_2\alpha^k. \tag{4.108}$$

Proof. (i) Let ϱ_0 satisfy (4.105). Comparing (4.80) with (4.68) and taking into account that $\varrho_k \geq \varrho_0$, we can check that all the assumptions of Lemma 4.13 are satisfied for $\mathbf{x} = \mathbf{x}^k$, $\boldsymbol{\lambda} = \boldsymbol{\lambda}^k$, and $\varrho = \varrho_k$. Substituting into (4.69) and using $\boldsymbol{\lambda}^{k+1} = \widetilde{\boldsymbol{\lambda}}$, we get

$$\|\boldsymbol{\lambda}^{k+1}-\overline{\boldsymbol{\lambda}}\| \leq 2{\varrho_k}^{-1}(\overline{\varrho}+\overline{\beta}_{\min}^{-1})\|\boldsymbol{\lambda}^k-\overline{\boldsymbol{\lambda}}\| \leq \alpha\|\boldsymbol{\lambda}^k-\overline{\boldsymbol{\lambda}}\|.$$

This proves (4.106).
(ii) Using the update rule of Step 2 of Algorithm 4.5, we get

$$\|\mathsf{B}\mathbf{x}^k-\mathbf{c}\| = \varrho_k^{-1}\|\boldsymbol{\lambda}^{k+1}-\boldsymbol{\lambda}^k\| \leq \varrho_k^{-1}(\|\boldsymbol{\lambda}^{k+1}-\overline{\boldsymbol{\lambda}}\|+\|\boldsymbol{\lambda}^k-\overline{\boldsymbol{\lambda}}\|),$$

and by (4.106), we get

$$\|\mathsf{B}\mathbf{x}^k-\mathbf{c}\| \leq \varrho_k^{-1}(\alpha^{k+1}+\alpha^k)\|\boldsymbol{\lambda}^0-\overline{\boldsymbol{\lambda}}\| \leq 2\alpha^k{\varrho_0}^{-1}\|\boldsymbol{\lambda}^0-\overline{\boldsymbol{\lambda}}\| = C_1\alpha^k.$$

This proves (4.107).
(iii) Observe that (4.107) and the condition (4.80) in the definition of Step 1 of Algorithm 4.5 imply that there is a constant C_1 such that

$$\|\mathsf{B}\mathbf{x}^k-\mathbf{c}\| \leq C_1\alpha^k \quad \text{and} \quad \|\mathbf{g}^k\| \leq C_1\alpha^k.$$

To finish the proof, it is enough to use Proposition 2.12 and simple manipulations. □

4.6.5 Optimality of the Outer Loop

Theorem 4.19 suggests that for homogeneous constraints, it is possible to give a rate of convergence of the feasibility error that does not depend on the constraint matrix B. To present explicitly this qualitatively new feature of Algorithm 4.5, at least as compared to the related Algorithm 4.4, let $\mathcal{T}$ denote any set of indices and assume that for any $t \in \mathcal{T}$ there is defined a problem

$$\text{minimize } f_t(\mathbf{x}) \text{ s.t. } \mathbf{x} \in \Omega_t \tag{4.109}$$

with $\Omega_t = \{\mathbf{x} \in \mathbb{R}^{n_t} : \mathsf{B}_t\mathbf{x} = \mathbf{o}\}$, $f_t(\mathbf{x}) = \frac{1}{2}\mathbf{x}^T\mathsf{A}_t\mathbf{x} - \mathbf{b}_t^T\mathbf{x}$, $\mathsf{A}_t \in \mathbb{R}^{n_t \times n_t}$ symmetric positive definite, $\mathsf{B}_t \in \mathbb{R}^{m_t \times n_t}$, and $\mathbf{b}_t, \mathbf{x} \in \mathbb{R}^{n_t}$. Our optimality result then reads as follows.

Theorem 4.21. *Let* $\{\mathbf{x}_t^k\}, \{\boldsymbol{\lambda}_t^k\}$, *and* $\{\varrho_{t,k}\}$ *be generated by Algorithm 4.5 for (4.109) with* $\|\mathbf{b}_t\| \geq \eta_t > 0$, $\beta > 1$, $M > 0$, $\varrho_{t,0} = \varrho_0 > 0$, $\boldsymbol{\lambda}_t^0 = \mathbf{o}$. *Let* $0 < a_{\min}$ *be a given constant. Finally, let the class of problems (4.109) satisfy*

$$a_{\min} \leq \lambda_{\min}(\mathsf{A}_t),$$

where $\lambda_{\min}(\mathsf{A}_t)$ *denotes the smallest eigenvalue of* A_t, *and denote*

$$a = (2+s)/(a_{\min}\varrho_0),$$

where $s \geq 0$ *is the smallest integer such that* $\beta^s\varrho_0 \geq M^2/a_{\min}$. *Then for each* $\varepsilon > 0$ *there are the indices* k_t, $t \in \mathcal{T}$, *such that*

$$k_t \le a/\varepsilon^2 + 1$$

and $\mathbf{x}_t^{k_t}$ *is an approximate solution of (4.109) satisfying*

$$\|\mathbf{g}_t(\mathbf{x}_t^{k_t}, \boldsymbol{\lambda}_t^{k_t}, \varrho_{t,k_t})\| \le M\varepsilon\|\mathbf{b}_t\| \quad \textit{and} \quad \|\mathsf{B}_t\mathbf{x}_t^{k_t}\| \le \varepsilon\|\mathbf{b}_t\|. \tag{4.110}$$

Proof. First notice that for any index j

$$\frac{j\varrho_0}{2}\min_{i\in\{1,\dots,j\}}\|\mathsf{B}_t\mathbf{x}_t^i\|^2 \le \sum_{i=1}^{j}\frac{\varrho_{t,i}}{2}\|\mathsf{B}_t\mathbf{x}_t^i\|^2 \le \sum_{i=1}^{\infty}\frac{\varrho_{t,i}}{2}\|\mathsf{B}_t\mathbf{x}_t^i\|^2. \tag{4.111}$$

Denoting by $L_t(\mathbf{x}, \boldsymbol{\lambda}, \varrho)$ the Lagrangian for problem (4.109), we get for any $\mathbf{x} \in \mathbb{R}^{n_t}$ and $\varrho \ge 0$

$$L_t(\mathbf{x}, \mathbf{o}, \varrho) = \frac{1}{2}\mathbf{x}^T(\mathsf{A}_t + \varrho\mathsf{B}_t^T\mathsf{B}_t)\mathbf{x} - \mathbf{b}_t^T\mathbf{x} \ge \frac{1}{2}a_{\min}\|\mathbf{x}\|^2 - \|\mathbf{b}_t\|\|\mathbf{x}\| \ge -\frac{\|\mathbf{b}_t\|^2}{2a_{\min}}.$$

If we substitute this inequality and $\mathbf{z} = \mathbf{z}_0^t = \mathbf{o}$ into (4.99) and use $\|\mathbf{b}_t\| \ge \eta_t$, we get for any $t \in \mathcal{T}$

$$\sum_{i=1}^{\infty}\frac{\varrho_{t,i}}{2}\|\mathsf{B}_t\mathbf{x}_t^i\|^2 \le \frac{\|\mathbf{b}_t\|^2}{2a_{\min}} + \frac{(1+s)\eta_t^2}{2a_{\min}} \le \frac{2+s}{2a_{\min}}\|\mathbf{b}_t\|^2 = \frac{a\varrho_0}{2}\|\mathbf{b}_t\|^2. \tag{4.112}$$

Combining the latter inequality with (4.111), we get

$$\min\{\|\mathsf{B}_t\mathbf{x}_t^i\|^2 : i = 1, \dots, k\} \le a\|\mathbf{b}_t\|^2/j. \tag{4.113}$$

Taking for j the smallest integer that satisfies $a/j \le \varepsilon^2$, so that

$$a/\varepsilon^2 \le j \le a/\varepsilon^2 + 1,$$

and denoting for any $t \in \mathcal{T}$ by $k_t \in \{1, \dots, j\}$ the index which minimizes $\{\|\mathsf{B}_t\mathbf{x}_t^i\| : i = 1, \dots, j\}$, we can use (4.113) to obtain

$$\|\mathsf{B}_t\mathbf{x}_t^{k_t}\|^2 = \min\{\|\mathsf{B}_t\mathbf{x}_t^i\|^2 : i = 1, \dots, j\} \le a\|\mathbf{b}_t\|^2/j \le \varepsilon^2\|\mathbf{b}_t\|^2.$$

Thus

$$\|\mathsf{B}_t\mathbf{x}_t^{k_t}\|^2 \le \varepsilon^2\|\mathbf{b}_t\|^2,$$

and, using the definition of Step 1 of Algorithm 4.5, we get also the inequality

$$\|\mathbf{g}_t(\mathbf{x}_t^{k_t}, \boldsymbol{\lambda}_t^{k_t}, \varrho_{t,k_t})\| \le M\|\mathsf{B}_t\mathbf{x}_t^{k_t}\| \le M\varepsilon\|\mathbf{b}_t\|.$$

□

Let us recall that

$$\|\mathbf{g}_t(\mathbf{x}_t^{k_t}, \boldsymbol{\lambda}_t^{k_t+1}, 0)\| = \|\mathbf{g}_t(\mathbf{x}_t^{k_t}, \boldsymbol{\lambda}_t^{k_t}, \varrho_{t,k_t})\|,$$

so that $(\mathbf{x}_t^{k_t}, \boldsymbol{\lambda}_t^{k_t+1})$ is an approximate KKT pair of problem (4.109). The assumption on homogeneity of the constraints was used to find $\mathbf{z}_0^t$ such that $f(\mathbf{z}_0^t)$ is uniformly bounded, in this case by zero.

4.6.6 Optimality of SMALE with Conjugate Gradients

We shall need the following simple lemma to prove optimality of the inner loop.

Lemma 4.22. *Let $\{\mathbf{x}^k\}, \{\boldsymbol{\lambda}^k\}$, and $\{\varrho_k\}$ be generated by Algorithm 4.5 for problem (4.1) with $\eta > 0$, $\beta > 1$, $M > 0$, $\varrho_0 > 0$, and $\boldsymbol{\lambda}^0 \in \mathbb{R}^m$. Let $\lambda_{\min}$ denote the least eigenvalue of A.*

Then for any $k \geq 0$

$$L(\mathbf{x}^k, \boldsymbol{\mu}^{k+1}, \varrho_{k+1}) - L(\mathbf{x}^{k+1}, \boldsymbol{\mu}^{k+1}, \varrho_{k+1}) \leq \frac{\eta^2}{2\lambda_{\min}} + \frac{\beta\varrho_k}{2}\|\mathsf{B}\mathbf{x}^k - \mathbf{c}\|^2. \quad (4.114)$$

Proof. Notice that by definition of the Lagrangian function and by the update rule (4.81)

$$\begin{aligned} L(\mathbf{x}^k, \boldsymbol{\lambda}^{k+1}, \varrho_{k+1}) &= L(\mathbf{x}^k, \boldsymbol{\lambda}^k, \varrho_k) + \varrho_k\|\mathsf{B}\mathbf{x}^k - \mathbf{c}\|^2 + \frac{\varrho_{k+1} - \varrho_k}{2}\|\mathsf{B}\mathbf{x}^k - \mathbf{c}\|^2 \\ &= L(\mathbf{x}^k, \boldsymbol{\lambda}^k, \varrho_k) + \frac{\varrho_{k+1} + \varrho_k}{2}\|\mathsf{B}\mathbf{x}^k - \mathbf{c}\|^2. \end{aligned}$$

After subtracting $L(\mathbf{x}^{k+1}, \boldsymbol{\lambda}^{k+1}, \varrho_{k+1})$ from both sides and observing that by (4.92)

$$L(\mathbf{x}^k, \boldsymbol{\lambda}^k, \varrho_k) - L(\mathbf{x}^{k+1}, \boldsymbol{\lambda}^{k+1}, \varrho_{k+1}) \leq \frac{\eta^2}{2\lambda_{\min}} - \frac{\varrho_k}{2}\|\mathsf{B}\mathbf{x}^k - \mathbf{c}\|^2,$$

we get

$$L(\mathbf{x}^k, \boldsymbol{\lambda}^{k+1}, \varrho_{k+1}) - L(\mathbf{x}^{k+1}, \boldsymbol{\lambda}^{k+1}, \varrho_{k+1}) \leq \frac{\eta^2}{2\lambda_{\min}} + \frac{\beta\varrho_k}{2}\|\mathsf{B}\mathbf{x}^k - \mathbf{c}\|^2.$$

□

Now we are ready to prove our main result concerning the inner loop.

Theorem 4.23. *Let $\{\mathbf{x}_t^k\}, \{\boldsymbol{\lambda}_t^k\}$, and $\{\varrho_{t,k}\}$ be generated by Algorithm 4.5 for (4.109) with $\|\mathbf{b}_t\| \geq \eta_t > 0$, $\beta > 1$, $M > 0$, $\varrho_{t,0} = \varrho_0 > 0$, $\boldsymbol{\lambda}_t^0 = \mathbf{o}$. Let $0 < a_{\min} < a_{\max}$ and $0 < B_{\max}$ be given constants. Let Step 1 be implemented by the conjugate gradient method which generates the iterates $\mathbf{x}_t^{k,0}, \mathbf{x}_t^{k,1}, \ldots, \mathbf{x}_t^{k,l} = \mathbf{x}_t^k$ starting from $\mathbf{x}_t^{k,0} = \mathbf{x}_t^{k-1}$ with $\mathbf{x}_t^{-1} = \mathbf{o}$, where $l = l(k,t)$ is the first index satisfying either*

$$\|g(\mathbf{x}_t^{k,l}, \boldsymbol{\lambda}_t^k, \varrho_k)\| \leq M\|\mathsf{B}_t\mathbf{x}_t^{k,l}\| \quad (4.115)$$

or

$$\|g(\mathbf{x}_t^{k,l}, \boldsymbol{\lambda}_t^k, \varrho_k)\| \leq \varepsilon M\|\mathbf{b}_t\|. \quad (4.116)$$

Finally, let the class of problems (4.109) satisfy

$$a_{\min} \le \lambda_{\min}(\mathsf{A}_t) \le \lambda_{\max}(\mathsf{A}_t) = \|\mathsf{A}_t\| \le a_{\max} \ \ and \ \ \|\mathsf{B}_t\| \le B_{\max}. \quad (4.117)$$

Then Algorithm 4.5 generates an approximate solution $\mathbf{x}_t^{k_t}$ *of any problem (4.109) which satisfies (4.110) at* $O(1)$ *matrix–vector multiplications by the Hessian of the augmented Lagrangian* L_t *for (4.109).*

Proof. Let $t \in \mathcal{T}$ be fixed and let us denote by $L_t(\mathbf{x}, \boldsymbol{\lambda}, \varrho)$ the augmented Lagrangian for problem (4.109), so that for any $\mathbf{x} \in \mathbb{R}^p$ and $\varrho \ge 0$

$$L_t(\mathbf{x}, \mathbf{o}, \varrho) = \frac{1}{2}\mathbf{x}^T(\mathsf{A}_t + \varrho \mathsf{B}_t^T \mathsf{B}_t)\mathbf{x} - \mathbf{b}_t^T\mathbf{x} \ge \frac{1}{2}a_{\min}\|\mathbf{x}\|^2 - \|\mathbf{b}_t\|\|\mathbf{x}\| \ge -\frac{\|\mathbf{b}_t\|^2}{2a_{\min}}.$$

Applying the latter inequality to (4.99) with $\mathbf{z}_0 = \mathbf{o}$ and $\boldsymbol{\lambda}_t^0 = \mathbf{o}$, we get, using the assumption $\|\mathbf{b}_t\| \ge \eta_t$, that for any $k \ge 0$

$$\begin{aligned} \frac{\varrho_{t,k}}{2}\|\mathsf{B}_t\mathbf{x}_t^k\|^2 &\le \sum_{i=1}^{\infty} \frac{\varrho_{t,i}}{2}\|\mathsf{B}_t\mathbf{x}_t^i\|^2 \le f(\mathbf{z}_0) - L(\mathbf{x}_t^0, \boldsymbol{\lambda}_t^0, \varrho_{t,0}) + (1+s)\frac{\eta_t^2}{2a_{\min}} \\ &\le (2+s)\|\mathbf{b}_t\|^2/(2a_{\min}), \end{aligned}$$

where $s \ge 0$ denotes the smallest integer such that $\beta^s \varrho_0 \ge M^2/a_{\min}$. Thus by (4.114)

$$\begin{aligned} L_t(\mathbf{x}_t^{k-1}, \boldsymbol{\lambda}_t^k, \varrho_{t,k}) - L_t(\mathbf{x}_t^k, \boldsymbol{\lambda}_t^k, \varrho_{t,k}) &\le \frac{\eta_t^2}{2a_{\min}} + \frac{\beta \varrho_{t,k-1}}{2}\|\mathsf{B}_t\mathbf{x}_t^{k-1}\|^2 \\ &\le (3+s)\beta\|\mathbf{b}_t\|^2/(2a_{\min}), \quad (4.118) \end{aligned}$$

and, since the minimizer $\overline{\mathbf{x}}_t^k$ of $L_t(\,.\,, \boldsymbol{\lambda}_t^k, \varrho_{t,k})$ satisfies (4.115) and is a possible choice for $\mathbf{x}_t^k$, also

$$L_t(\mathbf{x}_t^{k-1}, \boldsymbol{\lambda}_t^k, \varrho_{t,k}) - L_t(\overline{\mathbf{x}}_t^k, \boldsymbol{\lambda}_t^k, \varrho_{t,k}) \le (3+s)\beta\|\mathbf{b}_t\|^2/(2a_{\min}). \quad (4.119)$$

Denoting

$$a_1 = (3+s)\beta/a_{\min},$$

we can estimate the energy norm of the gradient by

$$\|\mathbf{g}_t(\mathbf{x}_t^{k,0}, \boldsymbol{\lambda}_t^k, \varrho_{t,k})\|^2_{\mathsf{A}_{t,k}^{-1}} = 2\big(L_t(\mathbf{x}_t^{k-1}, \boldsymbol{\lambda}_t^k, \varrho_{t,k}) - L_t(\overline{\mathbf{x}}_t^k, \boldsymbol{\lambda}_t^k, \varrho_{t,k})\big) \le a_1\|\mathbf{b}_t\|^2,$$

where

$$\mathsf{A}_{t,k} = \mathsf{A}_t + \frac{\varrho_{t,k}}{2}\mathsf{B}_t^T\mathsf{B}_t.$$

Since

$$a_{\min} \le \lambda_{\min}(\mathsf{A}_{t,k}) \le \|\mathsf{A}_{t,k}\| \le \|\mathsf{A}_t\| + \varrho_{t,k}\|\mathsf{B}_t\|^2 \le a_{\max} + \beta^s \varrho_0 B_{\max}^2,$$

we can also bound the spectral condition number $\kappa(\mathsf{A}_{t,k})$ of $\mathsf{A}_{t,k}$ by

$$K = \left(a_{\max} + \beta^s \varrho_0 B_{\max}^2\right)/a_{\min}.$$

Combining this bound with the estimate (3.21) which reads in our case

$$\|\mathbf{g}_t(\mathbf{x}_t^{k,l}, \boldsymbol{\lambda}_t^k, \varrho_{t,k})\|^2_{\mathsf{A}_{t,k}^{-1}} \leq 4\left(\frac{\sqrt{\kappa(\mathsf{A}_{t,k})}-1}{\sqrt{\kappa(\mathsf{A}_{t,k})}+1}\right)^{2l} \|\mathbf{g}_t(\mathbf{x}_t^{k,0}, \boldsymbol{\lambda}_t^k, \varrho_{t,k})\|^2_{\mathsf{A}_{t,k}^{-1}},$$

we get

$$\begin{aligned}\|\mathbf{g}_t(\mathbf{x}_t^{k,l}, \boldsymbol{\lambda}_t^k, \varrho_{t,k})\|^2 &\leq \frac{1}{a_{\min}}\|\mathbf{g}_t(\mathbf{x}_t^{k,l}, \boldsymbol{\lambda}_t^k, \varrho_{t,k})\|^2_{\mathsf{A}_{t,k}^{-1}} \leq \frac{4\sigma^{2l}}{a_{\min}}\|\mathbf{g}_t(\mathbf{x}_t^{k,0}, \boldsymbol{\lambda}_t^k, \varrho_{t,k})\|^2_{\mathsf{A}_{t,k}^{-1}} \\ &\leq \frac{4a_1}{a_{\min}}\sigma^{2l}\|\mathbf{b}_t\|^2,\end{aligned}$$

where

$$\sigma = \frac{\sqrt{K}-1}{\sqrt{K}+1} < 1.$$

It simply follows by the inner stop rule (4.116) that the number of the inner iterations is uniformly bounded by any index $l = l_{\max}$ which satisfies

$$\frac{4a_1}{a_{\min}}\sigma^{2l}\|\mathbf{b}_t\|^2 \leq \varepsilon^2\|\mathbf{b}_t\|^2 M^2.$$

To finish the proof, it is enough to combine this result with Theorem 4.21. □

We can observe optimality in the solution of more general classes of problems than those considered in Theorem 4.23 provided we can bound the number of iterations in the inner loop. For an example of optimality when $\|\mathsf{B}_t\|$ is not bounded see Sect. 4.8.2.

4.6.7 Solution of More General Problems

If A is positive definite only on the kernel of B, then we can use a suitable penalization to reduce such problem to the convex one. Using Lemma 1.3, it follows that there is $\overline{\varrho} > 0$ such that $\mathsf{A} + \overline{\varrho}\mathsf{B}^T\mathsf{B}$ is positive definite, so that we can apply our SMALE algorithm to the equivalent penalized problem

$$\min_{\mathbf{x}\in\Omega_E} f_{\overline{\varrho}}(\mathbf{x}), \tag{4.120}$$

where

$$f_{\overline{\varrho}}(\mathbf{x}) = \mathbf{x}^T(\mathsf{A} + \overline{\varrho}\mathsf{B}^T\mathsf{B})\mathbf{x} - \mathbf{b}^T\mathbf{x}.$$

Alternatively, we can modify the inner loop of SMALE so that it leaves the inner loop and increases the penalty parameter whenever the negative curvature is recognized. Let us point out that such modification does not guarantee optimality of the modified algorithm.

4.7 Implementation of Inexact Augmented Lagrangians

We shall complete the discussion of inexact augmented Lagrangian algorithms by a few hints concerning their implementation.

4.7.1 Stopping, Modification of Constraints, and Preconditioning

While implementing the inexact augmented Lagrangian algorithms of Sects. 4.4 and 4.6, a stopping criterion should be added not only after Step 1, but also into the procedure which generates $\mathbf{x}^k$ in Step 1. We use in our experiments the stopping criterion

$$\|\nabla L(\mathbf{x}^k, \boldsymbol{\lambda}^k, \varrho_k)\| \le \varepsilon_g \|\mathbf{b}\| \quad \text{and} \quad \|\mathsf{B}\mathbf{x}^k - \mathbf{c}\| \le \varepsilon_f \|\mathbf{b}\|.$$

The relative precisions ε_f and ε_g should be judiciously determined. Our stopping criterion in the inner conjugate gradient loop of SMALE reads

$$\|\mathbf{g}(\mathbf{y}^i, \boldsymbol{\lambda}^i, \varrho_i)\| \le \min\{M\|\mathsf{B}\mathbf{y}^i - \mathbf{c}\|, \eta\} \;\text{ or }\; \|\mathbf{g}(\mathbf{y}^i, \boldsymbol{\lambda}^i, \varrho_i)\| \le \min\{\varepsilon_g, M\varepsilon_f\}\|\mathbf{b}\|,$$

so that the inner loop is interrupted when either the solution or a new iterate $\mathbf{x}^k = \mathbf{y}^i$ is found.

Before applying the algorithms presented to problems with a well-conditioned Hessian A, we strongly recommend to rescale the equality constraints so that $\|\mathsf{A}\| \approx \|\mathsf{B}\|$. Taking into account the estimate of the rate of convergence like (4.69), it is also useful to orthonormalize or at least normalize the constraints. This approach has been successfully applied, e.g., in the FETI-DP-based solver for analysis of layered composites [137].

If the Hessian A is ill-conditioned and there is an approximation M of A that can be used as preconditioner, then we can use the preconditioning strategies introduced in the discussion on implementation of the penalty method in Sect. 4.2.6. The construction of the matrix M is typically problem dependent. We refer interested readers to the books by Axelsson [4], Saad [163], van der Vorst [178], or Chen [21].

Sometimes it is possible to exploit the structure of the problem for very efficient implementation of preconditioning. For example, it has been shown that it is possible to find multigrid preconditioners to the discretized Stokes problem so that the latter can be solved by SMALE with asymptotically linear complexity [144].

4.7.2 Initialization of Constants

Though all the inexact algorithms converge with $0 < \alpha < 1$, $\beta > 1$, $\eta > 0$, $\eta_0 > 0$, $\varrho_0 > 0$, $M > 0$, and $\boldsymbol{\lambda}^0 \in \mathbb{R}^m$, their choice affects the performance of the algorithms and should exploit available information. Here we give a few hints and heuristics that can be useful for their efficient implementation.

The parameter α is used only in the adaptive augmented Lagrangian algorithm 4.4. This parameter determines the final rate of convergence of approximations of the Lagrange multipliers in the outer loop; however, its small value can slow down the convergence in the inner loop via increasing the penalty parameter. We use $\alpha = 0.1$.

The parameter β is used by SMALE algorithm 4.5 and the adaptive augmented Lagrangian algorithm 4.4 to increase the penalty parameter. Our experience indicates that $\beta = 10$ is a reasonable choice.

The parameter η is used only by SMALE algorithm 4.5. It helps to avoid outer iterations that do not invoke the inner CG iterations; we use $\eta = 0.1\|\mathbf{b}\|$.

The parameter η_0 is used by Algorithm 4.4 to define the initial bound on the feasibility error which is used to control the update of the penalty parameter. The algorithm does not seem to be sensitive with respect to η_0; we use $\eta_0 = 0.1\|\mathbf{b}\|$.

The estimate (4.99) shows that a large value of the initial penalty parameter ϱ_0 guarantees fast convergence of the outer loop. By analysis of the penalty method in Sect. 4.2, it is even possible to find the solution in one outer iteration. At the same time, the large value of the penalty parameter slows down the rate of convergence of the conjugate gradient method in the inner loop, but the analysis of the conjugate gradient method in Sect. 4.2.6 based on the effective condition number of $\mathsf{A}_\varrho = \mathsf{A} + \varrho \mathsf{B}^T \mathsf{B}$ indicates that the slowdown need not be severe when the number of constraints is small, or when the constraints are close to orthogonal. If neither is the case and at least crude estimates of $\|\mathsf{A}\|$ and $\|\mathsf{B}\|$ are available, a simple strategy can be based on the observation that

$$\lambda_{\min}(\mathsf{A}) \leq \lambda_{\min}(\mathsf{A}_\varrho) \quad \text{and} \quad \|\mathsf{A}_\varrho\| \leq \|\mathsf{A}\| + \varrho\|\mathsf{B}\|^2,$$

so that

$$\varrho\|\mathsf{B}\|^2 \leq C\|\mathsf{A}\| \Rightarrow \kappa(\mathsf{A} + \varrho \mathsf{B}^T \mathsf{B}) \leq (C+1)\kappa(\mathsf{A}).$$

For example, choosing $\varrho_0 = 8 \times \|\mathsf{A}\|/\|\mathsf{B}\|^2$ seems to be a reasonable guess which results in $\kappa(\mathsf{A}_\varrho) \leq 9\kappa(\mathsf{A})$. Let us stress that the update of the penalty parameter should be considered as a safeguard that guarantees the convergence; we should always try to avoid invoking increase of the penalty parameter as the iterates with too small penalty parameters are inefficient.

The parameter M balances the weight of the cost function and the constraints. In our implementations we use

$$M = \varepsilon_g / \varepsilon_f.$$

Notice that by Lemma 4.18 small M can prevent the penalty parameter from increasing. We can even replace the update of the penalty parameter in Step 3 by the reduction of the parameter M using $M_{k+1} = M_k/\beta$ and obvious modifications of the rest of Algorithm 4.5. See also Sect. 6.11.

If there is no better guess of the initial approximation of $\boldsymbol{\lambda}^0$, we use $\boldsymbol{\lambda}^0 = \mathbf{o}$. Recall that using $\boldsymbol{\lambda}^0 \in \mathrm{Im}\mathsf{B}$ results in $\boldsymbol{\lambda}^k$ converging to the least square Lagrange multiplier $\boldsymbol{\lambda}_{\mathrm{LS}}$.

4.8 Numerical Experiments

Here we illustrate the performance of the exact Uzawa algorithm, the exact augmented Lagrangian algorithm, and SMALE Algorithm 4.5 on minimization of the cost functions $f_{\mathrm{L},h}$ and $f_{\mathrm{LW},h}$ introduced in Sect. 3.10 subject to ill-conditioned multipoint constraints. Let us recall that we refer to Algorithm 4.2 as the Uzawa algorithm when $\varrho = 0$, and as the augmented Lagrangian algorithm when $\varrho > 0$.

4.8.1 Uzawa, Exact Augmented Lagrangians, and SMALE

Let us start with minimization of the quadratic function $f_{\mathrm{L},h}$ defined by the discretization parameter h (see page 98) subject to the multipoint constraints which join the displacements of the node with the coordinates $(0, 1/3)$ with all the other nodes in the square $[h, 1/3] \times [1/3, 2/3]$. Let us recall that the Hessian $\mathsf{A}_{\mathrm{L},h}$ of $f_{\mathrm{L},h}$ is ill-conditioned with the spectral condition number $\kappa(\mathsf{A}_{\mathrm{L},h}) \approx h^{-2}$.

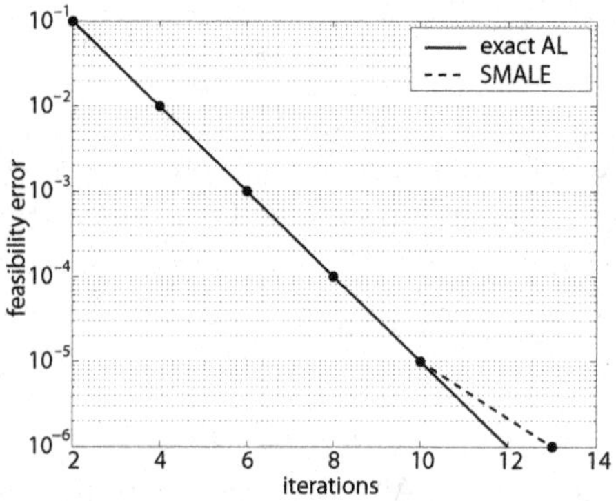

Fig. 4.3. Outer iterations of exact AL and SMALE algorithms

The graph of the relative feasibility error (vertical axis) against the numbers of outer iterations (horizontal axis) for exact augmented Lagrangians (exact AL) with $r_k = \varrho_k = 10$ and SMALE algorithm with $\varrho_0 = 10$ is in Fig. 4.3. The results were obtained with $h = 1/33$, which corresponds to $n = 1156$ unknowns and 131 multipliers. The inexact solution of auxiliary problems by SMALE has a small effect on the number of outer iterations. The SMALE algorithm required 964 CG iterations to reach the final precision. The same result was achieved by the original Uzawa algorithm with the optimal steplength after 3840 (!!!) iterations, each of them comprising direct solves of auxiliary linear problems. We conclude that even moderate regularization improves the convergence of the outer loop and the rate of convergence need not be slowed down by the inexact solution of auxiliary problems.

4.8.2 Numerical Demonstration of Optimality

To illustrate the optimality of SMALE for the solution of (4.1), let us consider the class of problems to minimize the quadratic function $f_{\mathrm{LW},h}$ (see page 99) subject to the multipoint constraints defined above. The class of problems can be given a mechanical interpretation associated to the expanding and partly stiff spring systems on Winkler's foundation. The spectrum of the Hessian $\mathsf{A}_{\mathrm{LW},h}$ of $f_{\mathrm{LW},h}$ is located in the interval $[2, 10]$. Thus the assumptions of Theorem 4.21 are satisfied and the number of outer iterations is bounded. Moreover, the rows of $\mathsf{B} \in \mathbb{R}^{m\times n}$ have a simple pattern given by

$$\mathsf{B}_{i*} = [0, \dots, 0, 1, 0, \dots, 0, -1, 0, \dots, 0], \quad i = 1, \dots, m.$$

It can be checked that $\mathsf{B}^T\mathsf{B}$ can be expressed as the sum of a matrix with the norm not exceeding four and a matrix of rank two. Using the reasoning of Sect. 4.2.6, we get that also the number of inner iterations is bounded.

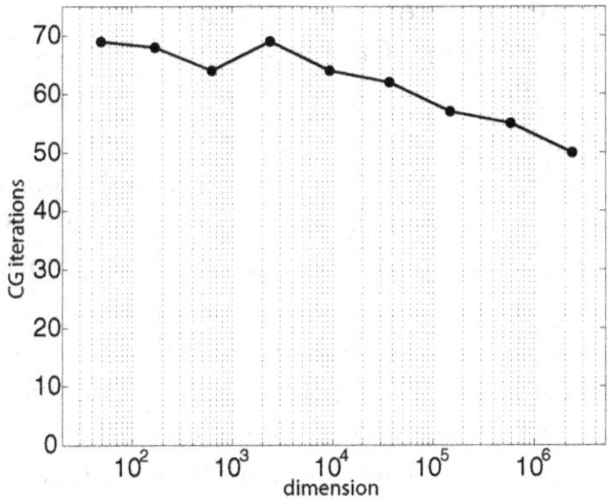

Fig. 4.4. Optimality of SMALE

In Fig. 4.4, on the vertical axis, we can see the numbers of the CG iterations k_n required to reduce the norm of the gradient and of the feasibility error to $10^{-6}\|\nabla f_{\mathrm{LW},h}(\mathbf{o})\|$ for the problems with the dimension n ranging from $n = 49$ to $n = 2362369$. The dimension n on the horizontal axis is in the logarithmic scale. We can see that k_n varies mildly with varying n, in agreement with Theorem 4.23 and the optimal property of CG. Moreover, since the cost of the matrix–vector multiplications is in our case proportional to the dimension n of the matrix $\mathsf{A}_{\mathrm{LW},h}$, it follows that the cost of the solution is also proportional to n. The number of outer iterations ranged from seven to ten.

The purpose of the above numerical experiment was just to illustrate the concept of optimality. For practical applications, it is necessary to combine SMALE with a suitable preconditioning. Application of SMALE with the multigrid preconditioning to development of in a sense optimal algorithm for the solution of the discretized Stokes problem is in Lukáš and Dostál [144].

4.9 Comments and References

The penalty method was exploited by a number of researchers to the solution of contact problems of elasticity [9, 108, 123, 125]. Theoretical results concerning the penalty method (e.g., Dostál [40], or Sect. 3.5 of Kikuchi and Oden [127]) yield that the norm of the approximation error depends on the condition number of the Hessian of the cost function. The analysis presented here generalizes the results of Dostál and Horák [65, 66]. The optimal feasibility estimate for the penalty methods (4.14) was used in development of a scalable algorithm for variational inequalities [65, 66]. The preconditioning preserving the gap in the spectrum was proposed in Dostál [44]. Reducing the spectrum of the penalized term to the one point, this preconditioning seems to be related to the constraint preconditioning for the saddle point systems introduced in nonlinear programming by Lukšan and Vlček [145]; see also Keller, Gould, and Wathen [126].

Augmented Lagrangian method was proposed independently by Powell [160] and Hestenes [116] for problems with a general cost function subject to general equality constraints. Comprehensive analysis of the augmented Lagrangian method (called the Lagrange multiplier method) including the asymptotically exact minimization of auxiliary problems was presented in the monograph by Bertsekas [11]. Applications to the solution of boundary value problems are discussed in Glowinski and Fortin [91] and Glowinski and Le Tallec [100]. Hager in [111, 113] obtained global convergence results for an algorithm of this type using inexact minimization in the solution of the auxiliary problems. In both papers the size of the optimality error was compared with the size of the feasibility error of the solution of the auxiliary problems trying to balance these quantities throughout the whole process. In [111] this comparison was used to decide whether the penalty parameter will be increased or not. In [113] it was used as a stopping criterion for the minimization of the auxiliary problems. The rate of convergence was free of any term due to inexact minimization when the least squares estimate of the Lagrange multipliers is used. Similar results for the linear update combined with the update of the penalty parameter that enforces a priori prescribed reduction of feasibility error were obtained by Dostál, Friedlander, and Santos [56] and Dostál, Friedlander, Santos, and Alesawi [58]. The same strategy was used by Conn, Gould, and Toint [26] for the solution of more general bound and equality constrained problems.

The SMALE algorithm was proposed by Dostál [46, 50]. The most attractive feature of this algorithm is a bound on the number of iterations which is independent of the constraint data. The bound has been obtained by a kind of global analysis; the result can hardly be obtained by analysis of one step of the algorithm. The algorithm has been combined with a multigrid preconditioning to develop in a sense optimal solver for the solution of a class of equality constrained problems arising from discretization of the Stokes problem; see Lukáš and Dostál [144].

Let us point out that our optimality results for the SMALE algorithm refer to the type of convergence which is known from the classical analysis of infinite series, but which is seldom exploited in numerical analysis. We shall call it the *sum bounding convergence of the second order* as it exploits the bound on the sum of the squares of errors. Though our sum bounding convergence does not guarantee even the linear rate of convergence, it is in our opinion rather a different characteristic of convergence than only a weaker one. For example, it does guarantee that the error bound for the following iterations is essentially reduced after any "bad" (here far from feasible) iteration, which is the property not guaranteed by more standard types of convergence. In our case, since we can control the upper bound by the penalty parameter, the sum bounding convergence offers an explanation to the fast convergence of the outer loop of SMALE which was observed in our numerical experiments [144].

5

Bound Constrained Minimization

We shall now be concerned with the *bound constrained problem* to find

$$\min_{\mathbf{x}\in\Omega_B} f(\mathbf{x}) \tag{5.1}$$

with $\Omega_B = \{\mathbf{x} \in \mathbb{R}^n : \ \mathbf{x} \geq \boldsymbol{\ell}\}$, $f(\mathbf{x}) = \frac{1}{2}\mathbf{x}^T\mathsf{A}\mathbf{x} - \mathbf{x}^T\mathbf{b}$, $\boldsymbol{\ell}$ and $\mathbf{b}$ given column n-vectors, and A an $n \times n$ symmetric positive definite matrix. To include the possibility that not all the components of $\mathbf{x}$ are constrained, we admit $\ell_i = -\infty$. Here we are again interested in large, sparse problems with a well-conditioned A, and in algorithms that can be used also for the solution of equality and inequality constrained problems. Such algorithms should be able to return an approximate solution at a cost proportional to the precision and to recognize an acceptable solution when it is found.

Our choice is the active set strategy with auxiliary problems solved approximately by the conjugate gradient method introduced in Sect. 3.5. It turns out that this type of algorithm can exploit effectively the specific structure of Ω_B, including the possibility to evaluate the projections in the Euclidean norm. We shall show that the resulting algorithm has an R-linear rate of convergence. If its parameters are chosen properly, the algorithm enjoys the finite termination property, even in the dual degenerate case with some active constraints corresponding to zero multipliers. We consider the finite termination property important, as it indicates that the algorithm does not suffer from undesirable oscillations and can exploit the superconvergence properties of the conjugate gradient method for linear problems.

As in the previous chapter, we first briefly review alternative algorithms for the solution of bound constrained problems. Then we introduce a basic active set algorithm and its modifications that are motivated by our effort to get the results on the rate of convergence in terms of bounds on the spectrum of the Hessian matrix A and on the finite termination. We restricted our attention to bound constrained problems because of their special structure which we exploit in the development of our algorithms. Let us recall that the problems with more general inequality constraints can be reduced to (5.1) by duality.

Zdeněk Dostál, *Optimal Quadratic Programming Algorithms*,
Springer Optimization and Its Applications, DOI 10.1007/978-0-387-84806-8_5,

Overview of algorithms

The *exact working (active) set method* of Sect. 5.3 reduces the solution of (5.1) to a sequence of unconstrained problems that are defined by the bounds which are assumed to be active at the solution. See also Algorithm 5.1. The performance of the algorithm is explained by the combinatorial arguments.

The *Polyak algorithm* is a variant of the working set method which solves the auxiliary linear problems by the conjugate gradient method. The active set is expanded whenever the unfeasible iterate is generated, typically by one index, but it is reduced only after the exact solution of an auxiliary unconstrained problem is found. The algorithm is described in Sect. 5.4. See Algorithm 5.2 for the formal description.

The *looking ahead Polyak algorithm* is based on observation that it is possible to recognize the incorrect active set before reaching the solution of the auxiliary unconstrained problem. The algorithm accepts inexact solutions of auxiliary unconstrained problems and preserves the finite termination property of the original Polyak algorithm. The algorithm is described in Sect. 5.5.2. See also Algorithm 5.3.

Even more relaxed solutions of the auxiliary unconstrained problems are accepted by the *easy re-release Polyak algorithm* of Sect 5.5.3. The algorithm preserves the finite termination property of the Polyak-type algorithms.

Unlike the Polyak-type algorithms, the *gradient projection with a fixed steplength* can typically add several indices to the active set in each step and it has established linear convergence in the bounds on the spectrum of the Hessian matrix. The algorithm is described in Sect. 5.6.3.

The *MPGP* (modified proportioning with gradient projections) algorithm of Sect. 5.7 uses the conjugate gradients to solve the auxiliary unconstrained problems with the precision controlled by the norm of violation of the Karush–Kuhn–Tucker conditions. The fixed steplength gradient projections are used to expand the active set. The basic scheme of MPGP is presented as Algorithm 5.6. The algorithm is proved to have an R-linear rate of convergence bounded in terms of the extreme eigenvalues of the Hessian matrix.

The *MPRGP* (modified proportioning with reduced gradient projections) algorithm of Sect. 5.8 is closely related to the MPGP algorithm, only the gradient projection step is replaced by the projection of the free gradient. The basic MPRGP scheme is presented as Algorithm 5.7. The R-linear rate of convergence is proved not only for the decrease of the cost function, but also for the norm of the projected gradient. The finite termination property is proved even for the problems with a dual degenerate solution.

The performance of MPGP and MPRGP can be improved by the *preconditioning* described in Sect. 5.10. The preconditioning in face improves the solution of the auxiliary unconstrained problems, while the preconditioning by the conjugate projector improves the convergence of the whole staff, including the nonlinear steps. The *monotonic MPRGP* and *semimonotonic MPRGP* algorithms which accept unfeasible iterations are described in Sect. 5.9.3.

5.1 Review of Alternative Methods

Before describing in detail the active set-based methods, let us briefly review alternative methods for the solution of the bound constrained problem (5.1).

Closely related to the active set strategy, various finite algorithms try to find $\mathbf{x} \in \mathbb{R}^n$ which solves the symmetric positive definite *LCP* (Linear Complementarity Problems)

$$\mathbf{g} = \mathsf{A}\mathbf{x} - \mathbf{b}, \quad \mathbf{x} \geq \mathbf{o}, \quad \mathbf{g} \geq \mathbf{o}, \quad \mathbf{x}^T\mathbf{g} = 0.$$

The LCP is equivalent to the minimization problem (5.1) with $\boldsymbol{\ell} = \mathbf{o}$. The algorithms are called finite as they find the solution in a finite number of steps; their analysis is based on the arguments of combinatorial nature. The most popular LCP solvers are probably *Lemke's algorithm* and *principal pivoting algorithm*, which reduce the LCP to the solution of a sequence of systems of linear equations in a way which is similar to the simplex method in linear programming. The solution of the auxiliary systems is typically implemented by LU-decompositions that are usually implemented by a rank one update. The result of the trial solve is used to improve a current approximation in order to reduce some characteristics of violation of the LCP conditions. These algorithms typically do not refer to the background minimization problems. The algorithms can be useful especially for more general LCP problems not considered here; see Cottle, Pang, and Stone [29].

Apart from the feasible active set methods presented in this chapter, it is possible to consider their unfeasible variants. For example, Kunisch and Rendl [139] proposed an iterative primal–dual algorithm which maintains the first-order optimality and complementarity conditions associated with (5.1) only; the feasibility is enforced by the update of the active set. The unfeasible methods are closely related to the *semismooth Newton method* applied to

$$\boldsymbol{\Phi}(\mathbf{x}) = \mathbf{o}, \quad \boldsymbol{\Phi}(\mathbf{x}) = \alpha^{-1}\big(\mathbf{x} - P_{\Omega_B}(\mathbf{x} - \alpha\nabla f(\mathbf{x}))\big), \quad \alpha > 0.$$

Hintermüller, Ito, and Kunisch [118] and Hintermüller, Kovtumenko, and Kunisch [119] describe the primal–dual semismooth Newton methods.

The bound constraints can be treated efficiently by the *interior point method*, which approximately minimizes the cost function modified by the parameterized barrier functions using Newton's method. The strong feature of the interior point methods is their capability to take into account all constraints, not only the active ones, at the cost of dealing with ill-conditioned problems. The performance of the interior point methods can exploit the sparsity pattern of the Hessian matrix A in the solution of auxiliary problems. There is a vast literature on this subject, see, e.g., the book by Wright [182] or the review paper by Forsgren, Gill, and Wright [90].

It is also possible to use the trust region-type methods that were developed to stabilize convergence of the Newton-type methods. We refer to Coleman and Lin [24, 25] for more details.

5.2 KKT Conditions and Related Inequalities

Since Ω_B is closed and convex and f is assumed to be strictly convex, the solution $\widehat{\mathbf{x}}$ of problem (5.1) exists and is necessarily unique by Proposition 2.5(i). Here we introduce some definitions and notations that enable us to exploit the special form of the KKT conditions in development of our algorithms. The KKT conditions fully determine the unique solution of (5.1).

By Proposition 2.18, the KKT conditions read

$$\mathsf{A}\widehat{\mathbf{x}} - \mathbf{b} \geq \mathbf{o} \quad \text{and} \quad (\mathsf{A}\widehat{\mathbf{x}} - \mathbf{b})^T(\widehat{\mathbf{x}} - \boldsymbol{\ell}) = 0,$$

or componentwise

$$\widehat{x}_i = \ell_i \Rightarrow \widehat{g}_i \geq 0 \quad \text{and} \quad \widehat{x}_i > \ell_i \Rightarrow \widehat{g}_i = 0, \quad i = 1, \dots, n, \tag{5.2}$$

where $\widehat{g}_i = [\mathsf{A}\widehat{\mathbf{x}} - \mathbf{b}]_i$. It may be observed that $\widehat{g}_i$ are the components of the vector of Lagrange multipliers for the bound constraints.

The KKT conditions (5.2) determine three important subsets of the set $\mathcal{N} = \{1, 2, \dots, n\}$ of all indices. The set of all indices for which $x_i = \ell_i$ is called an *active set* of $\mathbf{x}$. We denote it by $\mathcal{A}(\mathbf{x})$, so

$$\mathcal{A}(\mathbf{x}) = \{i \in \mathcal{N} : \ x_i = \ell_i\}.$$

Its complement

$$\mathcal{F}(\mathbf{x}) = \{i \in \mathcal{N} : \ x_i \neq \ell_i\}$$

and subsets

$$\mathcal{B}(\mathbf{x}) = \{i \in \mathcal{N} : \ x_i = \ell_i \text{ and } g_i > 0\}, \quad \mathcal{B}_0(\mathbf{x}) = \{i \in \mathcal{N} : \ x_i = \ell_i \text{ and } g_i \geq 0\}$$

are called a *free set*, a *binding set*, and a *weakly binding set*, respectively. Thus we can rewrite the KKT conditions in the form

$$\mathbf{g}_{\mathcal{A}}(\widehat{\mathbf{x}}) \geq \mathbf{o}_{\mathcal{A}} \quad \text{and} \quad \mathbf{g}_{\mathcal{F}}(\widehat{\mathbf{x}}) = \mathbf{o}_{\mathcal{F}}.$$

Using the subsets of $\mathcal{N}$, we can decompose the part of the gradient $\mathbf{g}(\mathbf{x}) = \mathsf{A}\mathbf{x} - \mathbf{b}$ which violates the KKT conditions into the *free gradient* $\boldsymbol{\varphi}$ and the *chopped gradient* $\boldsymbol{\beta}$ that are defined by

$$\begin{aligned} \varphi_i(\mathbf{x}) &= g_i(\mathbf{x}) \text{ for } i \in \mathcal{F}(\mathbf{x}), & \varphi_i(\mathbf{x}) &= 0 \text{ for } i \in \mathcal{A}(\mathbf{x}), \\ \beta_i(\mathbf{x}) &= 0 \text{ for } i \in \mathcal{F}(\mathbf{x}), & \beta_i(\mathbf{x}) &= g_i^-(\mathbf{x}) \text{ for } i \in \mathcal{A}(\mathbf{x}), \end{aligned}$$

where we have used the notation $g_i^- = \min\{g_i, 0\}$. Introducing the *projected gradient*

$$\mathbf{g}^P(\mathbf{x}) = \boldsymbol{\varphi}(\mathbf{x}) + \boldsymbol{\beta}(\mathbf{x}),$$

we can write the Karush–Kuhn–Tucker conditions (5.2) conveniently as

$$\mathbf{g}^P(\mathbf{x}) = \mathbf{o}. \tag{5.3}$$

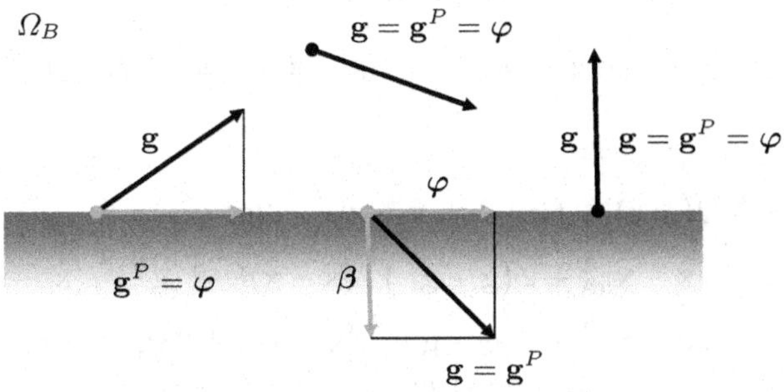

Fig. 5.1. Gradient splitting

Obviously $\beta(\mathbf{x})$ and $\varphi(\mathbf{x})$ are orthogonal and $-\beta(\mathbf{x})$ and $-\varphi(\mathbf{x})$ are feasible decrease directions of f at $\mathbf{x}$. See also Fig. 5.1.

If the dimension n of the bound constrained minimization problem (5.1) is large, it can be too ambitious to look for a solution which satisfies the gradient condition (5.3) exactly. A natural idea is to consider the weaker condition

$$\|\mathbf{g}^P(\mathbf{x})\| \leq \varepsilon, \tag{5.4}$$

but to require that the feasibility condition $\mathbf{x} \in \Omega_B$ is satisfied exactly. Notice that we are not able to check directly that we are near the solution as we do not know it, but we can easily evaluate (5.4). Thus the typical "solution" returned by iterative solvers is just $\mathbf{x}$ that satisfies the condition (5.4) with a small ε. The following lemma guarantees that any feasible vector $\mathbf{x}$ which satisfies (5.4) is near the solution.

Lemma 5.1. *Let $\widehat{\mathbf{x}}$ be the solution of (5.1) with a positive definite* A *and let* $\mathbf{g}^P = \mathbf{g}^P(\mathbf{x})$ *denote the projected gradient at* $\mathbf{x} \in \Omega_B$. *Then*

$$\|\mathbf{x} - \widehat{\mathbf{x}}\|_{\mathsf{A}}^2 \leq 2\big(f(\mathbf{x}) - f(\widehat{\mathbf{x}})\big) \leq \|\mathbf{g}^P\|_{\mathsf{A}^{-1}} \leq \lambda_{\min}^{-1}\|\mathbf{g}^P\|, \tag{5.5}$$

where $\lambda_{\min}$ *denotes the smallest eigenvalue of* A.

Proof. Let $\widehat{\mathcal{A}}$, $\widehat{\mathcal{F}}$, and $\widehat{\mathbf{g}}$ denote the active set, free set, and the gradient in the solution, respectively. Since $[\mathbf{x} - \widehat{\mathbf{x}}]_{\widehat{\mathcal{A}}} \geq \mathbf{o}_{\widehat{\mathcal{A}}}$, $\widehat{\mathbf{g}}_{\widehat{\mathcal{F}}} = \mathbf{o}_{\widehat{\mathcal{F}}}$, and $\widehat{\mathbf{g}} \geq \mathbf{o}$, we get

$$\begin{aligned} f(\mathbf{x}) - f(\widehat{\mathbf{x}}) &= \widehat{\mathbf{g}}^T(\mathbf{x} - \widehat{\mathbf{x}}) + \frac{1}{2}(\mathbf{x} - \widehat{\mathbf{x}})^T\mathsf{A}(\mathbf{x} - \widehat{\mathbf{x}}) \\ &= \widehat{\mathbf{g}}_{\widehat{\mathcal{A}}}^T[\mathbf{x} - \widehat{\mathbf{x}}]_{\widehat{\mathcal{A}}} + \frac{1}{2}\|\mathbf{x} - \widehat{\mathbf{x}}\|_{\mathsf{A}}^2 \geq \frac{1}{2}\|\mathbf{x} - \widehat{\mathbf{x}}\|_{\mathsf{A}}^2. \end{aligned}$$

This proves the left inequality of (5.5).

To prove the middle inequality, let $\mathcal{A} = \mathcal{A}(\mathbf{x})$ and $\mathcal{F} = \mathcal{F}(\mathbf{x})$ denote the active set and the free set of $\mathbf{x} \in \Omega_B$, respectively. Since

$$\mathbf{g}^P_{\mathcal{F}} = \mathbf{g}_{\mathcal{F}}, \quad [\widehat{\mathbf{x}} - \mathbf{x}]_{\mathcal{A}} \geq \mathbf{o}_{\mathcal{A}}, \quad \mathbf{g} = (\mathbf{g} - \mathbf{g}^P) + \mathbf{g}^P, \quad \text{and} \quad \mathbf{g} - \mathbf{g}^P \geq \mathbf{o},$$

we get

$$\begin{aligned}
0 \geq 2(f(\widehat{\mathbf{x}}) - f(\mathbf{x})) &= \|\widehat{\mathbf{x}} - \mathbf{x}\|^2_{\mathsf{A}} + 2\mathbf{g}^T(\widehat{\mathbf{x}} - \mathbf{x}) \\
&= \|\widehat{\mathbf{x}} - \mathbf{x}\|^2_{\mathsf{A}} + 2\left(\mathbf{g} - \mathbf{g}^P\right)^T(\widehat{\mathbf{x}} - \mathbf{x}) + 2\left(\mathbf{g}^P\right)^T(\widehat{\mathbf{x}} - \mathbf{x}) \\
&= \|\widehat{\mathbf{x}} - \mathbf{x}\|^2_{\mathsf{A}} + 2\left[\mathbf{g} - \mathbf{g}^P\right]^T_{\mathcal{A}}[\widehat{\mathbf{x}} - \mathbf{x}]_{\mathcal{A}} + 2\left(\mathbf{g}^P\right)^T(\widehat{\mathbf{x}} - \mathbf{x}) \\
&\geq \|\widehat{\mathbf{x}} - \mathbf{x}\|^2_{\mathsf{A}} + 2\left(\mathbf{g}^P\right)^T(\widehat{\mathbf{x}} - \mathbf{x}) \\
&\geq 2\left(\min_{\mathbf{y} \in \mathbb{R}^n} \frac{1}{2}\mathbf{y}^T\mathsf{A}\mathbf{y} + \left(\mathbf{g}^P\right)^T\mathbf{y}\right) = -(\mathbf{g}^P)^T\mathsf{A}^{-1}\mathbf{g}^P.
\end{aligned}$$

We used (2.11) in the last step. The middle inequality and the right inequality of (5.5) now follow by simple manipulations and (1.24), respectively. □

5.3 The Working Set Method with Exact Solutions

The basic idea of the *working set method*, or, as it is often called less correctly, the *active set method*, is to reduce the solution of an inequality constrained problem to the solution of a sequence of auxiliary equality constrained problems which are defined by a subset of the set $\mathcal{N} = \{1, \ldots, n\}$ of all indices of the constraints. This task would be very simple if we knew in advance which inequality constraints are active in the solution, as we could just replace the relevant inequalities by equalities, ignore the other inequalities, and solve the resulting equality constrained problem. As this is usually not the case, the working set method starts by making a guess which inequality constraints will be active in the solution, and if this guess turns out to be incorrect, it exploits the gradient and Lagrange multiplier information obtained by the trial minimization to define the next prediction.

5.3.1 Auxiliary Problems

If the working set method is applied to (5.1), it exploits the auxiliary equality constrained problems

$$\min_{\mathbf{y} \in \mathcal{W}_{\mathcal{I}}} f(\mathbf{y}), \tag{5.6}$$

where $\mathcal{I} \subseteq \mathcal{N}$ denotes the set of indices of bounds ℓ_i that are predicted to be active in the solution, and

$$\mathcal{W}_{\mathcal{I}} = \{\mathbf{y} : \; y_i = \ell_i, \; i \in \mathcal{I}\}.$$

The predicted set $\mathcal{I}$ of active bounds and $\mathcal{W}_\mathcal{I}$ are known as the *working set* and the *working face*, respectively. Since f is assumed to be strictly convex and $\mathcal{W}_\mathcal{I}$ is closed and convex, it follows by Proposition 2.5 that the auxiliary problem (5.6) has a unique solution $\widehat{\mathbf{y}}$.

Now observe that the equality constrained problem (5.6) can be reduced to an unconstrained problem in y_j, $j \notin \mathcal{I}$. To see its explicit form in the nontrivial cases $\mathcal{W}_\mathcal{I} \neq \{\boldsymbol{\ell}\}$ and $\mathcal{W}_\mathcal{I} \neq \mathbb{R}^n$, assume that $\emptyset \subsetneq \mathcal{J} \subsetneq \mathcal{N}$, and denote $\mathcal{J} = \mathcal{N} \setminus \mathcal{I}$, so that, after possibly rearranging the indices, we can write

$$\mathbf{y} = \begin{bmatrix} \mathbf{y}_\mathcal{I} \\ \mathbf{y}_\mathcal{J} \end{bmatrix}, \quad \mathbf{b} = \begin{bmatrix} \mathbf{b}_\mathcal{I} \\ \mathbf{b}_\mathcal{J} \end{bmatrix}, \quad \text{and} \quad \mathsf{A} = \begin{bmatrix} \mathsf{A}_{\mathcal{I}\mathcal{I}} & \mathsf{A}_{\mathcal{I}\mathcal{J}} \\ \mathsf{A}_{\mathcal{J}\mathcal{I}} & \mathsf{A}_{\mathcal{J}\mathcal{J}} \end{bmatrix}. \tag{5.7}$$

Thus for any $\mathbf{y} \in \mathbb{R}^n$

$$f(\mathbf{y}) = \frac{1}{2}\mathbf{y}_\mathcal{J}^T \mathsf{A}_{\mathcal{J}\mathcal{J}} \mathbf{y}_\mathcal{J} + \mathbf{y}_\mathcal{J}^T \mathsf{A}_{\mathcal{J}\mathcal{I}} \mathbf{y}_\mathcal{I} + \frac{1}{2}\mathbf{y}_\mathcal{I}^T \mathsf{A}_{\mathcal{I}\mathcal{I}} \mathbf{y}_\mathcal{I} - \mathbf{y}_\mathcal{J}^T \mathbf{b}_\mathcal{J} - \mathbf{y}_\mathcal{I}^T \mathbf{b}_\mathcal{I}.$$

Since $\mathbf{y} \in \mathcal{W}_\mathcal{I}$ if and only if $\mathbf{y}_\mathcal{I} = \boldsymbol{\ell}_\mathcal{I}$, we have for any $\mathbf{y} \in \mathcal{W}_\mathcal{I}$

$$f(\mathbf{y}) = f_\mathcal{J}(\mathbf{y}_\mathcal{J}) = \frac{1}{2}\mathbf{y}_\mathcal{J}^T \mathsf{A}_{\mathcal{J}\mathcal{J}} \mathbf{y}_\mathcal{J} - \mathbf{y}_\mathcal{J}^T (\mathbf{b}_\mathcal{J} - \mathsf{A}_{\mathcal{J}\mathcal{I}} \boldsymbol{\ell}_\mathcal{I}) + \frac{1}{2}\boldsymbol{\ell}_\mathcal{I}^T \mathsf{A}_{\mathcal{I}\mathcal{I}} \boldsymbol{\ell}_\mathcal{I} - \mathbf{b}_\mathcal{I}^T \boldsymbol{\ell}_\mathcal{I}.$$

Thus the solution $\widehat{\mathbf{y}}$ of (5.6) has the components $\widehat{\mathbf{y}}_\mathcal{I} = \boldsymbol{\ell}_\mathcal{I}$ and

$$\widehat{\mathbf{y}}_\mathcal{J} = \arg \min_{\mathbf{y}_\mathcal{J} \in \mathbb{R}^m} f_\mathcal{J}(\mathbf{y}_\mathcal{J}). \tag{5.8}$$

Since

$$\nabla f_\mathcal{J}(\mathbf{y}_\mathcal{J}) = \mathsf{A}_{\mathcal{J}\mathcal{J}} \mathbf{y}_\mathcal{J} - (\mathbf{b}_\mathcal{J} - \mathsf{A}_{\mathcal{J}\mathcal{I}} \boldsymbol{\ell}_\mathcal{I})$$

and $\nabla f_\mathcal{J}(\widehat{\mathbf{y}}_\mathcal{J}) = \mathbf{o}$, we get that $\widehat{\mathbf{y}}_\mathcal{J}$ satisfies

$$\mathsf{A}_{\mathcal{J}\mathcal{J}} \widehat{\mathbf{y}}_\mathcal{J} = \mathbf{b}_\mathcal{J} - \mathsf{A}_{\mathcal{J}\mathcal{I}} \boldsymbol{\ell}_\mathcal{I}. \tag{5.9}$$

We can check easily that (5.9) has a unique solution. Indeed, since $\mathsf{A}_{\mathcal{J}\mathcal{J}}$ is a submatrix of a positive definite matrix A, we get by Cauchy's interlacing inequalities (1.21) that $\mathsf{A}_{\mathcal{J}\mathcal{J}}$ is also positive definite. Alternatively, we can verify directly that $\mathsf{A}_{\mathcal{J}\mathcal{J}}$ is positive definite by observing that if $\mathbf{y}$ has the components $\mathbf{y}_\mathcal{I} = \mathbf{o}$ and $\mathbf{y}_\mathcal{J} \neq \mathbf{o}$, then $\mathbf{y} \neq \mathbf{o}$ and

$$\mathbf{y}_\mathcal{J}^T \mathsf{A}_{\mathcal{J}\mathcal{J}} \mathbf{y}_\mathcal{J} = \mathbf{y}^T \mathsf{A} \mathbf{y} > 0.$$

5.3.2 Algorithm

The *working set method with exact solutions* of auxiliary problems starts from an arbitrary $\mathbf{x}^0 \in \Omega_B$ and $\mathcal{I}^0 = \mathcal{B}_0(\mathbf{x}^0)$. Assuming that $\mathbf{x}^k$ is known, we first check if $\mathbf{x}^k$ is the solution of (5.1) by evaluating the KKT conditions

$$\mathbf{g}^P(\mathbf{x}^k) = \boldsymbol{\beta}(\mathbf{x}^k) + \boldsymbol{\varphi}(\mathbf{x}^k) = \mathbf{o}.$$

If this is not the case, we find the solution $\widehat{\mathbf{y}}$ of the auxiliary problem (5.6) by solving (5.9). There are two possibilities.

If $\widehat{\mathbf{y}} \in \Omega_B$, then we define the next iteration by the *feasible step*

$$\mathbf{x}^{k+1} = \widehat{\mathbf{y}}$$

and set $\mathcal{I}^{k+1} = \mathcal{B}_0(\mathbf{x}^{k+1})$. Notice that $f(\mathbf{x}^{k+1}) < f(\mathbf{x}^k)$ as $-\mathbf{g}^P(\mathbf{x}^k)$ is a feasible decrease direction of f at $\mathbf{x}^k$ with respect to $\mathcal{W}_I$.

In the other case, we define $\mathbf{x}^{k+1}$ by an *expansion step* so that

$$f(\mathbf{x}^{k+1}) \le f(\mathbf{x}^k) \quad \text{and} \quad \mathcal{A}(\mathbf{x}^{k+1}) \supsetneq \mathcal{I}^k, \tag{5.10}$$

and then set $\mathcal{I}^{k+1} = \mathcal{A}(\mathbf{x}^{k+1})$. The basic working set algorithm in the form that is convenient for analysis reads as follows.

Algorithm 5.1. The working set method with exact solutions.

Given a symmetric positive definite matrix $\mathsf{A} \in \mathbb{R}^{n\times n}$ *and n-vectors* $\mathbf{b}$, $\boldsymbol{\ell}$.
Step 0. {*Initialization.*}
  Choose $\mathbf{x}^0 \in \Omega_B$, *set* $\mathcal{I}^0 = \mathcal{B}_0(\mathbf{x}^0)$, $k = 0$
  while $\|\mathbf{g}^P(\mathbf{x}^k)\| > 0$
Step 1. {*Minimization in face* $\mathcal{W}_{\mathcal{I}^k}$. }
    $\widehat{\mathbf{y}} = \arg\ \min_{\mathbf{y}\in\mathcal{W}_{\mathcal{I}^k}} f(\mathbf{y})$
    if $\widehat{\mathbf{y}} \in \Omega_B$
Step 2. {*Feasible step.*}
      $\mathbf{x}^{k+1} = \widehat{\mathbf{y}}$
      $\mathcal{I}^{k+1} = \mathcal{B}_0(\mathbf{x}^{k+1})$
    else
Step 3. {*Expansion step.*}
      Set $\mathbf{x}^{k+1}$ *so that* $f(\mathbf{x}^{k+1}) \le f(\mathbf{x}^k)$ *and* $\mathcal{A}(\mathbf{x}^{k+1}) \supsetneq \mathcal{I}^k$
      $\mathcal{I}^{k+1} = \mathcal{A}(\mathbf{x}^{k+1})$
    end *if*
    $k = k + 1$
  end *while*
Step 4. {*Return solution.*}
    $\widehat{\mathbf{x}} = \mathbf{x}^k$

To implement the algorithm, we should specify the expansion step in more detail. For example, if $\mathbf{x}^k \in \Omega_B$ and

$$\mathbf{d} = \mathbf{x}^k - \widehat{\mathbf{y}},$$

we can observe that $-\mathbf{d}$ is a feasible decrease direction and that $f(\mathbf{x}^k - \alpha\mathbf{d})$ is a decreasing function of α for $\alpha \in [0, 1]$. Thus we can look for $\mathbf{x}^{k+1}$ in the form $\mathbf{x}^{k+1} = \mathbf{x}^k - \alpha\mathbf{d}$, $\alpha \in (0, 1]$. A possible choice of α is given by

$$\alpha_f = \arg\ \min_{\alpha\in(0,1]} \{f(\mathbf{x}^k - \alpha\mathbf{d}) :\ \mathbf{x}^k - \alpha\mathbf{d} \in \Omega_B\}, \tag{5.11}$$

which can be evaluated by using

$$\alpha_f = \min\{\alpha_m, 1\}, \quad \alpha_m = \min\{(x_i^k - \ell_i)/d_i : \ d_i > 0, \ i \in \mathcal{N}\}. \tag{5.12}$$

See also Fig. 5.2. Notice that if $\widehat{\mathbf{y}} \notin \Omega_B$, then the steplength α_f necessarily results in the expansion of the working set, typically by one index.

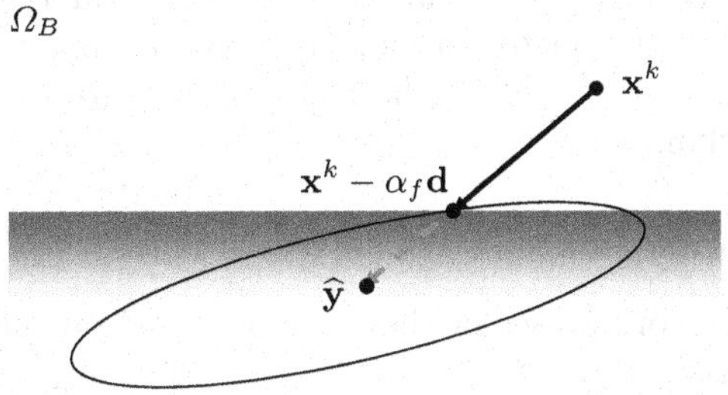

Fig. 5.2. Feasible steplength

This limitation may be overcome if we set $\mathbf{y} = \mathbf{x}^k - \alpha_f \mathbf{d}$ and define

$$\mathbf{x}^{k+1} = P_{\Omega_B}(\mathbf{y} - \alpha_p \mathbf{g}), \quad \alpha_p = \arg\min_{\alpha \geq 0} f\left(P_{\Omega_B}(\mathbf{y} - \alpha \mathbf{g})\right), \quad \mathbf{g} = \nabla f(\mathbf{y}),$$

where P_{Ω_B} is the Euclidean projection of Sect. 2.3.4. We prefer to use the gradient path, as the gradient defines a better local model of f than $\mathbf{d}$, though $-\mathbf{d}$ is the best global direction for minimization in the current working set. Figure 5.3 shows that α_f may be the best steplength for $\mathbf{d}$!

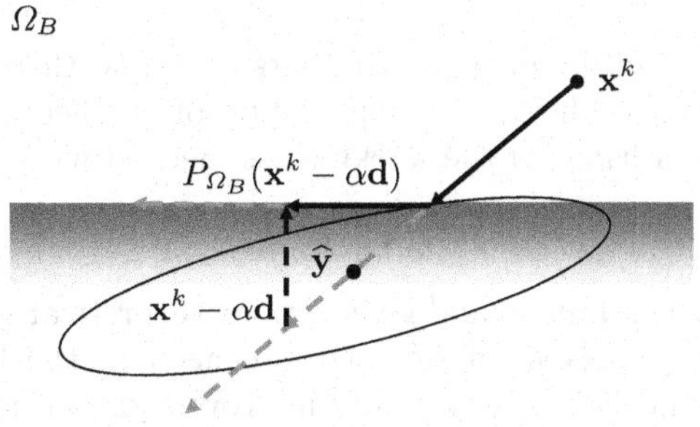

Fig. 5.3. Projected best unconstrained decrease path

To approximate α_p effectively, it is useful to notice that $f\left(P_{\Omega_B}(\mathbf{y}-\alpha\mathbf{g})\right)$ is a piecewise quadratic function because $P_{\Omega_B}(\mathbf{y}-\alpha\mathbf{g})$ is a linear mapping on any interval on which the active set of $P_{\Omega_B}(\mathbf{y}-\alpha\mathbf{g})$ is unchanged. We refer interested readers to Moré and Toraldo [153], Nocedal and Wright [155, Sect. 16.4], or to the discussion of the projected-gradient path in Conn, Gould, and Toint [28, Sect. 12.1.3]. We can also apply the fixed steplength reduced gradient projection which is described in Sect. 5.6.

The algorithm assumes by default that Step 1 is carried out by a direct method such as a matrix factorization, in which economies are possible by updating rather than recomputing the factorizations to account for gradual changes in the working set.

5.3.3 Finite Termination

The analysis of the working set method can be based on the following finite termination property.

Theorem 5.2. *Let Algorithm 5.1 be applied to find the solution $\widehat{\mathbf{x}}$ of (5.1) starting from $\mathbf{x}^0 \in \Omega_B$. Then there is k such that $\mathbf{x}^k = \widehat{\mathbf{x}}$.*

Proof. Since each expansion step adds at least one index into the working set, and the number of indices in the working set cannot exceed n, it follows that there are at most n consecutive expansion steps. Thus after each consecutive series of expansion steps, the algorithm either finds the solution of (5.1) and we are finished, or generates the next iterate, a feasible minimizer on the current face, by a feasible step. However, since $f(\mathbf{x}^k)$ is a nonincreasing sequence such that

$$f(\mathbf{x}^{k+1}) < f(\mathbf{x}^k)$$

whenever $\mathbf{x}^{k+1}$ is generated by a feasible step, it follows that no working set corresponding to an iterate generated by the feasible step can reappear. The number of different working sets being finite, we conclude that the working set method exploiting the exact solutions of auxiliary problems finds the solution of (5.1) in a finite number of steps. □

Since the number of different working sets is 2^n and there can be at most n expanding steps for each feasible step, the proof of Theorem 5.2 gives that the number N of iterations of the working set method with exact solution is bounded by

$$\overline{N} = n2^n. \tag{5.13}$$

This bound is very pessimistic and gives a poor theoretical support for practical computations, especially if we take into account the high cost of the iterations. The bound can be essentially improved for special problems. For example, if $\mathbf{x}^0 = \boldsymbol{\ell} = \mathbf{o}$ and the Hessian A of f is an M-matrix, then it is possible to show that Algorithm 5.1 generates only feasible steps and finds the solution in a number of iterations that does not exceed $n-p$, where p is the number of positive entries in $\mathbf{b}$. For more details see Diamond [32].

5.4 Polyak's Algorithm

If the auxiliary problems (5.6) are solved by the conjugate gradient method, it seems reasonable not to wait with the test of feasibility until their solution is found, but to modify the working set whenever unfeasible CG iteration is generated. This observation was enhanced in the *Polyak algorithm* [159], the starting point of our development of in a sense optimal algorithms.

5.4.1 Basic Algorithm

The new ingredient of the Polyak algorithm is that the minimization in face is replaced by a sequence of the *conjugate gradient steps* defined by

$$\mathbf{x}^k = \mathbf{x}^{k-1} - \alpha_{cg}\mathbf{p}^k, \tag{5.14}$$

where $\mathbf{p}^k$ denotes the recurrently constructed conjugate direction introduced in Sect. 3.2, and α_{cg} is the minimizer of $f(\mathbf{x}^{k-1} - \xi\mathbf{p}^k)$. The recurrence starts (or restarts) from $\mathbf{p}^{s+1} = \varphi(\mathbf{x}^s)$ whenever $\mathbf{x}^s$ is generated by the expansion step or $s = 0$. If $\mathbf{p}^k$ is known, then $\mathbf{p}^{k+1}$ is given by the formulae

$$\mathbf{p}^{k+1} = \varphi(\mathbf{x}^k) - \beta\mathbf{p}^k \quad \text{and} \quad \beta = \varphi(\mathbf{x}^k)^T\mathsf{A}\mathbf{p}^k/(\mathbf{p}^k)^T\mathsf{A}\mathbf{p}^k, \tag{5.15}$$

obtained by specialization of those introduced in Sect. 3.2. Let us recall that the conjugate directions $\mathbf{p}^{s+1}, \dots, \mathbf{p}^k$ that are generated by the recurrence (5.15) from the restart $\mathbf{x}^s$ are A-orthogonal, i.e., $(\mathbf{p}^i)^T\mathsf{A}\mathbf{p}^j = 0$ for any $i, j \in \{s+1, \dots, k\}$, $i \neq j$. Using the arguments of Sect. 3.1, it follows that

$$f(\mathbf{x}^k) = \min\left\{f(\mathbf{x}^s + \mathbf{y}) :\ \mathbf{y} \in \mathrm{Span}\{\mathbf{p}^{s+1}, \dots, \mathbf{p}^k\}\right\}. \tag{5.16}$$

The Polyak algorithm starts from an arbitrary feasible $\mathbf{x}^0$ by assigning $\mathcal{I}^0 = \mathcal{B}_0(\mathbf{x}^0)$ and initializing of the conjugate gradient loop (for details see Algorithm 5.2 or Sect. 3.2) for the minimization in $\mathcal{W}_{\mathcal{I}^0}$. Assuming that $\mathbf{x}^k$ is known, we first check if $\mathbf{x}^k$ solves either (5.1) or the auxiliary problem (5.6) by testing $\mathbf{g}^P(\mathbf{x}^k) = \mathbf{o}$ and $\varphi(\mathbf{x}^k) = \mathbf{o}$, respectively. If $\mathbf{g}^P(\mathbf{x}^k) = \mathbf{o}$, we are finished; if $\varphi(\mathbf{x}^k) = \mathbf{o}$, we reduce the working set to $\mathcal{I}^k = \mathcal{B}_0(\mathbf{x}^k)$ and initialize the conjugate gradient loop.

If the tests fail, we use the conjugate gradient step to define the trial iteration $\mathbf{y} = \mathbf{x}^k - \alpha_{cg}\mathbf{p}^{k+1}$. There are two possibilities. If $\mathbf{y}$ is feasible, then we set $\mathbf{x}^{k+1} = \mathbf{y}$. Otherwise we evaluate the feasible steplength by

$$\alpha_f = \arg\min_{\alpha \in (0, \alpha_{cg}]} \{f(\mathbf{x}^k - \alpha\mathbf{p}^{k+1}) :\ \mathbf{x}^k - \alpha\mathbf{p}^{k+1} \in \Omega_B\}, \tag{5.17}$$

set $\mathbf{x}^{k+1} = \mathbf{x}^k - \alpha_f\mathbf{p}^{k+1}$, expand the working set by $\mathcal{I}^{k+1} = \mathcal{A}(\mathbf{x}^{k+1})$, and finally initialize the new conjugate gradient loop.

The basic Polyak algorithm for the solution of strictly convex bound constrained quadratic programming problems takes the form shown by the following algorithm, where we omitted the indices of the vectors that are not referred to in what follows.

Algorithm 5.2. Polyak's algorithm.

Given a symmetric positive definite matrix $\mathsf{A} \in \mathbb{R}^{n\times n}$ *and n-vectors* $\mathbf{b}$, $\boldsymbol{\ell}$.
Step 0. {*Initialization.*}
  Choose $\mathbf{x}^0 \in \Omega_B$, *set* $\mathbf{g} = \mathsf{A}\mathbf{x}^0 - \mathbf{b}$, $\mathbf{p} = \mathbf{g}^P(\mathbf{x}^0)$, $k = 0$
  while $\|\mathbf{g}^P(\mathbf{x}^k)\| > 0$
   if $\|\varphi(\mathbf{x}^k)\| > 0$
Step 1. {*Trial conjugate gradient step.*}
    $\alpha_{cg} = \mathbf{g}^T\mathbf{p}/\mathbf{p}^T\mathsf{A}\mathbf{p}$, $\mathbf{y} = \mathbf{x}^k - \alpha_{cg}\mathbf{p}$
    $\alpha_f = \max\{\alpha : \mathbf{x}^k - \alpha\mathbf{p} \in \Omega_B\} = \min\{(x_i^k - \ell_i)/p_i : p_i > 0\}$
    if $\alpha_{cg} \le \alpha_f$
Step 2. {*Conjugate gradient step.*}
     $\mathbf{x}^{k+1} = \mathbf{y}$, $\mathbf{g} = \mathbf{g} - \alpha_{cg}\mathsf{A}\mathbf{p}$,
     $\beta = \varphi(\mathbf{y})^T\mathsf{A}\mathbf{p}/\mathbf{p}^T\mathsf{A}\mathbf{p}$, $\mathbf{p} = \varphi(\mathbf{y}) - \beta\mathbf{p}$
    else
Step 3. {*Expansion step.*}
     $\mathbf{x}^{k+1} = \mathbf{x}^k - \alpha_f\mathbf{p}$, $\mathbf{g} = \mathbf{g} - \alpha_f\mathsf{A}\mathbf{p}$, $\mathbf{p} = \varphi(\mathbf{x}^{k+1})$
    end *if*
   else
Step 4. {*Leaving the face after finding the minimizer.*}
    $\mathbf{d} = \beta(\mathbf{x}^k)$, $\alpha_{cg} = \mathbf{g}^T\mathbf{d}/\mathbf{d}^T\mathsf{A}\mathbf{d}$,
    $\mathbf{x}^{k+1} = \mathbf{x}^k - \alpha_{cg}\mathbf{d}$, $\mathbf{g} = \mathbf{g} - \alpha_{cg}\mathsf{A}\mathbf{d}$, $\mathbf{p} = \varphi(\mathbf{x}^{k+1})$
   end *if*
   $k = k + 1$
  end *while*
Step 5. {*Return solution.*}
  $\widehat{\mathbf{x}} = \mathbf{x}^k$

Our description of Algorithm 5.2 does not use explicitly the working sets; the information about the current working set is enhanced in the iterates $\mathbf{x}^k$ and the conjugate directions $\mathbf{p}^k$. Let us recall that the properties of the unconstrained conjugate gradient method are summarized in Theorem 3.1.

5.4.2 Finite Termination

Theorem 5.3. *Let Polyak's Algorithm 5.2 be applied to find the solution* $\widehat{\mathbf{x}}$ *of (5.1) starting from* $\mathbf{x}^0 \in \Omega_B$. *Then there is* k *such that* $\mathbf{x}^k = \widehat{\mathbf{x}}$.

Proof. First notice that by Theorem 3.1, there can be at most n consecutive conjugate gradient iterations before the minimizer in a face is found. If we remove all the iterates that are generated by Step 2 except the minimizers in the faces examined by the algorithm, which are used in Step 4 to generate the next iteration in the expanded face, we are left with the iterates that can be generated also by an implementation of Algorithm 5.1. The statement then follows by Theorem 5.2. □

The arguments of Sect. 5.3.3 can be used to show that the number of iterations of Polyak's algorithm is bounded by

$$\overline{N} = n^2 2^n. \tag{5.18}$$

Let us emphasize here that this bound is very pessimistic and can be improved, at least for special problems.

5.4.3 Characteristics of Polyak's Algorithm

The Polyak algorithm suffers from several drawbacks. The first one is related to an unpleasant consequence of application of the reduced conjugate gradient step with the steplength α_f defined by (5.17). Since the working set is typically expanded by one index only, there is a little chance that the number of iterations will be small when many indices of the binding set of the solution do not belong to $\mathcal{B}(\mathbf{x}^0)$. Another drawback concerns the basic approach combining the conjugate gradient method, which is now understood as an efficient iterative method for approximate solution of linear systems [4, 106, 163], and the finite termination strategy, which is based on combinatorial reasoning that requires exact solution of the auxiliary problems. Finally, as we have seen above, the combinatorial arguments give extremely poor bound on the number of iterations that are necessary to find the solution of (5.1). Though the bound (5.18) does not depend on the conditioning of A, it is rather poor and does not indicate why the algorithm should be efficient for the solution of well-conditioned problems.

5.5 Inexact Polyak's Algorithm

In this section we consider the variants of Polyak's algorithm which accept inexact solutions of auxiliary problems, but preserve the finite termination property.

5.5.1 Looking Ahead and Estimate

Let us first show that it is not necessary to solve the auxiliary problems (5.6) exactly in order to preserve the finite termination property of the Polyak algorithm. The key observation is that if $\mathbf{x}^{k+1} \in \Omega_B$ satisfies

$$f(\mathbf{x}^{k+1}) < \min\{f(\mathbf{x}) : \ \mathbf{x} \in \mathcal{W}_{\mathcal{I}}\}, \tag{5.19}$$

then the working set $\mathcal{I}$ cannot appear again as long as $\{f(\mathbf{x}^k)\}$ is nonincreasing. We shall use this simple observation to define both the precision control test and reduction of the active set.

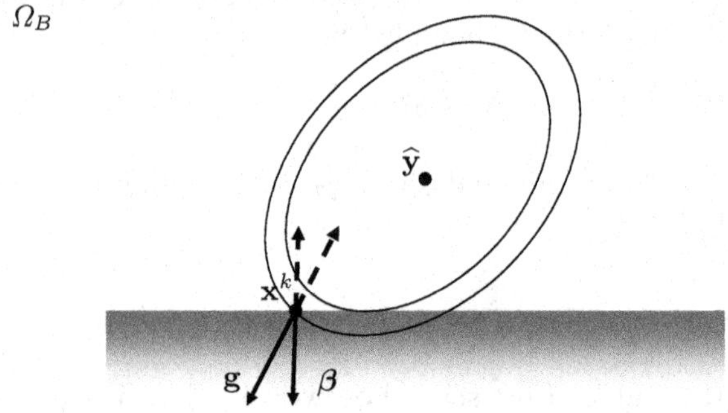

Fig. 5.4. Release directions at $\mathbf{x}^k$

Given $\mathbf{x}^k \in \mathcal{W}_\mathcal{I}$, we can try to find $\mathbf{x}^{k+1}$ which satisfies (5.19) in the form $\mathbf{x}^{k+1} = \mathbf{x}^k - \alpha\mathbf{d}$ with a given $\mathbf{d}$; if we are successful, we call $\mathbf{d}$ the *release direction of $\mathcal{W}_\mathcal{I}$ at $\mathbf{x}^k$*. The following lemma gives the conditions for $\mathbf{d}$, typically obtained from $\nabla f(\mathbf{x}^k)$ by reducing its components, to be a release direction. Such situation is depicted in Fig. 5.4 with $\mathbf{d} = \mathbf{g}(\mathbf{x}^k)$ and $\mathbf{d} = \boldsymbol{\beta}(\mathbf{x}^k)$.

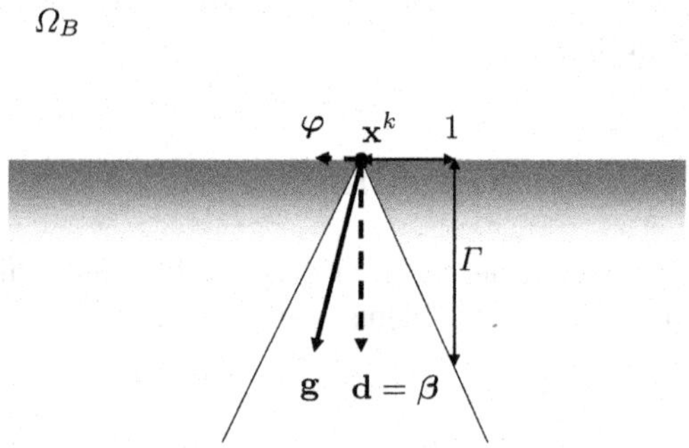

Fig. 5.5. The gradient and $\mathbf{d} = \boldsymbol{\beta}(\mathbf{x})$ that satisfy the release condition (5.20)

Lemma 5.4. *Let $\mathcal{I} = \mathcal{A}(\mathbf{x})$ and $\Gamma \geq \kappa(\mathsf{A})^{1/2}$, where $\kappa(\mathsf{A})^{1/2}$ denotes the spectral condition number of A. Denote $\mathbf{g} = \nabla f(\mathbf{x})$ and suppose that $\mathbf{d}$ satisfies*

$$\mathbf{g}^T\mathbf{d} \geq \|\mathbf{d}\|^2 \quad \textit{and} \quad \|\mathbf{d}\| > \Gamma\|\boldsymbol{\varphi}(\mathbf{x})\|. \tag{5.20}$$

Then the vector $\mathbf{y} = \mathbf{x} - \|\mathsf{A}\|^{-1}\mathbf{d}$ satisfies

$$f(\mathbf{y}) < \min\{f(\mathbf{x}) : \ \mathbf{x} \in \mathcal{W}_\mathcal{I}\}. \tag{5.21}$$

Proof. Let $\mathbf{x}$, Γ, and $\mathbf{d}$ satisfy the assumptions of Lemma 5.4 and notice that $\mathbf{g}^T\mathbf{d} \geq \|\mathbf{d}\|^2$ implies

$$f(\mathbf{y}) - f(\mathbf{x}) = \frac{1}{2}\|\mathsf{A}\|^{-2}\mathbf{d}^T\mathsf{A}\mathbf{d} - \|\mathsf{A}\|^{-1}\mathbf{d}^T\mathbf{g} \leq -\frac{1}{2}\|\mathsf{A}\|^{-1}\|\mathbf{d}\|^2. \tag{5.22}$$

Denoting $\mathcal{J} = \mathcal{F}(\mathbf{x})$, we have that $\|\mathbf{g}_\mathcal{J}\| = \|\varphi(\mathbf{x})\|$ and by the assumptions

$$\|\mathbf{d}\|^2 > \kappa(\mathsf{A})\|\mathbf{g}_\mathcal{J}\|^2. \tag{5.23}$$

Substituting (5.23) into (5.22) then yields

$$f(\mathbf{y}) - f(\mathbf{x}) < -\frac{1}{2}\|\mathsf{A}^{-1}\|\|\mathbf{g}_\mathcal{J}\|^2. \tag{5.24}$$

Now denote by $\bar{\mathbf{x}}$ and $\bar{\mathbf{g}}$ the minimizer of $f(\mathbf{x})$ on $\mathcal{W}_\mathcal{I}$ and the corresponding gradient vector, respectively. Direct computations yield

$$f(\mathbf{x}) - f(\bar{\mathbf{x}}) = \frac{1}{2}(\mathbf{x} - \bar{\mathbf{x}})^T\mathsf{A}(\mathbf{x} - \bar{\mathbf{x}}) + \bar{\mathbf{g}}^T(\mathbf{x} - \bar{\mathbf{x}}). \tag{5.25}$$

If we now rearrange the indices and take into account that

$$\bar{\mathbf{g}}_\mathcal{J} = \mathbf{o} \quad \text{and} \quad \mathbf{x}_\mathcal{I} = \bar{\mathbf{x}}_\mathcal{I},$$

we can further simplify the right-hand side of (5.25) to get

$$f(\mathbf{x}) - f(\bar{\mathbf{x}}) = \frac{1}{2}(\mathbf{x}_\mathcal{J} - \bar{\mathbf{x}}_\mathcal{J})^T\mathsf{A}_{\mathcal{J}\mathcal{J}}(\mathbf{x}_\mathcal{J} - \bar{\mathbf{x}}_\mathcal{J}). \tag{5.26}$$

To express $\mathbf{x}_\mathcal{J} - \bar{\mathbf{x}}_\mathcal{J}$ in terms of $\mathbf{g}_\mathcal{J}$, we can use the rearrangement (5.7) to get

$$\begin{bmatrix} \mathbf{g}_\mathcal{I} - \bar{\mathbf{g}}_\mathcal{I} \\ \mathbf{g}_\mathcal{J} \end{bmatrix} = \begin{bmatrix} \mathsf{A}_{\mathcal{I}\mathcal{I}} & \mathsf{A}_{\mathcal{I}\mathcal{J}} \\ \mathsf{A}_{\mathcal{J}\mathcal{I}} & \mathsf{A}_{\mathcal{J}\mathcal{J}} \end{bmatrix} \begin{bmatrix} \mathbf{o} \\ \mathbf{x}_\mathcal{J} - \bar{\mathbf{x}}_\mathcal{J} \end{bmatrix}. \tag{5.27}$$

In particular, since $\mathsf{A}_{\mathcal{J}\mathcal{J}}$ is also positive definite, it follows that

$$\mathbf{x}_\mathcal{J} - \bar{\mathbf{x}}_\mathcal{J} = \mathsf{A}_{\mathcal{J}\mathcal{J}}^{-1}\mathbf{g}_\mathcal{J}$$

and by (5.26)

$$f(\mathbf{x}) - f(\bar{\mathbf{x}}) = \frac{1}{2}\mathbf{g}_\mathcal{J}^T\mathsf{A}_{\mathcal{J}\mathcal{J}}^{-1}\mathbf{g}_\mathcal{J}. \tag{5.28}$$

Taking into account the interlacing properties of the spectra of principal submatrices of symmetric matrices (1.21), we get

$$\frac{1}{2}\mathbf{g}_\mathcal{J}^T\mathsf{A}_{\mathcal{J}\mathcal{J}}^{-1}\mathbf{g}_\mathcal{J} \leq \frac{1}{2}\|\mathsf{A}_{\mathcal{J}\mathcal{J}}^{-1}\|\|\mathbf{g}_\mathcal{J}\|^2 \leq \frac{1}{2}\|\mathsf{A}^{-1}\|\|\mathbf{g}_\mathcal{J}\|^2, \tag{5.29}$$

so that by (5.24) and (5.29)

$$\begin{aligned} f(\mathbf{y}) - f(\bar{\mathbf{x}}) &= \big(f(\mathbf{y}) - f(\mathbf{x})\big) + \big(f(\mathbf{x}) - f(\bar{\mathbf{x}})\big) \\ &< -\frac{1}{2}\|\mathsf{A}^{-1}\|\|\mathbf{g}_\mathcal{J}\|^2 + \frac{1}{2}\|\mathsf{A}^{-1}\|\|\mathbf{g}_\mathcal{J}\|^2 = 0. \end{aligned} \tag{5.30}$$

□

5.5.2 Looking Ahead Polyak's Algorithm

Using Lemma 5.4, we can now modify Polyak's algorithm so that it accepts approximate solution of the auxiliary problems and preserves its finite termination property. We only need to change the precision control of auxiliary problems. The *looking ahead Polyak algorithm* reads as follows.

Algorithm 5.3. Looking ahead Polyak's algorithm.

Given a symmetric positive definite matrix $\mathsf{A} \in \mathbb{R}^{n \times n}$, *n-vectors* $\mathbf{b}$, $\boldsymbol{\ell}$.
Step 0. {*Initialization.*}
  Choose $\mathbf{x}^0 \in \Omega_B$, $\Gamma \geq \kappa(\mathsf{A})^{1/2}$, *set* $\mathbf{g} = \mathsf{A}\mathbf{x}^0 - \mathbf{b}$, $\mathbf{p} = \mathbf{g}^P(\mathbf{x}^0)$, $k = 0$
  while $\|\mathbf{g}^P(\mathbf{x}^k)\| > 0$
   if $\Gamma\|\boldsymbol{\varphi}(\mathbf{x}^k)\| \geq \|\boldsymbol{\beta}(\mathbf{x}^k)\|$
Step 1. {*Trial conjugate gradient step.*}
    $\alpha_{cg} = \mathbf{g}^T\mathbf{p}/\mathbf{p}^T\mathsf{A}\mathbf{p}$, $\mathbf{y} = \mathbf{x}^k - \alpha_{cg}\mathbf{p}$
    $\alpha_f = \max\{\alpha : \mathbf{x}^k - \alpha\mathbf{p} \in \Omega_B\} = \min\{(x_i^k - \ell_i)/p_i : p_i > 0\}$
    if $\alpha_{cg} \leq \alpha_f$
Step 2. {*Conjugate gradient step.*}
     $\mathbf{x}^{k+1} = \mathbf{y}$, $\mathbf{g} = \mathbf{g} - \alpha_{cg}\mathsf{A}\mathbf{p}$,
     $\beta = \boldsymbol{\varphi}(\mathbf{y})^T\mathsf{A}\mathbf{p}/\mathbf{p}^T\mathsf{A}\mathbf{p}$, $\mathbf{p} = \boldsymbol{\varphi}(\mathbf{y}) - \beta\mathbf{p}$
    else
Step 3. {*Expansion step.*}
     $\mathbf{x}^{k+1} = \mathbf{x}^k - \alpha_f\mathbf{p}$, $\mathbf{g} = \mathbf{g} - \alpha_f\mathsf{A}\mathbf{p}$, $\mathbf{p} = \boldsymbol{\varphi}(\mathbf{x}^{k+1})$
    end *if*
   else
Step 4. {*Leaving the face in the release direction.*}
    $\mathbf{d} = \boldsymbol{\beta}(\mathbf{x}^k)$, $\alpha_{cg} = \mathbf{g}^T\mathbf{d}/\mathbf{d}^T\mathsf{A}\mathbf{d}$,
    $\mathbf{x}^{k+1} = \mathbf{x}^k - \alpha_{cg}\mathbf{d}$, $\mathbf{g} = \mathbf{g} - \alpha_{cg}\mathsf{A}\mathbf{d}$, $\mathbf{p} = \boldsymbol{\varphi}(\mathbf{x}^{k+1})$
   end *if*
   $k = k + 1$
  end *while*
Step 5. {*Return solution.*}
  $\widehat{\mathbf{x}} = \mathbf{x}^k$

To see that Algorithm 5.3 deserves its name, denote $\mathbf{d} = \boldsymbol{\beta}(\mathbf{x}^k)$ and assume that

$$\mathbf{x}^k \in \Omega_B, \quad \|\boldsymbol{\beta}(\mathbf{x}^k)\| > \Gamma\|\boldsymbol{\varphi}(\mathbf{x}^k)\|, \quad \text{and} \quad \Gamma \geq \kappa(\mathsf{A})^{1/2}, \tag{5.31}$$

so that $\mathbf{d}$ and Γ satisfy the assumptions of Lemma 5.4. Observing that α_{cg} minimizes $f(\mathbf{x}^k - \alpha\mathbf{d})$ with respect to α, we get for $\mathbf{x}^{k+1} = \mathbf{x}^k - \alpha_{cg}\mathbf{d}$ that

$$f(\mathbf{x}^{k+1}) \leq f(\mathbf{x}^k - \|\mathsf{A}\|^{-1}\mathbf{d}) < \min\{f(\mathbf{x}) : \mathbf{x} \in \mathcal{W}_{\mathcal{A}(\mathbf{x}^k)}\}.$$

Moreover, since $\mathbf{x}^k - \alpha\mathbf{d} \in \Omega_B$ for any $\alpha \geq 0$, we have $\mathbf{x}^{k+1} \in \Omega_B$. Thus the algorithm is able to recognize the face without the global solution before having a solution of the auxiliary problem, i.e., it "looks ahead".

The same reasoning as above can be carried out with $\mathbf{d} = \mathbf{g}^-(\mathbf{x}^k)$ or with some other nonzero vector $\mathbf{d}$ which satisfies the assumptions of Lemma 5.4. However, we found no significant evidence that there is a better choice than $\mathbf{d} = \beta(\mathbf{x}^k)$.

5.5.3 Easy Re-release Polyak's Algorithm

We can consider the relations like

$$\Gamma \|\varphi(\mathbf{x}^k)\| \geq \|\beta(\mathbf{x}^k)\|$$

for any $\Gamma > 0$. A reasonable choice is $\Gamma = 1$, as it seems natural to leave the current face when the norm of the chopped gradient starts to dominate the violation of the KKT conditions. The following *easy re-release Polyak's algorithm* enhances this observation by means of Lemma 5.4.

Algorithm 5.4. Easy re-release Polyak's algorithm.

Given a symmetric positive definite matrix $\mathsf{A} \in \mathbb{R}^{n \times n}$, *n-vectors* $\mathbf{b}$, $\boldsymbol{\ell}$.
Step 0. {*Initialization.*}
  Choose $\mathbf{x}^0 \in \Omega_B$, $\Gamma_M \geq \kappa(\mathsf{A})^{1/2}$, $0 \leq \Gamma_m \leq \Gamma_M$, *set* $\Gamma = \Gamma_M$, $k = 0$
  $\mathbf{g} = \mathsf{A}\mathbf{x}^0 - \mathbf{b}$, $\mathbf{p} = \mathbf{g}^P(\mathbf{x}^0)$
  while $\|\mathbf{g}^P(\mathbf{x}^k)\| > 0$
   if $\Gamma\|\varphi(\mathbf{x}^k)\| \geq \|\beta(\mathbf{x}^k)\|$
Step 1. {*Trial conjugate gradient step.*}
    $\alpha_{cg} = \mathbf{g}^T\mathbf{p}/\mathbf{p}^T\mathsf{A}\mathbf{p}$, $\mathbf{y} = \mathbf{x}^k - \alpha_{cg}\mathbf{p}$
    $\alpha_f = \max\{\alpha : \mathbf{x}^k - \alpha\mathbf{p} \in \Omega_B\} = \min\{(x_i^k - \ell_i)/p_i : p_i > 0\}$
    if $\alpha_{cg} \leq \alpha_f$
Step 2. {*Conjugate gradient step.*}
     $\mathbf{x}^{k+1} = \mathbf{y}$, $\mathbf{g} = \mathbf{g} - \alpha_{cg}\mathsf{A}\mathbf{p}$,
     $\beta = \varphi(\mathbf{y})^T\mathsf{A}\mathbf{p}/\mathbf{p}^T\mathsf{A}\mathbf{p}$, $\mathbf{p} = \varphi(\mathbf{y}) - \beta\mathbf{p}$
    else
Step 3. {*Expansion step.*}
     $\mathbf{x}^{k+1} = \mathbf{x}^k - \alpha_f\mathbf{p}$, $\mathbf{g} = \mathbf{g} - \alpha_f\mathsf{A}\mathbf{p}$, $\mathbf{p} = \varphi(\mathbf{x}^{k+1})$, $\Gamma = \Gamma_M$
    end *if*
   else
Step 4. {*Leaving the face in the release direction.*}
    $\mathbf{d} = \beta(\mathbf{x}^k)$, $\alpha_{cg} = \mathbf{g}^T\mathbf{d}/\mathbf{d}^T\mathsf{A}\mathbf{d}$,
    $\mathbf{x}^{k+1} = \mathbf{x}^k - \alpha_{cg}\mathbf{d}$, $\mathbf{g} = \mathbf{g} - \alpha_{cg}\mathsf{A}\mathbf{d}$, $\mathbf{p} = \varphi(\mathbf{x}^{k+1})$, $\Gamma = \Gamma_m$
   end *if*
   $k = k + 1$
  end *while*
Step 5. {*Return solution.*}
  $\widehat{\mathbf{x}} = \mathbf{x}^k$

Algorithm 5.4 uses the observations that we need not release the indices from the index set in one step and that the *release coefficient* Γ can change from iteration to iteration. The *easy re-release Polyak algorithm* starts with $\Gamma = \Gamma_M$, switches to $\Gamma = \Gamma_m$ when any index is released from the active set, and restores $\Gamma = \Gamma_M$ when the working set is expanded. Our experience [41] shows that Algorithm 5.4 is not very sensitive to the choice of Γ_m and works well with $\Gamma_m \approx 1$.

In what follows, we often use Step 4 of Algorithm 5.4 to release indices from the current active set. For any given $\Gamma > 0$, the iterates which satisfy $\|\boldsymbol{\beta}(\mathbf{x}^k)\| \leq \Gamma \|\boldsymbol{\varphi}(\mathbf{x}^k)\|$ are called *proportional*. The *proportioning step* sets $\mathbf{x}^{k+1} = \mathbf{x}^k - \alpha_{cg}\boldsymbol{\beta}(\mathbf{x}^k)$ in hope that the new iterate $\mathbf{x}^{k+1}$ is proportional.

5.5.4 Properties of Modified Polyak's Algorithms

Theorem 5.5. *Let the looking ahead Polyak Algorithm 5.3 or the easy re-release Polyak Algorithm 5.4 be applied to find the solution* $\widehat{\mathbf{x}}$ *of (5.1) starting from* $\mathbf{x}^0 \in \Omega_B$. *Then there is* k *such that* $\mathbf{x}^k = \widehat{\mathbf{x}}$.

Proof. First notice that the looking ahead Polyak Algorithm 5.3 generates the same iterates as the easy re-release Polyak Algorithm 5.4 provided $\Gamma_m = \Gamma_M$, so that it is enough to prove the statement for the latter algorithm. Since by Theorem 3.1 there can be at most n consecutive conjugate gradient iterations before the unconstrained minimizer is found, it follows that there can be at most n consecutive proportional conjugate gradient iterations. Moreover, since each proportioning step releases at least one index from the working set, which has at most n elements, we have that there can be at most n^2 iterations without an expansion step.

Now observe that the iterations start with $\Gamma = \Gamma_M$, that this value is reset by any expansion step, and that $\{f(\mathbf{x}^k)\}$ is nonincreasing. Since the chain of iterations with $\Gamma = \Gamma_M$ can be interrupted only after finding the iteration $\mathbf{x}^k$ which either solves (5.1), i.e., $\boldsymbol{\beta}(\mathbf{x}^k) = \boldsymbol{\varphi}(\mathbf{x}^k) = \mathbf{o}$, or is not proportional, i.e., satisfies $\|\boldsymbol{\beta}(\mathbf{x}^k)\| > \Gamma \|\boldsymbol{\varphi}(\mathbf{x}^k)\|$ with $\Gamma \geq \kappa(\mathsf{A})^{1/2}$, it follows by Lemma 5.4 that the associated active set $\mathcal{A}(\mathbf{x}^k)$ cannot be generated again in the following iterations. Since the number of all subsets of $\mathcal{N} = \{1, \ldots, n\}$ is bounded, and by Lemma 5.4 every iteration in the face with the solution is proportional when $\Gamma \geq \kappa(\mathsf{A})^{1/2}$, we conclude that the algorithm must generate $\mathbf{x}^k = \widehat{\mathbf{x}}$ in a finite number of steps. □

Our experience [41] indicates that our modifications of the Polyak algorithm outperform the original Polyak algorithm, but a little analysis shows that they suffer from many drawbacks described in Sect. 5.4.3. Moreover, their implementation requires an estimate of the condition number of A. The easy re-release Polyak algorithm with $\Gamma_m \approx 1$ usually outperforms the looking ahead Polyak algorithm as it can better avoid an "oversolve" of the auxiliary problems defined by the faces which do not contain the solution.

5.6 Gradient Projection Method

We shall now turn our attention to the iterative algorithms whose performance is substantiated by the convergence arguments. Instead of trying to find the exact solution of (5.1), these algorithms generate the iterates that steadily approach the solution until the KKT conditions are approximately satisfied. We start with a modification of the gradient method of Sect. 3.4 that uses the Euclidean projection P_{Ω_B} onto Ω_B to generate feasible iterates. The action of P_{Ω_B} is easy to calculate. As illustrated by Fig. 5.6, the components of the projection $P_{\Omega_B}(\mathbf{x})$ of $\mathbf{x}$ onto Ω_B are given by

$$[P_{\Omega_B}(\mathbf{x})]_i = \max\{\ell_i, x_i\}, \; i = 1, \ldots, n.$$

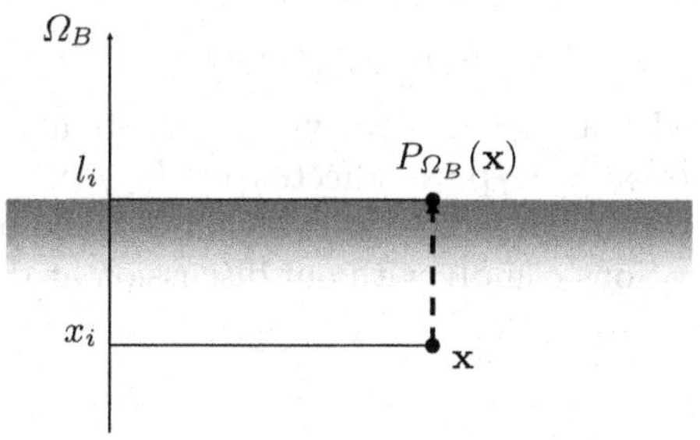

Fig. 5.6. Euclidean projection onto Ω_B

A typical step of the gradient projection method is in Fig. 5.7.

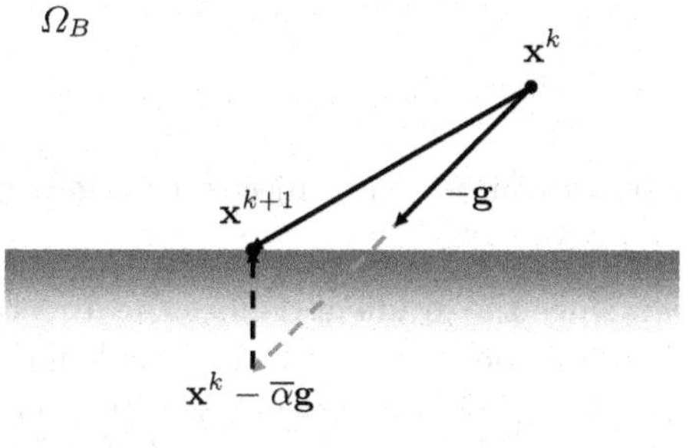

Fig. 5.7. Gradient projection step

5.6.1 Conjugate Gradient Versus Gradient Projections

Since the conjugate gradient is by Theorem 3.1 the best decrease direction which can be used to find the minimizer in the current Krylov space, probably the first idea how to plug the projection into the Polyak-type algorithms is to replace the reduced conjugate gradient step with the steplength α_f of (5.11) by the projected conjugate gradient step

$$\mathbf{x}^{k+1} = P_{\Omega_B}(\mathbf{x}^k - \alpha_{cg}\mathbf{p}^k).$$

However, if we examine Fig. 5.8, which depicts the 2D situation after the first conjugate gradient step, we can see that though the second conjugate gradient step finds the unconstrained minimizer $\mathbf{x}^k - \alpha_{cg}\mathbf{p}^k$, it can easily happen that

$$f(\mathbf{x}^k) < f(P_{\Omega_B}(\mathbf{x}^k - \alpha_{cg}\mathbf{p}^k)).$$

Figure 5.8 even suggests that it can happen for any $\alpha > \alpha_f$ that

$$f(P_{\Omega_B}(\mathbf{x}^k - \alpha\mathbf{p}^k)) > f\big(P_{\Omega_B}(\mathbf{x}^k - \alpha_f\mathbf{p}^k)\big).$$

Though Fig. 5.8 need not capture the typical situation when a small number of components of $\mathbf{x}^k - \alpha_f\mathbf{p}^k$ is affected by P_{Ω_B}, we conclude that the nice properties of the conjugate directions are guaranteed only in the feasible region. These observations comply with our discussion at the end of Sect. 3.5.

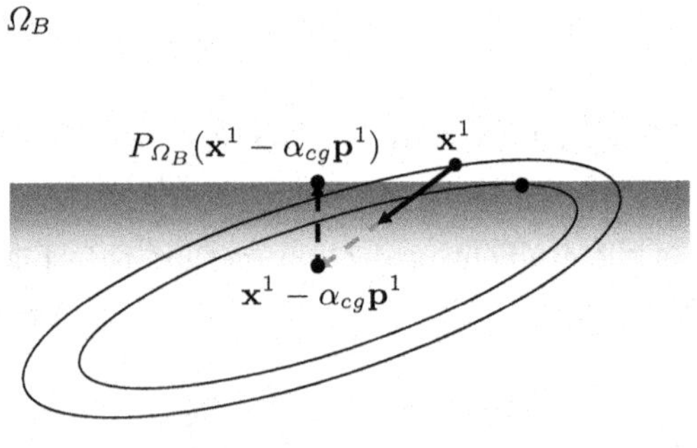

Fig. 5.8. Poor performance of the projected conjugate gradient step

On the other hand, since the gradient defines the direction of the steepest descent, it is natural to assume that for a small steplength the gradient perturbed by the projection P_{Ω_B} defines a decrease direction as in Fig. 5.9. We shall give a quantitative proof to this conjecture. In what follows, we restrict our attention to the analysis of the fixed steplength gradient iteration

$$\mathbf{x}^{k+1} = P_{\Omega_B}(\mathbf{x}^k - \overline{\alpha}\mathbf{g}^k), \tag{5.32}$$

where $\mathbf{g}^k = \nabla f(\mathbf{x}^k)$.

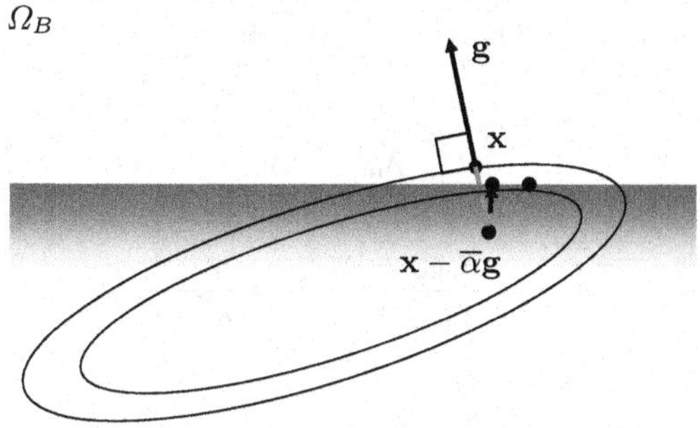

Fig. 5.9. Fixed steplength gradient step

5.6.2 Contraction in the Euclidean Norm

Which values of $\overline{\alpha}$ guarantee that the iterates defined by the fixed gradient projection step (5.32) approach the solution $\widehat{\mathbf{x}}$ in the Euclidean norm?

Proposition 5.6. *Let* $\mathbf{x} \in \Omega_B$ *and* $\mathbf{g} = \nabla f(\mathbf{x})$. *Then for any* $\overline{\alpha} > 0$

$$\|P_{\Omega_B}(\mathbf{x} - \overline{\alpha}\mathbf{g}) - \widehat{\mathbf{x}}\| \le \eta_E \|\mathbf{x} - \widehat{\mathbf{x}}\|, \tag{5.33}$$

where $\lambda_{\min}$, $\lambda_{\max}$ *are the extreme eigenvalues of* A *and*

$$\eta_E = \max\{|1 - \overline{\alpha}\lambda_{\min}|, |1 - \overline{\alpha}\lambda_{\max}|\}. \tag{5.34}$$

Proof. Since $\widehat{\mathbf{x}} \in \Omega_B$ and the projected gradient at the solution satisfies $\widehat{\mathbf{g}}^P = \mathbf{o}$, it follows that

$$P_{\Omega_B}(\widehat{\mathbf{x}} - \overline{\alpha}\widehat{\mathbf{g}}) = \widehat{\mathbf{x}}.$$

Using that the projection P_Ω is nonexpansive by Corollary 2.7, the formula $\mathbf{g}(\mathbf{x}) = \mathsf{A}\mathbf{x} - \mathbf{b}$, the relations between the norm of a symmetric matrix and its spectrum (1.23), and the observation that if λ_i are the eigenvalues of A, then $1 - \overline{\alpha}\lambda_i$ are the eigenvalues of $\mathsf{I} - \overline{\alpha}\mathsf{A}$ (see also (1.26)), we get

$$\begin{aligned}
\|P_{\Omega_B}(\mathbf{x} - \overline{\alpha}\mathbf{g}) - \widehat{\mathbf{x}}\| &= \|P_{\Omega_B}(\mathbf{x} - \overline{\alpha}\mathbf{g}) - P_{\Omega_B}(\widehat{\mathbf{x}} - \overline{\alpha}\widehat{\mathbf{g}})\| \\
&\le \|(\mathbf{x} - \overline{\alpha}\mathbf{g}) - (\widehat{\mathbf{x}} - \overline{\alpha}\widehat{\mathbf{g}})\| \\
&= \|\,(\mathbf{x} - \widehat{\mathbf{x}}) - \overline{\alpha}(\mathbf{g} - \widehat{\mathbf{g}})\| = \|\,(\mathbf{x} - \widehat{\mathbf{x}}) - \overline{\alpha}\mathsf{A}(\mathbf{x} - \widehat{\mathbf{x}})\| \\
&= \|(\mathsf{I} - \overline{\alpha}\mathsf{A})(\mathbf{x} - \widehat{\mathbf{x}})\| \\
&\le \max\{|1 - \overline{\alpha}\lambda_{\min}|, |1 - \overline{\alpha}\lambda_{\max}|\}\|\mathbf{x} - \widehat{\mathbf{x}}\|.
\end{aligned}$$

□

We call η_E the *coefficient of Euclidean contraction.* If $\overline{\alpha} \in (0, 2\|\mathsf{A}\|^{-1})$, then $\eta_E < 1$.

Using elementary arguments of Sect. 3.5.3, we get that the coefficient η_E of Euclidean contraction (5.34) is minimized by

$$\overline{\alpha}_E^{opt} = \frac{2}{\lambda_{\min} + \lambda_{\max}} \tag{5.35}$$

and

$$\eta_E^{opt} = \frac{\lambda_{\max} - \lambda_{\min}}{\lambda_{\max} + \lambda_{\min}} = \frac{\kappa - 1}{\kappa + 1}, \tag{5.36}$$

where

$$\kappa = \lambda_{\max}/\lambda_{\min}$$

denotes the spectral condition number of A.

If we compare our new estimate (5.36) of the contraction of the projected gradient step with the optimal steplength $\overline{\alpha}$ in the A-norm with the estimate (3.26) of the unconstrained gradient step with the optimal steplength α_{cg} in the A-norm norm, we find, a bit surprisingly, that they are the same. This might suggest to use the A-norm optimal steplength α_{cg} also in the projected gradient step.

Unfortunately, this strategy does not work. The counterexample of Fig. 5.10 shows that if $\mathbf{g} = \mathbf{g}(\mathbf{x})$ is the eigenvector corresponding to the smallest eigenvalue $\lambda_{\min}$, then the gradient projection step with the optimal conjugate gradient steplength

$$\alpha_{cg} = \|\mathbf{g}\|^2/\mathbf{g}^T\mathsf{A}\mathbf{g} = 1/\lambda_{\min}$$

generates the iterate which is worse than $\mathbf{x}$.

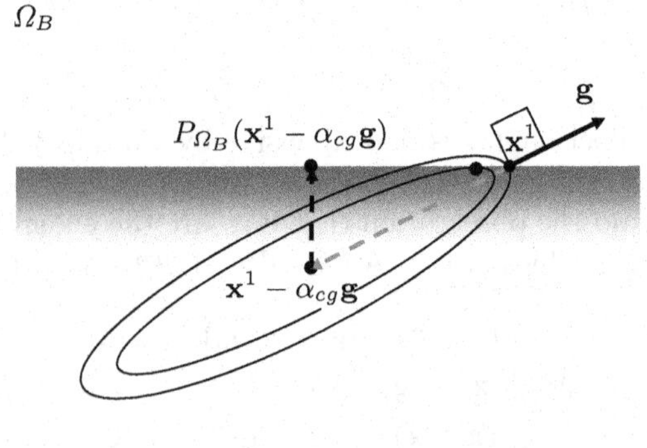

Fig. 5.10. Optimal unconstrained steplength may not be useful

Notice that the estimate (5.33) does not guarantee any bound on the decrease of the cost function. We study this topic in Sect. 5.6.5.

5.6.3 The Fixed Steplength Gradient Projection Method

Proposition 5.6 suggests that we can use the gradient projection with the fixed steplength to define an iterative algorithm with the rate of convergence in terms of bounds on the spectrum. To guarantee the convergence, the algorithm requires a computable upper bound on $\|\mathsf{A}\|$. Since A is assumed to be symmetric, it follows that $\|\mathsf{A}\|_1 = \|\mathsf{A}\|_\infty$ and, using (1.14), that $\|\mathsf{A}\| \leq \|\mathsf{A}\|_\infty$. Thus we can use $\|\mathsf{A}\|_\infty$ as the upper bound. The latter inequality can be obtained also from (1.24). More hints concerning effective evaluation of an upper bound on $\|\mathsf{A}\|$ can be found in Sect. 5.9.4. The gradient projection algorithm with the fixed steplength takes the following form.

Algorithm 5.5. Gradient projection method with the fixed steplength.

Given a symmetric positive definite matrix $\mathsf{A} \in \mathbb{R}^{n\times n}$ *and n-vectors* $\mathbf{b}$, $\boldsymbol{\ell}$.
Step 0. {*Initialization.*}
    Choose $\mathbf{x}^0 \in \Omega_B$, $\overline{\alpha} \in (0, 2\|\mathsf{A}\|^{-1})$, *set* $k = 0$
    while $\|\mathbf{g}^P(\mathbf{x}^k)\|$ *is not small*
Step 1. {*The gradient projection step.*}
        $\mathbf{x}^{k+1} = P_{\Omega_B}\left(\mathbf{x}^k - \overline{\alpha}\mathbf{g}(\mathbf{x}^k)\right)$
        $k = k + 1$
    end *while*
Step 2. {*Return (possibly inexact) solution.*}
    $\widetilde{\mathbf{x}} = \mathbf{x}^k$

We can use recurrently the estimate (5.33) of Proposition 5.6 to get for $k \geq 1$ that

$$\|\mathbf{x}^k - \widehat{\mathbf{x}}\| \leq \eta_E \|\mathbf{x}^{k-1} - \widehat{\mathbf{x}}\| \leq \cdots \leq \eta_E^k \|\mathbf{x}^0 - \widehat{\mathbf{x}}\|, \tag{5.37}$$

where $\eta_E < 1$ is the coefficient of Euclidean contraction defined by (5.34). It follows that Algorithm 5.5 generates the iterates $\mathbf{x}^k$ that *converge to the solution* $\widehat{\mathbf{x}}$ *of* (5.1) *in the Euclidean norm linearly* with the coefficient of contraction η_E. The iterates $\mathbf{x}^k$ converge in the A-norm only R-linearly with

$$\|\mathbf{x}^k - \widehat{\mathbf{x}}\|_{\mathsf{A}} \leq \eta_E^k \|\mathsf{A}\| \|\mathbf{x}^0 - \widehat{\mathbf{x}}\|. \tag{5.38}$$

Though the cost of a step of Algorithm 5.5 is comparable to that of the Polyak-type algorithms, the performance of these algorithms essentially differs. A nice feature of the gradient projection algorithm is the rate of convergence in terms of bounds on the spectrum. This can hardly be proved for the Polyak algorithm; when a component of the current iterate is near the bound and the corresponding component of the conjugate direction is large, then the feasible steplength α_f and the relative decrease of the cost function can be arbitrarily small. On the other hand, unlike the Polyak algorithm, Algorithm 5.5 is not able to exploit information from the previous steps in one face.

5.6.4 Quadratic Functions with Identity Hessian

Which values of $\overline{\alpha}$ guarantee that the cost function f decreases in each iterate defined by the fixed gradient projection step (5.32)? How much does f decrease when the answer is positive? To answer these questions, it is useful to carry out some analysis of a special quadratic function

$$F(\mathbf{x}) = \frac{1}{2}\mathbf{x}^T\mathbf{x} - \mathbf{c}^T\mathbf{x}, \quad \mathbf{x} \in \mathbb{R}^n, \tag{5.39}$$

which is defined by a fixed $\mathbf{c} \in \mathbb{R}^n, \ \mathbf{c} = [c_i]$. We shall also use

$$F(\mathbf{x}) = \sum_{i=1}^{n} F_i(x_i), \quad F_i(x_i) = \frac{1}{2}x_i^2 - c_i x_i, \quad \mathbf{x} = [x_i]. \tag{5.40}$$

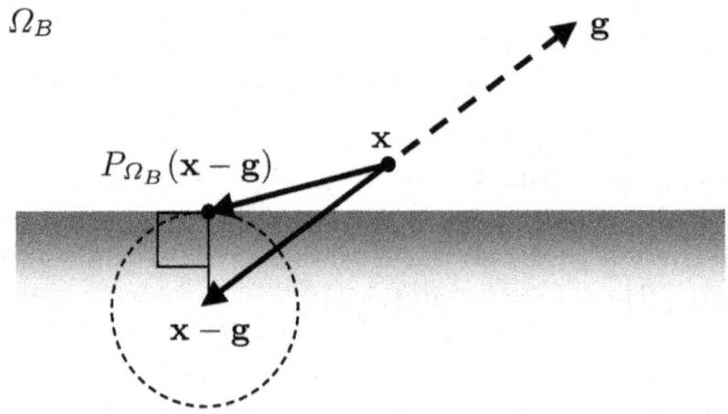

Fig. 5.11. Minimizer of F in Ω_B

The Hessian and the gradient of F are expressed, respectively, by

$$\nabla^2 F(\mathbf{x}) = \mathsf{I} \quad \text{and} \quad \mathbf{g} = \nabla F(\mathbf{x}) = \mathbf{x} - \mathbf{c}, \quad \mathbf{g} = [g_i]. \tag{5.41}$$

Thus $\mathbf{c} = \mathbf{x} - \mathbf{g}$ and for any $\mathbf{z} \in \mathbb{R}^n$

$$\|\mathbf{z} - \mathbf{c}\|^2 = \|\mathbf{z}\|^2 - 2\mathbf{c}^T\mathbf{z} + \|\mathbf{c}\|^2 = 2F(\mathbf{z}) + \|\mathbf{c}\|^2.$$

Since by Proposition 2.6 for any $\mathbf{z} \in \Omega_B$

$$\|\mathbf{z} - \mathbf{c}\| \geq \|P_{\Omega_B}(\mathbf{c}) - \mathbf{c}\|,$$

we get that for any $\mathbf{z} \in \Omega_B$

$$\begin{aligned} 2F(\mathbf{z}) &= \|\mathbf{z} - \mathbf{c}\|^2 - \|\mathbf{c}\|^2 \geq \|P_{\Omega_B}(\mathbf{c}) - \mathbf{c}\|^2 - \|\mathbf{c}\|^2 \\ &= 2F\left(P_{\Omega_B}(\mathbf{c})\right) = 2F\left(P_{\Omega_B}(\mathbf{x} - \mathbf{g})\right). \end{aligned} \tag{5.42}$$

We have thus proved that if $\mathbf{y} \in \Omega_B$, then, as illustrated in Fig. 5.11,

$$F\big(P_{\Omega_B}(\mathbf{x} - \mathbf{g})\big) \leq F(\mathbf{y}). \tag{5.43}$$

We are especially interested in the analysis of F along the projected-gradient path

$$p(\mathbf{x},\alpha) = P_{\Omega_B}(\mathbf{x} - \alpha\nabla F(\mathbf{x})) = \max\{\mathbf{x} - \alpha\mathbf{g}, \boldsymbol{\ell}\},$$

where the maximum is assumed to be carried out componentwise, $\alpha \geq 0$, and $\mathbf{x} \in \Omega_B$ is fixed. We shall often use that the projected-gradient path can be described by

$$p(\mathbf{x},\alpha) = P_{\Omega_B}(\mathbf{x} - \alpha\mathbf{g}) = \mathbf{x} - \alpha\widetilde{\mathbf{g}}(\alpha), \tag{5.44}$$

where $\widetilde{\mathbf{g}}(\alpha)$ denotes the *reduced gradient* whose components are defined by

$$\widetilde{g}_i(0) = 0 \quad \text{and} \quad \widetilde{g}_i(\alpha) = \min\{(x_i - \ell_i)/\alpha, g_i\} \;\text{ for }\; \alpha > 0.$$

A geometric illustration of the projected-gradient path is in Fig. 5.12.

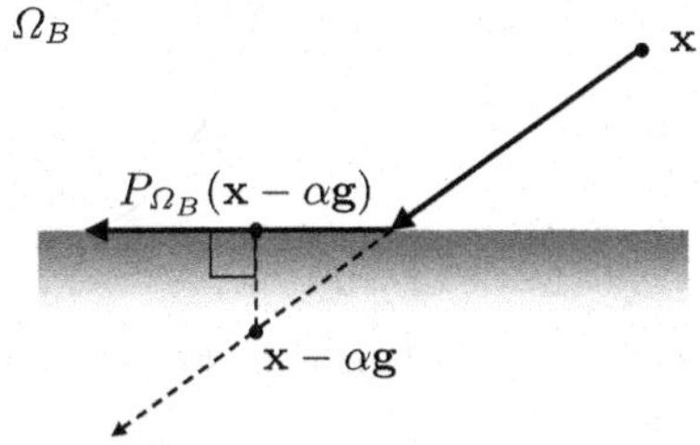

Fig. 5.12. Projected-gradient path

Due to the separability of F, the following analysis of a special case with F defined on $\mathbb{R}$ is important also in the general case.

Lemma 5.7. *Let $x, \ell, c \in \mathbb{R}$, $x \geq \ell$. Let F and g be defined by*

$$F(x) = \frac{1}{2}x^2 - cx \quad \text{and} \quad g = x - c.$$

Then for any $\delta \in [0,1]$

$$F\big(P_{\Omega_B}(x - (2-\delta)g)\big) \leq F\big(P_{\Omega_B}(x - \delta g)\big). \tag{5.45}$$

Proof. First assume that $x \geq l$ is fixed and denote

$$g = F'(x) = x - c, \quad \widetilde{g}(0) = 0, \quad \widetilde{g}(\alpha) = \min\{(x-\ell)/\alpha, g\}, \quad \alpha \neq 0.$$

For convenience, let us define

$$F\big(P_{\Omega_B}(x - \alpha g)\big) = F(x) + \Phi(\alpha), \quad \Phi(\alpha) = -\alpha\widetilde{g}(\alpha)g + \frac{\alpha^2}{2}\left(\widetilde{g}(\alpha)\right)^2, \quad \alpha \geq 0.$$

Moreover, using these definitions, it can be checked directly that Φ is defined explicitly by

$$\Phi(\alpha) = \begin{cases} \Phi_F(\alpha) & \text{for} \quad \alpha \in (-\infty, \overline{\xi}] \cap [0, \infty) \quad \text{or} \quad g \le 0, \\ \Phi_A(\alpha) & \text{for} \quad \alpha \in [\ \overline{\xi}, \infty) \cap [0, \infty) \quad \text{and} \quad g > 0, \end{cases}$$

where $\overline{\xi} = \infty$ if $g = 0$, $\overline{\xi} = (x - \ell)/g$ if $g \neq 0$,

$$\Phi_F(\alpha) = \left(-\alpha + \frac{\alpha^2}{2}\right) g^2, \quad \text{and} \quad \Phi_A(\alpha) = -g(x - \ell) + \frac{1}{2}(x - \ell)^2.$$

See also Fig. 5.13.

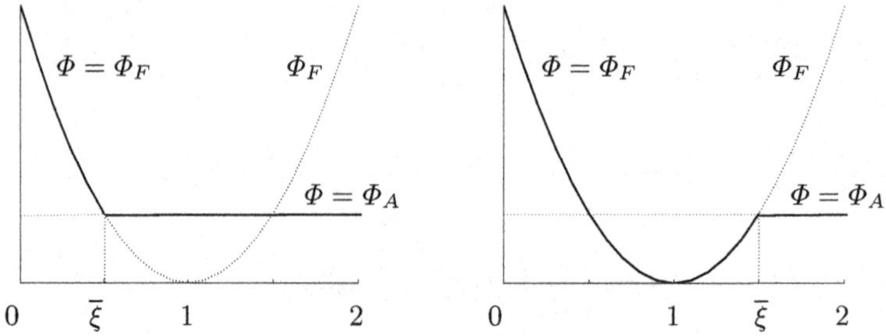

Fig. 5.13. Graphs of Φ for $\overline{\xi} < 1$ (left) and $\overline{\xi} > 1$ (right) when $g > 0$

It follows that for any α

$$\Phi_F(2 - \alpha) = \left(-(2 - \alpha) + \frac{(2 - \alpha)^2}{2}\right) g^2 = \Phi_F(\alpha), \tag{5.46}$$

and if $g \le 0$, then

$$\Phi(\alpha) = \Phi_F(\alpha) = \Phi_F(2 - \alpha) = \Phi(2 - \alpha).$$

Let us now assume that $g > 0$ and denote $\overline{\xi} = (x - \ell)/g$. Simple analysis shows that if $\overline{\xi} \in [0, 1]$, then Φ is nonincreasing on $[0, 2]$ and (5.45) is satisfied for $\alpha \in [0, 1]$. To finish the proof of (5.45), notice that if $1 < \overline{\xi}$, then

$$\Phi(\alpha) = \Phi_F(\alpha), \quad \alpha \in [0, 1], \quad \Phi(\alpha) \le \Phi_F(\alpha), \quad \alpha \in [1, 2],$$

so that we can use (5.46) to get that for $\alpha \in [0, 1]$

$$\Phi(2 - \alpha) \le \Phi_F(2 - \alpha) = \Phi_F(\alpha) = \Phi(\alpha).$$

□

The following property of F is essential in the analysis of the decrease of f along the projected-gradient path in the next subsection.

Corollary 5.8. *Let* $\mathbf{x}, \boldsymbol{\ell}, \mathbf{c} \in \mathbb{R}^n$, $\mathbf{x} \geq \boldsymbol{\ell}$. *Let* F *be defined by (5.39). Then for any* $\delta \in [0, 1]$

$$F\big(P_{\Omega_B}(\mathbf{x} - (2-\delta)\mathbf{g})\big) \leq F\big(P_{\Omega_B}(\mathbf{x} - \delta\mathbf{g})\big). \tag{5.47}$$

Proof. If $n = 1$, then the statement reduces to Lemma 5.7.

To prove the statement for $n > 1$, first observe that for any $\mathbf{y} \in \mathbb{R}$

$$[P_{\Omega_B}(\mathbf{y})]_i = \max\{y_i, \ell_i\}, \ i = 1, \dots, n.$$

It follows that P_{Ω_B} is separable and can be defined componentwise by the real functions

$$P_i(y) = \max\{y, \ell_i\}, \ i = 1, \dots, n.$$

Using the separable representation of F given by (5.40) and Lemma 5.7, we get

$$\begin{aligned} F\big(P_{\Omega_B}(\mathbf{x} - (2-\delta)\mathbf{g})\big) &= \sum_{i=1}^{n} F_i\big([P_{\Omega_B}(\mathbf{x} - (2-\delta)\mathbf{g})]_i\big) \\ &= \sum_{i=1}^{n} F_i\big(P_i(x_i - (2-\delta)g_i)\big) \\ &\leq \sum_{i=1}^{n} F_i\big(P_i(x_i - \delta g_i)\big) \\ &= F\big(P_{\Omega_B}(\mathbf{x} - \delta\mathbf{g})\big). \end{aligned}$$

□

5.6.5 Dominating Function and Decrease of the Cost Function

Now we are ready to give an estimate of the decrease of the cost function f in the iterates defined by the gradient projection step (5.32). The idea of the proof is to replace f by a suitable quadratic function F which dominates f and whose Hessian is the identity matrix.

Let us assume that $0 < \delta\|\mathsf{A}\| \leq 1$ and let $\mathbf{x} \in \Omega_B$ be arbitrary but fixed, so that we can define a quadratic function

$$F_\delta(\mathbf{y}) = \delta f(\mathbf{y}) + \frac{1}{2}(\mathbf{y} - \mathbf{x})^T(\mathsf{I} - \delta\mathsf{A})(\mathbf{y} - \mathbf{x}), \quad \mathbf{y} \in \mathbb{R}^n.$$

It is defined so that

$$F_\delta(\mathbf{x}) = \delta f(\mathbf{x}), \quad \nabla F_\delta(\mathbf{x}) = \delta\nabla f(\mathbf{x}) = \delta\mathbf{g}, \quad \text{and} \quad \nabla^2 F_\delta(\mathbf{y}) = \mathsf{I}. \tag{5.48}$$

Moreover, for any $\mathbf{y} \in \mathbb{R}^n$

$$\delta f(\mathbf{y}) \le F_\delta(\mathbf{y}). \tag{5.49}$$

It follows that

$$\delta f\left(P_{\Omega_B}(\mathbf{x} - \delta \mathbf{g})\right) - \delta f(\widehat{\mathbf{x}}) \le F_\delta\left(P_{\Omega_B}(\mathbf{x} - \delta \mathbf{g})\right) - \delta f(\widehat{\mathbf{x}}) \tag{5.50}$$

and

$$\nabla F_\delta(\mathbf{y}) = \delta \nabla f(\mathbf{y}) + (\mathsf{I} - \delta \mathsf{A})(\mathbf{y} - \mathbf{x}) = \mathbf{y} - (\mathbf{x} - \delta \mathbf{g}). \tag{5.51}$$

Using (5.43) and (5.48), we get that for any $\mathbf{z} \in \Omega_B$

$$F_\delta\left(P_{\Omega_B}(\mathbf{x} - \delta \mathbf{g})\right) \le F_\delta(\mathbf{z}). \tag{5.52}$$

The following lemma is due to Schöberl [165, 74].

Lemma 5.9. *Let* $\widehat{\mathbf{x}}$ *denote the unique solution of* (5.1), *let* $\lambda_{\min}$ *denote the smallest eigenvalue of* A, $\mathbf{g} = \nabla f(\mathbf{x})$, $\mathbf{x} \in \Omega_B$, *and* $\delta \in (0, \|\mathsf{A}\|^{-1}]$. *Then*

$$F_\delta\left(P_{\Omega_B}(\mathbf{x} - \delta \mathbf{g})\right) - \delta f(\widehat{\mathbf{x}}) \le \delta(1 - \delta \lambda_{\min})\left(f(\mathbf{x}) - f(\widehat{\mathbf{x}})\right). \tag{5.53}$$

Proof. Let us denote

$$[\widehat{\mathbf{x}}, \mathbf{x}] = \mathrm{Conv}\{\widehat{\mathbf{x}}, \mathbf{x}\} \quad \text{and} \quad \mathbf{d} = \widehat{\mathbf{x}} - \mathbf{x}.$$

Using (5.52),

$$[\widehat{\mathbf{x}}, \mathbf{x}] = \{\mathbf{x} + t\mathbf{d} : t \in [0, 1]\} \subseteq \Omega_B,$$

$0 < \lambda_{\min}\delta \le \|\mathsf{A}\|\delta \le 1$, and $\lambda_{\min}\|\mathbf{d}\|^2 \le \mathbf{d}^T\mathsf{A}\mathbf{d}$, we get

$$\begin{aligned}
F_\delta\left(P_{\Omega_B}(\mathbf{x} - \delta \mathbf{g})\right) - \delta f(\widehat{\mathbf{x}}) &= \min\{F_\delta(\mathbf{y}) - \delta f(\widehat{\mathbf{x}}) : \ \mathbf{y} \in \Omega_B\} \\
&\le \min\{F_\delta(\mathbf{y}) - \delta f(\widehat{\mathbf{x}}) : \ \mathbf{y} \in [\widehat{\mathbf{x}}, \mathbf{x}]\} \\
&= \min\{F_\delta(\mathbf{x} + t\mathbf{d}) - \delta f(\mathbf{x} + \mathbf{d}) : \ t \in [0, 1]\} \\
&= \min\{\delta t \mathbf{d}^T\mathbf{g} + \frac{t^2}{2}\|\mathbf{d}\|^2 - \delta \mathbf{d}^T\mathbf{g} - \frac{\delta}{2}\mathbf{d}^T\mathsf{A}\mathbf{d} : \ t \in [0, 1]\} \\
&\le \delta^2\lambda_{\min}\mathbf{d}^T\mathbf{g} + \frac{1}{2}\delta^2\lambda_{\min}^2\|\mathbf{d}\|^2 - \delta\mathbf{d}^T\mathbf{g} - \frac{\delta}{2}\mathbf{d}^T\mathsf{A}\mathbf{d} \\
&\le \delta^2\lambda_{\min}\mathbf{d}^T\mathbf{g} + \frac{1}{2}\delta^2\lambda_{\min}\mathbf{d}^T\mathsf{A}\mathbf{d} - \delta\mathbf{d}^T\mathbf{g} - \frac{\delta}{2}\mathbf{d}^T\mathsf{A}\mathbf{d} \\
&= \delta(\delta\lambda_{\min} - 1)(\mathbf{d}^T\mathbf{g} + \frac{1}{2}\mathbf{d}^T\mathsf{A}\mathbf{d}) \\
&= \delta(\delta\lambda_{\min} - 1)\left(f(\mathbf{x} + \mathbf{d}) - f(\mathbf{x})\right) \\
&= \delta(1 - \delta\lambda_{\min})\left(f(\mathbf{x}) - f(\widehat{\mathbf{x}})\right).
\end{aligned}$$

This proves (5.53). □

Proposition 5.10. *Let* $\widehat{\mathbf{x}}$ *denote the unique solution of* (5.1), $\mathbf{g} = \nabla f(\mathbf{x})$, $\mathbf{x} \in \Omega_B$, *and let* $\lambda_{\min}$ *denote the smallest eigenvalue of* A.

If $\overline{\alpha} \in (0, 2\|\mathsf{A}\|^{-1}]$, *then*

$$f\left(P_{\Omega_B}(\mathbf{x} - \overline{\alpha}\mathbf{g})\right) - f(\widehat{\mathbf{x}}) \leq \eta_f \left(f(\mathbf{x}) - f(\widehat{\mathbf{x}})\right), \tag{5.54}$$

where

$$\eta_f = 1 - \widehat{\alpha}\lambda_{\min} \tag{5.55}$$

is the cost function reduction coefficient and $\widehat{\alpha} = \min\{\overline{\alpha}, 2\|\mathsf{A}\|^{-1} - \overline{\alpha}\}$.

Proof. Let us first assume that $0 < \overline{\alpha}\|\mathsf{A}\| \leq 1$ and let $\mathbf{x} \in \Omega_B$ be arbitrary but fixed, so that we can use Lemma 5.9 with $\delta = \overline{\alpha}$ to get

$$F_{\overline{\alpha}}\left(P_{\Omega_B}(\mathbf{x} - \overline{\alpha}\mathbf{g})\right) - \overline{\alpha} f(\widehat{\mathbf{x}}) \leq \overline{\alpha}(1 - \overline{\alpha}\lambda_{\min})\left(f(\mathbf{x}) - f(\widehat{\mathbf{x}})\right). \tag{5.56}$$

In combination with (5.50), this proves (5.54) for $0 < \overline{\alpha} \leq \|\mathsf{A}\|^{-1}$.

To prove the statement for $\overline{\alpha} \in (\|A\|^{-1}, 2\|A\|^{-1}]$, let us first assume that $\|A\| = 1$ and let $\overline{\alpha} = 2 - \delta$, $\delta \in (0,1)$. Then F_1 dominates f and

$$\delta F_1(\mathbf{y}) \leq \delta F_1(\mathbf{y}) + \frac{1-\delta}{2}\|\mathbf{y} - \mathbf{x}\|^2 = F_\delta(\mathbf{y}). \tag{5.57}$$

Thus we can apply (5.49), Corollary 5.8, and the latter inequality to get

$$\begin{aligned}\delta f\big(P_\Omega(\mathbf{x} - \overline{\alpha}\mathbf{g})\big) &\leq \delta F_1\big(P_\Omega(\mathbf{x} - \overline{\alpha}\mathbf{g})\big) \leq \delta F_1\big(P_\Omega(\mathbf{x} - \delta\mathbf{g})\big) \\ &\leq F_\delta\big(P_\Omega(\mathbf{x} - \delta\mathbf{g})\big).\end{aligned}$$

Combining the latter inequalities with (5.56) for $\overline{\alpha} = \delta$, we get

$$\delta f\big(P_\Omega(\mathbf{x} - \overline{\alpha}\mathbf{g})\big) - \delta f(\widehat{\mathbf{x}}) \leq \delta(1 - \delta\lambda_{\min})\big((f(\mathbf{x}) - f(\widehat{\mathbf{x}})\big).$$

This proves the statement for $\overline{\alpha} \in (\|A\|^{-1}, 2\|A\|^{-1})$ and $\|A\| = 1$. To finish the proof, apply the last inequality divided by η to the function $\|A\|^{-1}f$ and recall that f and P_Ω are continuous. □

The estimate (5.54) gives the best value

$$\eta_f^{opt} = 1 - \kappa(\mathsf{A})^{-1}$$

for $\overline{\alpha} = \|\mathsf{A}\|^{-1}$ with $\kappa(\mathsf{A}) = \|\mathsf{A}\|\|\mathsf{A}^{-1}\|$. If $\overline{\alpha} \in (0, 2\|\mathsf{A}\|^{-1})$ and the iterates $\{\mathbf{x}^i\}$ are generated by Algorithm 5.5, we can use (5.54) to get for $k \geq 1$

$$f(\mathbf{x}^k) - f(\widehat{\mathbf{x}}) \leq \eta_f\left(f(\mathbf{x}^{k-1}) - f(\widehat{\mathbf{x}})\right) \leq \cdots \leq \eta_f^k\left(f(\mathbf{x}^0) - f(\widehat{\mathbf{x}})\right), \tag{5.58}$$

where $\eta_f < 1$ is given by (5.55). It follows by Lemma 5.1 that

$$\|\mathbf{x}^k - \widehat{\mathbf{x}}\|_A^2 \leq 2\left(f(\mathbf{x}^k) - f(\widehat{\mathbf{x}})\right) \leq 2\eta_f^k\left(f(\mathbf{x}^0) - f(\widehat{\mathbf{x}})\right) \leq 2\lambda_{\min}^{-1}\eta_f^k\|\mathbf{g}^P\|, \tag{5.59}$$

where $\mathbf{g}^P = \mathbf{g}^P(\mathbf{x}^0)$. The latter bound on the R-linear convergence in the energy norm is asymptotically worse than (5.38), but its right-hand side does not enhance the solution and can be effectively evaluated.

5.7 Modified Proportioning with Gradient Projections

In the previous sections, we learned that the solution of auxiliary problems in the active set algorithm for solving (5.1) can be implemented by the conjugate gradient method and we got the estimate (5.54) for the decrease of the cost function f in the gradient projection step with the fixed steplength $\overline{\alpha} \in (0, 2\|\mathsf{A}\|^{-1})$. Now we are ready to combine these observations in order to develop an *effective algorithm with the R-linear rate of convergence of f* that can be expressed in terms of bounds on the spectrum of the Hessian of f. The only difficulty which we must overcome is to ensure that the free gradient is always sufficiently large in the conjugate gradient iterations, since the conjugate gradient method reduces efficiently only the free gradient and is inefficient when the norm of the chopped gradient dominates the error of the KKT conditions. Using the methods of the next section, it is possible to prove for our new algorithm the finite termination for regular solution and the convergence, but not the R-linear convergence, of the projected gradient to zero in the general case. Here we restrict our attention to the R-linear convergence of the iterates in the energy norm.

5.7.1 MPGP Schema

The algorithm that we propose here exploits a user-defined constant $\Gamma > 0$, a test which is used to decide when to leave the face, and three types of steps.

The *conjugate gradient step*, defined as in Polyak's algorithm on page 165 by

$$\mathbf{x}^{k+1} = \mathbf{x}^k - \alpha_{cg}\mathbf{p}^{k+1}, \tag{5.60}$$

is used to carry out efficiently the minimization in the face $\mathcal{W}_\mathcal{I}$ given by $\mathcal{I} = \mathcal{A}(\mathbf{x}^s)$. We shall use in our proofs that by Theorem 3.1

$$f(\mathbf{x}^{k+1}) = \min\{f(\mathbf{x}^s + \mathbf{y}) : \mathbf{y} \in \mathrm{Span}\{\boldsymbol{\varphi}(\mathbf{x}^s), \dots, \boldsymbol{\varphi}(\mathbf{x}^k)\}\}. \tag{5.61}$$

The *gradient projection step* is defined by the gradient projection

$$\mathbf{x}^{k+1} = P_{\Omega_B}\big(\mathbf{x}^k - \overline{\alpha}\mathbf{g}(\mathbf{x}^k)\big) = \max\{\boldsymbol{\ell}, \mathbf{x}^k - \overline{\alpha}\mathbf{g}(\mathbf{x}^k)\} \tag{5.62}$$

with the fixed steplength. This step can both add and remove indices from the current working set. To describe the gradient projection step in the form suitable for our analysis, let us introduce, for any $\mathbf{x} \in \Omega_B$ and $\alpha > 0$, the *reduced free gradient* $\widetilde{\boldsymbol{\varphi}}_\alpha(\mathbf{x})$ with the entries

$$\widetilde{\varphi}_i = \widetilde{\varphi}_i(\mathbf{x}, \alpha) = \min\{(x_i - \ell_i)/\alpha, \varphi_i\}, \quad i \in \mathcal{N} = \{1, \dots, n\}. \tag{5.63}$$

Thus

$$P_{\Omega_B}\big(\mathbf{x} - \alpha\mathbf{g}(\mathbf{x})\big) = \mathbf{x} - \alpha\big(\widetilde{\boldsymbol{\varphi}}_\alpha(\mathbf{x}) + \boldsymbol{\beta}(\mathbf{x})\big). \tag{5.64}$$

If the steplength is equal to $\overline{\alpha}$ and the inequality

$$||\boldsymbol{\beta}(\mathbf{x}^k)||^2 \leq \Gamma^2 \widetilde{\boldsymbol{\varphi}}_{\overline{\alpha}}(\mathbf{x}^k)^T \boldsymbol{\varphi}(\mathbf{x}^k) \tag{5.65}$$

holds, then we call the iterate $\mathbf{x}^k$ *strictly proportional.* The test (5.65) is used to decide which components of the projected gradient $\mathbf{g}^P(\mathbf{x}^k)$ should be reduced in the next step. Notice that the right-hand side of (5.65) blends the information about the free gradient and its part that can be used in the gradient projection step.

The *proportioning step* is defined by

$$\mathbf{x}^{k+1} = \mathbf{x}^k - \alpha_{cg}\boldsymbol{\beta}(\mathbf{x}^k) \tag{5.66}$$

with the steplength α_{cg} that minimizes $f\left(\mathbf{x}^k - \alpha\boldsymbol{\beta}(\mathbf{x}^k)\right)$. It has been shown in Sect. 3.1 that the CG steplength α_{cg} that minimizes $f(\mathbf{x} - \alpha\mathbf{d})$ for a given $\mathbf{d}$ and $\mathbf{x}$ can be evaluated using the gradient $\mathbf{g} = \mathbf{g}(\mathbf{x}) = \nabla f(\mathbf{x})$ at $\mathbf{x}$ by

$$\alpha_{cg} = \alpha_{cg}(\mathbf{d}) = \frac{\mathbf{d}^T\mathbf{g}}{\mathbf{d}^T\mathsf{A}\mathbf{d}}. \tag{5.67}$$

The purpose of the proportioning step is to remove the indices of the components of the gradient $\mathbf{g}$ that violate the KKT conditions from the working set. Note that if $\mathbf{x}^k \in \Omega_B$, then

$$\mathbf{x}^{k+1} = \mathbf{x}^k - \alpha_{cg}\boldsymbol{\beta}(\mathbf{x}^k) \in \Omega_B.$$

Now we are ready to define the algorithm in the form that is convenient for analysis. For its implementation, see Sect. 5.9.

Algorithm 5.6. Modified proportioning with gradient projections (MPGP schema).

Given a symmetric positive definite matrix $\mathsf{A} \in \mathbb{R}^{n\times n}$ *and n-vectors* $\mathbf{b}$, $\boldsymbol{\ell}$. *Choose* $\mathbf{x}^0 \in \Omega_B$, $\overline{\alpha} \in (0, 2||\mathsf{A}||^{-1})$, *and* $\Gamma > 0$. *Set* $k = 0$. *For* $k \geq 0$ *and* $\mathbf{x}^k$ *known, choose* $\mathbf{x}^{k+1}$ *by the following rules:*

(i) If $\mathbf{g}^P(\mathbf{x}^k) = \mathbf{o}$, *set* $\mathbf{x}^{k+1} = \mathbf{x}^k$.

(ii) If $\mathbf{x}^k$ *is strictly proportional and* $\mathbf{g}^P(\mathbf{x}^k) \neq \mathbf{o}$, *try to generate* $\mathbf{x}^{k+1}$ *by the conjugate gradient step. If* $\mathbf{x}^{k+1} \in \Omega_B$, *then accept it, else generate* $\mathbf{x}^{k+1}$ *by the gradient projection step.*

(iii) If $\mathbf{x}^k$ *is not strictly proportional, define* $\mathbf{x}^{k+1}$ *by proportioning.*

We call our algorithm modified proportioning to distinguish it from earlier algorithms introduced independently by Friedlander and Martínez with their collaborators [94, 95, 96, 14, 33] and Dostál [41, 42]. These earlier algorithms applied the proportioning step when

$$||\boldsymbol{\beta}(\mathbf{x}^k)|| \leq \Gamma^2 ||\boldsymbol{\varphi}\left(\mathbf{x}^k\right)||.$$

5.7.2 Rate of Convergence

Now we are ready to prove the R-linear rate of convergence of MPGP in terms of bounds on the spectrum of the Hessian A for $\overline{\alpha} \in (0, 2\|\mathsf{A}\|^{-1})$.

Theorem 5.11. *Let* $\{\mathbf{x}^k\}$ *be generated by Algorithm 5.6 with* $\mathbf{x}^0 \in \Omega_B$, $\Gamma > 0$, *and* $\overline{\alpha} \in (0, 2\|\mathsf{A}\|^{-1}]$. *Then*

$$f(\mathbf{x}^{k+1}) - f(\widehat{\mathbf{x}}) \le \eta_\Gamma \left(f(\mathbf{x}^k) - f(\widehat{\mathbf{x}}) \right), \tag{5.68}$$

where $\widehat{\mathbf{x}}$ *denotes the unique solution of* (5.1),

$$\eta_\Gamma = 1 - \frac{\widehat{\alpha}\lambda_{\min}}{\vartheta + \vartheta \widehat{\Gamma}^2}, \qquad \widehat{\Gamma} = \max\{\Gamma, \Gamma^{-1}\}, \tag{5.69}$$

$$\vartheta = 2\max\{\overline{\alpha}\|\mathsf{A}\|, 1\}, \quad \widehat{\alpha} = \min\{\overline{\alpha}, 2\|\mathsf{A}\|^{-1} - \overline{\alpha}\}, \tag{5.70}$$

and $\lambda_{\min}$ *denotes the smallest eigenvalue of* A.

The error in the A*-norm is bounded by*

$$\|\mathbf{x}^k - \widehat{\mathbf{x}}\|_{\mathsf{A}}^2 \le 2\eta_\Gamma^k \left(f(\mathbf{x}^0) - f(\widehat{\mathbf{x}}) \right). \tag{5.71}$$

Proof. Since we have the estimate (5.54) for the gradient projection step with $\eta_f \le \eta_\Gamma$, it is enough to estimate the decrease of the cost function for the other two steps. Our main tools are (5.54) and the inequality

$$f\Big(P_{\Omega_B}\left(\mathbf{x}^k - \alpha \mathbf{g}(\mathbf{x}^k)\right)\Big) \ge f(\mathbf{x}^k) - \alpha\left(\widetilde{\varphi}_\alpha(\mathbf{x}^k)^T \varphi(\mathbf{x}^k) + \|\beta(\mathbf{x}^k)\|^2\right), \tag{5.72}$$

which is valid for any $\alpha \ge 0$ and can be obtained from the Taylor expansion

$$f(\mathbf{x} + \mathbf{d}) = f(\mathbf{x}) + \mathbf{d}^T\mathbf{g}(\mathbf{x}) + \frac{1}{2}\mathbf{d}^T\mathsf{A}\mathbf{d} \ge f(\mathbf{x}) + \mathbf{d}^T\mathbf{g}(\mathbf{x}) \tag{5.73}$$

by substituting

$$\mathbf{x} = \mathbf{x}^k, \quad \mathbf{d} = -\alpha\big(\widetilde{\varphi}_\alpha(\mathbf{x}^k) + \beta(\mathbf{x}^k)\big), \quad \text{and} \quad \mathbf{g} = \varphi(\mathbf{x}) + \beta(\mathbf{x}).$$

If $\mathbf{x}^{k+1}$ is generated by the conjugate gradient step (5.60), then by (5.61) and (5.67)

$$f(\mathbf{x}^{k+1}) \le f\left(\mathbf{x}^k - \alpha_{cg}\varphi(\mathbf{x}^k)\right) = f(\mathbf{x}^k) - \frac{1}{2}\frac{\|\varphi(\mathbf{x}^k)\|^4}{\varphi(\mathbf{x}^k)^T\mathsf{A}\varphi(\mathbf{x}^k)}$$
$$\le f(\mathbf{x}^k) - \frac{1}{2}\|\mathsf{A}\|^{-1}\|\varphi(\mathbf{x}^k)\|^2.$$

Taking into account $\widehat{\alpha} \le \|\mathsf{A}\|^{-1}$ and $\widetilde{\varphi}_i\varphi_i \le \varphi_i^2$, $i = 1, \dots, n$, we get

$$f(\mathbf{x}^{k+1}) \le f(\mathbf{x}^k) - \frac{1}{2}\|\mathsf{A}\|^{-1}\|\varphi(\mathbf{x}^k)\|^2 \le f(\mathbf{x}^k) - \frac{\widehat{\alpha}}{2}\widetilde{\varphi}_{\widehat{\alpha}}(\mathbf{x}^k)^T\varphi(\mathbf{x}^k). \tag{5.74}$$

Now observe that the conjugate gradient step is used only when $\mathbf{x}^k$ is strictly proportional, i.e.,

$$\|\boldsymbol{\beta}(\mathbf{x}^k)\|^2 \leq \Gamma^2 \widetilde{\varphi}_{\overline{\alpha}}(\mathbf{x}^k)^T \boldsymbol{\varphi}(\mathbf{x}^k).$$

Since $\widehat{\alpha} \leq \overline{\alpha}$ implies

$$\widetilde{\varphi}_{\overline{\alpha}}(\mathbf{x}^k)^T \boldsymbol{\varphi}(\mathbf{x}^k) \leq \widetilde{\varphi}_{\widehat{\alpha}}(\mathbf{x}^k)^T \boldsymbol{\varphi}(\mathbf{x}^k),$$

it follows that

$$\|\boldsymbol{\beta}(\mathbf{x}^k)\|^2 \leq \Gamma^2 \widetilde{\varphi}_{\widehat{\alpha}}(\mathbf{x}^k)^T \boldsymbol{\varphi}(\mathbf{x}^k). \tag{5.75}$$

After substituting (5.75) into (5.72) with $\alpha = \widehat{\alpha}$, we get

$$f\left(P_{\Omega_B}\left(\mathbf{x}^k - \widehat{\alpha}\mathbf{g}(\mathbf{x}^k)\right)\right) \geq f(\mathbf{x}^k) - \widehat{\alpha}(1+\Gamma^2)\widetilde{\varphi}_{\widehat{\alpha}}(\mathbf{x}^k)^T \boldsymbol{\varphi}(\mathbf{x}^k). \tag{5.76}$$

Thus for $\mathbf{x}^{k+1}$ generated by the conjugate gradient step, we get by elementary algebra and application of (5.76) that

$$\begin{aligned} f(\mathbf{x}^{k+1}) &\leq f(\mathbf{x}^k) - \frac{\widehat{\alpha}}{2}\widetilde{\varphi}_{\widehat{\alpha}}(\mathbf{x}^k)^T \boldsymbol{\varphi}(\mathbf{x}^k) \\ &= \frac{1}{2+2\Gamma^2}\Big(f(\mathbf{x}^k) - \widehat{\alpha}(1+\Gamma^2)\widetilde{\varphi}_{\widehat{\alpha}}(\mathbf{x}^k)^T \boldsymbol{\varphi}(\mathbf{x}^k) + (1+2\Gamma^2) f(\mathbf{x}^k)\Big) \\ &\leq \frac{1}{2+2\Gamma^2}\Big(f\left(P_{\Omega_B}\left(\mathbf{x}^k - \widehat{\alpha}\mathbf{g}(\mathbf{x}^k)\right)\right) + (1+2\Gamma^2) f(\mathbf{x}^k)\Big). \end{aligned}$$

After inserting $-f(\widehat{\mathbf{x}}) + f(\widehat{\mathbf{x}})$ into the last term and using (5.54) with simple manipulations, we get

$$\begin{aligned} f(\mathbf{x}^{k+1}) &\leq \frac{\eta_f + 1 + 2\Gamma^2}{2+2\Gamma^2} f(\mathbf{x}^k) + \frac{1-\eta_f}{2+2\Gamma^2} f(\widehat{\mathbf{x}}) \\ &= \frac{\eta_f + 1 + 2\Gamma^2}{2+2\Gamma^2}\left(f(\mathbf{x}^k) - f(\widehat{\mathbf{x}})\right) + f(\widehat{\mathbf{x}}). \end{aligned} \tag{5.77}$$

Let us finally assume that $\mathbf{x}^{k+1}$ is generated by the proportioning step (5.66), so that

$$\|\boldsymbol{\beta}(\mathbf{x}^k)\|^2 > \Gamma^2 \widetilde{\varphi}_{\overline{\alpha}}(\mathbf{x}^k)^T \boldsymbol{\varphi}(\mathbf{x}^k) \tag{5.78}$$

and

$$\begin{aligned} f(\mathbf{x}^{k+1}) = f\left(\mathbf{x}^k - \alpha_{cg}\boldsymbol{\beta}(\mathbf{x}^k)\right) &= f(\mathbf{x}^k) - \frac{1}{2}\frac{\|\boldsymbol{\beta}(\mathbf{x}^k)\|^4}{\boldsymbol{\beta}(\mathbf{x}^k)^T \mathsf{A}\boldsymbol{\beta}(\mathbf{x}^k)} \\ &\leq f(\mathbf{x}^k) - \frac{1}{2}\|\mathsf{A}\|^{-1}\|\boldsymbol{\beta}(\mathbf{x}^k)\|^2. \end{aligned}$$

Taking into account the definition of $\overline{\alpha}$ and ϑ, we get

$$\overline{\alpha}/\vartheta \leq \|\mathsf{A}\|^{-1}/2$$

and

$$f(\mathbf{x}^{k+1}) \le f(\mathbf{x}^k) - \frac{\overline{\alpha}}{\vartheta}\|\boldsymbol{\beta}(\mathbf{x}^k)\|^2, \tag{5.79}$$

where the right-hand side may be rewritten in the form

$$\begin{aligned} f(\mathbf{x}^k) - \frac{\overline{\alpha}}{\vartheta}\|\boldsymbol{\beta}(\mathbf{x}^k)\|^2 &= \frac{1}{\vartheta(1+\Gamma^{-2})}\left(f(\mathbf{x}^k) - \overline{\alpha}(1+\Gamma^{-2})\|\boldsymbol{\beta}(\mathbf{x}^k)\|^2\right) \\ &+ \frac{\vartheta + \vartheta\Gamma^{-2} - 1}{\vartheta(1+\Gamma^{-2})} f(\mathbf{x}^k). \end{aligned} \tag{5.80}$$

We can also substitute (5.78) into (5.72) to get

$$f\left(P_{\Omega_B}\left(\mathbf{x}^k - \overline{\alpha}\mathbf{g}(\mathbf{x}^k)\right)\right) > f(\mathbf{x}^k) - \overline{\alpha}(1+\Gamma^{-2})\|\boldsymbol{\beta}(\mathbf{x}^k)\|^2. \tag{5.81}$$

After substituting (5.81) into (5.80), using (5.79), (5.54) with $\mathbf{x} = \mathbf{x}^k$, and simple manipulations, we get

$$\begin{aligned} f(\mathbf{x}^{k+1}) &< \frac{1}{\vartheta + \vartheta\Gamma^{-2}} f\left(P_{\Omega_B}\left(\mathbf{x}^k - \overline{\alpha}\mathbf{g}(\mathbf{x}^k)\right)\right) + \frac{\vartheta + \vartheta\Gamma^{-2} - 1}{\vartheta + \vartheta\Gamma^{-2}} f(\mathbf{x}^k) \\ &= \frac{1}{\vartheta + \vartheta\Gamma^{-2}}\left(f\left(P_{\Omega_B}\left(\mathbf{x}^k - \overline{\alpha}\mathbf{g}(\mathbf{x}^k)\right)\right) - f(\widehat{\mathbf{x}})\right) \\ &+ \frac{1}{\vartheta + \vartheta\Gamma^{-2}} f(\widehat{\mathbf{x}}) + \frac{\vartheta + \vartheta\Gamma^{-2} - 1}{\vartheta + \vartheta\Gamma^{-2}} f(\mathbf{x}^k) \\ &\le \frac{\eta_f}{\vartheta + \vartheta\Gamma^{-2}}\left(f(\mathbf{x}^k) - f(\widehat{\mathbf{x}})\right) + \frac{1}{\vartheta + \vartheta\Gamma^{-2}} f(\widehat{\mathbf{x}}) + \frac{\vartheta + \vartheta\Gamma^{-2} - 1}{\vartheta + \vartheta\Gamma^{-2}} f(\mathbf{x}^k) \\ &= \frac{\eta_f + \vartheta + \vartheta\Gamma^{-2} - 1}{\vartheta + \vartheta\Gamma^{-2}}\left(f(\mathbf{x}^k) - f(\widehat{\mathbf{x}})\right) + f(\widehat{\mathbf{x}}). \end{aligned}$$

Comparing the last inequality with (5.77) and taking into account that by the definition $\Gamma \le \widehat{\Gamma}$, $\Gamma^{-1} \le \widehat{\Gamma}$, and $\vartheta \ge 2$, we obtain that the estimate

$$f(\mathbf{x}^{k+1}) - f(\widehat{\mathbf{x}}) \le \frac{\eta_f + \vartheta + \vartheta\Gamma^{-2} - 1}{\vartheta + \vartheta\Gamma^{-2}}\left(f(\mathbf{x}^k) - f(\widehat{\mathbf{x}})\right)$$

is valid for both the CG step and the proportioning step. The proof of (5.68) is completed by

$$\eta_\Gamma = \frac{\eta_f + \vartheta + \vartheta\Gamma^{-2} - 1}{\vartheta + \vartheta\Gamma^{-2}} = 1 - \frac{1-\eta_f}{\vartheta + \vartheta\Gamma^{-2}} = 1 - \frac{\widehat{\alpha}\lambda_{\min}}{\vartheta + \vartheta\widehat{\Gamma}^2}.$$

To get the error bound (5.71), notice that by Lemma 5.1

$$\|\mathbf{x}^k - \widehat{\mathbf{x}}\|_A^2 \le 2\left(f(\mathbf{x}^k) - f(\widehat{\mathbf{x}})\right) \le 2\eta_\Gamma^k\left(f(\mathbf{x}^0) - f(\widehat{\mathbf{x}})\right). \ \square \tag{5.82}$$

Theorem 5.11 gives the best bound on the rate of convergence for $\Gamma = \widehat{\Gamma} = 1$ in agreement with the heuristics that we should leave the face when the chopped gradient dominates the violation of the Karush–Kuhn–Tucker conditions. The formula for the best bound η_Γ^{opt} which corresponds to $\Gamma = 1$ and $\overline{\alpha} = \|\mathsf{A}\|^{-1}$ reads

$$\eta_\Gamma^{opt} = 1 - \kappa(\mathsf{A})^{-1}/4, \tag{5.83}$$

where $\kappa(\mathsf{A})$ denotes the spectral condition number of A.

5.8 Modified Proportioning with Reduced Gradient Projections

Even though the MPGP algorithm of the previous section combines the conjugate gradient method with the gradient projections in a way which enables to prove its linear rate of convergence that can be expressed in terms of bounds on the spectrum of the Hessian of f, there is still room for improvements. The reason is that the gradient projection at the same time adds and removes the indices from the active set, so the algorithm releases the indices from the active set rather randomly. The result is that MPGP may not exploit fully the self-preconditioning effect of the conjugate gradient method [168] and can suffer from the oscillations often attributed to the iterative active set methods. In this section we show that these drawbacks can be relieved if we replace the gradient projection step by the *free gradient projection* with a fixed steplength $\overline{\alpha}$. We show that the modified algorithm not only preserves the linear rate of convergence of the cost function, but *it has the finite termination property even for dual degenerate QP problems* with zero Lagrange multipliers corresponding to the active constraints and the *R-linear rate of convergence in the norm of projected gradient.*

5.8.1 MPRGP Schema

The algorithm that we propose here exploits a constant $\Gamma > 0$ defined by a user, a test to decide when to leave the face, and three types of steps. The test and two of the three steps, the conjugate gradient step and the proportioning step, are exactly those introduced in Sect. 5.7.1.

The gradient projection step is replaced by the *expansion step* defined by the free gradient projection

$$\mathbf{x}^{k+1} = P_{\Omega_B}\big(\mathbf{x}^k - \overline{\alpha}\boldsymbol{\varphi}(\mathbf{x}^k)\big) = \max\{\boldsymbol{\ell}, \mathbf{x}^k - \overline{\alpha}\boldsymbol{\varphi}(\mathbf{x}^k)\} \tag{5.84}$$

with the fixed steplength. This step expands the current working set. To describe it in the form suitable for analysis, let us recall, for any $\mathbf{x} \in \Omega_B$ and $\alpha > 0$, that the *reduced free gradient* $\widetilde{\boldsymbol{\varphi}}_\alpha(\mathbf{x})$ is defined by the entries

$$\widetilde{\varphi}_i = \widetilde{\varphi}_i(\mathbf{x}, \alpha) = \min\{(x_i - \ell_i)/\alpha, \varphi_i\}, \quad i \in \mathcal{N} = \{1, \dots, n\}, \tag{5.85}$$

so that

$$P_{\Omega_B}\big(\mathbf{x} - \alpha\boldsymbol{\varphi}(\mathbf{x})\big) = \mathbf{x} - \alpha\widetilde{\boldsymbol{\varphi}}_\alpha(\mathbf{x}). \tag{5.86}$$

Using the new notation, we can write also

$$P_{\Omega_B}\big(\mathbf{x} - \alpha\mathbf{g}(\mathbf{x})\big) = \mathbf{x} - \alpha\big(\widetilde{\boldsymbol{\varphi}}_\alpha(\mathbf{x}) + \boldsymbol{\beta}(\mathbf{x})\big). \tag{5.87}$$

Now we are ready to define the algorithm in the form that is convenient for analysis, postponing the discussion about implementation to the next section. Notice that we admit the fixed steplength $\overline{\alpha} = 2\|\mathsf{A}\|^{-1}$ which guarantees neither the contraction of the distance from the solution nor the decrease of the cost function in the expansion steps.

Algorithm 5.7. Modified proportioning with reduced gradient projections (MPRGP schema).

Given a symmetric positive definite matrix $\mathsf{A} \in \mathbb{R}^{n\times n}$ *and* n*-vectors* $\mathbf{b}$, $\boldsymbol{\ell}$. *Choose* $\mathbf{x}^0 \in \Omega_B$, $\overline{\alpha} \in (0, 2\|\mathsf{A}\|^{-1}]$, *and* $\Gamma > 0$. *Set* $k = 0$. *For* $k \geq 0$ *and* $\mathbf{x}^k$ *known, choose* $\mathbf{x}^{k+1}$ *by the following rules:*

(i) If $\mathbf{g}^P(\mathbf{x}^k) = \mathbf{o}$, *set* $\mathbf{x}^{k+1} = \mathbf{x}^k$.

(ii) If $\mathbf{x}^k$ *is strictly proportional and* $\mathbf{g}^P(\mathbf{x}^k) \neq \mathbf{o}$, *try to generate* $\mathbf{x}^{k+1}$ *by the conjugate gradient step. If* $\mathbf{x}^{k+1} \in \Omega_B$, *then accept it, else generate* $\mathbf{x}^{k+1}$ *by the expansion step.*

(iii) If $\mathbf{x}^k$ *is not strictly proportional, define* $\mathbf{x}^{k+1}$ *by proportioning.*

Proposition 5.12. *Let* $\{\mathbf{x}^k\}$ *be generated by Algorithm 5.7 with* $\mathbf{x}^0 \in \Omega_B$, $\Gamma > 0$, *and* $\overline{\alpha} \in (0, 2\|\mathsf{A}\|^{-1}]$. *Then* $\{\mathbf{x}^k\}$ *converges to the solution* $\{\widehat{\mathbf{x}}\}$ *and* $\{\mathbf{g}^P(\mathbf{x}^k)\}$ *converges to zero.*

Proof. MPRGP is a variant of the proportioning algorithm studied in [42]; it converges when each iterate $\mathbf{x}^{k+1}$ generated by the expansion step satisfies

$$f(\mathbf{x}^{k+1}) - f(\mathbf{x}^k) \leq 0.$$

This condition is satisfied by Proposition 5.10 for $\overline{\alpha} \in (0, 2\|\mathsf{A}\|^{-1}]$; the convergence is driven by the proportioning step, which is a spacer iteration (see, e.g., Bertsekas [12]). The second statement is an easy corollary of the identification lemma 5.17 and of the continuity of $\mathbf{g}(\mathbf{x})$. □

5.8.2 Rate of Convergence

The main tool of our analysis is the quadratic function

$$F(\mathbf{x}) = \frac{1}{2}\mathbf{x}^T\mathbf{x} - \mathbf{c}^T\mathbf{x} + d, \quad \mathbf{x}, \mathbf{c} \in \mathbb{R}^n, \quad \mathbf{c} = [c_i], \quad d \in \mathbb{R}, \tag{5.88}$$

and its properties similar to those developed in Sect. 5.6.5. In particular,

$$F(\mathbf{x}) = \sum_{i=1}^{n} F_i(x_i) + d, \quad F_i(x_i) = \frac{1}{2}x_i^2 - c_i x_i, \quad \mathbf{x} = [x_i]. \tag{5.89}$$

If $\mathbf{x} \in \mathbb{R}^n$ is arbitrary but fixed, we associate with f and $\delta \in (0, \|\mathsf{A}\|^{-1}]$ the quadratic function of the form (5.88)

$$F_\delta(\mathbf{y}) = \delta f(\mathbf{y}) + \frac{1}{2}(\mathbf{y} - \mathbf{x})^T(\mathsf{I} - \delta\mathsf{A})(\mathbf{y} - \mathbf{x}) \geq \delta f(\mathbf{y}). \tag{5.90}$$

It is defined so that

$$F_\delta(\mathbf{x}) = \delta f(\mathbf{x}), \quad \nabla F_\delta(\mathbf{x}) = \delta \nabla f(\mathbf{x}) = \delta\mathbf{g}, \quad \text{and} \quad \nabla^2 F_\delta(\mathbf{y}) = \mathsf{I}. \tag{5.91}$$

We need the following lemma which is analogous to Corollary 5.8.

Lemma 5.13. *Let* $\mathbf{x}, \boldsymbol{\ell}, \mathbf{c} \in \mathbb{R}^n$, $\mathbf{x} \geq \boldsymbol{\ell}$. *Let* F *be defined by* (5.88). *Then for any* $\delta \in [0,1]$

$$F\big(P_{\Omega_B}(\mathbf{x} - (2-\delta)\boldsymbol{\varphi}(\mathbf{x}))\big) \leq F\big(P_{\Omega_B}(\mathbf{x} - \delta\boldsymbol{\varphi}(\mathbf{x}))\big). \tag{5.92}$$

Proof. First recall that P_{Ω_B} is separable and can be defined componentwise by $P_i(y) = \max\{y, \ell_i\}$, $i = 1, \dots, n$, $y \in \mathbb{R}$. Denoting $\mathcal{F}$, $\mathcal{A}$, and g_i the free set of $\mathbf{x}$, the active set of $\mathbf{x}$, and the components of the gradient $\mathbf{g}(\mathbf{x})$, respectively, we can use the representation of F given by (5.89) and Lemma 5.7 to get

$$\begin{aligned}
F\big(P_{\Omega_B}(\mathbf{x} - (2-\delta)\boldsymbol{\varphi}(\mathbf{x}))\big) &= \sum_{i=1}^{n} F_i\big([P_{\Omega_B}(\mathbf{x} - (2-\delta)\boldsymbol{\varphi}(\mathbf{x}))]_i\big) + d \\
&= \sum_{i\in\mathcal{F}} F_i\big(P_i(x_i - (2-\delta)g_i)\big) + \sum_{i\in\mathcal{A}} F_i\big(P_i(x_i)\big) + d \\
&\leq \sum_{i\in\mathcal{F}} F_i\big(P_i(x_i - \delta g_i)\big) + \sum_{i\in\mathcal{A}} F_i\big(P_i(x_i)\big) + d \\
&= F\big(P_{\Omega_B}(\mathbf{x} - \delta\boldsymbol{\varphi}(\mathbf{x}))\big).
\end{aligned}$$

□

Now we are ready to prove the R-linear rate of convergence of MPRGP.

Theorem 5.14. *Let* $\{\mathbf{x}^k\}$ *be generated by Algorithm 5.7 with* $\mathbf{x}^0 \in \Omega_B$, $\Gamma > 0$, *and* $\overline{\alpha} \in (0, 2\|\mathsf{A}\|^{-1}]$. *Then*

$$f(\mathbf{x}^{k+1}) - f(\widehat{\mathbf{x}}) \leq \eta_\Gamma \left(f(\mathbf{x}^k) - f(\widehat{\mathbf{x}})\right), \tag{5.93}$$

where $\widehat{\mathbf{x}}$ *denotes a unique solution of* (5.1),

$$\eta_\Gamma = 1 - \frac{\widehat{\alpha}\lambda_{\min}}{\vartheta + \vartheta\widehat{\Gamma}^2}, \qquad \widehat{\Gamma} = \max\{\Gamma, \Gamma^{-1}\}, \tag{5.94}$$

$$\vartheta = 2\max\{\overline{\alpha}\|\mathsf{A}\|, 1\}, \quad \widehat{\alpha} = \min\{\overline{\alpha}, 2\|\mathsf{A}\|^{-1} - \overline{\alpha}\}, \tag{5.95}$$

and $\lambda_{\min}$ *denotes the smallest eigenvalue of* A. *The error in the* A*-norm is bounded by*

$$\|\mathbf{x}^k - \widehat{\mathbf{x}}\|_{\mathsf{A}}^2 \leq 2\eta_\Gamma^k \left(f(\mathbf{x}^0) - f(\widehat{\mathbf{x}})\right). \tag{5.96}$$

Proof. First observe that the only new type of iteration, as compared with MPGP of Sect. 5.7, is the expansion step. Moreover, the estimate (5.68) with η_Γ defined by (5.69) of Theorem 5.11 is the same as our estimate (5.93) with η_Γ defined by (5.94). Thus we can reduce our analysis to the expansion step. Our main tools are again (5.54) and the inequality

$$f\Big(P_{\Omega_B}\left(\mathbf{x}^k - \widehat{\alpha}\mathbf{g}(\mathbf{x}^k)\right)\Big) \geq f(\mathbf{x}^k) - \widehat{\alpha}\big(\widetilde{\varphi}_{\widehat{\alpha}}(\mathbf{x}^k)^T\boldsymbol{\varphi}(\mathbf{x}^k) + \|\boldsymbol{\beta}(\mathbf{x}^k)\|^2\big), \tag{5.97}$$

which can be obtained by the Taylor expansion and (5.87).

Let us first assume that $\|\mathsf{A}\| = 1$ and let $\mathbf{x}^{k+1}$ be generated by the expansion step (5.84). Using in sequence the definition of the dominating function (5.90) associated with $\mathbf{x} = \mathbf{x}^k$, Lemma 5.13, the assumption,

$\|\mathsf{A}\| = 1$ and $\widehat{\alpha} \le 1$ with (5.57), the Taylor expansion with (5.86), (5.91), $\|\widetilde{\varphi}_{\widehat{\alpha}}(\mathbf{x}^k)\|^2 \le \widetilde{\varphi}_{\widehat{\alpha}}(\mathbf{x}^k)^T\varphi(\mathbf{x}^k)$, and simple manipulations, we get

$$\begin{aligned}
\widehat{\alpha} f(\mathbf{x}^{k+1}) &\le \widehat{\alpha} F_1(\mathbf{x}^{k+1}) = \widehat{\alpha} F_1\big(P_{\Omega_B}\left(\mathbf{x}^k - \overline{\alpha}\varphi(\mathbf{x}^k)\right)\big) \\
&\le \widehat{\alpha} F_1\big(P_{\Omega_B}\left(\mathbf{x}^k - \widehat{\alpha}\varphi(\mathbf{x}^k)\right)\big) \le F_{\widehat{\alpha}}\big(P_{\Omega_B}\left(\mathbf{x}^k - \widehat{\alpha}\varphi(\mathbf{x}^k)\right)\big) \\
&= F_{\widehat{\alpha}}\left(\mathbf{x}^k\right) - \widehat{\alpha}^2\widetilde{\varphi}_{\widehat{\alpha}}(\mathbf{x}^k)^T\varphi(\mathbf{x}^k) + \frac{\widehat{\alpha}^2}{2}\|\widetilde{\varphi}_{\widehat{\alpha}}(\mathbf{x}^k)\|^2 \\
&\le F_{\widehat{\alpha}}\left(\mathbf{x}^k\right) - \frac{\widehat{\alpha}^2}{2}\widetilde{\varphi}_{\widehat{\alpha}}(\mathbf{x}^k)^T\varphi(\mathbf{x}^k) = \widehat{\alpha} f(\mathbf{x}^k) - \frac{\widehat{\alpha}^2}{2}\widetilde{\varphi}_{\widehat{\alpha}}(\mathbf{x}^k)^T\varphi(\mathbf{x}^k).
\end{aligned}$$

Thus

$$f(\mathbf{x}^{k+1}) \le f(\mathbf{x}^k) - \frac{\widehat{\alpha}}{2}\widetilde{\varphi}_{\widehat{\alpha}}(\mathbf{x}^k)^T\varphi(\mathbf{x}^k). \tag{5.98}$$

The expansion step is used only when $\mathbf{x}^k$ is strictly proportional, i.e.,

$$\|\boldsymbol{\beta}(\mathbf{x}^k)\|^2 \le \Gamma^2\widetilde{\varphi}_{\overline{\alpha}}(\mathbf{x}^k)^T\varphi(\mathbf{x}^k).$$

Since $\widehat{\alpha} \le \overline{\alpha}$ by the definition, it follows that

$$\widetilde{\varphi}_{\overline{\alpha}}(\mathbf{x}^k)^T\varphi(\mathbf{x}^k) \le \widetilde{\varphi}_{\widehat{\alpha}}(\mathbf{x}^k)^T\varphi(\mathbf{x}^k)$$

and

$$\|\boldsymbol{\beta}(\mathbf{x}^k)\|^2 \le \Gamma^2\widetilde{\varphi}_{\widehat{\alpha}}(\mathbf{x}^k)^T\varphi(\mathbf{x}^k). \tag{5.99}$$

After substituting (5.99) into (5.97), we get

$$f\big(P_{\Omega_B}\left(\mathbf{x}^k - \widehat{\alpha}\mathbf{g}(\mathbf{x}^k)\right)\big) \ge f(\mathbf{x}^k) - \widehat{\alpha}(1+\Gamma^2)\widetilde{\varphi}_{\widehat{\alpha}}(\mathbf{x}^k)^T\varphi(\mathbf{x}^k). \tag{5.100}$$

Thus for $\mathbf{x}^{k+1}$ generated by the expansion step, we get by elementary algebra and application of (5.100) that

$$\begin{aligned}
f(\mathbf{x}^{k+1}) &\le f(\mathbf{x}^k) - \frac{\widehat{\alpha}}{2}\widetilde{\varphi}_{\widehat{\alpha}}(\mathbf{x}^k)^T\varphi(\mathbf{x}^k) \\
&= \frac{1}{2+2\Gamma^2}\left(f(\mathbf{x}^k) - \widehat{\alpha}(1+\Gamma^2)\widetilde{\varphi}_{\widehat{\alpha}}(\mathbf{x}^k)^T\varphi(\mathbf{x}^k) + (1+2\Gamma^2)f(\mathbf{x}^k)\right) \\
&\le \frac{1}{2+2\Gamma^2}\left(f\left(P_{\Omega_B}\left(\mathbf{x}^k - \widehat{\alpha}\mathbf{g}(\mathbf{x}^k)\right)\right) + (1+2\Gamma^2)f(\mathbf{x}^k)\right).
\end{aligned}$$

Inserting $-f(\widehat{\mathbf{x}}) + f(\widehat{\mathbf{x}})$ into the last term and substituting (5.54) with $\mathbf{x} = \mathbf{x}^k$ and $\overline{\alpha} = \widehat{\alpha}$ into the last expression, we get

$$\begin{aligned}
f(\mathbf{x}^{k+1}) &\le \frac{\eta_f + 1 + 2\Gamma^2}{2+2\Gamma^2} f(\mathbf{x}^k) + \frac{1-\eta_f}{2+2\Gamma^2} f(\widehat{\mathbf{x}}) \\
&= \frac{\eta_f + 1 + 2\Gamma^2}{2+2\Gamma^2}\left(f(\mathbf{x}^k) - f(\widehat{\mathbf{x}})\right) + f(\widehat{\mathbf{x}}).
\end{aligned} \tag{5.101}$$

The proof of (5.93) for $\|\mathsf{A}\| = 1$ is completed by

$$\frac{\eta_f + 1 + 2\Gamma^2}{2 + 2\Gamma^2} = \frac{\eta_f - 1 + 2 + 2\Gamma^2}{2 + 2\Gamma^2} = 1 - \frac{1 - \eta_f}{2 + 2\Gamma^2} = 1 - \frac{\widehat{\alpha}\lambda_{\min}}{2 + 2\Gamma^2} \leq \eta_\Gamma .$$

To prove the general case, it is enough to apply the theorem to $h = \|\mathsf{A}\|^{-1} f$.

To get the error bound (5.96), notice that by Lemma 5.1

$$\|\mathbf{x}^k - \widehat{\mathbf{x}}\|_{\mathsf{A}}^2 \leq 2\left(f(\mathbf{x}^k) - f(\widehat{\mathbf{x}})\right) \leq 2\eta_\Gamma^k \left(f(\mathbf{x}^0) - f(\widehat{\mathbf{x}})\right). \tag{5.102}$$

□

The formula for the best bound η_Γ^{opt} is given by (5.83). Notice that the coefficient of the Euclidean contraction η_E defined by (5.34) is smaller than η_Γ and by (5.38) guarantees faster convergence in the energy norm. Does it follow that the gradient projection method is faster than MPRGP? The answer is *no.* We have got both estimates by the worst case analysis of just one step of each method. Such analysis at least partly enhances the improvement due to the long sequence of the same type of iterations of the projected gradient method, while this is not true in the case of MPRGP; the worst case assumes that the algorithm switches the types of iterations. The error in energy norm need not even decrease in one step of the gradient projection method.

5.8.3 Rate of Convergence of Projected Gradient

To use the MPRGP algorithm in the inner loops of other algorithms, we must be able to *recognize* when we are near the solution. There is a catch – though by Lemma 5.1 the latter can be tested by a norm of the projected gradient, Theorem 5.14 does not guarantee that such test is positive near the solution. The projected gradient is not a continuous function of the iterates! A large projected gradient near the solution is in Fig. 5.14. The R-linear convergence of the projected gradient is treated by the following theorem.

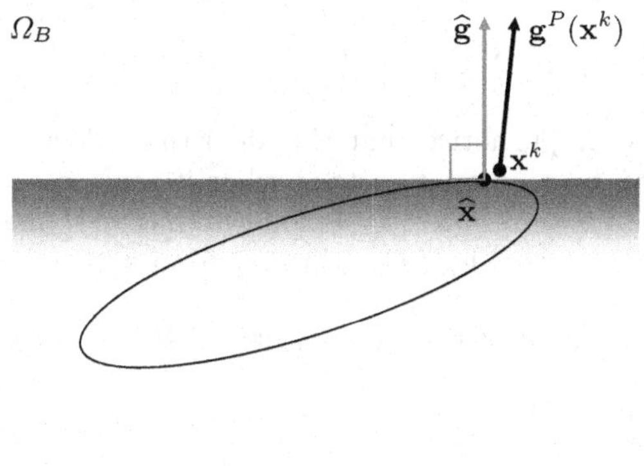

Fig. 5.14. Large projected gradient near the solution

Theorem 5.15. *Let $\{\mathbf{x}^k\}$ be generated by Algorithm 5.7 with $\mathbf{x}^0 \in \Omega_B$, $\Gamma > 0$, and $\overline{\alpha} \in (0, 2\|\mathsf{A}\|^{-1}]$. Let $\widehat{\mathbf{x}}$ denote the unique solution of (5.1) and let $\widehat{\Gamma}$, η_Γ, $\widehat{\alpha}$, and ϑ be those of Theorem 5.14.*

Then for any $k \geq 1$

$$\|\mathbf{g}^P(\mathbf{x}^{k+1})\|^2 \leq a_1 \eta_\Gamma^k \left(f(\mathbf{x}^0) - f(\widehat{\mathbf{x}})\right) \tag{5.103}$$

with

$$a_1 = \frac{38}{\widehat{\alpha}(1-\eta_\Gamma)} = \frac{38\vartheta(1+\widehat{\Gamma}^2)}{\widehat{\alpha}^2 \lambda_{\min}}. \tag{5.104}$$

Proof. First notice that it is enough to estimate separately $\boldsymbol{\beta}(\mathbf{x}^k)$ and $\boldsymbol{\varphi}(\mathbf{x}^k)$ as

$$\|\mathbf{g}^P(\mathbf{x}^k)\|^2 = \|\boldsymbol{\beta}(\mathbf{x}^k)\|^2 + \|\boldsymbol{\varphi}(\mathbf{x}^k)\|^2.$$

In particular, since $\widehat{\alpha} \leq \|\mathsf{A}^{-1}\|$, we have for any vector $\mathbf{d}$ which satisfies $\mathbf{d}^T\mathbf{g}(\mathbf{x}) \geq \|\mathbf{d}\|^2$

$$f(\mathbf{x}) - f(\mathbf{x} - \widehat{\alpha}\mathbf{d}) = \widehat{\alpha}\mathbf{d}^T\mathbf{g}(\mathbf{x}) - \frac{1}{2}\widehat{\alpha}^2\mathbf{d}^T\mathsf{A}\mathbf{d} \geq \frac{\widehat{\alpha}}{2}\|\mathbf{d}\|^2. \tag{5.105}$$

It follows that we can combine (5.105) with

$$\mathbf{x}^k - \widehat{\alpha}\boldsymbol{\beta}(\mathbf{x}^k) \geq \boldsymbol{\ell}$$

to estimate $\|\boldsymbol{\beta}(\mathbf{x}^k)\|$ by

$$\begin{aligned} f(\mathbf{x}^k) - f(\widehat{\mathbf{x}}) &= \left(f(\mathbf{x}^k) - f\left(\mathbf{x}^k - \widehat{\alpha}\boldsymbol{\beta}(\mathbf{x}^k)\right)\right) + \left(f\left(\mathbf{x}^k - \widehat{\alpha}\boldsymbol{\beta}(\mathbf{x}^k)\right) - f(\widehat{x})\right) \\ &\geq f(\mathbf{x}^k) - f\left(\mathbf{x}^k - \widehat{\alpha}\boldsymbol{\beta}(\mathbf{x}^k)\right) \geq \frac{\widehat{\alpha}}{2}\|\boldsymbol{\beta}(\mathbf{x}^k)\|^2. \end{aligned} \tag{5.106}$$

Applying (5.93), we get

$$\|\boldsymbol{\beta}(\mathbf{x}^k)\|^2 \leq \frac{2}{\widehat{\alpha}}\left(f(\mathbf{x}^k) - f(\widehat{\mathbf{x}})\right) \leq \frac{2\eta_\Gamma^k}{\widehat{\alpha}}\left(f(\mathbf{x}^0) - f(\widehat{\mathbf{x}})\right). \tag{5.107}$$

To estimate $\|\boldsymbol{\varphi}(\mathbf{x}^k)\|$, notice that the algorithm "does not know" about the components of the constraint vector $\boldsymbol{\ell}$ when it generates $\mathbf{x}^{k+1}$ unless their indices belong to $\mathcal{A}(\mathbf{x}^k)$ or $\mathcal{A}(\mathbf{x}^{k+1})$. It follows that $\mathbf{x}^{k+1}$ may be considered also as an iterate generated by Algorithm 5.7 from $\mathbf{x}^k$ for the problem

$$\text{minimize } f(\mathbf{x}) \quad \text{subject to} \quad x_i \geq \ell_i \text{ for } i \in \mathcal{A}(\mathbf{x}^k) \cup \mathcal{A}(\mathbf{x}^{k+1}). \tag{5.108}$$

If we denote

$$\overline{f}^k = \min\{f(\mathbf{x}) : \ x_i \geq \ell_i \text{ for } i \in \mathcal{A}(\mathbf{x}^k) \cup \mathcal{A}(\mathbf{x}^{k+1})\} \leq f(\widehat{\mathbf{x}})$$

and $\overline{\delta}_k = f(\widehat{\mathbf{x}}) - \overline{f}^k \geq 0$, we can use (5.93) to get

$$\overline{\delta}_k = f(\widehat{\mathbf{x}}) - \overline{f}^k \le f(\mathbf{x}^{k+1}) - \overline{f}^k \le \eta_\Gamma \left(f(\mathbf{x}^k) - \overline{f}^k \right)$$
$$= \eta_\Gamma \left(f(\mathbf{x}^k) - f(\widehat{\mathbf{x}}) \right) + \eta_\Gamma \overline{\delta}_k,$$

so that

$$\overline{\delta}_k \le \frac{\eta_\Gamma}{1-\eta_\Gamma} \left(f(\mathbf{x}^k) - f(\widehat{\mathbf{x}}) \right) \le \frac{\eta_\Gamma^{k+1}}{1-\eta_\Gamma} \left(f(\mathbf{x}^0) - f(\widehat{\mathbf{x}}) \right). \tag{5.109}$$

Now observe that the indices of the unconstrained components of the minimization problem (5.108) are those belonging to $\mathcal{I}^k = \mathcal{F}(\mathbf{x}^k) \cap \mathcal{F}(\mathbf{x}^{k+1})$ as

$$\mathcal{I}^k = \mathcal{F}(\mathbf{x}^k) \cap \mathcal{F}(\mathbf{x}^{k+1}) = (\mathcal{N} \setminus \mathcal{A}(\mathbf{x}^k)) \cap (\mathcal{N} \setminus \mathcal{A}(\mathbf{x}^{k+1}))$$
$$= \mathcal{N} \setminus \left(\mathcal{A}(\mathbf{x}^k) \cup \mathcal{A}(\mathbf{x}^{k+1}) \right).$$

It follows that if $\mathcal{I}^k$ is nonempty, then by the definition of $\overline{\delta}_k$ and (5.105)

$$\overline{\delta}_k \ge f(\widehat{\mathbf{x}}) - f\left(\widehat{\mathbf{x}} - \widehat{\alpha} \mathbf{g}_{\mathcal{I}^k}(\widehat{\mathbf{x}})\right) \ge \frac{\widehat{\alpha}}{2} \|\mathbf{g}_{\mathcal{I}^k}(\widehat{\mathbf{x}})\|^2. \tag{5.110}$$

For convenience, let us define $\mathbf{g}_{\mathcal{I}}(\mathbf{x}) = \mathbf{o}$ for any $\mathbf{x} \in \mathbb{R}^n$ and empty set $\mathcal{I} = \emptyset$. Then (5.110) remains valid for $\mathcal{I}^k = \emptyset$, so that we can combine it with (5.109) to get

$$\|\mathbf{g}_{\mathcal{I}^k}(\widehat{\mathbf{x}})\|^2 \le \frac{2}{\widehat{\alpha}} \overline{\delta}_k \le \frac{2\eta_\Gamma^{k+1}}{\widehat{\alpha}(1-\eta_\Gamma)} (f(\mathbf{x}^0) - f(\widehat{\mathbf{x}})). \tag{5.111}$$

Since our algorithm is defined so that either $\mathcal{I}^k = \mathcal{F}(\mathbf{x}^k) \subseteq \mathcal{F}(\mathbf{x}^{k+1})$ or $\mathcal{I}^k = \mathcal{F}(\mathbf{x}^{k+1}) \subseteq \mathcal{F}(\mathbf{x}^k)$, it follows that either

$$\|\mathbf{g}_{\mathcal{F}(\mathbf{x}^k)}(\widehat{\mathbf{x}})\|^2 = \|\mathbf{g}_{\mathcal{I}^k}(\widehat{\mathbf{x}})\|^2 \le \frac{2\eta_\Gamma^{k+1}}{\widehat{\alpha}(1-\eta_\Gamma)} (f(\mathbf{x}^0) - f(\widehat{\mathbf{x}}))$$
$$\le \frac{2\eta_\Gamma^{k}}{\widehat{\alpha}(1-\eta_\Gamma)} (f(\mathbf{x}^0) - f(\widehat{\mathbf{x}})) \tag{5.112}$$

or

$$\|\mathbf{g}_{\mathcal{F}(\mathbf{x}^{k+1})}(\widehat{\mathbf{x}})\|^2 = \|\mathbf{g}_{\mathcal{I}^k}(\widehat{\mathbf{x}})\|^2 \le \frac{2\eta_\Gamma^{k+1}}{\widehat{\alpha}(1-\eta_\Gamma)} (f(\mathbf{x}^0) - f(\widehat{\mathbf{x}})).$$

Using the same reasoning for $\mathbf{x}^{k-1}$ and $\mathbf{x}^k$, we conclude that the estimate (5.112) is valid for any $\mathbf{x}^k$ such that

$$\mathcal{F}(\mathbf{x}^{k-1}) \supseteq \mathcal{F}(\mathbf{x}^k) \quad \text{or} \quad \mathcal{F}(\mathbf{x}^k) \subseteq \mathcal{F}(\mathbf{x}^{k+1}). \tag{5.113}$$

Let us now recall that by Lemma 5.1 and (5.96)

$$\|\mathbf{g}(\mathbf{x}^k) - \mathbf{g}(\widehat{\mathbf{x}})\|^2 = \|\mathsf{A}(\mathbf{x}^k - \widehat{\mathbf{x}})\|^2 \le \|\mathsf{A}\| \|\mathbf{x}^k - \widehat{\mathbf{x}}\|_{\mathsf{A}}^2 \le 2\|\mathsf{A}\| \left(f(\mathbf{x}^k) - f(\widehat{\mathbf{x}}) \right)$$
$$\le \frac{2}{\widehat{\alpha}} \eta_\Gamma^k \left(f(\mathbf{x}^0) - f(\widehat{\mathbf{x}}) \right), \tag{5.114}$$

so that for any k satisfying the relations (5.113), we get

$$\begin{aligned}\|\boldsymbol{\varphi}(\mathbf{x}^k)\| &= \|\mathbf{g}_{\mathcal{F}(\mathbf{x}^k)}(\mathbf{x}^k)\| \le \|\mathbf{g}_{\mathcal{F}(\mathbf{x}^k)}(\mathbf{x}^k) - \mathbf{g}_{\mathcal{F}(\mathbf{x}^k)}(\widehat{\mathbf{x}})\| + \|\mathbf{g}_{\mathcal{F}(\mathbf{x}^k)}(\widehat{\mathbf{x}})\| \\ &\le \sqrt{\frac{2}{\widehat{\alpha}}\eta_\Gamma^k\left(f(x^0) - f(\widehat{\mathbf{x}})\right)} + \sqrt{\frac{2}{\widehat{\alpha}(1-\eta_\Gamma)}\eta_\Gamma^k\left(f(\mathbf{x}^0) - f(\widehat{\mathbf{x}})\right)} \\ &\le 2\sqrt{\frac{2}{\widehat{\alpha}(1-\eta_\Gamma)}\eta_\Gamma^k\left(f(\mathbf{x}^0) - f(\widehat{\mathbf{x}})\right)}.\end{aligned}$$

Combining the last inequality with (5.107), we get for any k satisfying the relations (5.113) that

$$\|\mathbf{g}^P(\mathbf{x}^k)\|^2 = \|\boldsymbol{\beta}(\mathbf{x}^k)\|^2 + \|\boldsymbol{\varphi}(\mathbf{x}^k)\|^2 \le \frac{10}{\widehat{\alpha}(1-\eta_\Gamma)}\eta_\Gamma^k\left(f(\mathbf{x}^0) - f(\widehat{\mathbf{x}})\right). \quad (5.115)$$

Now notice that the estimate (5.115) is valid for any iterate $\mathbf{x}^k$ which satisfies $\mathcal{F}(\mathbf{x}^{k-1}) \supseteq \mathcal{F}(\mathbf{x}^k)$, i.e., when $\mathbf{x}^k$ is generated by the conjugate gradient step or the expansion step. Thus it remains to estimate the projected gradient of the iterate $\mathbf{x}^k$ generated by the proportioning step. In this case

$$\mathcal{F}(\mathbf{x}^{k-1}) \subseteq \mathcal{F}(\mathbf{x}^k),$$

so that we can use the estimate (5.115) to get

$$\|\mathbf{g}^P(\mathbf{x}^{k-1})\| \le \sqrt{\frac{10}{\widehat{\alpha}(1-\eta_\Gamma)}\eta_\Gamma^{k-1}\left(f(\mathbf{x}^0) - f(\widehat{\mathbf{x}})\right)}. \quad (5.116)$$

Since the proportioning step is defined by $\mathbf{x}^k = \mathbf{x}^{k-1} - \alpha_{cg}\boldsymbol{\beta}(\mathbf{x}^{k-1})$, it follows that

$$\|\mathbf{g}_{\mathcal{F}(\mathbf{x}^k)}(\mathbf{x}^{k-1})\| = \|\mathbf{g}^P(\mathbf{x}^{k-1})\|.$$

Moreover, using the basic properties of the norm, we get

$$\begin{aligned}\|\boldsymbol{\varphi}(\mathbf{x}^k)\| &= \|\mathbf{g}_{\mathcal{F}(\mathbf{x}^k)}(\mathbf{x}^k)\| \le \|\mathbf{g}_{\mathcal{F}(\mathbf{x}^k)}(\mathbf{x}^k) - \mathbf{g}_{\mathcal{F}(\mathbf{x}^k)}(\mathbf{x}^{k-1})\| + \|\mathbf{g}_{\mathcal{F}(\mathbf{x}^k)}(\mathbf{x}^{k-1})\| \\ &\le \|\mathbf{g}(\mathbf{x}^k) - \mathbf{g}(\widehat{\mathbf{x}})\| + \|\mathbf{g}(\widehat{\mathbf{x}}) - \mathbf{g}(\mathbf{x}^{k-1})\| + \|\mathbf{g}^P(\mathbf{x}^{k-1})\|,\end{aligned}$$

and by (5.114) and (5.116)

$$\begin{aligned}\|\boldsymbol{\varphi}(\mathbf{x}^k)\| &\le \sqrt{\frac{2}{\widehat{\alpha}}\eta_\Gamma^k\left(f(\mathbf{x}^0) - f(\widehat{\mathbf{x}})\right)} + \sqrt{\frac{2}{\widehat{\alpha}}\eta_\Gamma^{k-1}\left(f(\mathbf{x}^0) - f(\widehat{\mathbf{x}})\right)} \\ &\quad + \sqrt{\frac{10}{\widehat{\alpha}(1-\eta_\Gamma)}\eta_\Gamma^{k-1}\left(f(\mathbf{x}^0) - f(\widehat{\mathbf{x}})\right)} \\ &\le (\sqrt{5}+2)\sqrt{\frac{2}{\widehat{\alpha}(1-\eta_\Gamma)}\eta_\Gamma^{k-1}\left(f(\mathbf{x}^0) - f(\widehat{\mathbf{x}})\right)}.\end{aligned}$$

Combining the last inequality with (5.107), we get by simple computation that

$$\|\mathbf{g}^P(\mathbf{x}^k)\|^2 = \|\varphi(\mathbf{x}^k)\|^2 + \|\beta(\mathbf{x}^k)\|^2 \le \frac{38}{\widehat{\alpha}(1-\eta_\Gamma)}\eta_\Gamma^{k-1}\left(f(\mathbf{x}^0) - f(\widehat{\mathbf{x}})\right).$$

Since the last estimate is obviously weaker than (5.115), it follows that (5.103) is valid for all indices k. □

The bound on the rate of convergence as given by (5.103) is rather poor. The reason is that it has been obtained by the worst case analysis of a general couple of consecutive iterations and does not reflect the structure of a longer chain of the same type of iterations. Recall that Fig. 5.14 shows that no bound can be obtained by the analysis of a single iteration!

5.8.4 Optimality

Theorems 5.14 and 5.15 give the bounds on the rates of convergence of the iterates and corresponding projected gradients that depend only on the bounds on the spectrum, but do not depend on the constraint vector $\boldsymbol{\ell}$. It simply follows that if we have a class of bound constrained problems with the spectrum of the Hessian of the cost function in an a priori fixed interval, then the rate of convergence of the MPRGP algorithm can be bounded uniformly for the whole class. To present explicitly this feature of Algorithm 5.7, let $\mathcal{T}$ denote any set of indices and assume that for any $t \in \mathcal{T}$ there is defined a problem

$$\text{minimize } f_t(\mathbf{x}) \text{ s.t. } \mathbf{x} \in \Omega_{B_t} \tag{5.117}$$

with $\Omega_{B_t} = \{\mathbf{x} \in \mathbb{R}^{n_t} : \mathbf{x} \ge \boldsymbol{\ell}_t\}$, $f_t(\mathbf{x}) = \frac{1}{2}\mathbf{x}^T\mathsf{A}_t\mathbf{x} - \mathbf{b}_t^T\mathbf{x}$, $\mathsf{A}_t \in \mathbb{R}^{n_t \times n_t}$ symmetric positive definite, and $\boldsymbol{\ell}_t \in \mathbb{R}^{n_t}$. Our optimality result then reads as follows.

Theorem 5.16. *Let $a_{\max} > a_{\min} > 0$ denote given constants and let $\{\mathbf{x}_t^k\}$ be generated by Algorithm 5.7 for the solution of the bound constrained problem (5.117) with $0 < \overline{\alpha} \le 2a_{\max}^{-1}$ and $\Gamma > 0$ starting from $\mathbf{x}_t^0 = \max\{\mathbf{o}, \boldsymbol{\ell}_t\}$. Let the class of problems (5.117) satisfy*

$$a_{\min} \le \lambda_{\min}(\mathsf{A}_t) \le \lambda_{\max}(\mathsf{A}_t) \le a_{\max},$$

where $\lambda_{\min}(\mathsf{A}_t)$ and $\lambda_{\max}(\mathsf{A}_t)$ denote respectively the smallest and the largest eigenvalues of A_t.

Then there are integers $\overline{k}$ and $\overline{\ell}$ such that for any $t \in \mathcal{T}$ and $\varepsilon > 0$

$$\|\mathbf{g}_t^P(\mathbf{x}_t^{\overline{k}})\| \le \varepsilon \|\mathbf{g}_t^P(\mathbf{x}_t^0)\|$$

and

$$f_t(\mathbf{x}_t^{\overline{\ell}}) - f_t(\widehat{\mathbf{x}}_t) \le \varepsilon\left(f_t(\mathbf{x}_t^0) - f(\widehat{\mathbf{x}}_t)\right).$$

Proof. First denote

$$\eta_\Gamma^t = 1 - \frac{\widehat{\alpha}\lambda_{\min}^t}{\vartheta + \vartheta\widehat{\Gamma}^2}, \qquad \overline{\eta}_\Gamma = 1 - \frac{\widehat{\alpha}a_{\min}}{\vartheta + \vartheta\widehat{\Gamma}^2},$$

$$a_1^t = \frac{38\vartheta(1+\widehat{\Gamma}^2)}{\widehat{\alpha}^2\lambda_{\min}^t}, \qquad \overline{a}_1 = \frac{38\vartheta(1+\widehat{\Gamma}^2)}{\widehat{\alpha}^2 a_{\min}},$$

where $\widehat{\Gamma} = \max\{\Gamma, \Gamma^{-1}\}$, so that

$$\eta_\Gamma^t \le \overline{\eta}_\Gamma < 1 \quad \text{and} \quad a_1^t \le \overline{a}_1.$$

Combining these estimates with Theorem 5.15 and inequality (5.5), we get for any $k \ge 1$

$$\|\mathbf{g}_t^P(\mathbf{x}^{k+1})\|^2 \le \overline{a}_1\overline{\eta}_\Gamma^k \left(f_t(\mathbf{x}_t^0) - f_t(\widehat{\mathbf{x}}_t)\right) \le \frac{\overline{a}_1}{a_{\min}}\overline{\eta}_\Gamma^k \|\mathbf{g}_t^P(\mathbf{x}_t^0)\|^2.$$

Similarly, using Theorem 5.14, we get

$$f_t(\mathbf{x}_t^k) - f_t(\widehat{\mathbf{x}}_t) \le \overline{\eta}_\Gamma^k \left(f_t(\mathbf{x}_t^0) - f(\widehat{\mathbf{x}}_t)\right).$$

To finish the proof, it is enough to take $\overline{k}$ and $\overline{\ell}$ so that

$$\frac{\overline{a}_1}{a_{\min}}\overline{\eta}_\Gamma^{\overline{k}-1} \le \varepsilon \quad \text{and} \quad \overline{\eta}_\Gamma^{\overline{\ell}} \le \varepsilon. \qquad \square$$

5.8.5 Identification Lemma and Finite Termination

Let us consider the conditions which guarantee that the MPRGP algorithm finds the solution $\widehat{\mathbf{x}}$ of (5.1) in a finite number of steps. There are at least two reasons to consider such results important. First the algorithm with the finite termination property is less likely to suffer from the oscillations that are often attributed to the working set-based algorithms as it is less likely to reexamine the working sets; if any working set reappears, it can happen "only" finitely many times. The second reason is that such algorithm is more likely to generate longer sequences of the conjugate gradient iterations. Thus the reduction of the cost function value is bounded by the "global" estimate (3.21), and finally switches to the conjugate gradient method, so that it can exploit its nice self-acceleration property [168]. It is difficult to enhance these characteristics of the algorithm into the rate of convergence as they cannot be obtained by the analysis of just one step of the method.

We first examine the finite termination of Algorithm 5.7 in a simpler case when the solution $\widehat{\mathbf{x}}$ of (5.1) is *regular*, i.e., the vector of Lagrange multipliers $\widehat{\lambda}$ of the solution satisfies the *strict complementarity condition* $\widehat{\lambda}_i > 0$ for $i \in \mathcal{A}(\widehat{\mathbf{x}})$. The proof is based on simple geometrical observations. For example, examining Fig. 5.15, it is easy to see that the free sets of the iterates $\mathbf{x}^k$ soon contain the free set of the solution $\widehat{\mathbf{x}}$. The formal analysis of such observations is a subject of the following identification lemma.

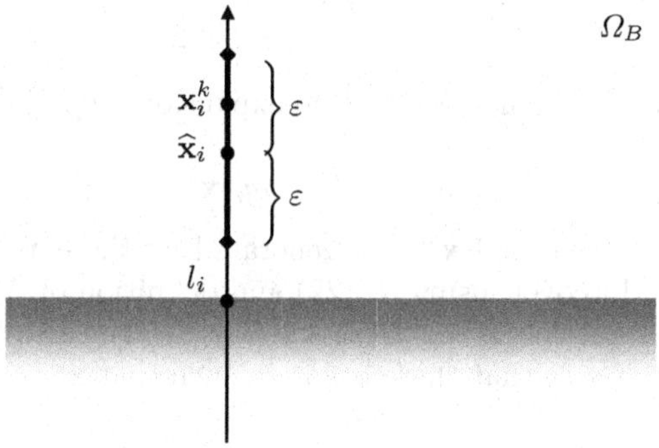

Fig. 5.15. Identification of the free set of the solution

Lemma 5.17. *Let* $\{\mathbf{x}^k\}$ *be generated by Algorithm 5.7 with* $\mathbf{x}^0 \in \Omega_B$, $\Gamma > 0$, *and* $\overline{\alpha} \in (0, 2\|\mathsf{A}\|^{-1}]$. *Then there is* k_0 *such that for* $k \geq k_0$

$$\mathcal{F}(\widehat{\mathbf{x}}) \subseteq \mathcal{F}(\mathbf{x}^k), \quad \mathcal{F}(\widehat{\mathbf{x}}) \subseteq \mathcal{F}(\mathbf{x}^k - \overline{\alpha}\widetilde{\varphi}(\mathbf{x}^k)), \quad \textit{and} \quad \mathcal{B}(\widehat{\mathbf{x}}) \subseteq \mathcal{B}(\mathbf{x}^k), \tag{5.118}$$

where $\widetilde{\varphi}(\mathbf{x}^k) = \widetilde{\varphi}_{\overline{\alpha}}(\mathbf{x}^k)$ *is defined by* (5.85).

Proof. Since (5.118) is trivially satisfied when there is $k = k_0$ such that $\mathbf{x}^k = \widehat{\mathbf{x}}$, we shall assume in what follows that $\mathbf{x}^k \neq \widehat{\mathbf{x}}$ for any $k \geq 0$. Let us denote $x_i^k = [\mathbf{x}^k]_i$ and $\widehat{x}_i = [\widehat{\mathbf{x}}]_i$, $i = 1, \ldots, n$.

Let us first assume that $\mathcal{F}(\widehat{\mathbf{x}}) \neq \emptyset$ and $\mathcal{B}(\widehat{\mathbf{x}}) \neq \emptyset$, so that we can define

$$\varepsilon = \min\{\widehat{x}_i - \ell_i : i \in \mathcal{F}(\widehat{\mathbf{x}})\} > 0 \quad \text{and} \quad \delta = \min\{g_i(\widehat{\mathbf{x}}) : i \in \mathcal{B}(\widehat{\mathbf{x}})\} > 0.$$

Since by Proposition 5.12 $\{\mathbf{x}^k\}$ converges to $\widehat{\mathbf{x}}$, there is k_0 such that for any $k \geq k_0$

$$g_i(\mathbf{x}^k) \leq \frac{\varepsilon}{4\overline{\alpha}} \text{ for } i \in \mathcal{F}(\widehat{\mathbf{x}}) \tag{5.119}$$

$$x_i^k \geq \ell_i + \frac{\varepsilon}{2} \text{ for } i \in \mathcal{F}(\widehat{\mathbf{x}}) \tag{5.120}$$

$$x_i^k \leq \ell_i + \frac{\overline{\alpha}\delta}{8} \text{ for } i \in \mathcal{B}(\widehat{\mathbf{x}}) \tag{5.121}$$

$$g_i(\mathbf{x}^k) \geq \frac{\delta}{2} \text{ for } i \in \mathcal{B}(\widehat{\mathbf{x}}). \tag{5.122}$$

In particular, for $k \geq k_0$, the first inclusion of (5.118) follows from (5.120), while the second inclusion follows from (5.119) and (5.120), as for $i \in \mathcal{F}(\widehat{\mathbf{x}})$

$$x_i^k - \overline{\alpha}\varphi_i(\mathbf{x}^k) = x_i^k - \overline{\alpha}g_i(\mathbf{x}^k) \geq \ell_i + \frac{\varepsilon}{2} - \frac{\overline{\alpha}\varepsilon}{4\overline{\alpha}} > \ell_i.$$

Let $k \geq k_0$ and observe that, by (5.121) and (5.122), for any $i \in \mathcal{B}(\widehat{\mathbf{x}})$

$$x_i^k - \overline{\alpha} g_i(\mathbf{x}^k) \le \ell_i + \frac{\overline{\alpha}\delta}{8} - \frac{\overline{\alpha}\delta}{2} < \ell_i,$$

so that if some $\mathbf{x}^{k+1}$ is generated by the expansion step (5.84), $k \ge k_0$, and $i \in \mathcal{B}(\widehat{\mathbf{x}})$, then

$$x_i^{k+1} = \max\{\ell_i, x_i^k - \overline{\alpha} g_i(\mathbf{x}^k)\} = \ell_i.$$

It follows that if $k \ge k_0$ and $\mathbf{x}^{k+1}$ is generated by the expansion step, then $\mathcal{B}(\mathbf{x}^{k+1}) \supseteq \mathcal{B}(\widehat{\mathbf{x}})$. Moreover, using (5.122) and definition of Algorithm 5.7, we can directly verify that if $\mathcal{B}(\mathbf{x}^k) \supseteq \mathcal{B}(\widehat{\mathbf{x}})$ and $k \ge k_0$, then also $\mathcal{B}(\mathbf{x}^{k+1}) \supseteq \mathcal{B}(\widehat{\mathbf{x}})$. Thus it remains to prove that there is $s \ge k_0$ such that $\mathbf{x}^s$ is generated by the expansion step.

Let us examine what can happen for $k \ge k_0$. First observe that we can never take the full CG step in the direction $\mathbf{p}^k = \varphi(\mathbf{x}^k)$. The reason is that

$$\alpha_{cg}(\mathbf{p}^k) = \frac{\varphi(\mathbf{x}^k)^T \mathbf{g}(\mathbf{x}^k)}{\varphi(\mathbf{x}^k)^T \mathsf{A} \varphi(\mathbf{x}^k)} = \frac{\|\varphi(\mathbf{x}^k)\|^2}{\varphi(\mathbf{x}^k)^T \mathsf{A} \varphi(\mathbf{x}^k)} \ge \|\mathsf{A}\|^{-1} \ge \frac{\overline{\alpha}}{2},$$

so that for $i \in \mathcal{F}(\mathbf{x}^k) \cap \mathcal{B}(\widehat{\mathbf{x}})$, by (5.121) and (5.122),

$$x_i^k - \alpha_{cg} p_i^k = x_i^k - \alpha_{cg} g_i(\mathbf{x}^k) \le x_i^k - \frac{\overline{\alpha}}{2} g_i(\mathbf{x}^k) \le \ell_i + \frac{\overline{\alpha}\delta}{8} - \frac{\overline{\alpha}\delta}{4} < \ell_i. \quad (5.123)$$

It follows by definition of Algorithm 5.7 that if $\mathbf{x}^k, k \ge k_0$, is generated by the proportioning step, then the following trial conjugate gradient step is not feasible, and $\mathbf{x}^{k+1}$ is necessarily generated by the expansion step.

To complete the proof, observe that Algorithm 5.7 can generate only a finite sequence of consecutive conjugate gradient iterates. Indeed, if there is neither proportioning step nor the expansion step for $k \ge k_0$, then it follows by the finite termination property of the conjugate gradient method that there is $l \le n$ such that $\varphi(\mathbf{x}^{k_0+l}) = \mathbf{o}$. Thus either $\mathbf{x}^{k_0+l} = \widehat{\mathbf{x}}$ and $\mathcal{B}(\mathbf{x}^k) = \mathcal{B}(\widehat{\mathbf{x}})$ for $k \ge k_0 + l$ by rule (i), or $\mathbf{x}^{k_0+l}$ is not strictly proportional, $\mathbf{x}^{k_0+l+1}$ is generated by the proportioning step, and $\mathbf{x}^{k_0+l+2}$ is generated by the expansion step. This completes the proof, as the cases $\mathcal{F}(\widehat{\mathbf{x}}) = \emptyset$ and $\mathcal{B}(\widehat{\mathbf{x}}) = \emptyset$ can be proved by a direct analysis of the above arguments. □

Proposition 5.18. *Let $\{\mathbf{x}^k\}$ be generated by Algorithm 5.7 with $\mathbf{x}^0 \in \Omega_B$, $\Gamma > 0$, and $\overline{\alpha} \in (0, 2\|\mathsf{A}\|^{-1}]$. Let the solution $\widehat{\mathbf{x}}$ satisfy the condition of strict complementarity, i.e., $\widehat{x}_i = \ell_i$ implies $g_i(\widehat{\mathbf{x}}) > 0$. Then there is $k \ge 0$ such that $\mathbf{x}^k = \widehat{\mathbf{x}}$.*

Proof. If $\widehat{\mathbf{x}}$ satisfies the condition of strict complementarity, then $\mathcal{A}(\widehat{\mathbf{x}}) = \mathcal{B}(\widehat{\mathbf{x}})$, and, by Lemma 5.17, there is $k_0 \ge 0$ such that for $k \ge k_0$ we have $\mathcal{F}(\mathbf{x}^k) = \mathcal{F}(\widehat{\mathbf{x}})$ and $\mathcal{B}(\mathbf{x}^k) = \mathcal{B}(\widehat{\mathbf{x}})$. Thus, for $k \ge k_0$, all $\mathbf{x}^k$ that satisfy $\widehat{\mathbf{x}} \ne \mathbf{x}^{k-1}$ are generated by the conjugate gradient steps and, by the finite termination property of the CG, there is $k \le k_0 + n$ such that $\mathbf{x}^k = \widehat{\mathbf{x}}$. □

5.8.6 Finite Termination for Dual Degenerate Solution

Our final goal is to prove the finite termination of Algorithm 5.7 when the solution of (5.1) does not satisfy the strict complementarity condition as in Fig. 5.16, where the iterations with different active sets are near the solution.

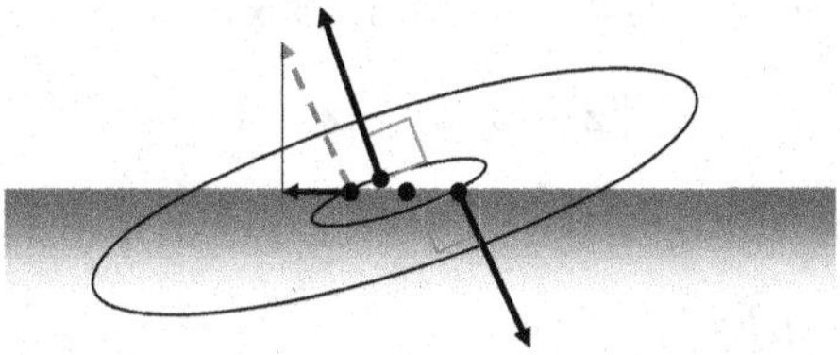

Fig. 5.16. Projected gradients near dual degenerate solution

Lemma 5.19. *Let* $\overline{\alpha} \in (0, 2\|\mathsf{A}\|^{-1}]$, $\mathbf{x} \in \Omega_B$, *and* $\mathbf{y} = \mathbf{x} - \overline{\alpha}\widetilde{\varphi}(\mathbf{x})$. *Then*

$$\|\varphi(\mathbf{y})\|^2 \leq 9\widetilde{\varphi}(\mathbf{x})^T\varphi(\mathbf{x}) \quad \textit{and} \quad \|\beta(\mathbf{y})\| \geq \|\beta(\mathbf{x})\| - 4\|\widetilde{\varphi}(\mathbf{x})\|, \tag{5.124}$$

where the reduced free gradient $\widetilde{\varphi}(\mathbf{x}) = \widetilde{\varphi}_{\overline{\alpha}}(\mathbf{x})$ *is defined by* (5.85).

Proof. First notice that $\mathcal{F}(\mathbf{y}) \subseteq \mathcal{F}(\mathbf{x})$. Since

$$\mathbf{g}(\mathbf{y}) = \mathbf{g}(\mathbf{x}) - \overline{\alpha}\mathsf{A}\widetilde{\varphi}(\mathbf{x}) \quad \text{and} \quad \widetilde{\varphi}_{\mathcal{F}(\mathbf{y})}(\mathbf{x}) = \varphi_{\mathcal{F}(\mathbf{y})}(\mathbf{x}) = \mathbf{g}_{\mathcal{F}(\mathbf{y})}(\mathbf{x}), \tag{5.125}$$

we get

$$\begin{aligned}\|\varphi(\mathbf{y})\| &= \|\mathbf{g}_{\mathcal{F}(\mathbf{y})}(\mathbf{y})\| = \|\mathbf{g}_{\mathcal{F}(\mathbf{y})}(\mathbf{x}) - \overline{\alpha}\,[\mathsf{A}\widetilde{\varphi}(\mathbf{x})]_{\mathcal{F}(\mathbf{y})}\,\| \\ &\leq \|\widetilde{\varphi}_{\mathcal{F}(\mathbf{y})}(\mathbf{x})\| + \overline{\alpha}\,\|\,[\mathsf{A}\widetilde{\varphi}(\mathbf{x})]_{\mathcal{F}(\mathbf{y})}\,\| \leq 3\|\widetilde{\varphi}(\mathbf{x})\|.\end{aligned}$$

Using the latter inequalities and the definition of $\widetilde{\varphi}(\mathbf{x})$, we get

$$\|\varphi(\mathbf{y})\|^2 \leq 9\|\widetilde{\varphi}(\mathbf{x})\|^2 \leq 9\widetilde{\varphi}(\mathbf{x})^T\varphi(\mathbf{x}).$$

To prove the second inequality of (5.124), denote

$$\mathcal{C} = \{i \in \mathcal{A}(\mathbf{x}) :\ g_i(\mathbf{x}) \leq 0\}$$

and notice that

$$\mathcal{A}(\mathbf{y}) \supseteq \mathcal{A}(\mathbf{x}) \supseteq \mathcal{C}.$$

Thus

$$\|\boldsymbol{\beta}(\mathbf{y})\| = \|\mathbf{g}_{\mathcal{A}(\mathbf{y})}(\mathbf{y})^-\| \geq \|\mathbf{g}_{\mathcal{C}}(\mathbf{y})^-\| = \|\left(\mathbf{g}_{\mathcal{C}}(\mathbf{x}) - \overline{\alpha}\,[\mathsf{A}\widetilde{\boldsymbol{\varphi}}(\mathbf{x})]_{\mathcal{C}}\right)^-\|$$
$$= \|\left(\boldsymbol{\beta}_{\mathcal{C}}(\mathbf{x}) - \overline{\alpha}\,[\mathsf{A}\widetilde{\boldsymbol{\varphi}}(\mathbf{x})]_{\mathcal{C}}\right)^-\|. \tag{5.126}$$

Using in sequence

$$\|\boldsymbol{\beta}_{\mathcal{C}}(\mathbf{x})\| = \|\boldsymbol{\beta}(\mathbf{x})\|, \quad \|\overline{\alpha}\,[\mathsf{A}\widetilde{\boldsymbol{\varphi}}(\mathbf{x})]_{\mathcal{C}}\| \leq 2\|\widetilde{\boldsymbol{\varphi}}(\mathbf{x})\|,$$

inequality (5.126), properties of the norm, $\boldsymbol{\beta}(\mathbf{x})^- = \boldsymbol{\beta}(\mathbf{x})$, and

$$\|\mathbf{z} - \mathbf{z}^-\| \leq \|\mathbf{z} - \mathbf{t}\|$$

for any $\mathbf{t}$ with nonpositive entries, we get

$$\|\boldsymbol{\beta}(\mathbf{x})\| - \|\widetilde{\boldsymbol{\varphi}}(\mathbf{x})\| - \|\boldsymbol{\beta}(\mathbf{y})\|$$
$$\leq \|\boldsymbol{\beta}_{\mathcal{C}}(\mathbf{x})\| - \frac{1}{2}\|\overline{\alpha}\,[\mathsf{A}\widetilde{\boldsymbol{\varphi}}(\mathbf{x})]_{\mathcal{C}}\| - \|\left(\boldsymbol{\beta}_{\mathcal{C}}(\mathbf{x}) - \overline{\alpha}\,[\mathsf{A}\widetilde{\boldsymbol{\varphi}}(\mathbf{x})]_{\mathcal{C}}\right)^-\|$$
$$\leq \|\boldsymbol{\beta}_{\mathcal{C}}(\mathbf{x}) - \frac{\overline{\alpha}}{2}\,[\mathsf{A}\widetilde{\boldsymbol{\varphi}}(\mathbf{x})]_{\mathcal{C}}\| - \|\left(\boldsymbol{\beta}_{\mathcal{C}}(\mathbf{x}) - \overline{\alpha}\,[\mathsf{A}\widetilde{\boldsymbol{\varphi}}(\mathbf{x})]_{\mathcal{C}}\right)^-\|$$
$$\leq \|\left(\boldsymbol{\beta}_{\mathcal{C}}(\mathbf{x}) - \overline{\alpha}\,[\mathsf{A}\widetilde{\boldsymbol{\varphi}}(\mathbf{x})]_{\mathcal{C}}\right) - \left(\boldsymbol{\beta}_{\mathcal{C}}(\mathbf{x}) - \overline{\alpha}\,[\mathsf{A}\widetilde{\boldsymbol{\varphi}}(\mathbf{x})]_{\mathcal{C}}\right)^-\| + \frac{\overline{\alpha}}{2}\|\,[\mathsf{A}\widetilde{\boldsymbol{\varphi}}(\mathbf{x})]_{\mathcal{C}}\|$$
$$\leq \|\left(\boldsymbol{\beta}_{\mathcal{C}}(\mathbf{x}) - \overline{\alpha}\,[\mathsf{A}\widetilde{\boldsymbol{\varphi}}(\mathbf{x})]_{\mathcal{C}}\right) - \boldsymbol{\beta}_{\mathcal{C}}(\mathbf{x})\| + \|\widetilde{\boldsymbol{\varphi}}(\mathbf{x})\| \leq 3\|\widetilde{\boldsymbol{\varphi}}(\mathbf{x})\|.$$

This proves the second inequality of (5.124). □

Corollary 5.20. *Let $\Gamma \geq 4$, $\overline{\alpha} \in (0, 2\|\mathsf{A}\|^{-1}]$, $\mathbf{x} \in \Omega_B$, and*

$$\Gamma^2 \widetilde{\boldsymbol{\varphi}}(\mathbf{x})^T \boldsymbol{\varphi}(\mathbf{x}) < \|\boldsymbol{\beta}(\mathbf{x})\|^2, \tag{5.127}$$

where the reduced free gradient $\widetilde{\boldsymbol{\varphi}}(\mathbf{x}) = \widetilde{\boldsymbol{\varphi}}_{\overline{\alpha}}(\mathbf{x})$ is defined by (5.85).
Then the vector $\mathbf{y} = \mathbf{x} - \overline{\alpha}\widetilde{\boldsymbol{\varphi}}(\mathbf{x})$ satisfies

$$\frac{\Gamma - 4}{3}\|\boldsymbol{\varphi}(\mathbf{y})\| < \|\boldsymbol{\beta}(\mathbf{y})\|. \tag{5.128}$$

Proof. Inequality (5.128) holds trivially for $\Gamma = 4$. For $\Gamma > 4$, using in sequence (5.124), $\|\widetilde{\varphi}(x)\|^2 \leq \widetilde{\boldsymbol{\varphi}}(\mathbf{x})^T \boldsymbol{\varphi}(\mathbf{x})$, twice (5.127), and (5.124), we get

$$\|\boldsymbol{\beta}(\mathbf{y})\| \geq \|\boldsymbol{\beta}(\mathbf{x})\| - 4\|\widetilde{\boldsymbol{\varphi}}(\mathbf{x})\| \geq \|\boldsymbol{\beta}(\mathbf{x})\| - 4\sqrt{\widetilde{\boldsymbol{\varphi}}^T(\mathbf{x})\boldsymbol{\varphi}(\mathbf{x})} > (1 - 4\Gamma^{-1})\|\boldsymbol{\beta}(\mathbf{x})\|$$
$$> (\Gamma - 4)\sqrt{\widetilde{\boldsymbol{\varphi}}^T(\mathbf{x})\boldsymbol{\varphi}(\mathbf{x})} \geq \frac{\Gamma - 4}{3}\|\boldsymbol{\varphi}(\mathbf{y})\|. \tag{5.129}$$

□

Theorem 5.21. *Let $\{\mathbf{x}^k\}$ denote the sequence generated by Algorithm 5.7 with*

$$\mathbf{x}^0 \in \Omega_B, \quad \Gamma \geq 3\left(\sqrt{\kappa(\mathsf{A})} + 4\right), \quad \text{and} \quad \overline{\alpha} \in (0, 2\|\mathsf{A}\|^{-1}]. \tag{5.130}$$

Then there is $k \geq 0$ such that $\mathbf{x}^k = \widehat{\mathbf{x}}$.

Proof. Let $\mathbf{x}^k$ be generated by Algorithm 5.7 and let Γ satisfy (5.130). Let k_0 be that of Lemma 5.17 and let $k \geq k_0$ be such that $\mathbf{x}^k$ is not strictly proportional, i.e., $\Gamma^2 \widetilde{\varphi}_{\overline{\alpha}}(\mathbf{x}^k)^T \varphi(\mathbf{x}^k) < \|\beta(\mathbf{x}^k)\|^2$. Then by Corollary 5.20 the vector $\mathbf{y} = \mathbf{x}^k - \overline{\alpha}\widetilde{\varphi}(\mathbf{x}^k)$ satisfies

$$\Gamma_1 \|\varphi(\mathbf{y})\| < \|\beta(\mathbf{y})\| \tag{5.131}$$

with

$$\Gamma_1 = (\Gamma - 4)/3 \geq \sqrt{\kappa(\mathsf{A})}.$$

Moreover, $\mathbf{y} \in \Omega_B$, and by Lemma 5.17 and definition of $\mathbf{y}$

$$\mathcal{A}(\widehat{\mathbf{x}}) \supseteq \mathcal{A}(\mathbf{y}) \supseteq \mathcal{A}(\mathbf{x}^k) \supseteq \mathcal{B}(\mathbf{x}^k) \supseteq \mathcal{B}(\widehat{\mathbf{x}}). \tag{5.132}$$

It follows by Lemma 5.4 that the vector $\mathbf{z} = \mathbf{y} - \|\mathsf{A}\|^{-1}\beta(\mathbf{y})$ satisfies

$$f(\mathbf{z}) < \min\{f(\mathbf{x}) : \ \mathbf{x} \in \mathcal{W}_{\mathcal{I}}\} \tag{5.133}$$

with $\mathcal{I} = \mathcal{A}(\mathbf{y})$. Since $\mathcal{I}$ satisfies by (5.132) $\mathcal{A}(\widehat{\mathbf{x}}) \supseteq \mathcal{I} \supseteq \mathcal{B}(\widehat{\mathbf{x}})$, we have also

$$f(\widehat{\mathbf{x}}) = \min\{f(\mathbf{x}) : \ \mathbf{x} \in \Omega_B\} = \min\{f(\mathbf{x}) : \ \mathbf{x} \in \mathcal{W}_{\mathcal{I}}\}. \tag{5.134}$$

However, $\mathbf{z} \in \Omega_B$, so that (5.134) contradicts (5.133). Thus all $\mathbf{x}^k$ are strictly proportional for $k \geq k_0$, so that

$$\mathcal{A}(\mathbf{x}^{k_0}) \subseteq \mathcal{A}(\mathbf{x}^{k_0+1}) \subseteq \dots.$$

Using the finite termination property of the conjugate gradient method, we conclude that there is $k \geq k_0$ such that $\widehat{\mathbf{x}} = \mathbf{x}^k$. □

Let us recall that the finite termination property of the MPRGP algorithm with a dual degenerate solution and

$$\overline{\alpha} \in (0, \|\mathsf{A}\|^{-1}]$$

has been proved for

$$\Gamma \geq 2\left(\sqrt{\kappa(\mathsf{A})} + 1\right).$$

For the details see Dostál [74].

5.9 Implementation of MPRGP with Optional Modifications

In this section, we describe Algorithm 5.7 in the form that is convenient for implementation. We include also some modifications that may be used to improve its performance. Implementation of Algorithm 5.6 is similar.

5.9.1 Expansion Step with Feasible Half-Step

To improve the efficiency of the expansion step, we can use the trial conjugate gradient direction $\mathbf{p}^k$ which is generated before the expansion step is invoked. We propose to generate first

$$\mathbf{x}^{k+\frac{1}{2}} = \mathbf{x}^k - \alpha_f \mathbf{p}^k \quad \text{and} \quad \mathbf{g}^{k+\frac{1}{2}} = \mathbf{g}^k - \alpha_f \mathsf{A}\mathbf{p}^k,$$

where the feasible steplength α_f for $\mathbf{p}^k$ is defined by

$$\alpha_f = \max\{\alpha : \ \mathbf{x}^k - \alpha \mathbf{p}^k \in \Omega_B\} = \min_{i=1,\dots,n} \{(x_i^k - \ell_i)/p_i^k, \ p_i^k > 0\},$$

and then define

$$\mathbf{x}^{k+1} = P_{\Omega_B}\left(\mathbf{x}^{k+\frac{1}{2}} - \overline{\alpha}\varphi(\mathbf{x}^{k+\frac{1}{2}})\right).$$

The half-step is illustrated in Fig. 5.17. Such modification does not require any additional matrix–vector multiplication and the estimate (5.93) remains valid as $f(\mathbf{x}^{k+\frac{1}{2}}) - f(\mathbf{x}^k) \le 0$ and

$$\begin{aligned} f(\mathbf{x}^{k+1}) - f(\widehat{\mathbf{x}}) &\le \eta_\Gamma \left((f(\mathbf{x}^{k+\frac{1}{2}}) - f(\mathbf{x}^k)) + f(\mathbf{x}^k) - f(\widehat{\mathbf{x}}) \right) \\ &\le \eta_\Gamma \left(f(\mathbf{x}^k) - f(\widehat{\mathbf{x}}) \right). \end{aligned}$$

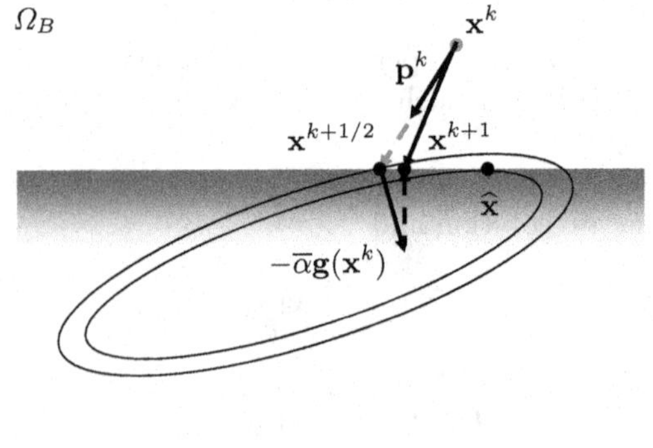

Fig. 5.17. Feasible half-step

5.9.2 MPRGP Algorithm

Now we are ready to give the details of the implementation of the MPRGP algorithm which was briefly described in the form suitable for analysis as Algorithm 5.7. To preserve readability, we do not distinguish generations of variables by indices unless it is convenient for further reference.

Algorithm 5.8. Modified proportioning with reduced gradient projections (MPRGP).

Given a symmetric positive definite matrix A *of the order* n, n*-vectors* $\mathbf{b}$, $\boldsymbol{\ell}$, $\Omega_B = \{\mathbf{x} : \mathbf{x} \geq \boldsymbol{\ell}\}$, $\mathbf{x}^0 \in \Omega_B$.

Step 0. {*Initialization.*}

  Choose $\Gamma > 0$, $\overline{\alpha} \in (0, 2\|\mathsf{A}\|^{-1}]$, *set* $k = 0$, $\mathbf{g} = \mathsf{A}\mathbf{x}^0 - \mathbf{b}$, $\mathbf{p} = \varphi(\mathbf{x}^0)$

  while $\|\mathbf{g}^P(\mathbf{x}^k)\|$ *is not small*

   if $\|\beta(\mathbf{x}^k)\|^2 \leq \Gamma^2 \widetilde{\varphi}(\mathbf{x}^k)^T \varphi(\mathbf{x}^k)$

Step 1. {*Proportional* $\mathbf{x}^k$. *Trial conjugate gradient step.*}

    $\alpha_{cg} = \mathbf{g}^T\mathbf{p}/\mathbf{p}^T\mathsf{A}\mathbf{p}$, $\mathbf{y} = \mathbf{x}^k - \alpha_{cg}\mathbf{p}$

    $\alpha_f = \max\{\alpha : \mathbf{x}^k - \alpha\mathbf{p} \in \Omega_B\} = \min\{(x_i^k - \ell_i)/p_i : p_i > 0\}$

    if $\alpha_{cg} \leq \alpha_f$

Step 2. {*Conjugate gradient step.*}

     $\mathbf{x}^{k+1} = \mathbf{y}$, $\mathbf{g} = \mathbf{g} - \alpha_{cg}\mathsf{A}\mathbf{p}$,

     $\beta = \varphi(\mathbf{y})^T\mathsf{A}\mathbf{p}/\mathbf{p}^T\mathsf{A}\mathbf{p}$, $\mathbf{p} = \varphi(\mathbf{y}) - \beta\mathbf{p}$

    else

Step 3. {*Expansion step.*}

     $\mathbf{x}^{k+\frac{1}{2}} = \mathbf{x}^k - \alpha_f\mathbf{p}$, $\mathbf{g} = \mathbf{g} - \alpha_f\mathsf{A}\mathbf{p}$

     $\mathbf{x}^{k+1} = P_{\Omega_B}(\mathbf{x}^{k+\frac{1}{2}} - \overline{\alpha}\varphi(\mathbf{x}^{k+\frac{1}{2}}))$

     $\mathbf{g} = \mathsf{A}\mathbf{x}^{k+1} - \mathbf{b}$, $\mathbf{p} = \varphi(\mathbf{x}^{k+1})$

    end if

   else

Step 4. {*Proportioning step.*}

    $\mathbf{d} = \beta(\mathbf{x}^k)$, $\alpha_{cg} = \mathbf{g}^T\mathbf{d}/\mathbf{d}^T\mathsf{A}\mathbf{d}$

    $\mathbf{x}^{k+1} = \mathbf{x}^k - \alpha_{cg}\mathbf{d}$, $\mathbf{g} = \mathbf{g} - \alpha_{cg}\mathsf{A}\mathbf{d}$, $\mathbf{p} = \varphi(\mathbf{x}^{k+1})$

   end if

   $k = k + 1$

  end while

Step 5. {*Return (possibly inexact) solution.*}

  $\widetilde{\mathbf{x}} = \mathbf{x}^k$

In our description, we denote by $\widetilde{\varphi}(\mathbf{x}) = \widetilde{\varphi}_{\overline{\alpha}}(\mathbf{x})$ the reduced free gradient defined by (5.85). Let us recall that by Proposition 5.12 the algorithm converges for any $\overline{\alpha} \in (0, 2\|\mathsf{A}\|^{-1}]$ and by Theorem 5.14 its R-linear rate of convergence is guaranteed for $\overline{\alpha} \in (0, 2\|\mathsf{A}\|^{-1})$.

5.9.3 Unfeasible MPRGP

The "global" bound on the rate of convergence of the CG method guaranteed by Theorem 3.2 indicates that MPRGP converges fast when it generates long chains of CG iterations. Thus it may be advantageous to continue the CG iterations when the trial CG step is unfeasible. The modification of MPRGP proposed here is based on the observation that the convergence of MPRGP is preserved when we insert between the last feasible iteration and the expansion step a finite number of unfeasible iterates as long as $\{f\left(P_{\Omega_B}(\mathbf{x}^k)\right)\}$ decreases. Thus if $f(P_{\Omega_B}(\mathbf{y})) \le f(P_{\Omega_B}(\mathbf{x}^k))$, we can define $\mathbf{x}^{k+1} = \mathbf{y}$ and continue the CG iterations; otherwise we generate $\mathbf{x}^{k+1}$ by the modified expansion step. The resulting *monotonic MPRGP* algorithm reads as follows.

Algorithm 5.9. Monotonic MPRGP.

Given a symmetric positive definite matrix A *of the order* n, n*-vectors* $\mathbf{b}$, $\boldsymbol{\ell}$, $\Omega_B = \{\mathbf{x} : \mathbf{x} \ge \boldsymbol{\ell}\}$, $\mathbf{x}^0 \in \Omega_B$.

Step 0. {*Initialization.*}
  Choose $\Gamma > 0$, $\overline{\alpha} \in (0, 2\|\mathsf{A}\|^{-1}]$, *set* $k = 0$, $\mathbf{g} = \mathsf{A}\mathbf{x}^0 - \mathbf{b}$, $\mathbf{p} = \varphi(\mathbf{x}^0)$
  while $\|\mathbf{g}^P(\mathbf{x}^k)\|$ *is not small*
   if $\|\beta(\mathbf{x}^k)\|^2 \le \Gamma^2 \widetilde{\varphi}(\mathbf{x}^k)^T \varphi(\mathbf{x}^k)$
    $\alpha_{cg} = \mathbf{g}^T\mathbf{p}/\mathbf{p}^T\mathsf{A}\mathbf{p}$, $\mathbf{y} = \mathbf{x}^k - \alpha_{cg}\mathbf{p}$
    while $f(P_{\Omega_B}(\mathbf{y})) \le f(P_{\Omega_B}(\mathbf{x}^k))$ **and**
     $\|\beta(\mathbf{x}^k)\|^2 \le \Gamma^2 \widetilde{\varphi}(\mathbf{x}^k)^T \varphi(\mathbf{x}^k)$ **and** $\|\mathbf{g}^P(\mathbf{x}^k)\|$ *not small*

Step 1. {*Conjugate gradient step.*}
     $\mathbf{x}^{k+1} = \mathbf{y}$, $\mathbf{g} = \mathbf{g} - \alpha_{cg}\mathsf{A}\mathbf{p}$,
     $\beta = \varphi(\mathbf{y})^T\mathsf{A}\mathbf{p}/\mathbf{p}^T\mathsf{A}\mathbf{p}$, $\mathbf{p} = \varphi(\mathbf{y}) - \beta\mathbf{p}$, $k = k + 1$

Step 2. {*Trial CG step for the next iteration of the CG loop.*}
     $\alpha_{cg} = \mathbf{g}^T\mathbf{p}/\mathbf{p}^T\mathsf{A}\mathbf{p}$, $\mathbf{y} = \mathbf{x}^k - \alpha_{cg}\mathbf{p}$
    end *while for CG loop*
   end *if*
   if $\mathbf{y} \notin \Omega_B$ **and** $\|\mathbf{g}^P(\mathbf{x}^k)\|$ *not small*

Step 3. {*Expansion step.*}
    $\mathbf{y} = P_{\Omega_B}(\mathbf{x}^k)$, $\mathbf{x}^{k+1} = P_{\Omega_B}(\mathbf{y} - \overline{\alpha}\varphi(\mathbf{y}))$
    $\mathbf{g} = \mathsf{A}\mathbf{x}^{k+1} - \mathbf{b}$, $\mathbf{p} = \varphi(\mathbf{x}^{k+1})$, $k = k + 1$
   else
    if $\|\beta(\mathbf{x}^k)\|^2 > \Gamma^2 \widetilde{\varphi}(\mathbf{x}^k)^T \varphi(\mathbf{x}^k)$ **and** $\|\mathbf{g}^P(\mathbf{x}^k)\|$ *not small*

Step 4. {*Proportioning step.*}
     $\mathbf{d} = \beta(\mathbf{x}^k)$, $\mathbf{g} = \mathsf{A}\mathbf{x}^k - \mathbf{b}$, $\alpha_{cg} = \mathbf{g}^T\mathbf{d}/\mathbf{d}^T\mathsf{A}\mathbf{d}$
     $\mathbf{x}^{k+1} = \mathbf{x}^k - \alpha_{cg}\mathbf{d}$, $\mathbf{g} = \mathbf{g} - \alpha_{cg}\mathsf{A}\mathbf{d}$, $\mathbf{p} = \varphi(\mathbf{x}^{k+1})$
     $k = k + 1$
    end if
   end if
  end while

Step 5. {*Return (possibly inexact) solution.*}
  $\widetilde{\mathbf{x}} = P_{\Omega_B}(\mathbf{x}^k)$

To see that the algorithm is well defined, namely, that $\mathbf{p} \neq \mathbf{o}$ in Step 1, it is enough to notice that this step is carried out when

$$\|\mathbf{g}^P(\mathbf{x}^k)\| > 0 \quad \text{and} \quad \|\boldsymbol{\beta}(\mathbf{x}^k)\|^2 \leq \Gamma^2 \widetilde{\varphi}(\mathbf{x}^k)^T \varphi(\mathbf{x}^k),$$

where $\widetilde{\varphi}(\mathbf{x}) = \widetilde{\varphi}_{\overline{\alpha}}(\mathbf{x})$ denotes the reduced free gradient defined by (5.85). Thus

$$\|\boldsymbol{\beta}(\mathbf{x}^k)\| + \|\varphi(\mathbf{x}^k)\| > 0 \quad \text{and} \quad \|\boldsymbol{\beta}(\mathbf{x}^k)\| \leq \Gamma \|\varphi(\mathbf{x}^k)\|.$$

It follows easily that $\varphi(\mathbf{x}^k) \neq \mathbf{o}$. Since

$$\|\mathbf{p}\| \geq \|\varphi(\mathbf{x}^k)\|,$$

we have $\mathbf{p} \neq \mathbf{o}$. If $\mathbf{x}^k$ is feasible, we can optionally implement the expansion step with the feasible half-step of Sect. 5.9.1. Notice that $\mathbf{x}^k$ is always feasible at the beginning of the outer loop.

Each unfeasible CG step of our implementation of the monotonic MPRGP algorithm requires two matrix–vector multiplications; the additional multiplication is necessary for evaluation of the test associated with the inner CG loop. To carry out the unfeasible CG step in one matrix–vector multiplication, we can use that for any $\mathbf{x}, \mathbf{d} \in \mathbb{R}^n$

$$f(\mathbf{x}+\mathbf{d}) = f(\mathbf{x}) + \mathbf{g}^T\mathbf{d} + \frac{1}{2}\mathbf{d}^T\mathsf{A}\mathbf{d} \leq f(\mathbf{x}) + \mathbf{g}^T\mathbf{d} + \frac{1}{2}\|\mathsf{A}\|\|\mathbf{d}\|^2$$

and

$$f(\mathbf{x}+\mathbf{d}) = f(\mathbf{x}) + \mathbf{g}^T\mathbf{d} + \frac{1}{2}\mathbf{d}^T\mathsf{A}\mathbf{d} \geq f(\mathbf{x}) + \mathbf{g}^T\mathbf{d}.$$

For example, if $\mathbf{x}^i$, $i = k, k+1, \ldots$, are generated in the inner CG loop of the monotonic MPRGP algorithm, $\mathbf{x}^k$ is feasible, and

$$\mathbf{d}^i = P_{\Omega_B}(\mathbf{y}^i) - \mathbf{x}^i, \quad \mathbf{g}^i = \mathbf{g}(\mathbf{y}^i), \quad i = k, k+1, \ldots,$$

where $\mathbf{y}^i$ is the trial CG iteration entering into the ith step, then we can use (3.9) to evaluate $f(\mathbf{y}^i)$ without additional matrix–vector multiplication,

$$f(\mathbf{y}^i) + (\mathbf{g}^i)^T\mathbf{d}^i + \|\mathsf{A}\|\|\mathbf{d}^i\|^2 \leq f(\mathbf{x}^k) \tag{5.135}$$

implies

$$f\left(P_{\Omega_B}(\mathbf{y}^i)\right) \leq f\left(\mathbf{x}^k\right),$$

and the unfeasible iterates $\mathbf{x}^{i+1} = \mathbf{y}^i$ which satisfy (5.135) can be accepted. Thus we can use (5.135) to modify the test at the beginning of the CG loop of Algorithm 5.9 so that the resulting *semimonotonic MPRGP* algorithm generates a *converging sequence of iterates that are evaluated at one matrix–vector multiplication.*

Using the lower bound on $f(\mathbf{x}^i)$, it is possible to develop a test applicable to unfeasible $\mathbf{x}^k$. The modifications presented in this section are closely related to the semismooth Newton methods.

5.9.4 Choice of Parameters

Our experience indicates that MPRGP is not sensitive to Γ as long as $\Gamma \approx 1$. Since $\Gamma = 1$ minimizes the upper bound on the rate of convergence and guarantees that the CG steps reduce directly the larger of the two components of the projected gradient, we can expect good efficiency with this value.

The choice of $\overline{\alpha}$ requires an estimate of $\|\mathsf{A}\|$. If the entries of A are available, we can use $\|\mathsf{A}\| \le \|\mathsf{A}\|_\infty$ to define $\overline{\alpha} = 2\|\mathsf{A}\|_\infty^{-1}$ which guarantees convergence. If this is not the case, or if $\|\mathsf{A}\|_\infty$ gives a poor upper bound on $\|\mathsf{A}\|$, then we can carry out a few, e.g., five, iterations of the following power method.

Algorithm 5.10. Power method for the estimate of $\|\mathsf{A}\|$.

Given a symmetric positive definite matrix $\mathsf{A} \in \mathbb{R}^{n \times n}$, *returns* $A \approx \|\mathbf{A}\|$.
Choose $\mathbf{x} \in \mathbb{R}^n$ *such that* $\mathbf{x} \neq \mathbf{o}$, $n_{\mathrm{it}} \ge 1$
for $i = 1, 2, \dots, n_{\mathrm{it}}$
$\mathbf{y} = \mathsf{A}\mathbf{x}$, $\mathbf{x} = \|\mathbf{y}\|^{-1}\mathbf{y}$
end for
$A = \|\mathsf{A}\mathbf{x}\|$

Alternatively, we can use the Lanczos method (see, e.g., Golub and van Loan [103]). We can conveniently enhance the Lanczos method into the conjugate gradient loop of the MPRGP algorithm by defining

$$\mathbf{q}_i = \|\varphi(\mathbf{x}^{s+i})\|^{-1}\varphi(\mathbf{x}^{s+i}), \quad i = 0, \dots, p,$$

where $\varphi(\mathbf{x}^s)$ and $\varphi(\mathbf{x}^{s+i})$ are the free gradients at respectively the initial and the ith iterate in one CG loop. Then we can estimate $\|\mathsf{A}\|$ by evaluation of the ℓ_∞-norm of the tridiagonal matrix

$$\mathsf{T} = \mathsf{Q}^T\mathsf{A}\mathsf{Q}, \quad \mathsf{Q} = [\mathbf{q}_0, \dots, \mathbf{q}_p].$$

Though these methods typically give only a lower bound A on the norm of $\|\mathsf{A}\|$, the choice like $\overline{\alpha} = 1.8A^{-1}$ is often sufficient in practice. The decrease of f can be achieved more reliably by initializing $\overline{\alpha} = 2(\mathbf{b}^T\mathsf{A}\mathbf{b})^{-1}\|\mathbf{b}\|^2$ and by inserting the following piece of code into the expansion step:

Algorithm 5.11. Modification of the steplength of the expansion step.

A piece of code to be inserted at the end of the expansion step of Algorithm 5.8.
if $f\left(P_{\Omega_B}(\mathbf{x}^{k+1})\right) > f(\mathbf{x}^k)$
$\overline{\alpha} = \overline{\alpha}/2$ *and repeat the expansion step*
end if

The modified algorithm can outperform that with $\overline{\alpha} = \|\mathsf{A}\|^{-1}$; the longer steps in an early stage of computations can be effective for identification of the solution. We observed a good performance with $\overline{\alpha}$ close to, but not greater than $2\|\mathsf{A}\|^{-1}$, near $\overline{\alpha}_E^{opt}$ which minimizes the coefficient η_E of the Euclidean contraction (5.36). Notice that Theorem 5.14 guarantees that the inserted loop of Algorithm 5.11 reduces the steplength in a small number of steps.

5.9.5 Dynamic Release Coefficient

The estimates given by Lemma 5.4 and Theorem 5.21 indicate that the value of $\Gamma = 1$, which gives the best upper bound on the rate of convergence of the MPRGP algorithm, may be too small to exclude repeated exploitation of any face. On the other hand, while discussing the original Polyak algorithm in Sect. 5.4, we have already expressed doubts that it is efficient to carry out the minimization in face to a high precision, especially in the early stage of computations, when we are far from the solution.

To accommodate these contradicting requirements, let us return to the description of the MPRGP algorithm in Sect. 5.8 and replace in its kth step the release coefficient Γ by Γ_k, so that it can change from iteration to iteration. For example, we shall now say that the iterate $\mathbf{x}^k$ is *strictly proportional* if

$$\|\boldsymbol{\beta}(\mathbf{x}^k)\|^2 \leq \Gamma_k^2 \widetilde{\boldsymbol{\varphi}}(\mathbf{x}^k)^T \boldsymbol{\varphi}(\mathbf{x}^k). \tag{5.136}$$

Repeating the arguments of the proof of Theorem 5.14, we can prove its following modification:

Theorem 5.22. *Let $\Gamma_{\max} \geq \Gamma_{\min}$ denote given positive numbers, let $\{\Gamma_i\}$ denote a given sequence such that $\Gamma_{\max} \geq \Gamma_k \geq \Gamma_{\min}$, let $\lambda_{\min}$ denote the smallest eigenvalue of A, and let $\{\mathbf{x}^k\}$ denote the sequence generated by Algorithm 5.7 with $\overline{\alpha} \in (0, 2\|\mathsf{A}\|^{-1}]$ and Γ replaced in the kth step by Γ_k.*

Then the error in the A-norm is bounded by

$$\|\mathbf{x}^k - \widehat{\mathbf{x}}\|_{\mathsf{A}}^2 \leq 2\eta_{\Gamma_1} \cdots \eta_{\Gamma_k} \left(f(\mathbf{x}^0) - f(\widehat{\mathbf{x}})\right) \leq 2\eta_\Gamma^k \left(f(\mathbf{x}^0) - f(\widehat{\mathbf{x}})\right), \tag{5.137}$$

where $\widehat{\mathbf{x}}$ denotes the unique solution of (5.1),

$$\eta_\Gamma = 1 - \frac{\overline{\alpha}\lambda_{\min}}{\vartheta + \vartheta \widehat{\Gamma}^2}, \quad \eta_{\Gamma_k} = 1 - \frac{\overline{\alpha}\lambda_{\min}}{\vartheta + \vartheta \widehat{\Gamma}_k^2}, \tag{5.138}$$

$$\vartheta = 2\max\{\overline{\alpha}/2, 1\}, \quad \widehat{\Gamma} = \max\{\Gamma_{\max}, \Gamma_{\min}^{-1}\}, \quad \widehat{\Gamma}_k = \max\{\Gamma_k, \Gamma_k^{-1}\}.$$

This definition opens room for implementation of heuristics that can be useful in some specific cases. Typically, the series of release coefficients $\{\Gamma_k\}$ is defined by a suitable function of $\|\mathbf{g}^P(\mathbf{x}^k)\|$. For example, specification

$$\overline{\alpha} = \|\mathsf{A}\|^{-1} \quad \text{and} \quad \Gamma_k = \begin{cases} 1 & \text{for} \quad \|\mathbf{g}^P(\mathbf{x}^k)\| \geq 10\varepsilon, \\ 2(\sqrt{\kappa(\mathsf{A})} + 1) & \text{for} \quad \|\mathbf{g}^P(\mathbf{x}^k)\| < 10\varepsilon \end{cases}$$

guarantees both favorable bound on the rate of convergence in the early stage of computation and the finite termination property.

5.10 Preconditioning

A natural way to improve the performance of the conjugate gradient-based methods is to apply the preconditioning described in Sect. 3.6. However, the application of preconditioning requires some care, as the *preconditioning transforms the variables, turning the bound constraints into more general inequality constraints.* In this section we present two strategies which preserve the bound constraints.

5.10.1 Preconditioning in Face

Probably the most straightforward preconditioning strategy which preserves the bound constraints is the preconditioning applied to the diagonal block $\mathsf{A}_{\mathcal{F}\mathcal{F}}$ of the Hessian matrix A in the conjugate gradient loop which minimizes the cost function f in the face defined by a free set $\mathcal{F}$. Such preconditioning requires that we are able to define for each diagonal block $\mathsf{A}_{\mathcal{F}\mathcal{F}}$ a regular matrix $\mathsf{M}(\mathcal{F})$ which satisfies the following two conditions. First, we require that $\mathsf{M}(\mathcal{F})$ approximates $\mathsf{A}_{\mathcal{F}\mathcal{F}}$ so that the convergence of the conjugate gradients method is significantly accelerated. The second condition requires that the solution of the system

$$\mathsf{M}(\mathcal{F})\mathbf{x} = \mathbf{y}$$

can be obtained easily. The preconditioners $\mathsf{M}(\mathcal{F})$ can be generated, e.g., by any of the methods described in Sect. 3.6.

Though the performance of the algorithm can be considerably improved by the preconditioning, *preconditioning in face does not result in the improved bound on rate of convergence.* The reason is that such preconditioning affects only the feasible conjugate gradient step, leaving the expansion and the proportioning steps without any preconditioning.

In probably the first application of preconditioning to the solution of bound constrained problems [157], O'Leary considered two simple methods which can be used to obtain the preconditioner for $\mathsf{A}_{\mathcal{F}\mathcal{F}}$ from the preconditioner M which approximates A, namely,

$$\mathsf{M}(\mathcal{F}) = \mathsf{M}_{\mathcal{F}\mathcal{F}} \quad \text{and} \quad \mathsf{M}(\mathcal{F}) = \mathsf{L}_{\mathcal{F}\mathcal{F}}\mathsf{L}_{\mathcal{F}\mathcal{F}}^T,$$

where L denotes the factor of the Cholesky factorization $\mathsf{M} = \mathsf{L}\mathsf{L}^T$. It can be proved that whichever method of the preconditioning is used, the convergence bound for the conjugate gradient algorithm applied to the subproblems is at least as good as that of the conjugate gradient method applied to the original matrix [157].

To describe the MPRGP algorithm with the preconditioning in face, let us assume that we are given the preconditioner $\mathsf{M}(\mathcal{F})$ for each set of indices $\mathcal{F}$, and let us denote $\mathcal{F}_k = \mathcal{F}(\mathbf{x}^k)$ and $\mathcal{A}_k = \mathcal{A}(\mathbf{x}^k)$ for each vector $\mathbf{x}^k \in \Omega_B$. To simplify the description of the algorithm, let M_k denote the preconditioner corresponding to the face defined by $\mathcal{F}_k$ padded with zeros so that

$$[\mathsf{M}_k]_{\mathcal{F}\mathcal{F}} = \mathsf{M}(\mathcal{F}_k), \quad [\mathsf{M}_k]_{\mathcal{A}\mathcal{A}} = \mathsf{O}, \quad [\mathsf{M}_k]_{\mathcal{A}\mathcal{F}} = [\mathsf{M}_k]_{\mathcal{F}\mathcal{A}}^T = \mathsf{O},$$

and recall that $\mathsf{M}_k^\dagger$ denotes the Moore–Penrose generalized inverse of M_k which is defined by

$$[\mathsf{M}_k^\dagger]_{\mathcal{F}\mathcal{F}} = \mathsf{M}(\mathcal{F}_k)^{-1}, \quad [\mathsf{M}_k^\dagger]_{\mathcal{A}\mathcal{A}} = \mathsf{O}, \quad [\mathsf{M}_k^\dagger]_{\mathcal{A}\mathcal{F}} = [\mathsf{M}_k^\dagger]_{\mathcal{F}\mathcal{A}}^T = \mathsf{O}.$$

In particular, it follows that

$$\mathsf{M}_k^\dagger \mathbf{g}(\mathbf{x}^k) = \mathsf{M}_k^\dagger \boldsymbol{\varphi}(\mathbf{x}^k).$$

The MPRGP algorithm with preconditioning in face reads as follows.

Algorithm 5.12. MPRGP with preconditioning in face.

Given a symmetric positive definite matrix A *of the order* n, n*-vectors* $\mathbf{b}$, $\boldsymbol{\ell}$, $\Omega_B = \{\mathbf{x} \in \mathbb{R}^n : \mathbf{x} \geq \boldsymbol{\ell}\}$; *choose* $\mathbf{x}^0 \in \Omega_B$, $\Gamma > 0$, $\overline{\alpha} \in (0, 2\|\mathsf{A}\|^{-1}]$, *and the rule which assigns to each* $\mathbf{x}^k \in \Omega_B$ *the preconditioner* M_k *which is SPD in the face defined by* $\mathcal{F}(\mathbf{x}^k)$.

Step 0. {*Initialization.*}

  Set $k = 0$, $\mathbf{g} = \mathsf{A}\mathbf{x}^0 - \mathbf{b}$, $\mathbf{z} = \mathsf{M}_0^\dagger \mathbf{g}$, $\mathbf{p} = \mathbf{z}$

  while $\|\mathbf{g}^P(\mathbf{x}^k)\|$ *is not small*

   if $\|\boldsymbol{\beta}(\mathbf{x}^k)\|^2 \leq \Gamma^2 \widetilde{\boldsymbol{\varphi}}(\mathbf{x}^k)^T \boldsymbol{\varphi}(\mathbf{x}^k)$

Step 1. {*Proportional* $\mathbf{x}^k$. *Trial conjugate gradient step.*}

    $\alpha_{cg} = \mathbf{z}^T \mathbf{g}/\mathbf{p}^T \mathsf{A}\mathbf{p}$, $\mathbf{y} = \mathbf{x}^k - \alpha_{cg}\mathbf{p}$

    $\alpha_f = \max\{\alpha : \mathbf{x}^k - \alpha\mathbf{p} \in \Omega_B\} = \min\{(x_i^k - \ell_i)/p_i : p_i > 0\}$

    if $\alpha_{cg} \leq \alpha_f$

Step 2. {*Conjugate gradient step.*}

    $\mathbf{x}^{k+1} = \mathbf{y}$, $\mathbf{g} = \mathbf{g} - \alpha_{cg}\mathsf{A}\mathbf{p}$, $\mathbf{z} = \mathsf{M}_k^\dagger \mathbf{g}$

    $\beta = \mathbf{z}^T \mathsf{A}\mathbf{p}/\mathbf{p}^T \mathsf{A}\mathbf{p}$, $\mathbf{p} = \mathbf{z} - \beta\mathbf{p}$

    else

Step 3. {*Expansion step.*}

    $\mathbf{x}^{k+\frac{1}{2}} = \mathbf{x}^k - \alpha_f\mathbf{p}$, $\mathbf{g} = \mathbf{g} - \alpha_f\mathsf{A}\mathbf{p}$

    $\mathbf{x}^{k+1} = P_{\Omega_B}\left(\mathbf{x}^{k+\frac{1}{2}} - \overline{\alpha}\boldsymbol{\varphi}(\mathbf{x}^{k+\frac{1}{2}})\right)$

    $\mathbf{g} = \mathsf{A}\mathbf{x}^{k+1} - \mathbf{b}$, $\mathbf{z} = \mathsf{M}_{k+1}^\dagger \mathbf{g}$, $\mathbf{p} = \mathbf{z}$

    end if

   else

Step 4. {*Proportioning step.*}

    $\mathbf{d} = \boldsymbol{\beta}(\mathbf{x}^k)$, $\alpha_{cg} = \mathbf{g}^T\mathbf{d}/\mathbf{d}^T\mathsf{A}\mathbf{d}$

    $\mathbf{x}^{k+1} = \mathbf{x}^k - \alpha_{cg}\mathbf{d}$, $\mathbf{g} = \mathbf{g} - \alpha_{cg}\mathsf{A}\mathbf{d}$, $\mathbf{z} = \mathsf{M}_{k+1}^\dagger\mathbf{g}$, $\mathbf{p} = \mathbf{z}$

   end if

   $k = k + 1$

  end while

Step 5. {*Return (possibly inexact) solution.*}

  $\widetilde{\mathbf{x}} = \mathbf{x}^k$

5.10.2 Preconditioning by Conjugate Projector

Let $1 \le m < n$ and let the vector of bounds satisfy

$$\ell_{m+1} = -\infty, \dots, \ell_n = -\infty,$$

so that problem (5.1) is only partially constrained and the feasible set can be described by

$$\Omega_B = \{\mathbf{x} \in \mathbb{R}^n : \mathbf{x}_{\mathcal{I}} \ge \boldsymbol{\ell}_{\mathcal{I}}\}, \quad \mathcal{I} = \{1, \dots, m\}. \tag{5.139}$$

Here we show that such partially constrained problems can be preconditioned by the conjugate projector of Sect. 3.7 and that *it is possible to give an improved bound on the rate of convergence of the preconditioned problem.*

Let us assume that $\mathcal{U}$ is the subspace spanned by the full column rank matrix $\mathsf{U} \in \mathbb{R}^{n \times p}$ of the form

$$\mathsf{U} = \begin{bmatrix} \mathsf{O} \\ \mathsf{V} \end{bmatrix}, \quad \mathsf{V} \in \mathbb{R}^{(n-m)\times p}.$$

As in Sect. 3.7.2, we decompose our partially constrained problem by means of the conjugate projectors

$$\mathsf{P} = \mathsf{U}(\mathsf{U}^T\mathsf{A}\mathsf{U})^{-1}\mathsf{U}^T\mathsf{A} \tag{5.140}$$

and $\mathsf{Q} = \mathsf{I} - \mathsf{P}$ onto $\mathcal{U}$ and $\mathcal{V} = \mathrm{Im}\mathsf{Q}$, respectively. Due to our special choice of U, we get that for any $\mathbf{x} \in \mathbb{R}^n$

$$[\mathsf{Q}\mathbf{x}]_{\mathcal{I}} = \mathbf{x}_{\mathcal{I}},$$

and that for any $\mathbf{y} \in \mathcal{U}$ and $\mathbf{z} \in \mathcal{V}$, $\mathbf{y} + \mathbf{z} \in \Omega_B$ if and only if $\mathbf{z} \in \Omega_B$. Using (3.32), (3.33), and the observations of Sect. 3.7.3, we thus get

$$\begin{aligned}
\min_{\mathbf{x}\in\Omega_B} f(\mathbf{x}) &= \min_{\substack{\mathbf{y}\in\mathcal{U},\ \mathbf{z}\in\mathcal{V} \\ \mathbf{y}+\mathbf{z}\in\Omega_B}} f(\mathbf{y}+\mathbf{z}) = \min_{\mathbf{y}\in\mathcal{U}} f(\mathbf{y}) + \min_{\mathbf{z}\in\mathcal{V}\cap\Omega_B} f(\mathbf{z}) \\
&= f(\mathbf{x}^0) + \min_{\mathbf{z}\in\mathcal{V}\cap\Omega_B} f(\mathbf{z}) = f(\mathbf{x}^0) + \min_{\substack{\mathbf{z}\in\mathsf{A}\mathcal{V} \\ \mathbf{z}_{\mathcal{I}}\ge\boldsymbol{\ell}_{\mathcal{I}}}} \frac{1}{2}\mathbf{z}^T\mathsf{Q}^T\mathsf{A}\mathsf{Q}\mathbf{z} - \mathbf{b}^T\mathsf{Q}\mathbf{z} \\
&= f(\mathbf{x}^0) + \min_{\substack{\mathbf{z}\in\mathsf{A}\mathcal{V} \\ \mathbf{z}_{\mathcal{I}}\ge\boldsymbol{\ell}_{\mathcal{I}}}} \frac{1}{2}\mathbf{z}^T\mathsf{Q}^T\mathsf{A}\mathsf{Q}\mathbf{z} + \left(\mathbf{g}^0\right)^T\mathbf{z},
\end{aligned}$$

where $\mathbf{x}^0 = \mathsf{P}\mathsf{A}^{-1}\mathbf{b}$ and $\mathbf{g}^0 = -\mathsf{Q}^T\mathbf{b}$. We have thus reduced our bound constrained problem (5.1) with the feasible set (5.139) to the problem

$$\min_{\substack{\mathbf{z}\in\mathsf{A}\mathcal{V} \\ \mathbf{z}_{\mathcal{I}}\ge\boldsymbol{\ell}_{\mathcal{I}}}} \frac{1}{2}\mathbf{z}^T\mathsf{Q}^T\mathsf{A}\mathsf{Q}\mathbf{z} + \left(\mathbf{g}^0\right)^T\mathbf{z}. \tag{5.141}$$

The following lemma shows that the above problem can be solved by the MPRGP algorithm.

Lemma 5.23. *Let $\mathbf{z}^1, \mathbf{z}^2, \ldots$ be generated by the MPRGP algorithm for the problem*

$$\min_{\mathbf{z}_\mathcal{I} \geq \boldsymbol{\ell}_\mathcal{I}} \frac{1}{2}\mathbf{z}^T \mathsf{Q}^T \mathsf{A} \mathsf{Q} \mathbf{z} + \left(\mathbf{g}^0\right)^T \mathbf{z} \tag{5.142}$$

starting from $\mathbf{z}^0 = P_{\Omega_B}\left(\mathbf{g}^0\right)$. Then $\mathbf{z}^k \in \mathsf{A}\mathcal{V},\ k = 0, 1, 2, \ldots$.

Proof. First observe that since $\mathsf{A}\mathcal{V}$ is orthogonal to $\mathcal{U}$ and $\dim \mathsf{A}\mathcal{V} = \dim \mathcal{V}$, it follows that $\mathsf{A}\mathcal{V}$ is the orthogonal complement of $\mathcal{U}$. Thus $\mathsf{A}\mathcal{V}$ is not only an invariant subspace of Q, but it is also an invariant subspace of P_{Ω_B}. Moreover, it also follows that $\mathsf{A}\mathcal{V}$ contains the set $\mathcal{V}_0 \subseteq \mathbb{R}^n$ of all the vectors of $\mathbb{R}^m$ padded with zeros,

$$\mathcal{V}_0 = \{\mathbf{x} \in \mathbb{R}^n : \ \mathbf{x}_\mathcal{J} = \mathbf{o},\ \mathcal{J} = \{m+1, \ldots, n\}\}.$$

More formally,

$$P_{\Omega_B}\left(\mathsf{A}\mathcal{V}\right) \subseteq \mathsf{A}\mathcal{V} \quad \text{and} \quad \mathcal{V}_0 \subseteq \mathsf{A}\mathcal{V}. \tag{5.143}$$

Let us now recall that by (3.33) $\mathbf{g}^0 \in \mathrm{Im}\mathsf{Q}^T$ and by (3.35) $\mathrm{Im}\mathsf{Q}^T = \mathsf{A}\mathcal{V}$, so that $\mathbf{g}^0 \in \mathsf{A}\mathcal{V}$. Using the definition of $\mathbf{z}^0$ and (5.143), we have $\mathbf{z}^0 \in \mathsf{A}\mathcal{V}$.

To finish the proof by induction, let us assume that $\mathbf{z}^k \in \mathsf{A}\mathcal{V}$. Since

$$\mathbf{g}^k = \mathsf{Q}^T \mathsf{A} \mathsf{Q} \mathbf{z}^k - \mathsf{Q}^T \mathbf{b} = \mathsf{A} \mathsf{Q} \mathbf{z}^k + \mathbf{g}^0,$$

we have $\mathbf{g}^k \in \mathsf{A}\mathcal{V}$. We shall use this simple observation to examine separately the three possible steps of the MPRGP algorithm of Sect. 5.8.1 that can be used to generate $\mathbf{z}^{k+1}$.

Let us first assume that $\mathbf{z}^{k+1}$ is generated by the proportioning step. Then

$$\mathbf{z}^{k+1} = \mathbf{z}^k - \alpha_{cg} \boldsymbol{\beta}(\mathbf{z}^k).$$

Using the definition of the chopped gradient, it is rather easy to check that $\boldsymbol{\beta}(\mathbf{z}^k) \in \mathcal{V}_0$. Since $\mathcal{V}_0 \subseteq \mathsf{A}\mathcal{V}$, $\mathsf{A}\mathcal{V}$ is a subspace of $\mathbb{R}^n$, and $\mathbf{z}^k \in \mathsf{A}\mathcal{V}$ by the assumptions, this proves that $\mathbf{z}^{k+1} \in \mathsf{A}\mathcal{V}$ when it is generated by the proportioning step.

Before examining the other two steps, observe that $\boldsymbol{\varphi}(\mathbf{z}^k) - \mathbf{g}^k \in \mathcal{V}_0$, so that

$$\boldsymbol{\varphi}(\mathbf{z}^k) = \left(\boldsymbol{\varphi}(\mathbf{z}^k) - \mathbf{g}^k\right) + \mathbf{g}^k \in \mathsf{A}\mathcal{V}.$$

Thus

$$\mathbf{z}^k - \alpha \boldsymbol{\varphi}(\mathbf{z}^k) \in \mathsf{A}\mathcal{V}$$

for any $\alpha \in \mathbb{R}$. Using the first inclusion of (5.143), we get that

$$P_{\Omega_B}\left(\mathbf{z}^k - \overline{\alpha} \boldsymbol{\varphi}(\mathbf{z}^k)\right) \in \mathsf{A}\mathcal{V}$$

for any $\overline{\alpha}$ of Algorithm 5.8. This proves that $\mathbf{z}^{k+1} \in \mathsf{A}\mathcal{V}$ for $\mathbf{z}^{k+1}$ generated by the expansion step. To finish the proof, observe that the conjugate direction $\mathbf{p}^k$ is either equal to $\boldsymbol{\varphi}(\mathbf{z}^k)$, or it is defined by the recurrence (see (5.15))

$\mathbf{p}^{k+1} = \varphi(\mathbf{z}^k) - \beta\mathbf{p}^k$ starting from the restart $\mathbf{p}^{s+1} = \varphi(\mathbf{z}^s)$. In any case, $\mathbf{p}^k \in \mathsf{A}\mathcal{V}$. Since we assume that $\mathbf{z}^k \in \mathsf{A}\mathcal{V}$ and the iterate $\mathbf{z}^{k+1}$ generated by the conjugate gradient step is a linear combination of $\mathbf{z}^k$ and $\mathbf{p}^k$, this completes the proof. □

It follows that we can obtain the correction $\widehat{\mathbf{z}}$ which solves the auxiliary problem by the standard MPRGP algorithm. Since the iterations are reduced to the subspace, the *projector preconditions all three types of steps* and we can give an improved bound on the rate of convergence. The solution $\widehat{\mathbf{x}}$ of the bound constrained problem (5.1) with the feasible set (5.139) can be expressed by $\widehat{\mathbf{x}} = \mathbf{x}^0 + \widehat{\mathbf{z}}$. For convenience of the reader, we give here the complete algorithm for the solution of the preconditioned problem (5.142).

Algorithm 5.13. MPRGP projection preconditioning correction.

Given a symmetric positive definite matrix A *of the order* n *and* $\mathbf{b}, \boldsymbol{\ell} \in \mathbb{R}^n$*; choose a full column rank matrix* $\mathsf{U} \in \mathbb{R}^{m\times n}$, $\mathbf{g}^0 = -\mathsf{Q}^T\mathbf{b}$, $\mathbf{x}^0 = \mathsf{P}\mathsf{A}^{-1}\mathbf{b}$, $\mathbf{z}^0 = P_{\Omega_B}(\mathbf{g}^0)$, $\Gamma > 0$, *and* $\overline{\alpha} \in (0, 2\|\mathsf{AQ}\|^{-1}]$, *where* P *is defined by (5.140) and* $\mathsf{Q} = \mathsf{I} - \mathsf{P}$.

Step 0. {*Initialization.*}
  Set $k = 0$, $\mathbf{g} = \mathsf{AQ}\mathbf{z}^0 + \mathbf{g}^0$, $\mathbf{p} = \varphi(\mathbf{z}^0)$
  while $\|\mathbf{g}^P(\mathbf{z}^k)\|$ *is not small*
   if $\|\beta(\mathbf{z}^k)\|^2 \le \Gamma^2\widetilde{\varphi}(\mathbf{z}^k)^T\varphi(\mathbf{z}^k)$

Step 1. {*Proportional* $\mathbf{z}^k$. *Trial conjugate gradient step.*}
   $\alpha_{cg} = \mathbf{g}^T\mathbf{p}/\mathbf{p}^T\mathsf{AQ}\mathbf{p}$, $\mathbf{y} = \mathbf{z}^k - \alpha_{cg}\mathbf{p}$
   $\alpha_f = \max\{\alpha : \mathbf{z}^k - \alpha\mathbf{p} \in \Omega_B\} = \min\{(z_i^k - \ell_i)/p_i : p_i > 0\}$
   if $\alpha_{cg} \le \alpha_f$

Step 2. {*Conjugate gradient step.*}
    $\mathbf{z}^{k+1} = \mathbf{y}$, $\mathbf{g} = \mathbf{g} - \alpha_{cg}\mathsf{AQ}\mathbf{p}$
    $\beta = \varphi(\mathbf{y})^T\mathsf{AQ}\mathbf{p}/\mathbf{p}^T\mathsf{AQ}\mathbf{p}$, $\mathbf{p} = \varphi(\mathbf{y}) - \beta\mathbf{p}$
   else

Step 3. {*Expansion step.*}
    $\mathbf{z}^{k+\frac{1}{2}} = \mathbf{z}^k - \alpha_f\mathbf{p}$, $\mathbf{g} = \mathbf{g} - \alpha_f\mathsf{AQ}\mathbf{p}$
    $\mathbf{z}^{k+1} = P_{\Omega_B}(\mathbf{z}^{k+\frac{1}{2}} - \overline{\alpha}\varphi(\mathbf{z}^{k+\frac{1}{2}}))$
    $\mathbf{g} = \mathsf{AQ}\mathbf{z}^{k+1} + \mathbf{g}^0$, $\mathbf{p} = \varphi(\mathbf{z}^{k+1})$
   end if
  else

Step 4. {*Proportioning step.*}
   $\mathbf{d} = \beta(\mathbf{z}^k)$, $\alpha_{cg} = \mathbf{g}^T\mathbf{d}/\mathbf{d}^T\mathsf{AQ}\mathbf{d}$
   $\mathbf{z}^{k+1} = \mathbf{z}^k - \alpha_{cg}\mathbf{d}$, $\mathbf{g} = \mathbf{g} - \alpha_{cg}\mathsf{AQ}\mathbf{d}$, $\mathbf{p} = \varphi(\mathbf{z}^{k+1})$
  end if
  $k = k + 1$
 end while

Step 5. {*Return (possibly inexact) solution.*}
 $\widetilde{\mathbf{x}} = \mathbf{z}^k + \mathbf{x}^0$

To describe the improved bound on the rate of convergence, let us denote, as in Sect. 3.7.4, the gap

$$\gamma = \|\mathsf{R}_{\mathsf{A}\mathcal{U}} - \mathsf{R}_{\mathcal{E}}\|$$

between $\mathsf{A}\mathcal{U}$ and the m-dimensional subspace $\mathcal{E}$ spanned by the eigenvectors corresponding to the m smallest eigenvalues

$$\lambda_{n-m+1} \geq \cdots \geq \lambda_{\min}$$

of A, so that the smallest nonzero eigenvalue $\overline{\lambda}_{\min}$ of $\mathsf{Q}^T\mathsf{A}\mathsf{Q}$ satisfies by Theorem 3.6

$$\overline{\lambda}_{\min} \geq \sqrt{(1-\gamma^2)\lambda_{n-m}^2 + \gamma^2\lambda_{\min}^2} \geq \lambda_{\min}. \tag{5.144}$$

Recall that by (3.36) and $\mathsf{A}\mathsf{Q} = \mathsf{Q}^T\mathsf{A}\mathsf{Q}$

$$\|\mathsf{A}\mathsf{Q}\| \leq \|\mathsf{A}\|.$$

Theorem 5.24. *Let $\{\mathbf{z}^k\}$ denote the sequence generated by Algorithm 5.7 for problem (5.142) with $\overline{\alpha} \in (0, 2\|\mathsf{A}\mathsf{Q}\|^{-1}]$ and $\Gamma > 0$ starting from $\mathbf{z}^0 = P_{\Omega_B}(\mathbf{g}^0)$. Let us denote*

$$f_{0,\mathsf{Q}}(\mathbf{z}) = \frac{1}{2}\mathbf{z}^T\mathsf{Q}^T\mathsf{A}\mathsf{Q}\mathbf{z} + (\mathbf{g}^0)^T\mathbf{z}.$$

Then

$$f_{0,\mathsf{Q}}(\mathbf{z}^{k+1}) - f_{0,\mathsf{Q}}(\widehat{\mathbf{z}}) \leq \eta_\Gamma \left(f_{0,\mathsf{Q}}(\mathbf{z}^k) - f_{0,\mathsf{Q}}(\widehat{\mathbf{z}})\right), \tag{5.145}$$

where

$$\eta_\Gamma = 1 - \frac{\widehat{\alpha}\overline{\lambda}_{\min}}{\vartheta + \vartheta\widehat{\Gamma}^2}, \qquad \widehat{\Gamma} = \max\{\Gamma, \Gamma^{-1}\}, \tag{5.146}$$

$$\vartheta = 2\max\{\overline{\alpha}\|\mathsf{A}\|, 1\}, \quad \widehat{\alpha} = \min\{\overline{\alpha}, 2\|\mathsf{A}\|^{-1} - \overline{\alpha}\}, \tag{5.147}$$

and $\overline{\lambda}_{\min}$ denote the least nonzero eigenvalue of $\mathsf{Q}^T\mathsf{A}\mathsf{Q}$ which satisfies (5.144).

Proof. It is enough to combine Theorem 5.14 with the bounds given by Theorem 3.6. □

The efficiency of preconditioning by conjugate projector depends on the choice of the matrix U whose columns span the subspace which should approximate an invariant subspace spanned by the eigenvectors which correspond to small eigenvalues of A. For the minimization problems arising from the discretization of variational inequalities, U is typically obtained by aggregation of variables using geometrical information or from the coarse discretization, as in the multigrid methods. A numerical example is given in the next section. For references on related topics see Sect. 5.12.

5.11 Numerical Experiments

Here we illustrate the performance of some CG-based algorithms for the bound constrained problem (5.1) on minimization of the cost functions $f_{\mathrm{L},h}$ and $f_{\mathrm{LW},h}$ introduced in Sect. 3.10 subject to bound constraints. All the computations are carried out with $\Gamma = 1$ and $\mathbf{x}^0 = \mathbf{o}$.

5.11.1 Polyak, MPRGP, and Preconditioned MPRGP

Let us first compare the performance of the CG-based algorithms presented in this chapter on minimization of the quadratic function $f_{\mathrm{L},h}$ defined by the discretization parameter h (see page 98) subject to the boundary obstacle ℓ defined by the upper part of the circle with the radius $R = 1$ and the center $S = (1, 0.5, -1.3)$. The boundary obstacle is placed under $\Gamma_c = 1 \times [0, 1]$. Our benchmark is described in more detail in Sect. 7.1; its solution is in Fig. 7.4. Recall that the Hessian $\mathsf{A}_{\mathrm{L},h}$ of $f_{\mathrm{L},h}$ is ill conditioned with $\kappa(\mathsf{A}_{\mathrm{L},h}) \approx h^{-2}$.

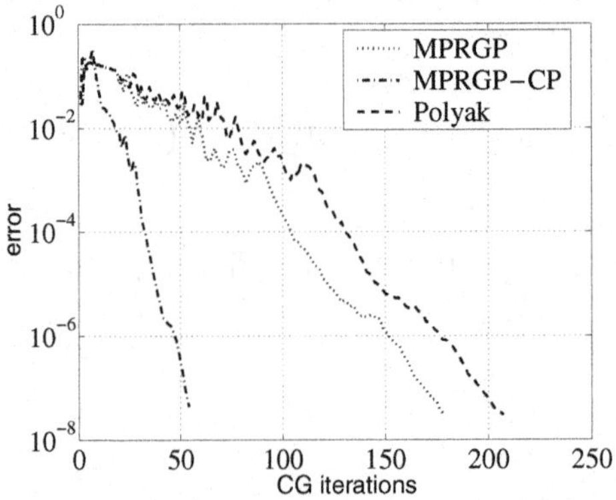

Fig. 5.18. Convergence of Polyak, MPRGP, and MPRGP–CP algorithms

The graph of the norm of the projected gradient (vertical axis) against the numbers of matrix–vector multiplications (horizontal axis) for Algorithm 5.2 (Polyak), Algorithm 5.8 (MPRGP), and MPRGP with preconditioning by the conjugate projector (MPRGP–CP) is in Fig. 5.18. The results were obtained with $h = 1/32$, which corresponds to $n = 1056$ unknowns. The conjugate projector was defined by the aggregation of variables in the squares with 8×8 variables as in Sect. 3.10.1, so that the matrix U has 16 columns. We can see not only that the MPRGP algorithm outperforms Polyak's algorithm, but also that the performance of MPRGP can be considerably improved by preconditioning. The difference between the Polyak and basic MPRGP algorithms is small due to the choice of $\boldsymbol{\ell}$ which makes identification of the active set easy; most iterations of both algorithms were CG steps. The picture can completely change for different $\boldsymbol{\ell}$ as documented in Dostál and Schöberl [74].

5.11.2 Numerical Demonstration of Optimality

To illustrate the concept of optimality, let us consider the class of problems to minimize the quadratic function $f_{\mathrm{LW},h}$ (see page 99) subject to the bound constraints defined by the obstacle as above. The class of problems can be given a mechanical interpretation associated to the expanding spring systems on Winkler's foundation. The spectrum of the Hessian $\mathsf{A}_{\mathrm{LW},h}$ of $f_{\mathrm{LW},h}$ is located in the interval $[2, 10]$. Moreover, $\boldsymbol{\ell} \leq \mathbf{o}$, so that the assumptions of Theorem 5.16 are satisfied.

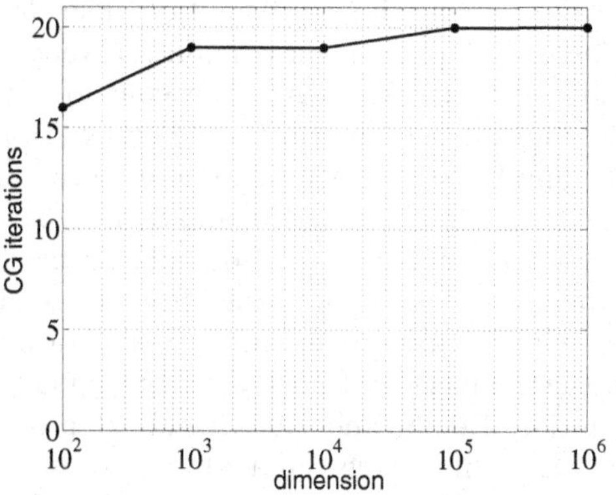

Fig. 5.19. Scalability of MPRGP algorithm

In Fig. 5.19, we can see the numbers of the CG iterations k_n (vertical axis) that were necessary to reduce the norm of the projected gradient by 10^{-6} for the problems with the dimension n ranging from 100 to 1000000. The dimension n on the horizontal axis is in the logarithmic scale. We can see that k_n varies mildly with varying n, in agreement with Theorem 5.16. Moreover, since the cost of the matrix–vector multiplications is in our case proportional to the dimension n of the matrix $\mathsf{A}_{\mathrm{LW},h}$, it follows that the cost of the solution is also proportional to n.

The purpose of the above numerical experiment was just to illustrate the concept of optimality. Realistic classes of problems arise from application of the discretization schemes, such as the finite element method, boundary element method, finite differences, etc., to the elliptic boundary variational inequalities, such as those arising in contact problems of elasticity, in combination with a suitable preconditioning scheme, such as FETI–DP or BETI–DP. An optimal algorithm for the solution of the class of problems arising from the finite element discretization of a model variational inequality with the FETI–DP preconditioning can be found in Chap. 7. More comprehensive related discussion and references can be found in the next section.

5.12 Comments and References

Since the conjugate gradient method was introduced in the celebrated paper by Hestenes and Stiefel [117] as a method for the solution of systems of linear equations, it seems that Polyak [159] was the first researcher who proposed to use the conjugate gradient method to minimize the quadratic cost function subject to bound constraints. Though Polyak assumed the auxiliary problems to be solved exactly, O'Leary [157] observed that this assumption can be replaced by refining the accuracy in the process of solution. In this way she managed to reduce the number of iterations to about a half as compared with the algorithm using the exact solution. The effective theoretically supported strategies for adaptive precision control were presented independently by Friedlander and Martínez with their collaborators [94, 95, 96, 14], and Dostál [41, 42]. Our exposition of inexact Polyak algorithms is based on Dostál [41, 43]. Comprehensive experiments and tests of heuristics can be found in Diniz-Ehrhardt, Gomes-Ruggiero, and Santos [34]. The research was not limited to the convex problems, see also Diniz-Ehrhardt et al. [33].

Many authors fought with an unpleasant consequence of the Polyak strategy which yields a lower bound on the number of iterations in terms of the difference between the numbers of the active constraints in the initial approximation and the solution. Dembo and Tulowitzski [30] proposed the conjugate gradient projection algorithm which could add and drop many constraints in an iteration. Later Yang and Tolle [183] further developed this algorithm with backtracking so that they were able to prove its finite termination property.

An important step forward was development of algorithms with a rigorous convergence theory. On the basis of the results of Calamai and Moré [20], Moré and Toraldo [153] proposed an algorithm that also exploits the conjugate gradients and projections, but its convergence is driven by the gradient projections with the steplength satisfying the sufficient decrease condition. The steplength is found, as in earlier algorithms, by possibly expensive backtracking. In spite of iterative basis of their algorithm, the authors proved that their algorithm preserved the finite termination property of the original algorithm provided the solution satisfies the strict complementarity condition. Friedlander, Martínez, Dostál, and their collaborators combined this result with inexact solution of auxiliary problems [94, 95, 96, 14, 33, 41, 42]. The concept of proportioning algorithm as presented here was introduced by Dostál in [42]. The convergence of the proportioning algorithm was driven by the proportioning step, leaving more room for the heuristic implementation of projections as compared with Moré and Toraldo [153]. The heuristics for implementation of the proportioning algorithm of Dostál [42] can be applied also to the MPRGP algorithm of Sect. 5.8.

The common drawbacks of all the above-mentioned strategies were possible backtracking in search of the gradient projection step and the lack of results on the rate of convergence. A key to further progress were the results by Schöberl [165, 166], who found the bound on the rate of convergence of the cost function in the energy norm for the gradient projection method with the fixed steplength $\overline{\alpha} \in (0, \|\mathsf{A}\|^{-1}]$ in terms of the spectral condition number of the Hessian matrix. It was observed later by Dostál [45] that this nice result can be plugged into the proportioning algorithm to get a similar result, but with the algorithm which can carry out more efficiently the unconstrained steps. The estimates were extended to $\overline{\alpha} \in (0, 2\|\mathsf{A}\|^{-1}]$ by Dostál [51] (gradient projection) and Dostál, Domorádová, and Sadowská [52] (MPRGP). In our exposition of the MPRGP algorithm, we follow Dostál and Schöberl [74], Dostál [51], and Dostál, Domorádová, and Sadowská [52]. Let us recall that the linear rate of convergence of the cost function for the gradient projection method was proved earlier even for more general problems by Luo and Tseng [146], but they did not make any attempt to specify the constants. Notice that the bound on the coefficient of contraction of the gradient projections in the Euclidean norm is a standard result [12]. The gradient projections were exploited also in the algorithms for more general bound constrained problems, see, e.g., Hager and Zhang [115]. Kučera [138] later modified the algorithm to the minimization of quadratic function subject to separated quadratic constraints.

The attempts to enhance unfeasible iterations into the active set-based methods are usually motivated by an effort to expand effectively the active set, especially in the early stage of computation. Of course, the problem is not how to expand the active set, but how to expand it properly. Our monotonic MPRGP algorithm introduced in Sect. 5.9.3 implements a natural heuristics that any decrease direction is acceptable when we want to expand the active set provided the decrease of the cost function in the unfeasible direction is not surpassed by the increase due to the projection to the feasible set. The algorithm can be considered as a special class of the semismooth Newton method with a globalization strategy. For the semismooth Newton algorithms, see, e.g., Hintermüller, Ito, and Kunisch [118] and Hintermüller, Kovtumenko, and Kunisch [119]. Recent application of Newton-type methods to the contact problem may be found in Hüeber, Stadler, and Wohlmuth [122].

The preconditioning in face was studied by O'Leary [157]. Kornhuber [131, 132] presented nice experimental results and convergence theory for the solution of quadratic programming problems arising from the discretization of boundary variational inequalities with multigrid preconditioning. See also Kornhuber and Krause [133] and Iontcheva and Vassilevski [124]. It turned out that the coarse grid should avoid the constrained variables as in our description of the preconditioning by a conjugate projector, see Domorádová and Dostál [36]. The first implementation of the latter idea can be found in Domorádová [35]. Dostál, Horák, and Stefanica combined the MPRGP algorithm with the FETI–DP domain decomposition method to develop a scalable algorithm for the solution of a boundary variational inequality [70]. For application to contact problems with friction see Dostál and Vondrák [75] and Dostál, Haslinger, and Kučera [63]. A discussion related to application of MPRGP in the cascade algorithm can be found in Braess [16].

Let us finish with a few comments on the bounds on the rates of convergence presented in Sect. 5.6 on the gradient projection method, in Sect. 5.7 on MPGP, and in Sect. 5.8 on MPRGP. Since the coefficient of the Euclidean contraction η_E and the coefficient η_f of the reduction of the cost function for the gradient projection step with the fixed steplength are smaller than the coefficient of reduction of the cost function η_Γ for MPGP and MPRGP, one can doubt superiority of the latter algorithms. However, such doubts are not substantiated. The point is that our estimates are based on the analysis of the worst case for isolated iterations and do not take into account the "global" performance of the conjugate gradient method, which dominates whenever a few consecutive conjugate gradient iterations are carried out; this feature of the CG method is captured by Theorem 3.2. Such global performance is partly captured by our finite termination results and, in the case of MPRGP, also by the result on the rate of convergence of the projected gradient.

6

Bound and Equality Constrained Minimization

We shall now combine the results of our previous investigation to develop efficient algorithms for the *bound and equality constrained problem*

$$\min_{\mathbf{x}\in\Omega_{BE}} f(\mathbf{x}), \tag{6.1}$$

where

$$\Omega_{BE} = \{\mathbf{x} \in \mathbb{R}^n : \mathsf{B}\mathbf{x} = \mathbf{o} \quad \text{and} \quad \mathbf{x} \geq \boldsymbol{\ell}\},$$

$f(\mathbf{x}) = \frac{1}{2}\mathbf{x}^T\mathsf{A}\mathbf{x} - \mathbf{x}^T\mathbf{b}$, $\mathbf{b}$ and $\boldsymbol{\ell}$ are given column n-vectors, A is an $n \times n$ symmetric positive definite matrix, and $\mathsf{B} \in \mathbb{R}^{m\times n}$. We consider similar assumptions as in previous chapters. In particular, we assume $\Omega_{BE} \neq \emptyset$ and admit dependent rows of B and $\ell_i = -\infty$. We assume that $\mathsf{B} \neq \mathsf{O}$ is not a full column rank matrix, so that $\mathrm{Ker}\mathsf{B} \neq \{\mathbf{o}\}$. Observe that more general quadratic programming problems can be reduced to (6.1) by duality, a suitable shift of variables, or modification of f.

If we compare the bound and equality constrained problem (6.1) with the bound constrained problem (5.1) of the previous chapter, we can see that the feasible set of (6.1) is more complicated than that of (5.1). For example, the evaluation of the Euclidean projection to the feasible set, one of the key ingredients of the algorithms of the previous section, is not tractable any more. The equality constraints complicate also the implementation of other ingredients of the algorithms developed in previous chapters.

The main idea of the algorithm that we develop in this chapter is to treat both sets of constraints, the equalities and the bound constraints, separately. This approach enables us to use the ingredients of the algorithms developed in the previous chapters, such as the precision control of the auxiliary problems and the update rule for the penalty parameter. We restrict our attention to the SMALBE (SemiMonotonic Augmented Lagrangian algorithm for Bound and Equality constraints) which will be proved to have similar optimality properties as SMALE of Sect. 4.6.1.

Zdeněk Dostál, *Optimal Quadratic Programming Algorithms*,
Springer Optimization and Its Applications, DOI 10.1007/978-0-387-84806-8_6,

6.1 Review of the Methods for Bound and Equality Constrained Problems

Probably the most simple way to reduce the bound and equality constrained quadratic programming problem (6.1) to the bound constrained one is to enhance the equality constraints into the objective function by adding a suitable term which penalizes the violation of the equality constraints, i.e., replacing (6.1) by

$$\min_{\mathbf{x}\geq\mathbf{o}} f_\varrho(\mathbf{x}), \quad f_\varrho(\mathbf{x}) = f(\mathbf{x}) + \frac{1}{2\varrho}\|\mathsf{B}\mathbf{x}\|^2.$$

The resulting bound constrained problem can then be solved by any algorithm for bound constrained problems introduced in Chap. 5. The penalty approximation of the equality constraints can be very efficient for well-conditioned problems; it was used in the development of scalable FETI-based algorithms for the solution of elliptic boundary variational inequalities (see Dostál and Horák [66, 65]). As for the equality constrained problems, we can get the approximation of the Lagrange multipliers

$$\widetilde{\boldsymbol{\lambda}} = \varrho\mathsf{B}\mathbf{x}.$$

If only a part of variables is bound constrained, then it may be possible to reduce the bound and equality constrained problem (6.1) to the bound constrained problem by eliminating some variables. Then most of the methods reviewed in Sect. 5.1 can be adapted to the solution of problem (6.1).

A special structure of (6.1) can be exploited by a modification of the original Polyak algorithm 5.2 provided the dimension of problem (6.1) is not too large and we are able to find a feasible initial approximation $\mathbf{x}^0$. The conjugate gradient iterates are forced to stay in the feasible region by means of orthogonal projectors to the intersection of the current working face and Ker$\mathsf{B}$. The Polyak scheme can be modified for the solution of more general equality and inequality constrained problems; the details can be found, e.g., in Pshenichny and Danilin [161, Chap. III]. Since the modified Polyak algorithm is not compatible with application of nonlinear projectors and effective adaptive precision control, there are no results on its rate of convergence.

Problem (6.1) can also be reduced to the problem with quadratic equality constraints by observing that the inequality $g(\mathbf{x}) \leq 0$ is equivalent to $g(\mathbf{x}) + t^2 = 0$ (see [11]). Though simple and straightforward from the theoretical point of view, the latter approach does not seem to be able to exploit fully the specific structure of quadratic programming problems as it transforms the strictly convex problems with linear constraints to more general problems with the quadratic constraints.

6.2 SMALBE Algorithm for Bound and Equality Constraints

The basic idea of the augmented Lagrangian method for the equality constraints as presented in Sect. 4.3 was based on general observations which remain valid even if the inner auxiliary problems are minimized with respect to some other constraints. This suggests to modify the augmented Lagrangian method so that the bound constrained problems are solved in the inner loop. The idea goes back to Conn, Gould, and Toint [26, 27], who proposed an algorithm for the solution of more general problems that generates the Lagrange multipliers for the equality constraints in the outer loop while solving the auxiliary bound constrained problems in the inner loop.

6.2.1 KKT Conditions and Projected Gradient

Since Ω_{BE} is closed and convex and f is assumed to be strictly convex, the solution of problem (6.1) exists and is necessarily unique by Proposition 2.5. The special form of the feasible set Ω_{BE} enables us to use a specific form of the KKT conditions (2.81) which fully determine the unique solution of (6.1). Enhancing the equality constraints into the Lagrangian

$$L(\mathbf{x}, \boldsymbol{\lambda}, \varrho) = \frac{1}{2}\mathbf{x}^T \mathsf{A}\mathbf{x} - \mathbf{x}^T \mathbf{b} + \mathsf{B}^T \boldsymbol{\lambda} + \frac{\varrho}{2}\|\mathsf{B}\mathbf{x}\|^2,$$

we can easily express the KKT conditions for (6.1) by means of its gradient

$$\mathbf{g} = \mathbf{g}(\mathbf{x}, \boldsymbol{\lambda}, \varrho) = \nabla L(\mathbf{x}, \boldsymbol{\lambda}, \varrho) = (\mathsf{A} + \varrho \mathsf{B}^T \mathsf{B})\mathbf{x} - \mathbf{b} + \mathsf{B}^T \boldsymbol{\lambda}.$$

Using (2.81), we get that a feasible vector $\mathbf{x} \in \Omega_{BE}$ is the solution of (6.1) if and only if

$$\mathbf{g} \geq \mathbf{o} \quad \text{and} \quad \mathbf{g}^T(\mathbf{x} - \boldsymbol{\ell}) = 0,$$

or equivalently

$$\mathbf{g}^P = \mathbf{o}. \tag{6.2}$$

6.2.2 SMALBE Algorithm

The following algorithm is a modification of the SMALE algorithm of Sect. 4.6.1. The only difference is that the SMALBE algorithm solves the bound constrained problems in the inner loop with the precision controlled by the Euclidean norm of the projected gradients. As compared with the original algorithm proposed by Conn, Gould, and Toint [26, 27], the SMALBE algorithm differs in the adaptive precision control introduced by Dostál, Friedlander, and Santos [57] and in the control of the regularization parameter ϱ that was introduced by Dostál [49]. The complete SMALBE algorithm reads as follows.

In Step 1 we can use any algorithm for minimizing the strictly convex quadratic function subject to bound constraints as long as it guarantees the

Algorithm 6.1. Semimonotonic augmented Lagrangians (SMALBE).

Given a symmetric positive definite matrix $\mathsf{A} \in \mathbb{R}^{n\times n}$, $\mathsf{B} \in \mathbb{R}^{m\times n}$, *n-vectors* $\mathbf{b}$, $\boldsymbol{\ell}$.
Step 0. {*Initialization.*}
Choose $\eta > 0,\ \beta > 1,\ M > 0,\ \varrho_0 > 0,\ \boldsymbol{\lambda}^0 \in \mathbb{R}^m$
for $k = 0, 1, 2, \ldots$
Step 1. {*Inner iteration with adaptive precision control.*}
Find $\mathbf{x}^k \geq \boldsymbol{\ell}$ *such that*

$$\|\mathbf{g}^P(\mathbf{x}^k, \boldsymbol{\lambda}^k, \varrho_k)\| \leq \min\{M\|\mathsf{B}\mathbf{x}^k\|, \eta\} \tag{6.3}$$

Step 2. {*Updating the Lagrange multipliers.*}

$$\boldsymbol{\lambda}^{k+1} = \boldsymbol{\lambda}^k + \varrho_k \mathsf{B}\mathbf{x}^k \tag{6.4}$$

Step 3. {*Update* ϱ *provided the increase of the Lagrangian is not sufficient.*}
if $k > 0$ *and*

$$L(\mathbf{x}^k, \boldsymbol{\lambda}^k, \varrho_k) < L(\mathbf{x}^{k-1}, \boldsymbol{\lambda}^{k-1}, \varrho_{k-1}) + \frac{\varrho_k}{2}\|\mathsf{B}\mathbf{x}^k\|^2 \tag{6.5}$$

$\varrho_{k+1} = \beta\varrho_k$
else
$\varrho_{k+1} = \varrho_k$
end if
end for

convergence of the projected gradient to zero, such as the MPGP of Sect. 5.7 the MPRGP algorithm of Sect. 5.8. The next lemma shows that Algorithm 6.1 is well defined, that is, any algorithm for the solution of the auxiliary problem required in Step 1 which guarantees the convergence of the projected gradient to zero generates either $\mathbf{x}^k$ which satisfies (6.3) in a finite number of steps or approximations which converge to the solution of (6.1). It is also clear that there is no hidden enforcement of exact solution in (6.3), and typically inexact solutions of the auxiliary unconstrained problems are obtained in Step 1. Notice that it is not enough to guarantee the convergence of the algorithm in the inner loop. Since the projected gradient is not a continuous function, it is necessary to guarantee that also the projected gradient converges to the zero vector!

Lemma 6.1. *Let* $M > 0$, $\boldsymbol{\lambda} \in \mathbb{R}^m$, $\eta > 0$, *and* $\varrho \geq 0$ *be given. Let* $\{\mathbf{y}^k\}$ *denote any sequence such that* $\mathbf{y}^k \geq \boldsymbol{\ell}$ *and* $\mathbf{g}^P(\mathbf{y}^k, \boldsymbol{\lambda}, \varrho)$ *converges to the zero vector. Then* $\{\mathbf{y}^k\}$ *either converges to the unique solution* $\widehat{\mathbf{x}}$ *of problem (6.1), or there is an index* k *such that*

$$\|\mathbf{g}^P(\mathbf{y}^k, \boldsymbol{\lambda}, \varrho)\| \leq \min\{M\|\mathsf{B}\mathbf{y}^k\|, \eta\}. \tag{6.6}$$

Proof. First notice that the sequence $\{\mathbf{y}^k\}$ converges by Lemma 5.1 to the solution $\widehat{\mathbf{y}}$ of the problem

$$\min_{\mathbf{x} \geq \boldsymbol{\ell}} L(\mathbf{y}^k, \boldsymbol{\lambda}, \varrho).$$

If (6.6) does not hold for any k, then $\|\mathbf{g}^P(\mathbf{y}^k, \boldsymbol{\lambda}, \varrho)\| > M\|\mathsf{B}\mathbf{y}^k\|$ for any k. Since $\mathbf{g}^P(\mathbf{y}^k, \boldsymbol{\lambda}, \varrho)$ converges to the zero vector by the assumption, it follows that $\|\mathsf{B}\mathbf{y}^k\|$ converges to zero. Thus $\mathsf{B}\widehat{\mathbf{y}} = \mathbf{o}$ and

$$\mathbf{g}^P(\widehat{\mathbf{y}}, \boldsymbol{\lambda}, \varrho_k) = \mathbf{o}.$$

It follows that $\widehat{\mathbf{y}}$ satisfies the KKT conditions (6.2) and $\widehat{\mathbf{y}} = \widehat{\mathbf{x}}$. □

Lemma 6.1 shows that it is necessary to include the stop criterion into the procedure which implements the inner loop. See Sect. 6.10 for implementation of the stopping criterion.

6.3 Inequalities Involving the Augmented Lagrangian

In this section we establish basic inequalities that relate the bound on the norm of the projected gradient $\mathbf{g}^P$ of the augmented Lagrangian L to the values of L. These inequalities are similar to those of Sect. 4.6.2 and will be the key ingredients in the proof of convergence and other analysis concerning Algorithm 6.1. We shall derive them similarly as in Sect. 4.6.2 using the observation that for any $\mathbf{x} \geq \boldsymbol{\ell}$ and $\mathbf{y} \geq \boldsymbol{\ell}$

$$(\mathbf{y} - \mathbf{x})^T \mathbf{g}(\mathbf{x}, \boldsymbol{\lambda}, \varrho) \geq (\mathbf{y} - \mathbf{x})^T \mathbf{g}^P(\mathbf{x}, \boldsymbol{\lambda}, \varrho).$$

Lemma 6.2. *Let* $\mathbf{x}, \mathbf{y}, \boldsymbol{\ell} \in \mathbb{R}^n$, $\mathbf{x} \geq \boldsymbol{\ell}$, $\mathbf{y} \geq \boldsymbol{\ell}$, $\boldsymbol{\lambda} \in \mathbb{R}^m$, $\varrho > 0$, $\eta > 0$, *and* $M > 0$. *Let* $\lambda_{\min}$ *denote the least eigenvalue of* A *and* $\widetilde{\boldsymbol{\lambda}} = \boldsymbol{\lambda} + \varrho\mathsf{B}\mathbf{x}$.
(i) If

$$\|\mathbf{g}^P(\mathbf{x}, \boldsymbol{\lambda}, \varrho)\| \leq M\|\mathsf{B}\mathbf{x}\|, \tag{6.7}$$

then

$$L(\mathbf{y}, \widetilde{\boldsymbol{\lambda}}, \varrho) \geq L(\mathbf{x}, \boldsymbol{\lambda}, \varrho) + \frac{1}{2}\left(\varrho - \frac{M^2}{\lambda_{\min}}\right)\|\mathsf{B}\mathbf{x}\|^2 + \frac{\varrho}{2}\|\mathsf{B}\mathbf{y}\|^2. \tag{6.8}$$

(ii) If

$$\|\mathbf{g}^P(\mathbf{x}, \boldsymbol{\lambda}, \varrho)\| \leq \eta, \tag{6.9}$$

then

$$L(\mathbf{y}, \widetilde{\boldsymbol{\lambda}}, \varrho) \geq L(\mathbf{x}, \boldsymbol{\lambda}, \varrho) + \frac{\varrho}{2}\|\mathsf{B}\mathbf{x}\|^2 + \frac{\varrho}{2}\|\mathsf{B}\mathbf{y}\|^2 - \frac{\eta^2}{2\lambda_{\min}}. \tag{6.10}$$

(iii) If (6.9) holds and $\mathbf{z}_0 \in \Omega_{BE}$, *then*

$$L(\mathbf{x}, \boldsymbol{\lambda}, \varrho) \leq f(\mathbf{z}_0) + \frac{\eta^2}{2\lambda_{\min}}. \tag{6.11}$$

Proof. Let us denote $\boldsymbol{\delta} = \mathbf{y} - \mathbf{x}$ and $\mathsf{A}_\varrho = \mathsf{A} + \varrho \mathsf{B}^T\mathsf{B}$. Using

$$L(\mathbf{x}, \widetilde{\boldsymbol{\lambda}}, \varrho) = L(\mathbf{x}, \boldsymbol{\lambda}, \varrho) + \varrho\|\mathsf{B}\mathbf{x}\|^2 \quad \text{and} \quad \mathbf{g}(\mathbf{x}, \widetilde{\boldsymbol{\lambda}}, \varrho) = \mathbf{g}(\mathbf{x}, \boldsymbol{\lambda}, \varrho) + \varrho\mathsf{B}^T\mathsf{B}\mathbf{x},$$

we get

$$\begin{aligned} L(\mathbf{y}, \widetilde{\boldsymbol{\lambda}}, \varrho) &= L(\mathbf{x}, \widetilde{\boldsymbol{\lambda}}, \varrho) + \boldsymbol{\delta}^T\mathbf{g}(\mathbf{x}, \widetilde{\boldsymbol{\lambda}}, \varrho) + \frac{1}{2}\boldsymbol{\delta}^T\mathsf{A}_\varrho\boldsymbol{\delta} \\ &= L(\mathbf{x}, \boldsymbol{\lambda}, \varrho) + \boldsymbol{\delta}^T\mathbf{g}(\mathbf{x}, \boldsymbol{\lambda}, \varrho) + \frac{1}{2}\boldsymbol{\delta}^T\mathsf{A}_\varrho\boldsymbol{\delta} + \varrho\boldsymbol{\delta}^T\mathsf{B}^T\mathsf{B}\mathbf{x} + \varrho\|\mathsf{B}\mathbf{x}\|^2 \\ &\geq L(\mathbf{x}, \boldsymbol{\lambda}, \varrho) + \boldsymbol{\delta}^T\mathbf{g}^P(\mathbf{x}, \boldsymbol{\lambda}, \varrho) + \frac{1}{2}\boldsymbol{\delta}^T\mathsf{A}_\varrho\boldsymbol{\delta} + \varrho\boldsymbol{\delta}^T\mathsf{B}^T\mathsf{B}\mathbf{x} + \varrho\|\mathsf{B}\mathbf{x}\|^2 \\ &\geq L(\mathbf{x}, \boldsymbol{\lambda}, \varrho) + \boldsymbol{\delta}^T\mathbf{g}^P(\mathbf{x}, \boldsymbol{\lambda}, \varrho) + \frac{\lambda_{\min}}{2}\|\boldsymbol{\delta}\|^2 + \frac{\varrho}{2}\|\mathsf{B}\boldsymbol{\delta}\|^2 + \varrho\boldsymbol{\delta}^T\mathsf{B}^T\mathsf{B}\mathbf{x} \\ &\quad +\varrho\|\mathsf{B}\mathbf{x}\|^2. \end{aligned}$$

Noticing that

$$\frac{\varrho}{2}\|\mathsf{B}\mathbf{y}\|^2 = \frac{\varrho}{2}\|\mathsf{B}(\boldsymbol{\delta} + \mathbf{x})\|^2 = \varrho\boldsymbol{\delta}^T\mathsf{B}^T\mathsf{B}\mathbf{x} + \frac{\varrho}{2}\|\mathsf{B}\boldsymbol{\delta}\|^2 + \frac{\varrho}{2}\|\mathsf{B}\mathbf{x}\|^2,$$

we get

$$\begin{aligned} L(\mathbf{y}, \widetilde{\boldsymbol{\lambda}}, \varrho) \geq & L(\mathbf{x}, \boldsymbol{\lambda}, \varrho) + \boldsymbol{\delta}^T\mathbf{g}^P(\mathbf{x}, \boldsymbol{\lambda}, \varrho) \\ & + \frac{\lambda_{\min}}{2}\|\boldsymbol{\delta}\|^2 + \frac{\varrho}{2}\|\mathsf{B}\mathbf{x}\|^2 + \frac{\varrho}{2}\|\mathsf{B}\mathbf{y}\|^2. \end{aligned} \tag{6.12}$$

Using (6.7) and simple manipulations then yields

$$\begin{aligned} L(\mathbf{y}, \widetilde{\boldsymbol{\lambda}}, \varrho) &\geq L(\mathbf{x}, \boldsymbol{\lambda}, \varrho) - M\|\boldsymbol{\delta}\|\|\mathsf{B}\mathbf{x}\| + \frac{\lambda_{\min}}{2}\|\boldsymbol{\delta}\|^2 + \frac{\varrho}{2}\|\mathsf{B}\mathbf{x}\|^2 + \frac{\varrho}{2}\|\mathsf{B}\mathbf{y}\|^2 \\ &= L(\mathbf{x}, \boldsymbol{\lambda}, \varrho) + \left(\frac{\lambda_{\min}}{2}\|\boldsymbol{\delta}\|^2 - M\|\boldsymbol{\delta}\|\|\mathsf{B}\mathbf{x}\| + \frac{M^2\|\mathsf{B}\mathbf{x}\|^2}{2\lambda_{\min}}\right) \\ &\quad -\frac{M^2\|\mathsf{B}\mathbf{x}\|^2}{2\lambda_{\min}} + \frac{\varrho}{2}\|\mathsf{B}\mathbf{x}\|^2 + \frac{\varrho}{2}\|\mathsf{B}\mathbf{y}\|^2 \\ &\geq L(\mathbf{x}, \boldsymbol{\lambda}, \varrho) + \frac{1}{2}\left(\varrho - \frac{M^2}{\lambda_{\min}}\right)\|\mathsf{B}\mathbf{x}\|^2 + \frac{\varrho}{2}\|\mathsf{B}\mathbf{y}\|^2. \end{aligned}$$

This proves (i).

If we assume that (6.9) holds, then by (6.12)

$$\begin{aligned} L(\mathbf{y}, \widetilde{\boldsymbol{\lambda}}, \varrho) &\geq L(\mathbf{x}, \boldsymbol{\lambda}, \varrho) - \|\boldsymbol{\delta}\|\eta + \frac{\lambda_{\min}}{2}\|\boldsymbol{\delta}\|^2 + \frac{\varrho}{2}\|\mathsf{B}\mathbf{x}\|^2 + \frac{\varrho}{2}\|\mathsf{B}\mathbf{y}\|^2 \\ &\geq L(\mathbf{x}, \boldsymbol{\lambda}, \varrho) + \frac{\varrho}{2}\|\mathsf{B}\mathbf{x}\|^2 + \frac{\varrho}{2}\|\mathsf{B}\mathbf{y}\|^2 \ - \frac{\eta^2}{2\lambda_{\min}}. \end{aligned}$$

This proves (ii).

Finally, let $\widehat{\mathbf{z}}$ denote the solution of the auxiliary problem

$$\text{minimize} \;\; L(\mathbf{z}, \boldsymbol{\lambda}, \varrho) \;\; \text{s.t.} \;\; \mathbf{z} \geq \boldsymbol{\ell}, \tag{6.13}$$

let $\mathbf{z}_0 \in \Omega_{BE}$ so that $\mathsf{B}\mathbf{z}_0 = \mathbf{o}$, and let $\widehat{\boldsymbol{\delta}} = \widehat{\mathbf{z}} - \mathbf{x}$. If (6.9) holds, then

$$\begin{aligned} 0 &\geq L(\widehat{\mathbf{z}}, \boldsymbol{\lambda}, \varrho) - L(\mathbf{x}, \boldsymbol{\lambda}, \varrho) = \widehat{\boldsymbol{\delta}}^T \mathbf{g}(\mathbf{x}, \boldsymbol{\lambda}, \varrho) + \frac{1}{2} \widehat{\boldsymbol{\delta}}^T \mathsf{A}_\varrho \widehat{\boldsymbol{\delta}} \\ &\geq \widehat{\boldsymbol{\delta}}^T \mathbf{g}^P(\mathbf{x}, \boldsymbol{\lambda}, \varrho) + \frac{1}{2} \widehat{\boldsymbol{\delta}}^T \mathsf{A}_\varrho \widehat{\boldsymbol{\delta}} \geq -\|\widehat{\boldsymbol{\delta}}\| \eta + \frac{1}{2} \lambda_{\min} \|\widehat{\boldsymbol{\delta}}\|^2 \geq -\frac{\eta^2}{2\lambda_{\min}}. \end{aligned}$$

Since $L(\widehat{\mathbf{z}}, \boldsymbol{\lambda}, \varrho) \leq L(\mathbf{z}_0, \boldsymbol{\lambda}, \varrho) = f(\mathbf{z}_0)$, we conclude that

$$L(\mathbf{x}, \boldsymbol{\lambda}, \varrho) \leq L(\mathbf{x}, \boldsymbol{\lambda}, \varrho) - L(\widehat{\mathbf{z}}, \boldsymbol{\lambda}, \varrho) + f(\mathbf{z}_0) \leq f(\mathbf{z}_0) + \frac{\eta^2}{2\lambda_{\min}}.$$

□

6.4 Monotonicity and Feasibility

Now we shall translate the results on the relations that are satisfied by the augmented Lagrangian into the relations concerning the iterates generated by Algorithm 6.1 (SMALBE).

Lemma 6.3. *Let $\{\mathbf{x}^k\}, \{\boldsymbol{\lambda}^k\}$, and $\{\varrho_k\}$ be generated by Algorithm 6.1 for the solution of (6.1) with $\eta > 0$, $\beta > 1$, $M > 0$, $\varrho_0 > 0$, and $\boldsymbol{\lambda}^0 \in \mathbb{R}^m$. Let $\lambda_{\min}$ denote the least eigenvalue of the Hessian A of the quadratic function f.*
(i) If $k \geq 0$ and

$$\varrho_k \geq M^2/\lambda_{\min}, \tag{6.14}$$

then

$$L(\mathbf{x}^{k+1}, \boldsymbol{\lambda}^{k+1}, \varrho_{k+1}) \geq L(\mathbf{x}^k, \boldsymbol{\lambda}^k, \varrho_k) + \frac{\varrho_{k+1}}{2} \|\mathsf{B}\mathbf{x}^{k+1}\|^2. \tag{6.15}$$

(ii) For any $k \geq 0$

$$\begin{aligned} L(\mathbf{x}^{k+1}, \boldsymbol{\lambda}^{k+1}, \varrho_{k+1}) \geq L(\mathbf{x}^k, \boldsymbol{\lambda}^k, \varrho_k) &+ \frac{\varrho_k}{2} \|\mathsf{B}\mathbf{x}^k\|^2 \\ &+ \frac{\varrho_{k+1}}{2} \|\mathsf{B}\mathbf{x}^{k+1}\|^2 - \frac{\eta^2}{2\lambda_{\min}}. \end{aligned} \tag{6.16}$$

(iii) For any $k \geq 0$ and $\mathbf{z}_0 \in \Omega_{BE}$

$$L(\mathbf{x}^k, \boldsymbol{\lambda}^k, \varrho_k) \leq f(\mathbf{z}_0) + \frac{\eta^2}{2\lambda_{\min}}. \tag{6.17}$$

Proof. In Lemma 6.2, let us substitute $\mathbf{x} = \mathbf{x}^k, \boldsymbol{\lambda} = \boldsymbol{\lambda}^k, \varrho = \varrho_k$, and $\mathbf{y} = \mathbf{x}^{k+1}$, so that inequality (6.7) holds by (6.3), and by (6.4) $\widetilde{\boldsymbol{\lambda}} = \boldsymbol{\lambda}^{k+1}$.

If (6.14) holds, we get by (6.8) that

$$L(\mathbf{x}^{k+1}, \boldsymbol{\lambda}^{k+1}, \varrho_k) \geq L(\mathbf{x}^k, \boldsymbol{\lambda}^k, \varrho_k) + \frac{\varrho_k}{2}\|\mathsf{B}\mathbf{x}^{k+1}\|^2. \tag{6.18}$$

To prove (6.15), it is enough to add

$$\frac{\varrho_{k+1} - \varrho_k}{2}\|\mathsf{B}\mathbf{x}^{k+1}\|^2 \tag{6.19}$$

to both sides of (6.18) and to notice that

$$L(\mathbf{x}^{k+1}, \boldsymbol{\lambda}^{k+1}, \varrho_{k+1}) = L(\mathbf{x}^{k+1}, \boldsymbol{\lambda}^{k+1}, \varrho_k) + \frac{\varrho_{k+1} - \varrho_k}{2}\|\mathsf{B}\mathbf{x}^{k+1}\|^2. \tag{6.20}$$

Since by the definition of Step 1 of Algorithm 6.1

$$\|\mathbf{g}^P(\mathbf{x}^k, \boldsymbol{\lambda}^k, \varrho_k)\| \leq \eta,$$

we can apply the same substitution as above to Lemma 6.2(ii) to get

$$\begin{aligned} L(\mathbf{x}^{k+1}, \boldsymbol{\lambda}^{k+1}, \varrho_k) \geq \ & L(\mathbf{x}^k, \boldsymbol{\lambda}^k, \varrho_k) \\ & + \frac{\varrho_k}{2}\|\mathsf{B}\mathbf{x}^k\|^2 + \frac{\varrho_k}{2}\|\mathsf{B}\mathbf{x}^{k+1}\|^2 - \frac{\eta^2}{2\lambda_{\min}}. \end{aligned} \tag{6.21}$$

After adding the nonnegative expression (6.19) to both sides of (6.21) and using (6.20), we get (6.16). Similarly, inequality (6.17) results from application of the substitution to Lemma 6.2(iii). □

Theorem 6.4. *Let $\{\mathbf{x}^k\}, \{\boldsymbol{\lambda}^k\}$, and $\{\varrho_k\}$ be generated by Algorithm 6.1 for the solution of (6.1) with $\eta > 0$, $\beta > 1$, $M > 0$, $\varrho_0 > 0$, and $\boldsymbol{\lambda}^0 \in \mathbb{R}^m$. Let $\lambda_{\min}$ denote the least eigenvalue of the Hessian A of the cost function f, and let $s \geq 0$ denote the smallest integer such that $\beta^s \varrho_0 \geq M^2/\lambda_{\min}$. Then the following statements hold.*
(i) The sequence $\{\varrho_k\}$ is bounded and

$$\varrho_k \leq \beta^s \varrho_0. \tag{6.22}$$

(ii) If $\mathbf{z}_0 \in \Omega_{BE}$, then

$$\sum_{k=1}^{\infty} \frac{\varrho_k}{2}\|\mathsf{B}\mathbf{x}^k\|^2 \leq f(\mathbf{z}_0) - L(\mathbf{x}^0, \boldsymbol{\lambda}^0, \varrho_0) + (1+s)\frac{\eta^2}{2\lambda_{\min}}. \tag{6.23}$$

Proof. Let $s \geq 0$ denote the smallest integer such that $\beta^s \varrho_0 \geq M^2/\lambda_{\min}$ and let $\mathcal{I} \subseteq \{1, 2, \dots\}$ denote the possibly empty set of the indices k_i such that $\varrho_{k_i} > \varrho_{k_i - 1}$. Using Lemma 6.3(i), $\varrho_{k_i} = \beta \varrho_{k_i - 1} = \beta^i \varrho_0$ for $k_i \in \mathcal{I}$, and

$\beta^s \varrho_0 \geq M^2/\lambda_{\min}$, we conclude that there is no k such that $\varrho_k > \beta^s \varrho_0$. Thus $\mathcal{I}$ has at most s elements and (6.22) holds.

By the definition of Step 3, if $k > 0$, then either $k \notin \mathcal{I}$ and

$$\frac{\varrho_k}{2}\|\mathsf{B}\mathbf{x}^k\|^2 \leq L(\mathbf{x}^k, \boldsymbol{\lambda}^k, \varrho_k) - L(\mathbf{x}^{k-1}, \boldsymbol{\lambda}^{k-1}, \varrho_{k-1}),$$

or $k \in \mathcal{I}$ and by (6.16)

$$\begin{aligned}\frac{\varrho_k}{2}\|\mathsf{B}\mathbf{x}^k\|^2 &\leq \frac{\varrho_{k-1}}{2}\|\mathsf{B}\mathbf{x}^{k-1}\|^2 + \frac{\varrho_k}{2}\|\mathsf{B}\mathbf{x}^k\|^2 \\ &\leq L(\mathbf{x}^k, \boldsymbol{\lambda}^k, \varrho_k) - L(\mathbf{x}^{k-1}, \boldsymbol{\lambda}^{k-1}, \varrho_{k-1}) + \frac{\eta^2}{2\lambda_{\min}}.\end{aligned}$$

Summing up the appropriate cases of the last two inequalities for $k = 1, \ldots, n$ and taking into account that $\mathcal{I}$ has at most s elements, we get

$$\sum_{k=1}^{n} \frac{\varrho_k}{2}\|\mathsf{B}\mathbf{x}^k\|^2 \leq L(\mathbf{x}^n, \boldsymbol{\lambda}^n, \varrho_n) - L(\mathbf{x}^0, \boldsymbol{\lambda}^0, \varrho_0) + s\frac{\eta^2}{2\lambda_{\min}}. \tag{6.24}$$

To get (6.23), it is enough to replace $L(\mathbf{x}^n, \boldsymbol{\lambda}^n, \varrho_n)$ by the upper bound (6.17).

□

6.5 Boundedness

The first step toward the proof of convergence of our SMALBE Algorithm 6.1 is to show that the iterates $\mathbf{x}^k$ are bounded.

Lemma 6.5. *Let $\{\mathbf{x}^k\}$, $\{\boldsymbol{\lambda}^k\}$, and $\{\varrho_k\}$ be generated by Algorithm 6.1 for the solution of (6.1) with $\eta > 0$, $\beta > 1$, $M > 0$, $\varrho_0 > 0$, and $\boldsymbol{\lambda}^0 \in \mathbb{R}^m$. Then the sequence $\{\mathbf{x}^k\}$ is bounded.*

Proof. Since there is only a finite number of different subsets $\mathcal{F}$ of the set of all indices $\mathcal{N} = \{1, \ldots, n\}$, and $\{\mathbf{x}^k\}$ is bounded if and only if $\{\mathbf{x}^k_{\mathcal{F}(\mathbf{x}^k)}\}$ is bounded, we can restrict our attention to the analysis of infinite subsequences $\{\mathbf{x}^k_{\mathcal{F}} : \mathcal{F}(\mathbf{x}^k) = \mathcal{F}\}$ that are defined by the nonempty subsets $\mathcal{F}$ of $\mathcal{N}$.

Let $\mathcal{F} \subseteq \mathcal{N}$, $\mathcal{F} \neq \emptyset$, let $\{\mathbf{x}^k : \mathcal{F}(\mathbf{x}^k) = \mathcal{F}\}$ be infinite, and denote $\mathcal{A} = \mathcal{N} \setminus \mathcal{F}$. Using Lemma 6.3(i), we get that there is an integer k_0 such that $\varrho_k = \varrho_{k_0}$ for $k \geq k_0$. Thus, for $k \geq k_0$, we can denote $\mathsf{H} = \mathsf{A} + \varrho_k \mathsf{B}^T\mathsf{B}$, so that

$$\mathbf{g}^k = \mathbf{g}(\mathbf{x}^k, \boldsymbol{\lambda}^k, \varrho_k) = \mathsf{H}\mathbf{x}^k + \mathsf{B}^T\boldsymbol{\lambda}^k - \mathbf{b},$$

and

$$\begin{bmatrix}\mathsf{H}_{\mathcal{F}\mathcal{F}} & \mathsf{B}^T_{*\mathcal{F}} \\ \mathsf{B}_{*\mathcal{F}} & \mathsf{O}\end{bmatrix}\begin{bmatrix}\mathbf{x}^k_{\mathcal{F}} \\ \boldsymbol{\lambda}^k\end{bmatrix} = \begin{bmatrix}\mathbf{g}^k_{\mathcal{F}} + \mathbf{b}_{\mathcal{F}} - \mathsf{H}_{\mathcal{F}\mathcal{A}}\boldsymbol{\ell}_{\mathcal{A}} \\ \mathsf{B}_{*\mathcal{F}}\mathbf{x}^k_{\mathcal{F}}\end{bmatrix}. \tag{6.25}$$

Since

$$\mathsf{B}_{*\mathcal{F}}\mathbf{x}_{\mathcal{F}}^k = \mathsf{B}\mathbf{x}^k - \mathsf{B}_{*\mathcal{A}}\boldsymbol{\ell}_{\mathcal{A}}, \quad \|\mathbf{g}_{\mathcal{F}}^k\| = \|\mathbf{g}_{\mathcal{F}}(\mathbf{x}^k, \boldsymbol{\lambda}^k, \varrho_k)\| \le \|\mathbf{g}^P(\mathbf{x}^k \boldsymbol{\lambda}^k, \varrho_k)\|,$$

and both $\|\mathbf{g}^P(\mathbf{x}^k, \boldsymbol{\lambda}^k, \varrho_k)\|$ and $\|\mathsf{B}\mathbf{x}^k\|$ converge to zero by the definition of $\mathbf{x}^k$ in Step 1 of Algorithm 6.1 and (6.23), the right-hand side of (6.25) is bounded. Using Lemma 2.11, we get that the matrix of the system (6.25) is regular when $\mathsf{B}_{*\mathcal{F}}$ is a full row rank matrix. Thus both $\mathbf{x}^k$ and $\boldsymbol{\lambda}^k$ are bounded provided the matrix of the system (6.25) is regular.

If $\mathsf{B}_{*\mathcal{F}} \in \mathbb{R}^{m\times s}$ is not a full row rank matrix, then its rank r satisfies $r < m$, and by the singular value decomposition formula (1.28) there are orthogonal matrices $\mathsf{U} = [\mathbf{u}_1, \dots, \mathbf{u}_m] \in \mathbb{R}^{m\times m}$, $\mathsf{V} = [\mathbf{v}_1, \dots, \mathbf{v}_s] \in \mathbb{R}^{s\times s}$, and the diagonal matrix $\mathsf{S} = [s_{ij}] \in \mathbb{R}^{m\times s}$ with the nonzero diagonal entries $s_{11} > 0, \dots, s_{rr} > 0$ such that $\mathsf{B}_{*\mathcal{F}} = \mathsf{U}\mathsf{S}\mathsf{V}^T$. Thus, taking $\hat{\mathsf{U}} = [\mathbf{u}_1, \dots, \mathbf{u}_r]$, $\hat{\mathsf{D}} = \mathrm{diag}(s_{11}, \dots, s_{rr})$, and $\hat{\mathsf{V}} = [\mathbf{v}_1, \dots, \mathbf{v}_r]$, we have $\mathsf{B}_{*\mathcal{F}} = \hat{\mathsf{U}}\hat{\mathsf{D}}\hat{\mathsf{V}}^T$, and we can define a full row rank matrix

$$\hat{\mathsf{B}}_{*\mathcal{F}} = \hat{\mathsf{D}}\hat{\mathsf{V}}^T = \hat{\mathsf{U}}^T\mathsf{B}_{*\mathcal{F}}$$

that satisfies $\hat{\mathsf{B}}_{*\mathcal{F}}^T\hat{\mathsf{B}}_{*\mathcal{F}} = \mathsf{B}_{*\mathcal{F}}^T\mathsf{B}_{*\mathcal{F}}$ and $\|\hat{\mathsf{B}}_{*\mathcal{F}}\mathbf{x}_{\mathcal{F}}\| = \|\mathsf{B}_{*\mathcal{F}}\mathbf{x}_{\mathcal{F}}\|$ for any vector $\mathbf{x}$. We shall assign to any $\boldsymbol{\lambda} \in \mathbb{R}^m$ the vector

$$\hat{\boldsymbol{\lambda}} = \hat{\mathsf{U}}^T\boldsymbol{\lambda}$$

so that $\hat{\mathsf{B}}_{\mathcal{F}}^T\hat{\boldsymbol{\lambda}} = \mathsf{B}_{\mathcal{F}}^T\boldsymbol{\lambda}$. Substituting the latter identity into (6.25), we get

$$\begin{bmatrix} \mathsf{H}_{\mathcal{F}\mathcal{F}} & \hat{\mathsf{B}}_{*\mathcal{F}}^T \\ \mathsf{B}_{*\mathcal{F}} & \mathsf{O} \end{bmatrix} \begin{bmatrix} \mathbf{x}_{\mathcal{F}}^k \\ \hat{\boldsymbol{\lambda}}^k \end{bmatrix} = \begin{bmatrix} \mathbf{g}_{\mathcal{F}}^k + \mathbf{b}_{\mathcal{F}} - \mathsf{H}_{\mathcal{F}\mathcal{A}}\boldsymbol{\ell}_{\mathcal{A}} \\ \mathsf{B}_{*\mathcal{F}}\mathbf{x}_{\mathcal{F}}^k \end{bmatrix}.$$

Since $\mathsf{B}_{*\mathcal{F}} = \hat{\mathsf{U}}\hat{\mathsf{D}}\hat{\mathsf{V}}^T = \hat{\mathsf{U}}\hat{\mathsf{B}}_{*\mathcal{F}}$ and $\hat{\mathsf{U}}$ is a full column rank matrix, the latter system is equivalent to the system

$$\begin{bmatrix} \mathsf{H}_{\mathcal{F}\mathcal{F}} & \hat{\mathsf{B}}_{*\mathcal{F}}^T \\ \hat{\mathsf{B}}_{*\mathcal{F}} & \mathsf{O} \end{bmatrix} \begin{bmatrix} \mathbf{x}_{\mathcal{F}}^k \\ \hat{\boldsymbol{\lambda}}^k \end{bmatrix} = \begin{bmatrix} \mathbf{g}_{\mathcal{F}}^k + \mathbf{b}_{\mathcal{F}} - \mathsf{H}_{\mathcal{F}\mathcal{A}}\boldsymbol{\ell}_{\mathcal{A}} \\ \hat{\mathsf{B}}_{*\mathcal{F}}\mathbf{x}_{\mathcal{F}}^k \end{bmatrix} \tag{6.26}$$

with a regular matrix. The right-hand side of (6.26) being bounded due to $\|\hat{\mathsf{B}}_{*\mathcal{F}}\mathbf{x}_{\mathcal{F}}^k\| = \|\mathsf{B}_{*\mathcal{F}}\mathbf{x}_{\mathcal{F}}^k\|$, we conclude that the set $\{\mathbf{x}_{\mathcal{F}}^k : \mathcal{F}(\mathbf{x}^k) = \mathcal{F}\}$ is bounded. □

The next step is to prove that $\boldsymbol{\lambda}^k$ are either bounded or closely related to a bounded sequence of auxiliary Lagrange multipliers that are not generated by the algorithm. We split our proof into several steps to cope with the difficulties that arise from admitting dependent rows of the constraint matrix B.

Lemma 6.6. *Let $\{\mathbf{z}^k\}$ denote a bounded sequence, let $\mathsf{B} \in \mathbb{R}^{m\times n}$ denote a full row rank matrix, and let there be a sequence $\{\boldsymbol{\zeta}^k\}$ such that $\mathsf{B}^T\boldsymbol{\zeta}^k \ge \mathbf{z}^k$. Then there is a bounded sequence $\{\hat{\boldsymbol{\zeta}}^k\}$ such that $\mathsf{B}^T\hat{\boldsymbol{\zeta}}^k \ge \mathbf{z}^k$.*

Proof. Let us denote $\mathbf{e} = (1, 1, \ldots, 1)^T$ and consider for a given integer k a linear programming problem of the form

$$\min\{\mathbf{e}^T \mathsf{B}^T \boldsymbol{\xi} : \mathsf{B}^T \boldsymbol{\xi} \geq \mathbf{z}^k\} \tag{6.27}$$

with B and $\mathbf{z}^k$ of the lemma. Since $\boldsymbol{\zeta}^k$ satisfies $\mathsf{B}^T \boldsymbol{\zeta}^k \geq \mathbf{z}^k$, it follows that problem (6.27) is feasible. Moreover, observing that for any feasible $\boldsymbol{\xi}$

$$\mathbf{e}^T \mathsf{B}^T \boldsymbol{\xi} = \mathbf{e}^T (\mathsf{B}^T \boldsymbol{\xi} - \mathbf{z}^k) + \mathbf{e}^T \mathbf{z}^k \geq \mathbf{e}^T \mathbf{z}^k,$$

we conclude that problem (6.27) is also bounded from below, so that it has a solution $\boldsymbol{\xi}^k$. Using the results of the duality theory of linear programming presented in Sect. 2.7, it follows that the dual problem

$$\max\{\boldsymbol{\eta}^T \mathbf{z}^k : \boldsymbol{\eta} \geq 0 \ \text{ and } \ \mathsf{B}\boldsymbol{\eta} = \mathbf{e}\} \tag{6.28}$$

is feasible and bounded from above, so that it attains its solution $\boldsymbol{\eta}^k$ at a vertex of the convex boundary of the feasible set of dual problem (6.28) and

$$(\boldsymbol{\eta}^k)^T \mathbf{z}^k = \mathbf{e}^T \mathsf{B}^T \boldsymbol{\xi}^k.$$

Since the number of the vertices is finite, it follows that there is only a finite number of different $\boldsymbol{\eta}^k$, so that, as $\{\mathbf{z}^k\}$ is bounded, there is a constant c such that $\mathbf{e}^T \mathsf{B}^T \boldsymbol{\xi}^k = (\boldsymbol{\eta}^k)^T \mathbf{z}^k \leq c$ for any integer k. Thus

$$\begin{aligned} \|\mathsf{B}^T \boldsymbol{\xi}^k\|_1 &\leq \|\mathsf{B}^T \boldsymbol{\xi}^k - \mathbf{z}^k\|_1 + \|\mathbf{z}^k\|_1 = \mathbf{e}^T (\mathsf{B}^T \boldsymbol{\xi}^k - \mathbf{z}^k) + \|\mathbf{z}^k\|_1 \\ &\leq \mathbf{e}^T \mathsf{B}^T \boldsymbol{\xi}^k + 2\|\mathbf{z}^k\|_1 \leq c + 2\|\mathbf{z}^k\|_1. \end{aligned}$$

Since $\{\mathbf{z}^k\}$ is bounded and B^T is a full column rank matrix, also the vectors $\boldsymbol{\xi}^k$ are bounded and $\hat{\boldsymbol{\zeta}}^k = \boldsymbol{\xi}^k$ satisfies the statement of the lemma. □

Lemma 6.7. *Let $\{\mathbf{x}^k\}, \{\boldsymbol{\lambda}^k\}$, and $\{\varrho_k\}$ be generated by Algorithm 6.1 for the solution of (6.1) with $\eta > 0$, $\beta > 1$, $M > 0$, $\varrho_0 > 0$, and $\boldsymbol{\lambda}^0 \in \mathbb{R}^m$. Then there is a bounded sequence $\hat{\boldsymbol{\lambda}}^k$ such that*

$$\mathbf{g}^P(\mathbf{x}^k, \hat{\boldsymbol{\lambda}}^k, \varrho_k) = \mathbf{g}^P(\mathbf{x}^k, \boldsymbol{\lambda}^k, \varrho_k). \tag{6.29}$$

Proof. Let $\mathcal{B} \subset \mathcal{N}$, $\mathcal{B} \neq \emptyset$, $\mathcal{B} \neq \mathcal{N}$ be such that $\{\mathbf{x}^k : \mathcal{B}_0(\mathbf{x}^k, \boldsymbol{\lambda}^k, \varrho^k) = \mathcal{B}\}$ is infinite, where

$$\mathcal{B}_0(\mathbf{x}^k, \boldsymbol{\lambda}^k, \varrho^k) = \{i \in \mathcal{A}(\mathbf{x}^k) : \ g_i(\mathbf{x}^k, \boldsymbol{\lambda}^k, \varrho^k) \geq 0\}$$

denotes the *weakly binding set* of $\mathbf{x}^k$, and denote $\mathcal{C} = \mathcal{N} \setminus \mathcal{B}$. Using a variant of the Gramm–Schmidt orthogonalization process, we can find a regular matrix R such that

$$\begin{bmatrix} \mathsf{B}_{*\mathcal{C}}^T \\ \mathsf{B}_{*\mathcal{B}}^T \end{bmatrix} \mathsf{R} = \begin{bmatrix} \mathsf{P} & \mathsf{O} & \mathsf{O} \\ \mathsf{Q} & \mathsf{T} & \mathsf{O} \end{bmatrix},$$

where P and T are full column rank matrices. Thus decomposing properly $\mathsf{R}^{-1}\boldsymbol{\lambda}^k$ into the blocks $\mathsf{R}^{-1}\boldsymbol{\lambda}^k = (\boldsymbol{\xi}^k, \boldsymbol{\zeta}^k, \boldsymbol{\nu}^k)^T$, we get

$$\begin{bmatrix} \mathsf{B}_{*\mathcal{C}}^T \\ \mathsf{B}_{*\mathcal{B}}^T \end{bmatrix} \boldsymbol{\lambda}^k = \begin{bmatrix} \mathsf{B}_{*\mathcal{C}}^T \\ \mathsf{B}_{*\mathcal{B}}^T \end{bmatrix} \mathsf{R}\mathsf{R}^{-1}\boldsymbol{\lambda}^k = \begin{bmatrix} \mathsf{P} & \mathsf{O} & \mathsf{O} \\ \mathsf{Q} & \mathsf{T} & \mathsf{O} \end{bmatrix} \begin{bmatrix} \boldsymbol{\xi}^k \\ \boldsymbol{\zeta}^k \\ \boldsymbol{\nu}^k \end{bmatrix} = \begin{bmatrix} \mathsf{P} & \mathsf{O} & \mathsf{O} \\ \mathsf{Q} & \mathsf{T} & \mathsf{O} \end{bmatrix} \begin{bmatrix} \boldsymbol{\xi}^k \\ \boldsymbol{\zeta}^k \\ \mathbf{o} \end{bmatrix}. \quad (6.30)$$

Using Theorem 6.4(i), we get that there is an integer k_0 such that $\varrho_k = \varrho_{k_0}$ for $k \geq k_0$. Let us denote $\mathsf{H} = \mathsf{A} + \varrho_{k_0}\mathsf{B}^T\mathsf{B}$ and $\mathbf{g}^k = \mathbf{g}(\mathbf{x}^k, \boldsymbol{\lambda}^k, \varrho_k)$, so that for $k \geq k_0$

$$\mathsf{B}^T\boldsymbol{\lambda}^k = \mathbf{b} + \mathbf{g}^k - \mathsf{H}\mathbf{x}^k. \quad (6.31)$$

Substituting into (6.30) and using (6.31), we get that for $k \geq k_0$

$$\mathsf{B}_{*\mathcal{C}}^T\boldsymbol{\lambda}^k = \mathsf{P}\boldsymbol{\xi}^k = \mathsf{b}_\mathcal{C} + \mathbf{g}_\mathcal{C}^k - \mathsf{H}_{\mathcal{C}*}\mathbf{x}^k.$$

Since P is a full column rank matrix, $\|\mathbf{g}_\mathcal{C}^k\| = \|\mathbf{g}^P(\mathbf{x}^k, \boldsymbol{\lambda}^k, \varrho_k)\|$, and both $\mathbf{x}^k$ and $\mathbf{g}^P(\mathbf{x}^k, \boldsymbol{\lambda}^k, \varrho_k)$ are bounded, it follows that $\boldsymbol{\xi}^k$ is bounded, too. Moreover, for $k \geq k_0$ and $\mathcal{B} = \mathcal{B}_0(\mathbf{x}^k, \boldsymbol{\lambda}^k, \varrho^k)$, we get

$$\mathsf{B}_{*\mathcal{B}}^T\boldsymbol{\lambda}^k = \mathsf{Q}\boldsymbol{\xi}^k + \mathsf{T}\boldsymbol{\zeta}^k = \mathbf{b}_\mathcal{B} + \mathbf{g}_\mathcal{B}^k - \mathsf{H}_{\mathcal{B}*}\mathbf{x}^k \geq \mathbf{b}_\mathcal{B} - \mathsf{H}_{\mathcal{B}*}\mathbf{x}^k,$$

that is,

$$\mathsf{T}\boldsymbol{\zeta}^k \geq \mathbf{b}_\mathcal{B} - \mathsf{H}_{\mathcal{B}*}\mathbf{x}^k - \mathsf{Q}\boldsymbol{\xi}^k.$$

Since we have just shown that $\boldsymbol{\xi}^k$ are bounded, and $\mathbf{x}^k$ are bounded due to Lemma 6.5, we can apply Lemma 6.6 to get bounded sequence $\hat{\boldsymbol{\zeta}}^k$ such that

$$\mathsf{T}\hat{\boldsymbol{\zeta}}^k \geq \mathbf{b}_\mathcal{B} - \mathsf{H}_{\mathcal{B}*}\mathbf{x}^k - \mathsf{Q}\boldsymbol{\xi}^k. \quad (6.32)$$

Let us now define for $k \geq k_0$ a bounded sequence

$$\hat{\boldsymbol{\lambda}}^k = \mathsf{R}\begin{bmatrix} \boldsymbol{\xi}^k \\ \hat{\boldsymbol{\zeta}}^k \\ \mathbf{o} \end{bmatrix},$$

so that by (6.30)

$$\mathsf{B}_{*\mathcal{C}}^T\hat{\boldsymbol{\lambda}}^k = \mathsf{B}_{*\mathcal{C}}^T\mathsf{R}\begin{bmatrix} \boldsymbol{\xi}^k \\ \hat{\boldsymbol{\zeta}}^k \\ \mathbf{o} \end{bmatrix} = \mathsf{P}\boldsymbol{\xi}^k = \mathsf{B}_{*\mathcal{C}}^T\boldsymbol{\lambda}^k$$

and

$$\mathsf{B}_{*\mathcal{B}}^T\hat{\boldsymbol{\lambda}}^k = \mathsf{B}_{*\mathcal{B}}^T\mathsf{R}\begin{bmatrix} \boldsymbol{\xi}^k \\ \hat{\boldsymbol{\zeta}}^k \\ \mathbf{o} \end{bmatrix} = \mathsf{Q}\boldsymbol{\xi}^k + T\hat{\boldsymbol{\zeta}}^k.$$

If we use (6.32) and the latter equation, we get

$$\mathbf{g}_\mathcal{B}(\mathbf{x}^k, \hat{\boldsymbol{\lambda}}^k, \varrho_k) = \mathsf{H}_{\mathcal{B}*}\mathbf{x}^k - \mathbf{b}_\mathcal{B} + \mathsf{B}_{*\mathcal{B}}^T\hat{\boldsymbol{\lambda}}^k = \mathsf{H}_{\mathcal{B}*}\mathbf{x}^k - \mathbf{b}_\mathcal{B} + \mathsf{Q}\boldsymbol{\xi}^k + T\hat{\boldsymbol{\zeta}}^k \geq 0.$$

Recalling that we assume that $\mathcal{B}_0(\mathbf{x}^k, \boldsymbol{\lambda}^k, \varrho^k) = \mathcal{B}$, the last equation together with

$$\mathbf{g}_{\mathcal{C}}^k = \mathbf{g}_{\mathcal{C}}(\mathbf{x}^k, \boldsymbol{\lambda}^k, \varrho_k) = \mathbf{g}_{\mathcal{C}}^P(\mathbf{x}^k, \boldsymbol{\lambda}^k, \varrho_k)$$

yields

$$\mathbf{g}^P(\mathbf{x}^k, \boldsymbol{\lambda}^k, \varrho_k) = \mathbf{g}^P(\mathbf{x}^k, \hat{\boldsymbol{\lambda}}^k, \varrho_k).$$

If $\mathcal{B} = \emptyset$ or $\mathcal{B} = \mathcal{N}$ are such that $\{\mathbf{x}^k : \mathcal{B}_0(\mathbf{x}^k, \boldsymbol{\lambda}^k, \varrho^k) = \mathcal{B}\}$ is infinite, we can find the multipliers $\hat{\boldsymbol{\lambda}}^k$ that satisfy the statement of our lemma by specializing the above arguments. Since there is only a finite number of different subsets $\mathcal{B}$ of $\mathcal{N}$, we have shown that there are the multipliers $\hat{\boldsymbol{\lambda}}^k$ that satisfy the statement of our lemma for all k except possibly a finite number of indices for which we shall define $\hat{\boldsymbol{\lambda}}^k = \boldsymbol{\lambda}^k$. This completes the proof. □

6.6 Convergence

Now we are ready to prove the main convergence result of this chapter. It turns out that the convergence of the Lagrange multipliers requires additional assumptions. To describe them effectively, let $\mathcal{F} = \mathcal{F}(\widehat{\mathbf{x}})$ denote the free set of the unique solution $\widehat{\mathbf{x}}$. The solution $\widehat{\mathbf{x}}$ is a *regular solution* of (6.1) if $\mathsf{B}_{*\mathcal{F}}$ is a full row rank matrix, and $\widehat{\mathbf{x}}$ is a *range regular solution* of (6.1) if $\mathrm{Im}\mathsf{B} = \mathrm{Im}\mathsf{B}_{*\mathcal{F}}$.

Theorem 6.8. *Let $\{\mathbf{x}^k\}, \{\boldsymbol{\lambda}^k\}$, and $\{\varrho_k\}$ be generated by Algorithm 6.1 for the solution of (6.1) with $\eta > 0$, $\beta > 1$, $M > 0$, $\varrho_0 > 0$, and $\boldsymbol{\lambda}^0 \in \mathbb{R}^m$. Then the following statements hold.*
(i) The sequence $\{\mathbf{x}^k\}$ converges to the solution $\widehat{\mathbf{x}}$ of (6.1).
(ii) If the solution $\widehat{\mathbf{x}}$ of (6.1) is regular, then $\{\boldsymbol{\lambda}^k\}$ converges to the uniquely determined vector $\widehat{\boldsymbol{\lambda}}$ of Lagrange multipliers of (6.1). Moreover, if ϱ_0 is sufficiently large, then the convergence of both the Lagrange multipliers and the feasibility error is linear.
(iii) If the solution $\widehat{\mathbf{x}}$ of (6.1) is range regular, then $\{\boldsymbol{\lambda}^k\}$ converges to the vector

$$\overline{\boldsymbol{\lambda}} = \boldsymbol{\lambda}_{\mathrm{LS}} + (\mathsf{I} - \mathsf{P})\boldsymbol{\lambda}^0,$$

where P is the orthogonal projector onto $\mathrm{Im}\mathsf{B} = \mathrm{Im}\mathsf{B}_{*\mathcal{F}}$, and $\boldsymbol{\lambda}_{\mathrm{LS}}$ is the least square Lagrange multiplier of (6.1).

Proof. Let $\hat{\boldsymbol{\lambda}}^k$ denote the sequence of Lemma 6.7 so that it satisfies

$$\mathbf{g}^P(\mathbf{x}^k, \boldsymbol{\lambda}^k, \varrho_k) = \mathbf{g}^P(\mathbf{x}^k, \hat{\boldsymbol{\lambda}}^k, \varrho_k).$$

Since both $\mathbf{x}^k$ and $\hat{\boldsymbol{\lambda}}^k$ are bounded, it follows that there is a cluster point $(\overline{\mathbf{x}}, \overline{\boldsymbol{\lambda}})$ of the sequence $(\mathbf{x}^k, \hat{\boldsymbol{\lambda}}^k)$. Using Theorem 6.4(i), we get that there is k_0 such that $\varrho_k = \varrho_{k_0}$ for $k \geq k_0$. Moreover, by Theorem 6.4(ii) and the definition of Step 1 of Algorithm 6.1, $\mathsf{B}\overline{\mathbf{x}} = \mathbf{o}$ and

$$\mathbf{g}^P(\bar{\mathbf{x}}, \bar{\boldsymbol{\lambda}}, \varrho_{k_0}) = \mathbf{g}^P(\bar{\mathbf{x}}, \bar{\boldsymbol{\lambda}}, 0) = \mathbf{o}.$$

Since $\bar{\mathbf{x}} \geq \boldsymbol{\ell}$, $\bar{\mathbf{x}}$ is the solution of (6.1). The solution $\widehat{\mathbf{x}}$ of (6.1) being unique, it follows that $\mathbf{x}^k$ converges to $\bar{\mathbf{x}} = \widehat{\mathbf{x}}$.

Let k_0 be as above, and let us denote $\mathcal{F} = \mathcal{F}(\widehat{\mathbf{x}})$ and $\mathsf{H} = \mathsf{A} + \varrho_{k_0}\mathsf{B}^T\mathsf{B}$. Since we have just proved that $\{\mathbf{x}^k\}$ converges to $\widehat{\mathbf{x}}$, there is $k_1 \geq k_0$ such that $\mathcal{F} \subseteq \mathcal{F}\{\mathbf{x}^k\}$ for $k \geq k_1$ and

$$\mathbf{g}_{\mathcal{F}}(\mathbf{x}^k, \boldsymbol{\lambda}^k, \varrho_k) = \mathsf{H}_{\mathcal{F}*}\mathbf{x}^k - \mathbf{b}_{\mathcal{F}} + \mathsf{B}^T_{*\mathcal{F}}\boldsymbol{\lambda}^k$$

converges to zero. It follows that the sequence

$$\mathsf{B}^T_{*\mathcal{F}}\boldsymbol{\lambda}^k = \mathbf{b}_{\mathcal{F}} - \mathsf{H}_{\mathcal{F}*}\mathbf{x}^k + \mathbf{g}_{\mathcal{F}}(\mathbf{x}^k, \boldsymbol{\lambda}^k, \varrho_k)$$

is bounded. Moreover, if $\overline{\boldsymbol{\lambda}}$ is any vector of Lagrange multipliers, then

$$\mathbf{b} = \mathsf{H}\widehat{\mathbf{x}} + \mathsf{B}^T\overline{\boldsymbol{\lambda}}$$

and

$$\mathsf{B}^T_{*\mathcal{F}}(\boldsymbol{\lambda}^k - \overline{\boldsymbol{\lambda}}) = -\mathsf{H}_{\mathcal{F}*}(\mathbf{x}^k - \widehat{\mathbf{x}}) + \mathbf{g}_{\mathcal{F}}(\mathbf{x}^k, \boldsymbol{\lambda}^k, \varrho_k) \tag{6.33}$$

converges to zero.

If the solution $\widehat{\mathbf{x}}$ of (6.1) is regular, then $\mathsf{B}^T_{*\mathcal{F}}$ is a full column rank matrix and there is the unique Lagrange multiplier $\widehat{\boldsymbol{\lambda}}$ for problem (6.1). Moreover, since $\boldsymbol{\lambda}^k - \widehat{\boldsymbol{\lambda}} \in \mathbb{R}^m = \mathrm{Im}\mathsf{B}_{*\mathcal{F}}$, it follows by (1.34) that

$$\|\mathsf{B}^T_{*\mathcal{F}}(\boldsymbol{\lambda}^k - \widehat{\boldsymbol{\lambda}})\| \geq \overline{\sigma}^{\mathcal{F}}_{\min}\|\boldsymbol{\lambda}^k - \widehat{\boldsymbol{\lambda}}\|,$$

where $\overline{\sigma}^{\mathcal{F}}_{\min}$ denotes the least nonzero singular value of $\mathsf{B}_{*\mathcal{F}}$. The convergence of the right-hand side of (6.33) to zero thus implies that $\boldsymbol{\lambda}^k$ converges to $\widehat{\boldsymbol{\lambda}}$. The proof of linear convergence of the Lagrange multipliers and the feasibility error for large ϱ_0 is technical and can be found in Dostál, Friedlander, and Santos [57, Theorems 5.2 and 5.5].

Let us now assume that the solution $\widehat{\mathbf{x}}$ of (6.1) is only range regular, so that $\mathrm{Im}\mathsf{B}_{*\mathcal{F}} = \mathrm{Im}\mathsf{B}$, and let $\mathsf{Q} = \mathsf{I} - \mathsf{P}$ denote the orthogonal projector onto $\mathrm{Ker}\mathsf{B}^T = \mathrm{Ker}\mathsf{B}^T_{*\mathcal{F}}$. Using $\mathsf{P} + \mathsf{Q} = \mathsf{I}$, $\mathsf{B}^T\mathsf{Q} = \mathsf{O}$, and (1.34), we get

$$\begin{aligned}\|\mathsf{B}^T_{*\mathcal{F}}(\boldsymbol{\lambda}^k - \overline{\boldsymbol{\lambda}})\| &= \|\mathsf{B}^T_{*\mathcal{F}}(\mathsf{P} + \mathsf{Q})(\boldsymbol{\lambda}^k - \overline{\boldsymbol{\lambda}})\| = \|\mathsf{B}^T_{*\mathcal{F}}(\mathsf{P}\boldsymbol{\lambda}^k - \mathsf{P}\overline{\boldsymbol{\lambda}})\| \\ &\geq \overline{\sigma}^{\mathcal{F}}_{\min}\|\mathsf{P}\boldsymbol{\lambda}^k - \mathsf{P}\overline{\boldsymbol{\lambda}}\|.\end{aligned}$$

Thus $\mathsf{P}\boldsymbol{\lambda}^k$ converges to $\mathsf{P}\overline{\boldsymbol{\lambda}}$, where $\overline{\boldsymbol{\lambda}}$ is a vector of Lagrange multipliers for (6.1). Since

$$\boldsymbol{\lambda}^k = \boldsymbol{\lambda}^0 + \varrho_0\mathsf{B}\mathbf{x}^0 + \cdots + \varrho_k\mathsf{B}\mathbf{x}^k$$

with $\mathsf{B}\mathbf{x}^k \in \mathrm{Im}\mathsf{B}$, we get

$$\boldsymbol{\lambda}^k = (\mathsf{P} + \mathsf{Q})\boldsymbol{\lambda}^k = \mathsf{Q}\boldsymbol{\lambda}^0 + \mathsf{P}\boldsymbol{\lambda}^k.$$

Observing that $\overline{\boldsymbol{\lambda}} = \boldsymbol{\lambda}_{\mathrm{LS}} + \mathsf{Q}\boldsymbol{\lambda}^0$ is a Lagrange multiplier for (6.1), and that $\mathsf{P}\overline{\boldsymbol{\lambda}} = \boldsymbol{\lambda}_{\mathrm{LS}}$, we get

$$\|\boldsymbol{\lambda}^k - \overline{\boldsymbol{\lambda}}\| = \|\mathsf{Q}\boldsymbol{\lambda}^0 + \mathsf{P}\boldsymbol{\lambda}^k - (\boldsymbol{\lambda}_{\mathrm{LS}} + \mathsf{Q}\boldsymbol{\lambda}^0)\| = \|\mathsf{P}\boldsymbol{\lambda}^k - \mathsf{P}\overline{\boldsymbol{\lambda}}\|.$$

Since the right-hand side converges to zero, we conclude that $\boldsymbol{\lambda}^k$ converges to $\overline{\boldsymbol{\lambda}}$, which completes the proof of (iii). □

6.7 Optimality of the Outer Loop

Theorem 6.4 suggests that it is possible to give an independent of B upper bound on the number of outer iterations of Algorithm 6.1 that are necessary to achieve a prescribed feasibility error for a class of problems like (6.1). To present explicitly this qualitatively new feature of Algorithm 6.1, at least as compared to the related algorithms [57], let $\mathcal{T}$ denote any set of indices and let for any $t \in \mathcal{T}$ be defined a problem

$$\text{minimize } f_t(\mathbf{x}) \text{ s.t. } \mathbf{x} \in \Omega_t \tag{6.34}$$

with $\Omega_t = \{\mathbf{x} \in \mathbb{R}^{n_t} : \mathsf{B}_t\mathbf{x} = \mathbf{o} \text{ and } \mathbf{x} \geq \boldsymbol{\ell}_t\}$, $f_t(\mathbf{x}) = \frac{1}{2}\mathbf{x}^T\mathsf{A}_t\mathbf{x} - \mathbf{b}_t^T\mathbf{x}$, $\mathsf{A}_t \in \mathbb{R}^{n_t \times n_t}$ symmetric positive definite, $\mathsf{B}_t \in \mathbb{R}^{m_t \times n_t}$, and $\mathbf{b}_t, \boldsymbol{\ell}_t \in \mathbb{R}^{n_t}$. To simplify our exposition, we assume that the bound constraints $\boldsymbol{\ell}_t$ are not positive so that $\mathbf{o} \in \Omega_t$. Our optimality result reads as follows.

Theorem 6.9. *Let $\{\mathbf{x}_t^k\}$, $\{\boldsymbol{\lambda}_t^k\}$, and $\{\varrho_{t,k}\}$ be generated by Algorithm 6.1 for (6.34) with $\|\mathbf{b}_t\| \geq \eta_t > 0$, $\beta > 1$, $M > 0$, $\varrho_{t,0} = \varrho_0 > 0$, and $\boldsymbol{\lambda}_t^0 = \mathbf{o}$. Let there be an $a_{\min} > 0$ such that the least eigenvalue $\lambda_{\min}(\mathsf{A}_t)$ of the Hessian A_t of the quadratic function f_t satisfies*

$$\lambda_{\min}(\mathsf{A}_t) \geq a_{\min},$$

let $s \geq 0$ denote the smallest integer such that $\beta^s\varrho_0 \geq M^2/a_{\min}$, and denote

$$a = \frac{2+s}{a_{\min}\varrho_0}.$$

Then for each $\varepsilon > 0$ there are indices k_t, $t \in \mathcal{T}$, such that

$$k_t \leq a/\varepsilon^2 + 1 \tag{6.35}$$

and $\mathbf{x}_t^{k_t}$ is an approximate solution of (4.109) satisfying

$$\|\mathbf{g}^P(\mathbf{x}_t^{k_t}, \boldsymbol{\lambda}_t^{k_t}, \varrho_{t,k_t})\| \leq M\varepsilon\|\mathbf{b}_t\| \quad \text{and} \quad \|\mathsf{B}_t\mathbf{x}_t^{k_t}\| \leq \varepsilon\|\mathbf{b}_t\|. \tag{6.36}$$

Proof. First notice that for any index j

$$\frac{\varrho_0 j}{2}\min\{\|\mathsf{B}_t\mathbf{x}_t^i\|^2 : i=1,\dots,j\} \le \sum_{i=1}^{j}\frac{\varrho_{t,i}}{2}\|\mathsf{B}_t\mathbf{x}_t^i\|^2 \le \sum_{i=1}^{\infty}\frac{\varrho_{t,i}}{2}\|\mathsf{B}_t\mathbf{x}_t^i\|^2. \tag{6.37}$$

Denoting by $L_t(\mathbf{x},\boldsymbol{\lambda},\varrho)$ the augmented Lagrangian for problem (6.34), we get for any $\mathbf{x}\in\mathbb{R}^{n_t}$ and $\varrho\ge 0$

$$L_t(\mathbf{x},\mathbf{o},\varrho) = \frac{1}{2}\mathbf{x}^T(\mathsf{A}_t+\varrho\mathsf{B}_t^T\mathsf{B}_t)\mathbf{x}-\mathbf{b}_t^T\mathbf{x} \ge \frac{1}{2}a_{\min}\|\mathbf{x}\|^2-\|\mathbf{b}_t\|\|\mathbf{x}\| \ge -\frac{\|\mathbf{b}_t\|^2}{2a_{\min}}.$$

If we substitute this inequality and $\mathbf{z}_0=\mathbf{o}$ into (6.23) and use the assumption $\|\mathbf{b}_t\|\ge\eta_t$, we get

$$\sum_{i=1}^{\infty}\frac{\varrho_{t,i}}{2}\|\mathsf{B}_t\mathbf{x}_t^i\|^2 \le \frac{\|\mathbf{b}_t\|^2}{2a_{\min}}+(1+s)\frac{\eta^2}{2a_{\min}} \le \frac{(2+s)\|\mathbf{b}_t\|^2}{2a_{\min}}. \tag{6.38}$$

Using (6.37) and (6.38), we get

$$\frac{\varrho_0 j}{2}\min\{\|\mathsf{B}_t\mathbf{x}_t^i\|^2 : i=1,\dots,j\} \le \frac{(2+s)}{2a_{\min}\varepsilon^2}\varepsilon^2\|\mathbf{b}_t\|^2.$$

Taking for j the least integer that satisfies $a/j\le\varepsilon^2$, so that

$$a/\varepsilon^2 \le j \le a/\varepsilon^2+1,$$

and denoting for any $t\in\mathcal{T}$ by $k_t\in\{1,\dots,j\}$ the index which minimizes $\{\|\mathsf{B}_t\mathbf{x}_t^i\| : i=1,\dots,j\}$, we can use the last inequality with simple manipulations to obtain

$$\|\mathsf{B}_t\mathbf{x}_t^{k_t}\|^2 = \min\{\|\mathsf{B}_t\mathbf{x}_t^i\|^2 : i=1,\dots,j\} \le \frac{a}{j\varepsilon^2}\varepsilon^2\|\mathbf{b}_t\|^2 \le \varepsilon^2\|\mathbf{b}_t\|^2.$$

The inequality

$$M^{-1}\|\mathbf{g}^P(\mathbf{x}_t^{k_t},\boldsymbol{\lambda}_t^{k_t},\varrho_{t,k_t})\| \le \|\mathsf{B}_t\mathbf{x}_t^{k_t}\| \le \varepsilon\|\mathbf{b}_t\|$$

results easily from the definition of Step 1 of Algorithm 6.1. □

Having proved that there is a bound on the number of outer iterations of SMALBE that is necessary to get an approximate solution, it remains to bound the number of inner iterations. In the next section, we consider implementation of the inner loop by the CG algorithm and give sufficient conditions which guarantee that the number of inner iterations is bounded.

6.8 Optimality of the Inner Loop

We need the following simple lemma to prove optimality of the inner loop implemented by the CG algorithm.

Lemma 6.10. *Let* $\{\mathbf{x}^k\}$, $\{\boldsymbol{\lambda}^k\}$, *and* $\{\varrho_k\}$ *be generated by Algorithm 6.1 for the solution of (6.1) with* $\eta > 0$, $\beta > 1$, $M > 0$, $\varrho_0 > 0$, *and* $\boldsymbol{\lambda}^0 \in \mathbb{R}^m$. *Let* $0 < a_{\min} \leq \lambda_{\min}(\mathsf{A})$, *where* $\lambda_{\min}(\mathsf{A})$ *denotes the least eigenvalue of the Hessian* A *of the quadratic function* f. *Then for any* $k \geq 0$

$$L(\mathbf{x}^k, \boldsymbol{\lambda}^{k+1}, \varrho_{k+1}) - L(\mathbf{x}^{k+1}, \boldsymbol{\lambda}^{k+1}, \varrho_{k+1}) \leq \frac{\eta^2}{2a_{\min}} + \frac{\beta\varrho_k}{2}\|\mathsf{B}\mathbf{x}^k\|^2. \tag{6.39}$$

Proof. Notice that by the definition of the Lagrangian function

$$\begin{aligned} L(\mathbf{x}^k, \boldsymbol{\lambda}^{k+1}, \varrho_{k+1}) &= L(\mathbf{x}^k, \boldsymbol{\lambda}^k, \varrho_k) + \varrho_k\|\mathsf{B}\mathbf{x}^k\|^2 + \frac{\varrho_{k+1} - \varrho_k}{2}\|\mathsf{B}\mathbf{x}^k\|^2 \\ &= L(\mathbf{x}^k, \boldsymbol{\lambda}^k, \varrho_k) + \frac{\varrho_{k+1} + \varrho_k}{2}\|\mathsf{B}\mathbf{x}^k\|^2, \end{aligned}$$

so that by (6.16)

$$\begin{aligned} L(\mathbf{x}^k, \boldsymbol{\lambda}^{k+1}, \varrho_{k+1}) - L(\mathbf{x}^{k+1}, \boldsymbol{\lambda}^{k+1}, \varrho_{k+1}) &= L(\mathbf{x}^k, \boldsymbol{\lambda}^k, \varrho_k) - L(\mathbf{x}^{k+1}, \boldsymbol{\lambda}^{k+1}, \varrho_{k+1}) \\ &\quad + \frac{\varrho_{k+1} + \varrho_k}{2}\|\mathsf{B}\mathbf{x}^k\|^2 \\ &\leq \frac{\eta^2}{2a_{\min}} + \frac{\beta\varrho_k}{2}\|\mathsf{B}\mathbf{x}^k\|^2. \end{aligned}$$

□

Now we are ready to prove the main result of this chapter, the optimality of Algorithm 6.1 (SMALBE) in terms of matrix–vector multiplications, provided Step 1 is implemented by Algorithm 5.8 (MPRGP).

Theorem 6.11. *Let*

$$0 < a_{\min} < a_{\max} \quad \text{and} \quad 0 < c_{\max}$$

be given constants and let the class of problems (6.34) satisfy

$$a_{\min} \leq \lambda_{\min}(\mathsf{A}_t) \leq \lambda_{\max}(\mathsf{A}_t) \leq a_{\max} \quad \text{and} \quad \|\mathsf{B}_t\| \leq c_{\max}. \tag{6.40}$$

Let $\{\mathbf{x}_t^k\}, \{\boldsymbol{\lambda}_t^k\}$, *and* $\{\varrho_{t,k}\}$ *be generated by Algorithm 6.1 (SMALBE) for (6.34) with*

$$\|\mathbf{b}_t\| \geq \eta_t > 0,\ \beta > 1,\ M > 0,\ \varrho_{t,0} = \varrho_0 > 0, \quad \text{and} \quad \boldsymbol{\lambda}_t^0 = \mathbf{o}.$$

Let $s \geq 0$ *denote the smallest integer such that*

$$\beta^s \varrho_0 \geq M^2/a_{\min},$$

and let Step 1 of Algorithm 6.1 be implemented by Algorithm 5.8 (MPRGP) with the parameters $\Gamma > 0$ *and* $\overline{\alpha} \in (0, 2(a_{\max} + \beta^s \varrho_0 c_{\max}^2)^{-1}]$ *to generate the iterates* $\mathbf{x}_t^{k,0}, \mathbf{x}_t^{k,1}, \ldots, \mathbf{x}_t^{k,l} = \mathbf{x}_t^k$ *for the solution of (6.34) starting from* $\mathbf{x}_t^{k,0} = \mathbf{x}_t^{k-1}$ *with* $\mathbf{x}_t^{-1} = \mathbf{o}$*, where* $l = l_{t,k}$ *is the first index satisfying*

$$\|g^P(\mathbf{x}_t^{k,l}, \boldsymbol{\lambda}_t^k, \varrho_{t,k})\| \leq M\|\mathsf{B}_t\mathbf{x}_t^{k,l}\| \tag{6.41}$$

or

$$\|g^P(\mathbf{x}_t^{k,l}, \boldsymbol{\lambda}_t^k, \varrho_{t,k})\| \leq M\varepsilon\|\mathbf{b}_t\|. \tag{6.42}$$

Then Algorithm 6.1 generates an approximate solution $\mathbf{x}_t^{k_t}$ *of any problem (6.34) which satisfies*

$$k_t \leq j, \;\; \|\mathbf{g}^P(\mathbf{x}_t^{k_t}, \boldsymbol{\lambda}_t^{k_t}, \varrho_{t,k_t})\| \leq M\varepsilon\|\mathbf{b}_t\|, \;\; \textit{and} \;\; \|\mathsf{B}_t\mathbf{x}_t^{k_t}\| \leq \varepsilon\|\mathbf{b}_t\| \tag{6.43}$$

at $O(1)$ *matrix–vector multiplications by the Hessian of the augmented Lagrangian* L_t *for (6.34).*

Proof. Let $t \in \mathcal{T}$ be fixed and let us denote by $L_t(\mathbf{x}, \boldsymbol{\lambda}, \varrho)$ the augmented Lagrangian for problem (6.34), so that for any $\mathbf{x} \in \mathbb{R}^{n_t}$ and $\varrho \geq 0$

$$L_t(\mathbf{x}, \mathbf{o}, \varrho) = \frac{1}{2}\mathbf{x}^T(\mathsf{A}_t + \varrho \mathsf{B}_t^T\mathsf{B}_t)\mathbf{x} - \mathbf{b}_t^T\mathbf{x} \geq \frac{1}{2}a_{\min}\|\mathbf{x}\|^2 - \|\mathbf{b}_t\|\|\mathbf{x}\| \geq -\frac{\|\mathbf{b}_t\|^2}{2a_{\min}}.$$

Applying the latter inequality to (6.23) with $\mathbf{z}_0 = \mathbf{o}$ and using the assumption $\eta_t \leq \|\mathbf{b}_t\|$, we get

$$\begin{aligned}
\frac{\varrho_{t,k}}{2}\|\mathsf{B}_t\mathbf{x}_t^k\|^2 &\leq \sum_{i=1}^{\infty}\frac{\varrho_{t,i}}{2}\|\mathsf{B}_t\mathbf{x}_t^i\|^2 \\
&\leq f(\mathbf{z}_0) - L(\mathbf{x}_t^0, \boldsymbol{\lambda}_t^0, \varrho_{t,0}) + (1+s)\frac{\eta_t^2}{2a_{\min}} \\
&= (2+s)\frac{\|\mathbf{b}_t\|^2}{2a_{\min}}
\end{aligned}$$

for any $k \geq 0$. Thus by (6.39)

$$\begin{aligned}
L_t(\mathbf{x}_t^{k-1}, \boldsymbol{\lambda}_t^k, \varrho_{t,k}) - L_t(\mathbf{x}_t^k, \boldsymbol{\lambda}_t^k, \varrho_{t,k}) &\leq \frac{\eta_t^2}{2a_{\min}} + \frac{\beta\varrho_{t,k-1}}{2}\|\mathsf{B}_t\mathbf{x}_t^{k-1}\|^2 \\
&\leq (3+s)\frac{\beta\|\mathbf{b}_t\|^2}{2a_{\min}}
\end{aligned}$$

and, since the minimizer $\overline{\mathbf{x}}_t^k$ of $L_t(\mathbf{x}, \boldsymbol{\lambda}_t^k, \varrho_{t,k})$ subject to $\mathbf{x} \geq \boldsymbol{\ell}_t$ satisfies (6.3) and is a possible choice for $\mathbf{x}_t^k$, also that

$$L_t(\mathbf{x}_t^{k-1}, \boldsymbol{\lambda}_t^k, \varrho_{t,k}) - L_t(\overline{\mathbf{x}}_t^k, \boldsymbol{\lambda}_t^k, \varrho_{t,k}) \leq (3+s)\frac{\beta\|\mathbf{b}_t\|^2}{2a_{\min}}. \tag{6.44}$$

Using Theorem 5.15, we get that the MPRGP algorithm 5.8 used to implement Step 1 of Algorithm 6.1 (SMALBE) starting from $\mathbf{x}_t^{k,0} = \mathbf{x}_t^{k-1}$ generates $\mathbf{x}_t^{k,l}$ satisfying

$$\begin{aligned}\|g_t^P(\mathbf{x}_t^{k,l}, \boldsymbol{\lambda}_t^k, \varrho_{t,k})\|^2 &\le a_1 \eta_\Gamma^l \left(L_t(\mathbf{x}_t^{k-1}, \boldsymbol{\lambda}_t^k, \varrho_{t,k}) - L_t(\overline{\mathbf{x}}_t^k, \boldsymbol{\lambda}_t^k, \varrho_{t,k})\right)\\ &\le a_1(3+s)\frac{\beta\|\mathbf{b}_t\|^2}{2a_{\min}}\eta_\Gamma^l,\end{aligned}$$

where

$$a_1 = \frac{38}{\widehat{\alpha}(1-\eta_\Gamma)}, \qquad \eta_\Gamma = 1 - \frac{\widehat{\alpha} a_{\min}}{\vartheta + \vartheta \widehat{\Gamma}^2}, \qquad \vartheta = 2\max\{\overline{\alpha}\|\mathsf{A}\|, 1\},$$
$$\widehat{\Gamma} = \max\{\Gamma, \Gamma^{-1}\}, \qquad \widehat{\alpha} = \min\{\overline{\alpha}, 2\|\mathsf{A}\|^{-1} - \overline{\alpha}\}.$$

It simply follows by the inner stop rule (6.42) that the number l of the inner iterations in Step 1 is uniformly bounded by an index $l_{\max}$ which satisfies

$$a_1(3+s)\frac{\beta\|\mathbf{b}_t\|^2}{2a_{\min}}\eta_\Gamma^{l_{\max}} \le M^2\varepsilon^2\|\mathbf{b}_t\|^2.$$

To finish the proof, it is enough to combine this result with Theorem 6.9. □

The assumption that $\|\mathsf{B}_t\|$ is bounded is essential; it guarantees that it is possible to choose the steplength $\overline{\alpha}$ bounded away from zero.

6.9 Solution of More General Problems

If A is positive definite only on the kernel of B, then we can use a suitable penalization to reduce such problem to the convex one. Using Lemma 1.3, it follows that there is $\overline{\varrho} > 0$ such that $\mathsf{A} + \overline{\varrho}\mathsf{B}^T\mathsf{B}$ is positive definite, so that we can apply our SMALBE algorithm to the equivalent penalized problem

$$\min_{\mathbf{x}\in\Omega_{BE}} f_{\overline{\varrho}}(\mathbf{x}), \tag{6.45}$$

where

$$f_{\overline{\varrho}}(\mathbf{x}) = \frac{1}{2}\mathbf{x}^T(\mathsf{A} + \overline{\varrho}\mathsf{B}^T\mathsf{B})\mathbf{x} - \mathbf{b}^T\mathbf{x}.$$

If A is a positive semidefinite matrix which is positive definite on the kernel of B, then we can use by Lemma 1.2 any $\overline{\varrho} > 0$.

Alternatively, we can modify the inner loop of SMALBE so that it leaves the inner loop and increases the penalty parameter whenever the negative curvature is recognized. Let us point out that such modification does not guarantee optimality of the modified algorithm.

6.10 Implementation

Let us give here a few hints that can be helpful for effective implementation of SMALBE with the inner loop implemented by MPRGP. See also Sect. 4.7.

The purpose and choice of the parameter η are the same as described in Sect. 4.7.

The basic strategy for initialization of the parameters M, β, and ϱ_0 is more complicated than that described in Sect. 4.7. The reason is that Theorem 5.14 guarantees the R-linear convergence of the inner loop only for the steplength

$$\overline{\alpha} \in (0, 2/\|\mathsf{A} + \varrho_k \mathsf{B}^T \mathsf{B}\|],$$

but fast convergence of the outer loop requires large values of the penalty parameter ϱ_k. Thus it is necessary to find a reasonable compromise between the two contradicting requirements. See also Sect. 4.7.2.

Since the choice of ϱ_0 should not be too large but should satisfy

$$\varrho_0 \geq M^2/\lambda_{\min}, \tag{6.46}$$

we recommend to use the values of β that do not cause a big "overshoot", such as $\beta \approx 2$.

The formula (6.46) shows that the values of ϱ_k are related to M. Given ϱ_0, we can achieve that (6.46) is satisfied by choosing

$$M \leq \sqrt{\lambda_{\min} \varrho_0}.$$

Implementation of the inner loop

On entering the inner loop, we recommend to choose

$$\overline{\alpha} \in (\|\mathsf{A} + \varrho_k \mathsf{B}^T \mathsf{B}\|^{-1}, 2\|\mathsf{A} + \varrho_k \mathsf{B}^T \mathsf{B}\|^{-1})$$

to guarantee the R-linear convergence of the inner loop and fast expansion of the active set. Our experience shows the best performance with $\overline{\alpha}$ slightly less than $2/\|\mathsf{A} + \varrho_k \mathsf{B}^T \mathsf{B}\|$; see Sect. 5.9.4 for more discussion on the choice of $\overline{\alpha}$. The parameter Γ can be determined by the heuristics described in Sect. 5.9.4. Recall that $\Gamma \approx 1$ is a good choice.

Lemma 6.1 shows that it is necessary to include the stop criterion not only after Step 1, but also in the procedure which generates $\mathbf{x}^k$ in Step 1. For example, we use in our experiments the stopping criterion

$$\|\mathbf{g}^P(\mathbf{x}^k, \boldsymbol{\lambda}^k, \varrho)\| \leq \varepsilon_g \|\mathbf{b}\| \quad \text{and} \quad \|\mathsf{B}\mathbf{x}^k\| \leq \varepsilon_f \|\mathbf{b}\|, \quad \varepsilon_f = \varepsilon_g/M,$$

and our stopping criterion of the inner loop of MPRGP reads

$$\|\mathbf{g}^P(\mathbf{y}^i, \boldsymbol{\lambda}^i, \varrho_i)\| \leq \min\{M\|\mathsf{B}\mathbf{y}^i\|, \eta\} \text{ or } \|\mathbf{g}^P(\mathbf{y}^i, \boldsymbol{\lambda}^i, \varrho_i)\| \leq \min\{\varepsilon_g, M\varepsilon_f\}\|\mathbf{b}\|.$$

6.11 SMALBE–M

To get a better control of the penalty parameter, we can observe that by Lemma 6.3 small M can prevent the penalty parameter from increasing. It follows that we can modify the SMALBE algorithm so that it updates M and keeps the penalty parameter constant as indicated in Sect. 4.7.2. For convenience of the reader, we present here the complete modified algorithm which we call SMALBE–M.

Algorithm 6.2. SMALBE with modification of M (SMALBE–M).

Given a symmetric positive definite matrix $\mathsf{A} \in \mathbb{R}^{n\times n}$, $\mathsf{B} \in \mathbb{R}^{m\times n}$, *n-vectors* $\mathbf{b}$, $\boldsymbol{\ell}$.
Step 0. {*Initialization.*}
Choose $\eta > 0$, $\beta > 1$, $M_0 > 0$, $\varrho > 0$, $\boldsymbol{\lambda}^0 \in \mathbb{R}^m$
for $k = 0, 1, 2, \ldots$
Step 1. {*Inner iteration with adaptive precision control.*}
Find $\mathbf{x}^k \geq \boldsymbol{\ell}$ *such that*

$$\|\mathbf{g}^P(\mathbf{x}^k, \boldsymbol{\lambda}^k, \varrho)\| \leq \min\{M_k\|\mathsf{B}\mathbf{x}^k\|, \eta\} \tag{6.47}$$

Step 2. {*Updating the Lagrange multipliers.*}

$$\boldsymbol{\lambda}^{k+1} = \boldsymbol{\lambda}^k + \varrho\mathsf{B}\mathbf{x}^k \tag{6.48}$$

Step 3. {*Update M provided the increase of the Lagrangian is not sufficient.*}
if $k > 0$ *and*

$$L(\mathbf{x}^k, \boldsymbol{\lambda}^k, \varrho) < L(\mathbf{x}^{k-1}, \boldsymbol{\lambda}^{k-1}, \varrho) + \frac{\varrho}{2}\|\mathsf{B}\mathbf{x}^k\|^2 \tag{6.49}$$

$M_{k+1} = M_k/\beta$
else
$M_{k+1} = M_k$
end if
end for

The SMALBE–M algorithm has similar properies as the original SMALBE algorithm. In particular, if we choose ϱ_0 and M for the algorithm SMALBE and ϱ and M_0 for the algorithm SMALBE–M such that $\varrho_0 = \varrho$, $M = M_0$, and

$$\varrho_0 \geq M^2/\lambda_{\min},$$

then SMALBE and SMALBE–M will generate $\varrho_k = \varrho$ and $M_k = M$, respectively. Thus if the other parameters of both algorithms are initiated by the same values, the algorithms will generate exactly the same iterates.

6.12 Numerical Experiments

Here we illustrate the performance of Algorithm 6.1 on minimization of the functions $f_{L,h}$ and $f_{LW,h}$ introduced in Sect. 3.10 subject to the multipoint constraints and the bound constraints used in our previous numerical tests. Numerical experiments were carried out with the values of basic parameters equal to

$$M = 1, \quad \Gamma = 1, \quad \text{and} \quad \varrho = 10.$$

The norm of feasibility error is the norm of violation of the equality constraints. The bound constraints are satisfied in each iteration exactly.

6.12.1 Balanced Reduction of Feasibility and Gradient Errors

Let us first show how SMALBE balances the norm of the feasibility error with the norm of the projected gradient on minimization of the quadratic function $f_{L,h}$ defined by the discretization parameter h (see page 98) subject to the multipoint equality constraints introduced in Sect. 4.8.1 and the bound constraints defined in Sect. 5.11.1. The solution given in Fig. 6.1 illustrates also the solution of the benchmarks in Sects. 4.8.1 and 5.11.1.

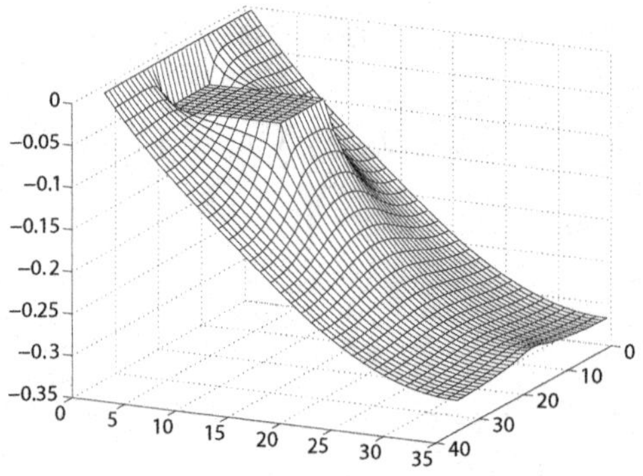

Fig. 6.1. Development of the norms of the projected gradient and feasibility error

The graph of the norms of the projected gradient and the feasibility error (vertical axis) in inner iterations for SMALBE is given in Fig. 6.2. The results were obtained with $h = 1/33$, which corresponds to $n = 1156$ unknowns and 131 equality constraints. We can see that the decrease of the norm of the projected gradient is linear and is balanced with the norm of the feasibility error. Let us recall that the Hessian $\mathsf{A}_{L,h}$ of $f_{L,h}$ is ill-conditioned with the spectral condition number $\kappa(\mathsf{A}_{L,h}) \approx h^{-2}$.

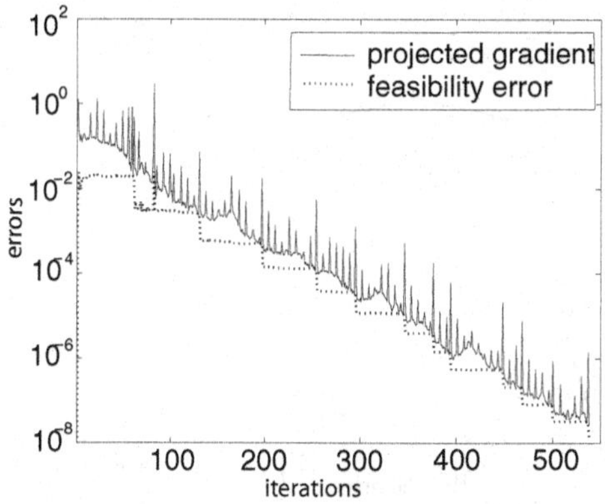

Fig. 6.2. Development of the norms of the projected gradient and feasibility error

6.12.2 Numerical Demonstration of Optimality

To illustrate optimality of SMALBE, we consider the class of well-conditioned problems to minimize the quadratic function $f_{\mathrm{LW},h}$ (see page 99) defined by the discretization parameter h subject to the orthonormal multipoint and bound constraints which describe the same feasible set as in Sect. 6.12.1.

The orthogonalization was used to comply with the assumptions of Theorem 6.11. The matrices B of the equality constraints were obtained by the Gramm–Schmidt process applied to the matrices of the multipoint constraints of Sect. 4.8.1. The class of problems can be given a mechanical interpretation associated to the expanding and partly stiff spring systems on Winkler's foundation and an obstacle. The spectrum of the Hessian $\mathsf{A}_{\mathrm{LW},h}$ of $f_{\mathrm{LW},h}$ is located in the interval $[2, 10]$. Moreover $\|\mathsf{B}\| \le 1$ and $\boldsymbol{\ell} \le \mathbf{o}$ (see Sect. 5.11.2), so that the assumptions of Theorem 6.11 are satisfied.

In Fig. 6.3, we can see the numbers of the CG iterations k_n (vertical axis) that were necessary to reduce the norm of the projected gradient and of the feasibility error to $10^{-6}\|\nabla f_{\mathrm{LW},h}(\mathbf{o})\|$ for the problems with the dimension n ranging from 49 to 9409. The dimension n is on the horizontal axis. We can see that k_n varies mildly with varying n, in agreement with Theorem 6.11. The number of outer iterations was decreasing from 11 for $n = 49$ to 7 for $n = 9409$.

The purpose of the above numerical experiment was just to illustrate the concept of optimality. Similar experiments can be found in Dostál [47, 50]. For practical applications, it is necessary to combine SMALBE with a suitable preconditioning. Application of SMALBE with the FETI domain decomposition method to development of in a sense optimal algorithm for the solution of a semicoercive variational inequality is in Chap. 8.

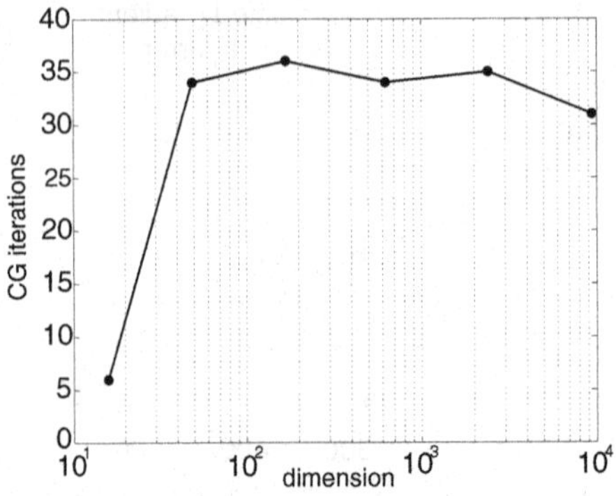

Fig. 6.3. Optimality of SMALBE for a class of well-conditioned problems

6.13 Comments and References

This chapter is based on our research whose starting point was the algorithm introduced by Conn, Gould, and Toint [26]; they adapted the augmented Lagrangian method of Powell [160] and Hestenes [116] to the solution of problems with a nonlinear cost function subject to nonlinear equality constraints and bound constraints. Conn, Gould, and Toint proved that the potentially troublesome penalty parameter ϱ_k is bounded and the algorithm converges to a solution also with asymptotically exact solutions of auxiliary problems [26]. Moreover, they used this algorithm to develop the package LANCELOT [27] for the solution of more general nonlinear optimization problems. More references can be found in their comprehensive book on trust region methods [28]. The inexact augmented Lagrangian method for more general QP problems with the precision control of the auxiliary subproblems by filter were proposed by Friedlander and Leyfer [97].

Our SMALBE algorithm differs from the original algorithm in two points. The first one is the adaptive precision control introduced for the bound and equality constrained problems by Dostál, Friedlander, and Santos [57]. These authors also proved the basic convergence results for the problems with a regular solution, including linear convergence of both the Lagrange multipliers and the feasibility error for a large initial penalty parameter ϱ_0.

The second modification, the update rule of SMALBE for the penalty parameter ϱ_k which is increased until there is a sufficient monotonic increase of $L(x^k \mu^k, \varrho_k)$, was first published by Dostál [49]. The convergence analysis included the optimality of the outer loop and the bound on the penalty parameter; however, the first optimality results for the bound and equality constrained problems were proved by Dostál and Horák for the penalty method [66, 67].

The optimality of SMALBE with the auxiliary problems solved by MPRGP was proved in Dostál [48]; the generalization of the results achieved earlier for the penalty method was based on a well-known observation that the basic augmented Lagrangian algorithm can be considered as a variant of the penalty method (see, e.g., Bertsekas [12, Sect. 4.4)] or Sect. 4.3). Both the optimal penalty method and SMALBE were used in the development of scalable FETI or BETI-based algorithms for the solution of boundary variational inequalities such as those describing the equilibrium of a system of elastic bodies in mutual contact, see, e.g., Dostál and Horák [66, 65, 64], Dostál [48], Bouchala, Dostál, and Sadowská [18, 17, 19], Dostál, Horák, and Stefanica [73], and Dostál et al. [76]. Applications to the contact problems with friction in 2D or 3D can be found in Dostál, Haslinger, and Kučera [63] and Dostál et al. [69].

Part III

Applications to Variational Inequalities

7

Solution of a Coercive Variational Inequality by FETI–DP Method

Numerical experiments in Chap. 5 demonstrated the capability of algorithms with the rate of convergence in bounds on the spectrum to solve special classes of bound constrained problems with optimal, i.e., asymptotically linear, complexity. There is a natural question whether there are effective methods which can reduce the solution of some real-world problems to these special classes.

To give an example of such method, we present here the one which can be used to reduce the coercive variational inequality which describes the equilibrium of a system of 2D elastic bodies in mutual contact to the class of bound constrained QP problems with uniformly bounded spectrum of the Hessian matrix. Let us recall that a contact problem is called coercive if all the bodies are fixed along the part of the boundary in a way which excludes their rigid body motion. To simplify our exposition, we restricted our attention to the solution of a scalar variational inequality governed by the Laplace operator.

Our main tool is a variant of the *finite element tearing and interconnecting (FETI) method*, which was originally proposed by Farhat and Roux [86, 87] as a parallel solver for the problems described by elliptic partial differential equations. The basic idea of FETI is to decompose the domain into non-overlapping subdomains that are "glued" by equality constraints. The variant that we consider here is the FETI–DP method proposed for linear problems by Farhat et al. [83]; it assumes that the subdomains are not completely separated, but remain joined at some nodes that are called *corners* as in Fig. 7.2. After eliminating the primal variables from the KKT conditions for the minimum of the discretized energy function subject to the bound and equality constraints by solving nonsingular local problems, the original problem is reduced to a small, relatively well conditioned bound constrained quadratic programming problem in the Lagrange multipliers.

Though not discovered in this way, the FETI-based methods for linear elliptic problems can be considered as a successful application of the duality theory to the convex QP problems. Here we use the standard duality theory for coercive equality and inequality constrained problems as described in Sect. 2.6.4.

Zdeněk Dostál, *Optimal Quadratic Programming Algorithms*,
Springer Optimization and Its Applications, DOI 10.1007/978-0-387-84806-8_7,

7.1 Model Coercive Variational Inequality

Let $\Omega = (0,1) \times (0,1)$ denote an open domain with the boundary Γ and its three parts $\Gamma_u = \{0\} \times [0,1]$, $\Gamma_f = [0,1] \times \{0,1\}$, and $\Gamma_c = \{1\} \times [0,1]$. The parts Γ_u, Γ_f, and Γ_c are called respectively the Dirichlet boundary, the Neumann boundary, and the contact boundary. On the contact boundary Γ_c, let us define the obstacle ℓ by the upper part of the circle with the radius $R = 1$ and the center $S = (1, 0.5, -1.3)$.

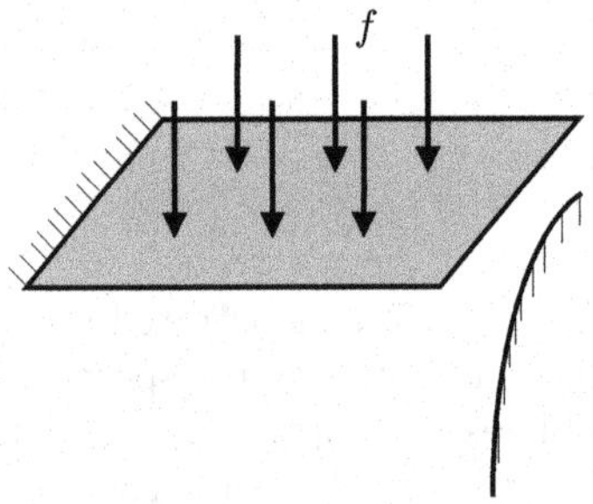

Fig. 7.1. Coercive model problem

Let $H^1(\Omega)$ denote the Sobolev space of the first order in the space $L^2(\Omega)$ of functions on Ω whose squares are integrable in the Lebesgue sense, let

$$\mathcal{K} = \{u \in H^1(\Omega) : \ u = 0 \ \text{ on } \ \Gamma_u \quad \text{and} \quad \ell \leq u \ \text{ on } \ \Gamma_c\},$$

and let us define for any $u \in H^1(\Omega)$

$$f(u) = \frac{1}{2}\int_\Omega \|\nabla u(x)\|^2 \mathrm{d}\Omega + \int_\Omega u \mathrm{d}\Omega.$$

Thus we can define the continuous problem to find

$$\min_{u \in \mathcal{K}} f(u). \tag{7.1}$$

Since the Dirichlet conditions are prescribed on the part Γ_u of the boundary with the positive measure, the cost function f is coercive, which guarantees the existence and uniqueness of the solution by Proposition 2.5.

The solution can be interpreted as the displacement of the membrane under the traction defined by the unit density. The membrane is fixed on Γ_u, not allowed to penetrate the obstacle on Γ_c, and pulled horizontally in the direction of the outer normal by the forces with the unit density along Γ_f. See also Fig. 7.1. We used the discretized problem (7.1) as a benchmark in Sect. 5.11.1.

7.2 FETI–DP Domain Decomposition and Discretization

The first step in our domain decomposition method is to partition the domain Ω into p square subdomains with the sides $H = 1/q$, $q > 1$, $p = q^2$. We call H the *decomposition parameter.* The continuity of the global solution in Ω is enforced by the "gluing" conditions $u^i(\mathbf{X}) = u^j(\mathbf{X})$ that should be satisfied for any point $\mathbf{X}$ on the interface Γ^{ij} of Ω^i and Ω^j except crosspoints. We call a common *crosspoint* either a corner that belongs to four subdomains, or a corner that belongs to two subdomains and is located on Γ. An important feature for developing FETI–DP type algorithms is that a single degree of freedom is considered at each crosspoint, while two degrees of freedom are introduced at all the other matching nodes across subdomain edges. Thus the body is decomposed into the subdomains that are joined in the corners as in Fig. 7.2.

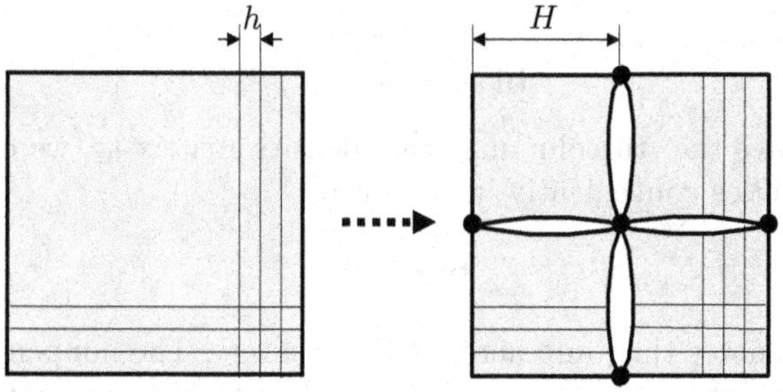

Fig. 7.2. FETI–DP domain decomposition and crosspoints

After modifying appropriately the definition of problem (7.1), introducing regular grids in the subdomains Ω^i with the *discretization parameter* h that match across the interfaces Γ^{ij} of Ω^i and Ω^j, keeping in mind that the crosspoints are global, and using the Lagrangian finite element discretization, we get the discretized version of problem (7.1) with auxiliary domain decomposition in the form

$$\min \frac{1}{2}\mathbf{x}^T \mathsf{A}\mathbf{x} - \mathbf{b}^T\mathbf{x} \quad \text{s.t.} \quad \mathsf{B}_{\mathcal{I}*}\mathbf{x} \le \mathbf{c}_{\mathcal{I}} \quad \text{and} \quad \mathsf{B}_{\mathcal{E}*}\mathbf{x} = \mathbf{o}. \tag{7.2}$$

We assume that the nodes that are not the crosspoints are indexed contiguously in the subdomains, so that Hessian matrix $\mathsf{A} \in \mathbb{R}^{n\times n}$ in (7.2) has the form

$$\mathsf{A} = \begin{bmatrix} \mathsf{A}_{r1} & \mathsf{O} & \dots & \mathsf{O} & \mathsf{A}_{c1} \\ \mathsf{O} & \mathsf{A}_{r2} & \dots & \mathsf{O} & \mathsf{A}_{c2} \\ . & . & \dots & . & . \\ \mathsf{O} & \mathsf{O} & \dots & \mathsf{A}_{rp} & \mathsf{A}_{cp} \\ \mathsf{A}_{c1}^T & \mathsf{A}_{c2}^T & \dots & \mathsf{A}_{cp}^T & \mathsf{A}_{cc} \end{bmatrix}$$

with the band matrices $\mathsf{A}_{\mathrm{r}i}$. Since the diagonal blocks can be interpreted as the stiffness matrices of the subdomains that are fixed at least in corners, $\mathsf{A}_{\mathrm{r}i}$ are positive definite. We refer to the points that are not crosspoints as *reminders*; the subscripts c and r refer to the crosspoints and reminders, respectively. We assume that the Dirichlet conditions are enhanced in A by deleting the corresponding rows and columns. The vector $\mathbf{b} \in \mathbb{R}^n$ represents the discrete analog of the linear term $b(u)$.

The full rank submatrices $\mathsf{B}_{\mathcal{I}*}$ and $\mathsf{B}_{\mathcal{E}*}$ of a matrix $\mathsf{B} \in \mathbb{R}^{m\times n}$ describe the discretized nonpenetration and gluing conditions, respectively. The rows of $\mathsf{B}_{\mathcal{E}*}$ are filled with zeros except 1 and -1 in positions that correspond to the nodes with the same coordinates on the subdomain interfaces. If $\mathbf{b}_i$ denotes a row of $\mathsf{B}_{\mathcal{E}*}$, then $\mathbf{b}_i$ has just two nonzero entries, 1 and -1. The continuity of the solution across the interface in the nodes with indices i, j (see Fig. 7.3) is enforced by the equalities

$$x_i = x_j.$$

Denoting

$$\mathbf{b}_k = (\mathbf{s}_i - \mathbf{s}_j)^T,$$

where $\mathbf{s}_i$ denotes the ith column of the identity matrix I_n, we can write the "gluing" equalities conveniently in the form

$$\mathbf{b}_k \mathbf{x} = 0,$$

so that $\mathbf{b}_k\mathbf{x}$ denotes the jump across the boundary. The nonpenetration condition $x_i \geq \ell_i$ that should be satisfied for the variables corresponding to the nodes on Γ_c, is implemented by $\mathbf{b}_i\mathbf{x} \leq -\ell_i$ with $\mathbf{b}_i = -\mathbf{s}_i^T$. The coordinates $-\ell_i$ are assembled into the vector $\mathbf{c}_{\mathcal{I}}$.

i 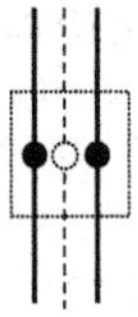 j

Fig. 7.3. "Gluing" along subdomain interface

Our next step is to reduce the problem to the subdomain interfaces and Γ_c by the duality theory. To this end, let us denote the Lagrange multipliers associated with the inequality and equality constraints of problem (7.2) by $\lambda_{\mathcal{E}}$ and $\lambda_{\mathcal{I}}$, respectively, and assume that the rows of B are ordered in such a way that

$$\boldsymbol{\lambda} = \begin{bmatrix} \boldsymbol{\lambda}_{\mathcal{I}} \\ \boldsymbol{\lambda}_{\mathcal{E}} \end{bmatrix}, \quad \mathbf{c} = \begin{bmatrix} \mathbf{c}_{\mathcal{I}} \\ \mathbf{o}_{\mathcal{E}} \end{bmatrix}, \quad \text{and} \quad \mathsf{B} = \begin{bmatrix} \mathsf{B}_{\mathcal{I}} \\ \mathsf{B}_{\mathcal{E}} \end{bmatrix}.$$

Since we formed B in such a way that it is a full rank matrix with orthogonal rows, we can use Proposition 2.21 to get that the Lagrange multipliers $\boldsymbol{\lambda}$ for problem (7.1) solve the dual problem

$$\max\ \Theta(\boldsymbol{\lambda}) \quad \text{s.t.} \quad \boldsymbol{\lambda}_{\mathcal{I}} \geq \mathbf{o},$$

where $\Theta(\boldsymbol{\lambda})$ is the dual function. Changing the signs of Θ and discarding the constant term, we get that the Lagrange multipliers $\boldsymbol{\lambda}$ solve the bound constrained problem

$$\min\ \theta(\boldsymbol{\lambda}) \quad \text{s.t.} \quad \boldsymbol{\lambda}_{\mathcal{I}} \geq \mathbf{o}, \tag{7.3}$$

where θ and the standard FETI notation are defined by

$$\theta(\boldsymbol{\lambda}) = \frac{1}{2}\boldsymbol{\lambda}^T\mathsf{F}\boldsymbol{\lambda} - \boldsymbol{\lambda}^T\mathbf{d}, \quad \mathsf{F} = \mathsf{B}\mathsf{A}^{-1}\mathsf{B}^T, \quad \mathbf{d} = \boldsymbol{\lambda}^T\mathsf{B}\mathsf{A}^{-1}\mathbf{b} - \mathbf{c}.$$

Notice that using the block and band structure of A, we can effectively evaluate $\mathsf{A}^{-1}\mathbf{y}$ for any $\mathbf{y} \in \mathbb{R}^n$ in two steps. Indeed, using the Cholesky decomposition described in Sect. 1.5, we can eliminate the reminders, reducing the unknowns to the corners. In the next step, we decompose the small Schur complement matrix which is associated with the crosspoint variables. However, the implementation of this procedure is a bit tricky and not directly related to the quadratic programming, the main topic of this book. We refer interested readers to Dostál, Horák, and Stefanica [70] or to the Ph.D. thesis of Horák [121]. If the dimension of the blocks A_{ri} is uniformly bounded, then the computational cost increases nearly proportionally with p. Moreover, the time that is necessary for the decomposition $\mathsf{A} = \mathsf{L}\mathsf{L}^T$ and evaluation of $(\mathsf{L}^{-1})^T\mathsf{L}^{-1}\mathbf{y}$ can be reduced nearly proportionally by parallel implementation.

The preconditioning effect of the FETI–DP duality transformation is formulated in the following proposition.

Proposition 7.1. *Let $\mathsf{F}_{H,h}$ denote the Hessian of the reduced dual function θ of (7.3) defined by the decomposition parameter H and the discretization parameter h.*

Then there are constants $C_1 > 0$ and $C_2 > 0$ independent of h and H such that

$$C_1 \leq \lambda_{\min}(\mathsf{F}_{H,h}) \quad \textit{and} \quad \lambda_{\max}(\mathsf{F}_{H,h}) = \|\mathsf{F}_{H,h}\| \leq C_2\left(\frac{H}{h}\right)^2. \tag{7.4}$$

Proof. See [70]. □

Proposition 7.1 shows that the FETI–DP procedure reduces the conditioning of the Hessian of discretized energy from $O(h^{-2})$ to $O(H^2/h^2)$.

7.3 Optimality

To show that Algorithm 5.8 is optimal for the solution of problem (or a class of problems) (7.3), let us introduce new notation that complies with that used to define the class of problems (5.117) introduced in Sect. 5.8.4.

We use

$$\mathcal{T} = \{(H,h) \in \mathbb{R}^2 : H \le 1,\ 0 < 2h \le H, \quad \text{and} \quad H/h \in \mathbb{N}\}$$

as the set of indices, where $\mathbb{N}$ denotes the set of all positive integers. Given a constant $C \ge 2$, we define a subset $\mathcal{T}_C$ of $\mathcal{T}$ by

$$\mathcal{T}_C = \{(H,h) \in \mathcal{T} :\ H/h \le C\}.$$

For any $t \in \mathcal{T}$, we define

$$\mathsf{A}_t = \mathsf{F}, \quad \mathbf{b}_t = \mathbf{d}, \quad \boldsymbol{\ell}_{t,\mathcal{I}} = \mathbf{o}_{\mathcal{I}}, \quad \text{and} \quad \boldsymbol{\ell}_{t,\mathcal{E}} = -\infty$$

by the vectors and matrices generated with the discretization and decomposition parameters H and h, respectively, so problem (7.3) with the fixed discretization and decomposition parameters h and H is equivalent to the problem

$$\text{minimize}\ \ f_t(\boldsymbol{\lambda}_t)\ \ \text{s.t.}\ \ \boldsymbol{\lambda}_t \ge \boldsymbol{\ell}_t \tag{7.5}$$

with $t = (H,h)$, $f_t(\boldsymbol{\lambda}) = \frac{1}{2}\boldsymbol{\lambda}^T \mathbf{A}_t \boldsymbol{\lambda} - \mathbf{b}_t^T \boldsymbol{\lambda}$. Using these definitions, we obtain

$$\|\boldsymbol{\ell}_t^+\| = 0, \tag{7.6}$$

where for any vector $\mathbf{v} = [v_i]$, $\mathbf{v}^+$ denotes the vector with the entries $v_i^+ = \max\{v_i, 0\}$. Moreover, it follows by Proposition 7.1 that for any $C \ge 2$, there are the constants $a_{\max}^C > a_{\min}^C > 0$ such that for any $t \in \mathcal{T}_C$

$$a_{\min}^C \le \lambda_{\min}(\mathsf{A}_t) \le \lambda_{\max}(\mathsf{A}_t) \le a_{\max}^C, \tag{7.7}$$

where $\lambda_{\min}(\mathsf{A}_t)$ and $\lambda_{\max}(\mathsf{A}_t)$ denote the extreme eigenvalues of A_t.

Our optimality result for a model coercive boundary variational inequality then reads as follows.

Theorem 7.2. *Let $C \ge 2$ and $\varepsilon > 0$ denote given constants, let $\{\boldsymbol{\lambda}_t^k\}$ be generated by Algorithm 5.8 (MPRGP) for the solution of (7.5) with the parameters $\Gamma > 0$ and $\overline{\alpha} \in (0, a_{\max}^{-1}]$, starting from $\boldsymbol{\lambda}_t^0 = \max\{\mathbf{o}, \boldsymbol{\ell}_t\}$.*

Then an approximate solution $\lambda_t^{k_t}$ of any problem (7.5) which satisfies

$$\|\mathbf{g}_t^P(\boldsymbol{\lambda}^{k_t})\| \le \varepsilon \|\mathbf{g}_t^P(\boldsymbol{\lambda}_t^0)\|$$

and

$$a_{\min}^C \|\boldsymbol{\lambda}^{k_t} - \widehat{\boldsymbol{\lambda}}_t\| \le f_t(\boldsymbol{\lambda}_t^{\overline{\ell}}) - f_t(\widehat{\boldsymbol{\lambda}}_t) \le \varepsilon \left(f_t(\boldsymbol{\lambda}_t^0) - f(\widehat{\boldsymbol{\lambda}}_t) \right)$$

is generated at $O(1)$ matrix–vector multiplications by A_t for any $t \in \mathcal{T}_C$.

Proof. The class of problems (7.5) with $t \in \mathcal{T}_C$ satisfies the assumptions of Theorem 5.16. □

7.4 Numerical Experiments

In this section we illustrate numerical scalability of MPRGP Algorithm 5.8 on the class of problems arising from application of the FETI–DP method to our boundary variational inequality (7.1). The domain Ω was partitioned into identical squares with the side $H \in \{1/2, 1/4, 1/8\}$. The squares were then discretized by the regular grid with the stepsize h. The solution for $H = 1/4$ and $h = 1/4$ is in Fig. 7.4.

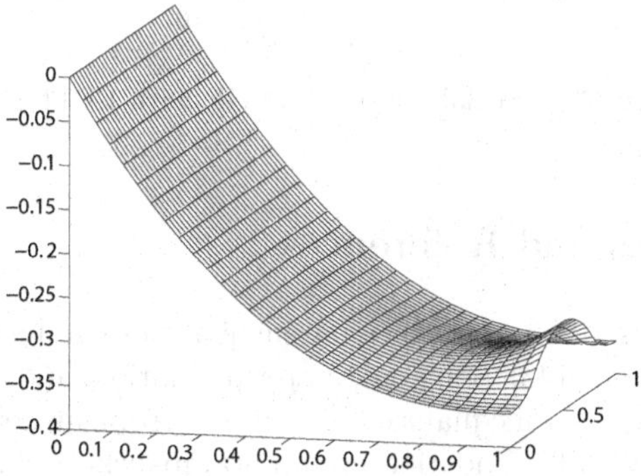

Fig. 7.4. Solution of the coercive model problem (7.1)

The computations were performed with parameters $\Gamma = 1$, $\overline{\alpha} \approx 1/\|\mathsf{A}\|$, and $\boldsymbol{\lambda}^0 = \mathbf{o}$. The stopping criterion in the conjugate gradient iteration was

$$\|\mathbf{g}^P(\boldsymbol{\lambda}^k)\|/\|\mathbf{g}^P(\boldsymbol{\lambda}^0)\| < 10^{-6}.$$

For each H, we chose $h = H/16$, so that ratio H/h was fixed to $H/h = 16$ and the meshes matched across the interface of each couple of neighboring subdomains. Selected results of the computations for varying values of $H \in \{1/8, 1/32, 1/64\}$ and $h = H/16$ are in Fig. 7.5. The primal dimension n is on the horizontal axis; the computation was carried out for primal dimension $n \in \{1156, 4624, 18496\}$ with corresponding dual dimensions $m \in \{93, 425, 1809\}$. The key point is that the number of the conjugate gradient iterations for a fixed ratio H/h varies very moderately with the increasing number of subdomains. This indicates that the unspecified constants in Theorem 7.2 are not very large and we can observe numerical scalability in practical computations. For more numerical experiments with the solution of coercive problems see Dostál, Horák, and Stefanica [70].

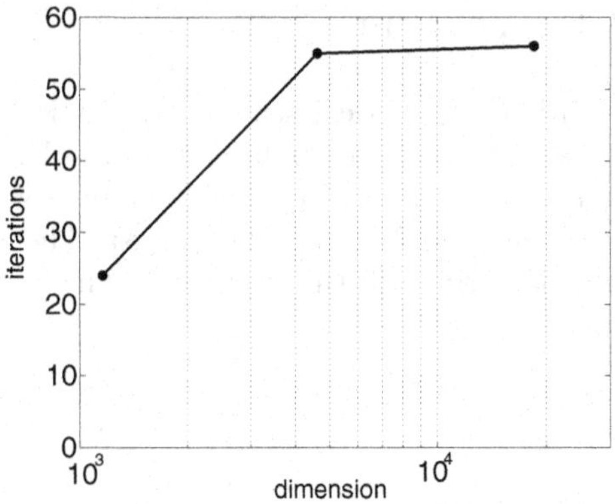

Fig. 7.5. Scalability of MPRGP with FETI–DP

7.5 Comments and References

More problems described by variational inequalities can be found in the book by Lions and Duvaut [143]. Solvability, approximation, and classical numerical methods for variational inequalities or contact problems are discussed in the books by Glowinski [99], Kinderlehrer and Stampaccia [128], Glowinski, Lions, and Trèmoliéres [101], Hlaváček et al. [120], or Eck, Jarušek, and Krbec [80]. The formulation and alternative algorithms for the solution contact problems of elasticity are in Kikuchi and Oden [127], Laursen [142], or Wriggers [181].

Probably the first theoretical results concerning development of scalable algorithms for coercive problems were proved by Schöberl [165, 166]. Our first proof of numerical scalability of an algorithm for the solution of a coercive variational inequality used optimal penalty in dual FETI problem [66]. The proof of Proposition 7.1 is due to D. Stefanica [70]. The optimality was proved also for multidomain coercive problems [70] and for the FETI–DP solution of coercive problems with nonpenetration mortar conditions on contact interface [71]. For more details of mortar implementation of constraints we refer to Wohlmuth [179]. Numerical evidence of scalability of a different approach combining FETI–DP with a Newton-type algorithm for 3D contact problems was given in Avery et al. [2]. See also Dostál et al. [76]. The performance of the method can be further improved by enforcing zero averages of primal variables on the interfaces of subdomains as used by Klawonn and Rheinbach [129] or by preconditioning of linear auxiliary problems by standard preconditioners described, e.g., in Tosseli and Widlund [175]. Preconditioning of linear step was successfully applied by Avery et al. [2].

It should be noted that the effort to develop scalable solvers for coercive variational inequalities was not restricted to FETI. For example, using the ideas related to Mandel [147], Kornhuber [132], Kornhuber and Krause [133], and Krause and Wohlmuth [135] gave an experimental evidence of numerical scalability and the convergence theory for the algorithm based on monotone multigrid. Badea, Tai, and Wang [6] proved linear rate of convergence in terms of the decomposition parameter and overlap for the Schwarz domain decomposition method which assumes exact solution of subdomain problems. See also Zeng and Zhou [185], Tai and Tseng [173], Tarvainen [174], and references therein.

A readable introduction into the formulation and implementation of the FETI methods, including FETI–DP, can be found in Kruis [136]. Let us stress that our goal here is only to illustrate the optimality of MPRGP Algorithm 5.8 on the problem whose structure is the same as that of important real-world problems.

8

Solution of a Semicoercive Variational Inequality by TFETI Method

To give an example of a class of bound and equality constrained problems with uniformly bounded spectrum arising in important applications, let us consider the solution of the discretized elliptic semicoercive variational inequalities, such as those describing the equilibrium of a system of elastic bodies in mutual unilateral frictionless contact in case that some bodies are not sufficiently fixed along the boundary. The presence of "floating" bodies is a considerable complication as the corresponding stiffness matrices are singular. To simplify our exposition, we again restrict our attention to a model variational inequality governed by the Laplace operator on 2D domains.

Our main tool is a variant of the classical FETI method called total FETI (TFETI), which was proposed independently by Dostál, Horák, and Kučera [68] and Of (all floating FETI) [156] as a parallel solver for the problems described by elliptic partial differential equations. The TFETI method differs from the original FETI method in the way which is used to implement the Dirichlet boundary conditions. While the FETI method assumes that the subdomains inherit the Dirichlet boundary conditions from the original problem, TFETI uses the Lagrange multipliers to "glue" the subdomains to the boundary whenever the Dirichlet boundary conditions are prescribed. Such approach simplifies the implementation as all the stiffness matrices of the subdomains have typically a priori known kernels and can be treated in the same way. Moreover, the kernels can be used for effective evaluation of the action of a generalized inverse by means of Lemma 1.1. The procedure can be naturally combined with the preconditioning by the "natural coarse grid" introduced by Farhat, Mandel, and Roux [85]. This preconditioning results in the class of problems with the condition number of the regular part of the Hessian matrix bounded by CH/h, where C, H, and h are a constant, decomposition, and discretization parameters, respectively. This compares favorably with the estimate CH^2/h^2 of Proposition 7.1 for non-preconditioned FETI–DP. Here we use the duality theory of Sect. 2.6.5 and Theorem 6.11 to modify the TFETI method for the solution of variational inequalities.

Zdeněk Dostál, *Optimal Quadratic Programming Algorithms*,
Springer Optimization and Its Applications, DOI 10.1007/978-0-387-84806-8_8,

8.1 Model Semicoercive Variational Inequality

Let $\Omega = \Omega^1 \cup \Omega^2$, where $\Omega^1 = (0,1) \times (0,1)$ and $\Omega^2 = (1,2) \times (0,1)$ denote open domains with boundaries Γ^1, Γ^2 and their parts Γ_u^i, Γ_f^i, and Γ_c^i formed by the sides of Ω^i, $i = 1, 2$.

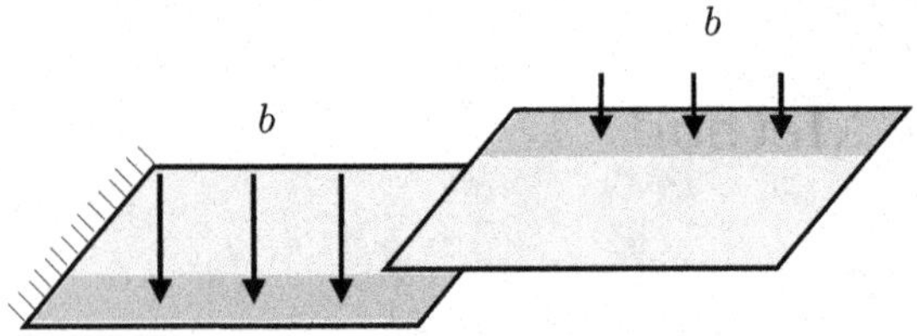

Fig. 8.1. Semicoercive model problem

Let $H^1(\Omega^i)$, $i = 1, 2$, denote the Sobolev space of the first order in the space $L^2(\Omega^i)$ of the functions on Ω^i whose squares are integrable in the sense of Lebesgue. Let

$$\mathcal{V}^i = \left\{ v^i \in H^1(\Omega^i) : v^i = 0 \quad \text{on} \quad \Gamma_u^i \right\}$$

denote the closed subspaces of $H^1(\Omega^i)$, $i = 1, 2$, and let

$$\mathcal{V} = \mathcal{V}^1 \times \mathcal{V}^2 \qquad \text{and} \qquad \mathcal{K} = \left\{ (v^1, v^2) \in \mathcal{V} : v^2 - v^1 \geq 0 \quad \text{on} \quad \Gamma_c \right\}$$

denote the closed subspace and the closed convex subset of

$$\mathcal{H} = H^1(\Omega^1) \times H^1(\Omega^2),$$

respectively. The relations on the boundaries are in terms of traces. We shall define on $\mathcal{H}$ the symmetric bilinear form

$$a(u, v) = \sum_{i=1}^{2} \int_{\Omega^i} \left(\frac{\partial u^i}{\partial x} \frac{\partial v^i}{\partial x} + \frac{\partial u^i}{\partial y} \frac{\partial v^i}{\partial y} \right) d\Omega$$

and the linear form

$$b(v) = \sum_{i=1}^{2} \int_{\Omega^i} b^i v^i d\Omega,$$

where $b^i \in L^2(\Omega^i)$, $i = 1, 2$ are the restrictions of

$$b(x, y) = \begin{cases} -3 \text{ for } (x, y) \in (0, 1) \times [0.75, 1), \\ \ \ 0 \text{ for } (x, y) \in (0, 1) \times [0, 0.75) \quad \text{and} \quad (x, y) \in (1, 2) \times [0.25, 1). \\ -1 \text{ for } (x, y) \in (1, 2) \times [0, 0.25). \end{cases}$$

Denoting for each $u \in \mathcal{H}$

$$f(u) = \frac{1}{2}a(u,u) - b(u) = \frac{1}{2}\sum_{i=1}^{2}\int_{\Omega^i} \|\nabla u^i\|^2 d\Omega - \sum_{i=1}^{2}\int_{\Omega^i} b^i v^i d\Omega,$$

we can define the continuous problem to find

$$\min_{u \in \mathcal{K}} f(u). \tag{8.1}$$

The solution of the model problem can be interpreted as the displacement of two membranes under the traction b as in Fig. 8.1. The left edge of the right membrane is not allowed to penetrate below the right edge of the left membrane. Notice that only the left membrane is fixed on the outer edge and the right membrane has no prescribed displacement, so that

$$\Gamma_u^1 = \{(0,y) \in \mathbb{R}^2 : y \in [0,1]\}, \quad \Gamma_u^2 = \emptyset.$$

Even though the form a is only semicoercive, the form b is still coercive due to the choice of b so that it has a unique solution [120, 99].

8.2 TFETI Domain Decomposition and Discretization

In our definition of the problem, we have so far used only the natural decomposition of the spatial domain Ω into Ω^1 and Ω^2. To enable efficient application of the domain decomposition methods, we decompose each Ω^i into subdomains $\Omega^{i1}, \ldots, \Omega^{ip}$, $p > 1$, as in Fig. 8.2.

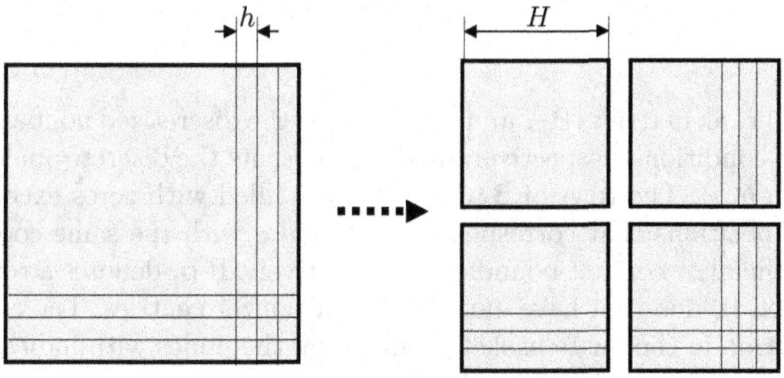

Fig. 8.2. Domain decomposition and discretization

The continuity of the global solution in Ω^1 and Ω^2 is enforced by the "gluing" conditions $u^{ij}(\mathbf{X}) = u^{ik}(\mathbf{X})$ that should be satisfied for any point $\mathbf{X}$ on the interface $\Gamma^{ij,ik}$ of Ω^{ij} and Ω^{ik}.

After modifying appropriately the definition of problem (8.1), introducing regular grids in the subdomains Ω^{ij} that match across the interfaces $\Gamma^{ij,kl}$, indexing contiguously the nodes and entries of corresponding vectors in the subdomains, and using the Lagrangian finite element discretization, we get the discretized version of problem (8.1) with auxiliary domain decomposition that reads

$$\min \frac{1}{2}\mathbf{x}^T \mathsf{A}\mathbf{x} - \mathbf{b}^T\mathbf{x} \quad \text{s.t.} \quad \mathsf{B}_{\mathcal{I}*}\mathbf{x} \le \mathbf{o} \quad \text{and} \quad \mathsf{B}_{\mathcal{E}*}\mathbf{x} = \mathbf{o}. \tag{8.2}$$

In (8.2), the Hessian matrix

$$\mathsf{A} = \begin{bmatrix} \mathsf{A}_1 & \mathsf{O} & \dots & \mathsf{O} \\ \mathsf{O} & \mathsf{A}_2 & \dots & \mathsf{O} \\ . & . & \dots & . \\ \mathsf{O} & \mathsf{O} & \dots & \mathsf{A}_{2p} \end{bmatrix}$$

is a block diagonal positive semidefinite stiffness matrix. The diagonal blocks A_i are the local stiffness matrices of the subdomains with the same kernel; for $j = 1, 2$ and $k = 1, \dots, p$, the matrix $\mathsf{A}_{p(j-1)+k}$ corresponds to the subdomain Ω^{jk}. If the nodes in each subdomain are ordered columnwise, the blocks A_i are band matrices.

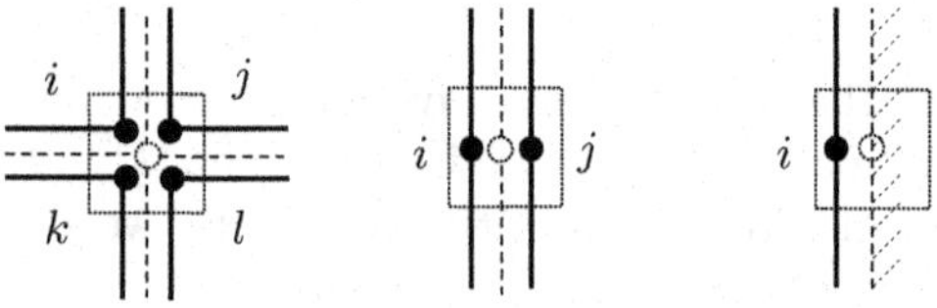

Fig. 8.3. Three types of constraints

The full rank matrices $\mathsf{B}_{\mathcal{I}*}$ and $\mathsf{B}_{\mathcal{E}*}$ describe the discretized nonpenetration and gluing conditions, respectively, and $\mathbf{b}$ represents the discrete analog of the linear term $b(u)$. The rows of $\mathsf{B}_{\mathcal{E}*}$ and $\mathsf{B}_{\mathcal{I}*}$ are filled with zeros except 1 and -1 in the positions that correspond to the nodes with the same coordinates on the artificial or contact boundaries, respectively. If $\mathbf{b}_i$ denotes a row of $\mathsf{B}_{\mathcal{E}*}$ or $\mathsf{B}_{\mathcal{I}*}$, then $\mathbf{b}_i$ does not have more than four nonzero entries. The continuity of the solution in the "wire basket" comprising the nodes with indices i, j, k, l (see Fig. 8.3 left) is enforced by the equalities

$$x_i = x_j, \quad x_k = x_l, \quad x_i + x_j = x_k + x_l,$$

which can be expressed by the vectors

$$\mathbf{b}_{ij} = (\mathbf{s}_i - \mathbf{s}_j)^T, \ \mathbf{b}_{kl} = (\mathbf{s}_k - \mathbf{s}_l)^T, \ \mathbf{b}_{ijkl} = (\mathbf{s}_i + \mathbf{s}_j - \mathbf{s}_k - \mathbf{s}_l)^T,$$

where $\mathbf{s}_i$ denotes the ith column of the identity matrix I_n. The continuity of the solution across the subdomains interface (see Fig. 8.3 middle) is implemented by $\mathbf{b}_{ij}\mathbf{x} = 0$ as in the FETI–DP method discussed in Sect. 7.2, so that $\mathbf{b}_{ij}\mathbf{x}$ denotes the jump across the boundary, and the Dirichlet boundary condition (se Fig. 8.3 right) $x_i = 0$ is implemented by the row $\mathbf{b}_i = \mathbf{s}_i^T$.

Our next step is to simplify the problem, in particular to replace the general inequality constraints $\mathsf{B}_{\mathcal{I}}\mathbf{x} \le \mathbf{o}$ by the nonnegativity constraints using the duality theory. To this end, let us denote the Lagrange multipliers associated with the inequality and equality constraints of problem (8.2) by $\boldsymbol{\lambda}_{\mathcal{I}}$ and $\boldsymbol{\lambda}_{\mathcal{E}}$, respectively, and assume that the rows of B are ordered in such a way that

$$\boldsymbol{\lambda} = \begin{bmatrix} \boldsymbol{\lambda}_{\mathcal{I}} \\ \boldsymbol{\lambda}_{\mathcal{E}} \end{bmatrix} \quad \text{and} \quad \mathsf{B} = \begin{bmatrix} \mathsf{B}_{\mathcal{I}} \\ \mathsf{B}_{\mathcal{E}} \end{bmatrix}.$$

We formed B in such a way that it is a full rank matrix. Finally, let R denote the full column rank matrix whose columns span Ker$\mathsf{A}$. Then we can use Proposition 2.22 to get that the Lagrange multipliers $\boldsymbol{\lambda}$ for problem (8.2) solve the constrained dual problem

$$\max\ \Theta(\boldsymbol{\lambda}) \quad \text{s.t.} \quad \boldsymbol{\lambda}_{\mathcal{I}} \ge \mathbf{o} \quad \text{and} \quad \mathsf{R}^T(\mathbf{b} - \mathsf{B}^T\boldsymbol{\lambda}) = \mathbf{o},$$

where $\Theta(\boldsymbol{\lambda})$ is the dual function. Changing the signs of Θ and discarding the constant term, we get that the Lagrange multipliers $\boldsymbol{\lambda}$ solve the bound and equality constrained problem

$$\min\ \theta(\boldsymbol{\lambda}) \quad \text{s.t.} \quad \boldsymbol{\lambda}_{\mathcal{I}} \ge \mathbf{o} \quad \text{and} \quad \mathsf{R}^T(\mathbf{b} - \mathsf{B}^T\boldsymbol{\lambda}) = \mathbf{o}, \tag{8.3}$$

where

$$\theta(\lambda) = \frac{1}{2}\boldsymbol{\lambda}^T\mathsf{B}\mathsf{A}^+\mathsf{B}^T\boldsymbol{\lambda} - \boldsymbol{\lambda}^T\mathsf{B}\mathsf{A}^+\mathbf{b}$$

and A^+ is any symmetric positive semidefinite generalized inverse. In our computations, we use the generalized inverse $\mathsf{A}^\#$ defined by (1.7).

Notice that using the block diagonal and band structure of A together with

$$\text{Ker}\mathsf{A}_i = [1, \ldots, 1]^T, \quad i = 1, \ldots, 2p,$$

we can effectively evaluate $\mathsf{A}^\#\mathbf{y}$ for any $\mathbf{y} \in \mathbb{R}^n$. Indeed, using the Cholesky decomposition described in Sect. 1.5, we get the lower triangular band matrices L_i such that $\mathsf{A}_i = \mathsf{L}_i\mathsf{L}_i^T$. Since $\mathsf{A}_i^\# = (\mathsf{L}_i^\#)^T\mathsf{L}_i^\#$ and

$$\mathsf{A}^\# = \text{diag}(\mathsf{A}_1^\#, \mathsf{A}_2^\#, \ldots, \mathsf{A}_{2p}^\#),$$

we get

$$\mathsf{A}^\#\mathbf{y} = \sum_{i=1}^{2p} \mathsf{A}_i\mathbf{y}_i = \sum_{i=1}^{2p} (\mathsf{L}_i^\#)^T(\mathsf{L}_i^\#\mathbf{y}_i),$$

where we assume that the decomposition $\mathbf{y}^T = [\mathbf{y}_1^T, \mathbf{y}_2^T, \ldots, \mathbf{y}_{2p}^T]$ complies with the block structure of A. If the dimension of the blocks A_i is uniformly bounded, then the computational cost increases nearly proportionally to p. Moreover, the time that is necessary for the decomposition $\mathsf{A} = \mathsf{L}\mathsf{L}^T$ and evaluation of $(\mathsf{L}^\#)^T\mathsf{L}^\#\mathbf{y}$ can be reduced nearly proportionally by parallel implementation.

8.3 Natural Coarse Grid

Even though problem (8.3) is much more suitable for computations than (8.2), further improvement may be achieved by adapting some simple observations and the results of Farhat, Mandel, and Roux [85]. Let us denote

$$\mathsf{F} = \mathsf{B}\mathsf{A}^{+}\mathsf{B}^{T}, \qquad \widetilde{\mathbf{d}} = \mathsf{B}\mathsf{A}^{T}\mathbf{b},$$
$$\widetilde{\mathsf{G}} = \mathsf{R}^{T}\mathsf{B}^{T}, \qquad \widetilde{\mathbf{e}} = \mathsf{R}^{T}\mathbf{b},$$

and let T denote a regular matrix that defines orthonormalization of the rows of $\widetilde{\mathsf{G}}$ so that the matrix

$$\mathsf{G} = \mathsf{T}\widetilde{\mathsf{G}}$$

has orthonormal rows. After denoting

$$\mathbf{e} = \mathsf{T}\widetilde{\mathbf{e}},$$

problem (8.3) reads

$$\min \ \frac{1}{2}\boldsymbol{\lambda}^{T}\mathsf{F}\boldsymbol{\lambda} - \boldsymbol{\lambda}^{T}\widetilde{\mathbf{d}} \quad \text{s.t.} \quad \boldsymbol{\lambda}_{\mathcal{I}} \geq \mathbf{o} \quad \text{and} \quad \mathsf{G}\boldsymbol{\lambda} = \mathbf{e}. \tag{8.4}$$

Next we shall transform the problem of minimization on the subset of the affine space to that on the subset of the vector space by looking for the solution of (8.4) in the form $\boldsymbol{\lambda} = \boldsymbol{\mu} + \widetilde{\boldsymbol{\lambda}}$, where $\mathsf{G}\widetilde{\boldsymbol{\lambda}} = \mathbf{e}$. The following lemma shows that we can even find $\widetilde{\boldsymbol{\lambda}}$ such that $\widetilde{\boldsymbol{\lambda}}_{\mathcal{I}} = \mathbf{o}$.

Lemma 8.1. *Let B be such that the negative entries of $\mathsf{B}_{\mathcal{I}}$ are in the columns that correspond to the nodes in the floating subdomain Ω^2. Then there is $\widetilde{\boldsymbol{\lambda}}_{\mathcal{I}}$ such that $\widetilde{\boldsymbol{\lambda}}_{\mathcal{I}} \geq \mathbf{o}$ and $\widetilde{\mathsf{G}}\widetilde{\boldsymbol{\lambda}} = \widetilde{\mathbf{e}}$.*

Proof. See [65]. □

To carry out the transformation, substitute $\boldsymbol{\lambda} = \boldsymbol{\mu} + \widetilde{\boldsymbol{\lambda}}$ to get

$$\frac{1}{2}\boldsymbol{\lambda}^{T}\mathsf{F}\boldsymbol{\lambda} - \boldsymbol{\lambda}^{T}\widetilde{\mathbf{d}} = \frac{1}{2}\boldsymbol{\mu}^{T}\mathsf{F}\boldsymbol{\mu} - \boldsymbol{\mu}^{T}(\widetilde{\mathbf{d}} - \mathsf{F}\widetilde{\boldsymbol{\lambda}}) + \frac{1}{2}\widetilde{\boldsymbol{\lambda}}^{T}\mathsf{F}\widetilde{\boldsymbol{\lambda}} - \widetilde{\boldsymbol{\lambda}}^{T}\widetilde{\mathbf{d}}.$$

After returning to the old notation, problem (8.4) is reduced to

$$\min \frac{1}{2}\boldsymbol{\lambda}^{T}\mathsf{F}\boldsymbol{\lambda} - \boldsymbol{\lambda}^{T}\mathbf{d} \quad \text{s.t.} \quad \mathsf{G}\boldsymbol{\lambda} = \mathbf{o} \quad \text{and} \quad \boldsymbol{\lambda}_{\mathcal{I}} \geq -\widetilde{\boldsymbol{\lambda}}_{\mathcal{I}} \tag{8.5}$$

with $\mathbf{d} = \widetilde{\mathbf{d}} - \mathsf{F}\widetilde{\boldsymbol{\lambda}}$ and $\widetilde{\boldsymbol{\lambda}}_{\mathcal{I}} \geq \mathbf{o}$.

Our final step is based on the observation that (8.5) is equivalent to

$$\min \frac{1}{2}\boldsymbol{\lambda}^{T}(\mathsf{PFP} + \overline{\varrho}\mathsf{Q})\boldsymbol{\lambda} - \boldsymbol{\lambda}^{T}\mathsf{P}\mathbf{d} \quad \text{s.t.} \quad \mathsf{G}\boldsymbol{\lambda} = \mathbf{o} \quad \text{and} \quad \boldsymbol{\lambda}_{\mathcal{I}} \geq -\widetilde{\boldsymbol{\lambda}}_{\mathcal{I}}, \tag{8.6}$$

where $\overline{\varrho}$ is an arbitrary positive constant and

$$\mathsf{Q} = \mathsf{G}^T\mathsf{G} \qquad \text{and} \qquad \mathsf{P} = \mathsf{I} - \mathsf{Q}$$

denote the orthogonal projectors on the image space of $G^\top$ and on the kernel of G, respectively. The regularization term is introduced in order to simplify the reference to the results of quadratic programming that assume regularity of the Hessian matrix of the quadratic form. Problem (8.6) turns out to be a suitable starting point for development of an efficient algorithm for variational inequalities due to the following classical estimates of the extreme eigenvalues.

Theorem 8.2. *There are constants $C_1 > 0$ and $C_2 > 0$ independent of the discretization parameter h and the decomposition parameter H such that*

$$C_1 \ \leq \ \lambda_{\min}(\mathsf{PFP}|\mathrm{Im}\mathsf{P}) \quad \textit{and} \quad \lambda_{\max}(\mathsf{PFP}|\mathrm{Im}\mathsf{P}) \leq ||\mathsf{PFP}|| \ \leq \ C_2\frac{H}{h},$$

where $\lambda_{\min}$ and $\lambda_{\max}$ denote the corresponding extremal eigenvalues of corresponding matrices.

Proof. See Theorem 3.2 of Farhat, Mandel, and Roux [85]. Let us point out that the statement of Theorem 3.2 of Farhat, Mandel and Roux [85] gives only an upper bound on the spectral condition number $\kappa(\mathsf{PFP}|\mathrm{Im}\mathsf{P})$, but the reasoning that precedes and substantiates their estimate proves both bounds of (8.2). □

8.4 Optimality

To show that Algorithm 6.1 with the inner loop implemented by Algorithm 5.8 is optimal for the solution of problem (or a class of problems) (8.6), let us introduce new notation that complies with that used to define the class of problems (6.34) introduced in Sect. 6.7.

As in Chap. 7, we use

$$\mathcal{T} = \{(H,h) \in \mathbb{R}^2 : H \leq 1,\ 0 < 2h \leq H, \ \text{ and } \ H/h \in \mathbb{N}\}$$

as the set of indices, where $\mathbb{N}$ denotes the set of all positive integers. Given a constant $C \geq 2$, we shall define a subset $\mathcal{T}_C$ of $\mathcal{T}$ by

$$\mathcal{T}_C = \{(H,h) \in \mathcal{T} : \ H/h \leq C\}.$$

For any $t \in \mathcal{T}$, we shall define

$$\begin{aligned} \mathsf{A}_t &= \mathsf{PFP} + \overline{\varrho}\mathsf{Q}, \quad & \mathsf{b}_t &= \mathsf{Pd} \\ \mathsf{B}_t &= \mathsf{G}, & \boldsymbol{\ell}_{t,\mathcal{I}} &= -\widetilde{\boldsymbol{\lambda}}_{\mathcal{I}} \ \text{ and } \ \boldsymbol{\ell}_{t,\mathcal{E}} = -\infty \end{aligned}$$

by the vectors and matrices generated with the discretization and decomposition parameters H and h, respectively, so that problem (8.6) is equivalent to the problem

$$\text{minimize } f_t(\boldsymbol{\lambda}_t) \text{ s.t. } \mathsf{C}_t\boldsymbol{\lambda}_t = \mathbf{o} \text{ and } \boldsymbol{\lambda}_t \geq \boldsymbol{\ell}_t \tag{8.7}$$

with $f_t(\boldsymbol{\lambda}) = \frac{1}{2}\boldsymbol{\lambda}^T\mathsf{A}_t\boldsymbol{\lambda} - \mathbf{b}_t^T\boldsymbol{\lambda}$. Using these definitions, Lemma 8.1, and $\mathsf{G}\mathsf{G}^T = \mathsf{I}$, we obtain

$$\|\mathsf{B}_t\| \leq 1 \text{ and } \|\boldsymbol{\ell}_t^+\| = 0, \tag{8.8}$$

where for any vector $\mathbf{v} = [v_i]$, $\mathbf{v}^+$ denotes the vector with the entries $v_i^+ = \max\{v_i, 0\}$. Moreover, it follows by Theorem 8.2 that for any $C \geq 2$ there are constants $a_{\max}^C > a_{\min}^C > 0$ such that

$$a_{\min}^C \leq \lambda_{\min}(\mathsf{A}_t) \leq \lambda_{\max}(\mathsf{A}_t) \leq a_{\max}^C \tag{8.9}$$

for any $t \in \mathcal{T}_C$. As above, we denote by $\lambda_{\min}(\mathsf{A}_t)$ and $\lambda_{\max}(\mathsf{A}_t)$ the extreme eigenvalues of A_t. Our optimality result for a model semicoercive boundary variational inequality then reads as follows.

Theorem 8.3. *Let $C \geq 2$ denote a given constant, let $\{\boldsymbol{\lambda}_t^k\}$, $\{\boldsymbol{\mu}_t^k\}$, and $\{\varrho_{t,k}\}$ be generated by Algorithm 6.1 (SMALBE) for (8.7) with $\|\mathbf{b}_t\| \geq \eta_t > 0$, $\beta > 1$, $M > 0$, $\varrho_{t,0} = \varrho_0 > 0$, $\varepsilon > 0$, and $\boldsymbol{\mu}_t^0 = \mathbf{o}$. Let $s \geq 0$ denote the smallest integer such that $\beta^s\varrho_0 \geq M^2/a_{\min}$ and assume that Step 1 of Algorithm 6.1 is implemented by means of Algorithm 5.8 (MPRGP) with parameters $\Gamma > 0$ and $\overline{\alpha} \in (0, (a_{\max} + \beta^s\varrho_0)^{-1}]$, so that it generates the iterates*

$$\boldsymbol{\lambda}_t^{k,0}, \boldsymbol{\lambda}_t^{k,1}, \ldots, \boldsymbol{\lambda}_t^{k,l} = \boldsymbol{\lambda}_t^k$$

for the solution of (8.7) starting from $\boldsymbol{\lambda}_t^{k,0} = \boldsymbol{\lambda}_t^{k-1}$ with $\boldsymbol{\lambda}_t^{-1} = \mathbf{o}$, where $l = l_{t,k}$ is the first index satisfying

$$\|\mathbf{g}^P(\boldsymbol{\lambda}_t^{k,l}, \boldsymbol{\mu}_t^k, \varrho_{t,k})\| \leq M\|\mathsf{B}_t\boldsymbol{\lambda}_t^{k,l}\| \tag{8.10}$$

or

$$\|\mathbf{g}^P(\boldsymbol{\lambda}_t^{k,l}, \boldsymbol{\mu}_t^k, \varrho_{t,k})\| \leq \varepsilon M\|\mathbf{b}_t\|. \tag{8.11}$$

Then for any $t \in \mathcal{T}_C$ and problem (8.7), an approximate solution $\boldsymbol{\lambda}_t^{k_t}$ which satisfies

$$\|\mathbf{g}^P(\boldsymbol{\lambda}_t^{k_t}, \boldsymbol{\mu}_t^{k_t}, \varrho_{t,k_t})\| \leq \varepsilon M\|\mathbf{b}_t\| \quad \text{and} \quad \|\mathsf{B}_t\boldsymbol{\lambda}_t^{k_t}\| \leq \varepsilon\|\mathbf{b}_t\| \tag{8.12}$$

is generated at $O(1)$ matrix–vector multiplications by the Hessian of the augmented Lagrangian L_t for (8.7) and

$$\varrho_{t,k} \leq \beta^s\varrho_0.$$

Proof. Notice that we assume that the constant C is fixed, so all the assumptions of Theorem 6.11 (i.e., the inequalities (8.8) and (8.9)) are satisfied for the set of indices $\mathcal{T}_C$. Thus to complete the proof, it is enough to apply Theorem 6.11. □

Since the cost of a matrix–vector multiplication by the Hessian of the augmented Lagrangian L_t is proportional to the number of the dual variables, Theorem 8.3 proves numerical scalability of Algorithm 6.1 (SMALBE) for (8.7) provided the inner bound constrained minimization is implemented by means of Algorithm 5.8 (MPRGP). The parallel scalability follows directly from the discussion at the end of Sect. 8.2. We shall illustrate these features numerically in the next section.

8.5 Numerical Experiments

In this section we illustrate numerical scalability of SMALBE Algorithm 6.1 on the class of problems arising from application of the TFETI method described above to our boundary variational inequality (8.1). The domain Ω was first partitioned into identical squares with the side

$$H \in \{1/2, 1/4, 1/8, 1/16\}.$$

The square subdomains were then discretized by regular grids with the discretization parameter $h = H/64$, so that the discretized problems have the primal dimension

$$n \in \{33282, 133128, 532512, 21300048\}$$

and the dual dimension

$$m \in \{258, 1545, 7203, 30845\}.$$

The computations were performed with the parameters

$$M = 1, \quad \varrho_0 = 30, \quad \Gamma = 1, \quad \text{and} \quad \varepsilon = 10^{-4}.$$

The stopping criterion was

$$\|\mathbf{g}_t^P(\boldsymbol{\lambda}^k)\| \le 10^{-4}\|\mathbf{b}_t\| \quad \text{and} \quad \|\mathsf{B}_t\boldsymbol{\lambda}^k\| \le 10^{-4}\|\mathbf{b}_t\|.$$

Algorithm 6.1 with the solution of auxiliary bound constrained problem by Algorithm 5.8 was implemented in C exploiting PETSc [7]. Using Theorem 8.3, we get that the number of iterations that are necessary to find the approximate solution is bounded provided H/h is bounded. The solution for $H = 1/4$ and $h = 1/4$ is in Fig. 8.4.

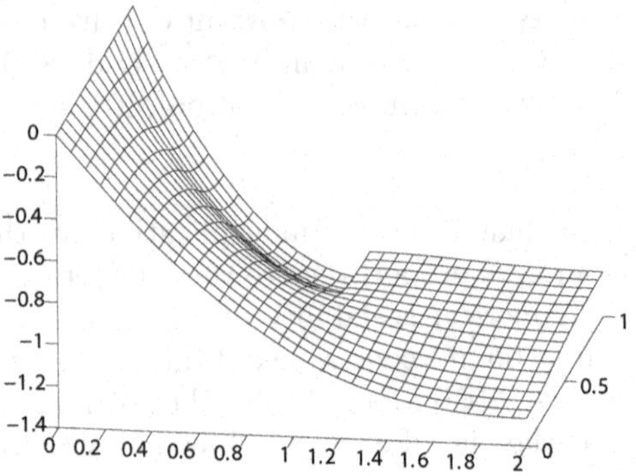

Fig. 8.4. Solution of the model semicoercive problem

The results of computations are in Fig. 8.5. We can see that the numbers of the conjugate gradient iterations (on vertical axis) which correspond to $H/h = 64$ vary very moderately with the dimension of the problem in agreement with Theorem 8.3, so that the cost of computations increases nearly linearly. The algorithm shares its parallel scalability with FETI; see, e.g., Dostál and Horák [64]. We conclude that it is possible to observe numerical scalability and that SMALBE with the inner loop implemented by MPRGP can be an efficient solver for semicoercive variational inequalities.

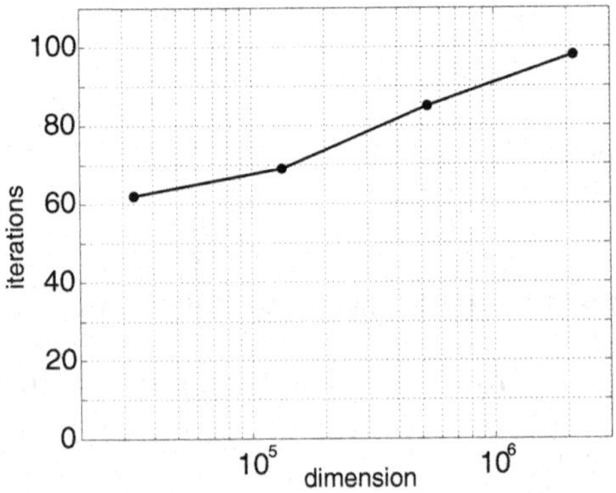

Fig. 8.5. Scalability of SMALBE with TFETI for the semicoercive problem with $H/h = 64$

More results of numerical experiments can be found in Dostál [49]. See also Dostál and Horák [64]. Applications to the contact problems of elasticity are in Dostál et al. [76].

8.6 Comments and References

For solvability and approximation theory for semicoercive variational inequalities see the references in Sect. 7.5. See also Proposition 2.16.

The linear augmented Lagrangians were often used in engineering algorithms to implement active constraints as in Simo and Laursen [167]. The first application of the nonlinear augmented Lagrangians with adaptive precision control in combination with FETI to the solution of variational inequalities and contact problems seems to be in Dostál, Friedlander, and Santos [55] and Dostál, Gomes, and Santos [60, 61]. Applications to 3D frictionless contact problems with preconditioning of linear step are, e.g., in Dostál et al. [53] and Dostál, Gomes, and Santos [59, 62]. Experimental evidence of scalability of the algorithm with the inner loop implemented by the proportioning [42] was given in Dostál and Horák [64]. Applications to the contact shape optimization are, e.g., in Dostál, Vondrák, and Rasmussen [77].

The method presented in this chapter solves both coercive and semicoercive problems. Our first proof of numerical scalability of an algorithm for the solution of a semicoercive variational inequality used the optimal penalty in dual FETI problem [65]. Optimality of outer loop was proved in Dostál [67]; the theory was completed in Dostál [48]. In particular, it was proved that the relative feasibility error of the solution of the FETI problem with a given penalty parameter can be bounded independently of the discretization parameter. The results presented here for a scalar semicoercive variational inequality can be extended, including the theoretical results, to the solution of 2D or 3D multibody contact problems of elasticity, including 2D problems with a given (Tresca) friction [63] and an approximation of 3D ones [69]. The scalability was proved also for the problems discretized by the BETI (boundary element tearing and interconnecting) method of Langer and Steinbach [141]; see Bouchala, Dostál, and Sadowská [18, 17, 19]. See also Sadowská [164].

There is an interesting corollary of our theory. If we are given a class of contact problems which involves bodies of comparable shape, so that the regular part of their spectrum is contained in a given positive interval, then Theorem 8.3 implies that *there is a bound, independent of a number of the bodies, on the number of iterations that are necessary to approximate the solution to a given precision.* The linear auxiliary problems can be preconditioned by the FETI preconditioners. For comprehensive review of domain decomposition methods with many references see, e.g., Toselli and Widlund [175].

Some methods reported in Sect. 7.5 can be naturally adopted for the solution of semicoercive problems. This concerns especially the active set-based algorithms with multigrid solvers of linear problems (see, e.g., Krause [134]) and the algorithm proposed by Schöberl. A FETI-based algorithm for coercive and semicoercive contact problems was proposed by Dureisseix and Farhat [78]. These authors gave experimental evidence of scalability of their algorithms.

References

1. Arrow, K.J., Hurwitz, L., Uzawa, H.: Studies in Nonlinear Programming. Stanford University Press, Stanford (1958)
2. Avery, P., Rebel, G., Lesoinne, M., Farhat, C.: A numerically scalable dual–primal substructuring method for the solution of contact problems – part I: the frictionless case. Comput. Methods Appl. Mech. Eng. **193**, 2403–2426 (2004)
3. Axelsson, O.: A class of iterative methods for finite element equations. Comput. Methods Appl. Mech. Eng. **9**, 127–137 (1976)
4. Axelsson, O.: Iterative Solution Methods. Cambridge University Press, Cambridge (1994)
5. Axelsson, O., Lindskøg, G.: On the rate of convergence of the preconditioned conjugate gradient method. Numer. Math. **48**, 499–523 (1986)
6. Badea, L., Tai, X.-C., Wang, J.: Convergence rate analysis of a multiplicative Schwarz method for variational inequalities. SIAM J. Numer. Anal. **41**, 3, 1052–1073 (2003)
7. Balay, S., Gropp, W., McInnes, L.C., Smith, B.: PETSc 2.0 Users Manual. Argonne National Laboratory, http://www.mcs.anl.gov/petsc/
8. Bazaraa, M.S., Shetty, C.M., Sherali, H.D.: Nonlinear Programming, Theory and Algorithms. Second Edition. Wiley, New York (1993)
9. Belytschko, T., Neal, K.O.: Contact–impact by the pinball algorithm with penalty and Lagrangian methods. Int. J. Num. Methods Eng. **31**, 3, 547–572 (1991)
10. Benzi, M., Golub, G.H., Liesen, J.: Numerical solution of saddle point problems. Acta Numer. 1–137 (2005)
11. Bertsekas, D.P.: Constrained Optimization and Lagrange Multiplier Methods. Academic Press, London (1982)
12. Bertsekas, D.P.: Nonlinear Optimization. Athena Scientific, Belmont (1999)
13. Bertsekas, D.P., Tsitsiklis, J.N.: Introduction to Linear Optimization. Athena Scientific, Belmont (1997)
14. Bielschowski, R.H., Friedlander, A., Gomes, F.A.M., Martínez, J.M., Raydan, M.: An adaptive algorithm for bound constrained quadratic minimization. Invest. Oper. **7**, 67–102 (1997)

15. Blum, E., Oettli, W.: Direct proof of the existence theorem in quadratic programming. Oper. Res. **20**, 165–167 (1971)
16. Blum, H., Braess, D., Suttmeier, F.T.: A cascadic multigrid algorithm for variational inequalities. Comput. Vis. Sci. **7**, 3–4, 153–157 (2004)
17. Bouchala, J., Dostál, Z., Sadowská, M.: Duality based algorithms for the solution of multidomain variational inequalities discretized by BEM. Submitted (2007)
18. Bouchala, J., Dostál, Z., Sadowská, M.: Scalable BETI for Variational Inequalities. In: U. Langer et al. (eds.) Domain Methods in Science and Engineering XVII. Springer, Lecture Notes in Computational Science and Engineering (LNCSE) **60**, Berlin, 167–174 (2008)
19. Bouchala, J., Dostál, Z., Sadowská, M.: Theoretically Supported Scalable BETI Method for Variational Inequalities. Computing **82**, 53–75 (2008)
20. Calamai, P.H., Moré, J.J.: Projected gradient methods for linearly constrained problems. Math. Program. **39**, 93–116 (1987)
21. Chen, Ke: Matrix Preconditioning Techniques and Applications. Cambridge University Press, Cambridge (2005)
22. Chvátal, V.: Linear Programming. W.H. Freeman and Company, New York (1983)
23. Coleman, T.F., Hulbert, L.A.: A globally and superlinearly convergent algorithm for convex quadratic programs with simple bounds. SIAM J. Optim. **3**, 298–321 (1993)
24. Coleman, T.F., Lin, Y.: An interior trust region approach for nonlinear minimization subject to bounds. SIAM J. Optim. **6**, 418–445 (1996)
25. Coleman, T.F., Liu, J.G.: An interior Newton method for quadratic programming. Math. Program. **85**, 491–523 (1999)
26. Conn, A.R., Gould, N.I.M., Toint, Ph.L.: A globally convergent augmented Lagrangian algorithm for optimization with general constraints and simple bounds. SIAM J. Numer. Anal. **28**, 545–572 (1991)
27. Conn, A.R., Gould, N.I.M., Toint, Ph.L.: LANCELOT: A FORTRAN Package for Large Scale Nonlinear Optimization (Release A), No. 17 in Springer Series in Computational Mathematics, Springer–Verlag, New York (1992)
28. Conn, A.R., Gould, N.I.M., Toint, Ph.L.: Trust Region Methods. SIAM, Philadelphia (2000)
29. Cottle, R., Pang, J., Stone, R.: The Linear Complementarity Problems. Academic Press, New York (1992)
30. Dembo, R.S., Tulowitzski, U.: On minimization of a quadratic function subject to box constraints. Working Paper No. 71, Series B, School of Organization and Management, Yale University, New Haven (1983)
31. Demmel, J.W.: Applied Numerical Linear Algebra. SIAM, Philadelphia (1997)
32. Diamond, M.A.: The solution of a quadratic programming problem using fast methods to solve systems of linear equations. Int. J. Syst. Sci. **5**, 131–136 (1974)
33. Diniz-Ehrhardt, M.A., Dostál, Z., Gomes-Ruggiero, M.A., Martínez, J.M., Santos, S.A.: Nonmonotone strategy for minimization of quadratics with simple constraints. Appl. Math. **46**, 5, 321–338 (2001)

34. Diniz-Ehrhardt, M.A., Gomes-Ruggiero, M.A., Santos, S.A.: Numerical analysis of the leaving-face criterion in bound-constrained quadratic minimization. Optim. Methods Software **15**, 45–66 (2001)
35. Domorádová, M.: Projector Preconditioning for the Solution of Large Scale Bound Constrained Problems. M.Sc. Thesis, FEECS VŠB–Technical University of Ostrava, Ostrava (2006)
36. Domorádová, M., Dostál, Z.: Projector Preconditioning for Partially Bound Constrained Quadratic Optimization. Numer. Linear Algebra Appl. **14**, 10, 791–806 (2007)
37. Dorn, W.S.: Duality in Quadratic Programming. Quart. Appl. Math. **18**, 155–162 (1960)
38. Dorn, W.S.: Self-dual quadratic programs. J. Soc. Ind. Appl. Math. **9**, 51–54 (1961)
39. Dostál, Z.: Conjugate gradient method with preconditioning by projector. Int. J. Comput. Math. **23**, 315–323 (1988)
40. Dostál, Z.: On the Penalty Approximation of Quadratic Programming Problem. Kybernetika **27**, 2, 151–154 (1991)
41. Dostál, Z.: Directions of Large Decrease and Quadratic Programming. In: Marek, I. (ed.) Software and Algorithms of Numerical Mathematics 1993. University of West Bohemia, Charles University, and Union of Czech Mathematicians and Physicists, 9–22 (1993)
42. Dostál, Z.: Box constrained quadratic programming with proportioning and projections. SIAM J. Optim. **7**, 3, 871–887 (1997)
43. Dostál, Z.: Inexact solution of auxiliary problems in Polyak type algorithms. Acta Univ. Palacki. Olomuc. Fac. Rerum Nat. Math. **38**, 25–30 (1999)
44. Dostál, Z.: On preconditioning and penalized matrices. Numer. Linear Algebra Appl. **6**, 109–114 (1999)
45. Dostál, Z.: A proportioning based algorithm for bound constrained quadratic programming with the rate of convergence. Numer. Algorithms **34**, 2–4, 293-302 (2003)
46. Dostál, Z.: Semi-monotonic inexact augmented Lagrangians for quadratic programming with equality constraints. Optim. Methods Software **20**, 6, 715–727 (2005)
47. Dostál, Z.: Optimal algorithms for large sparse quadratic programming problems with uniformly bounded spectrum. In: Di Pillo, G., Roma, M. (eds.) Large-Scale Nonlinear Optimization. Nonconvex Optimization and Its Applications, **83**, Springer, Berlin, 83–93 (2006)
48. Dostál, Z.: An optimal algorithm for bound and equality constrained quadratic programming problems with bounded spectrum. Computing **78**, 311–328 (2006)
49. Dostál, Z.: Inexact semi-monotonic augmented Lagrangians with optimal feasibility convergence for quadratic programming with simple bounds and equality constraints. SIAM J. Numer. Anal. **45**, 2, 500-513 (2007)
50. Dostál, Z.: An optimal algorithm for a class of equality constrained quadratic programming problems with bounded spectrum. Comput. Optim. Appl. **38**, 1, 47–59 (2007)
51. Dostál, Z.: On the decrease of a quadratic function along the projected–gradient path. Accepted in ETNA (2008)

52. Dostál, Z., Domorádová, M., Sadowská, M.: Superrelaxation in minimizing quadratic functions subject to bound constraints. Submitted (2008)
53. Dostál, Z., Friedlander, A.,Gomes, F.A.M., Santos, S.A.: Preconditioning by projectors in the solution of contact problems: a parallel implementation. Ann. Oper. Res. **117**, 117–129 (2002)
54. Dostál, Z., Friedlander, A., Santos, S.A.: Adaptive precision control in quadratic programming with simple bounds and/or equalities. In: De Leone, R., Murli, A., Pardalos P.M., Toraldo, G. (eds.) High Performance Software for Non-linear Optimization. Kluwer, Applied Optimization, **24**, 161–173 (1998)
55. Dostál, Z., Friedlander, A., Santos, S.A.: Solution of contact problems of elasticity by FETI domain decomposition. In: Mandel, J., Farhat, C., Cai, X.–C. (eds.) Domain Decomposition Methods 10. AMS, Providence, Contemporary Mathematics, **218**, 82–93 (1998)
56. Dostál, Z., Friedlander, A., Santos, S.A.: Augmented Lagrangians with adaptive precision control for quadratic programming with equality constraints. Comput. Optim. Appl. **14**, 37–53 (1999)
57. Dostál, Z., Friedlander, A., Santos, S.A.: Augmented Lagrangians with adaptive precision control for quadratic programming with simple bounds and equality constraints. SIAM J. Optim. **13**, 1120–1140 (2003)
58. Dostál, Z., Friedlander, A., Santos S.A., Alesawi, K.: Augmented Lagrangians with adaptive precision control for quadratic programming with equality constraints: corrigendum and addendum. Comput. Optim. Appl. **23**, 1, 127–133 (2002)
59. Dostál, Z., Gomes, F.A.M., Santos, S.A.: Duality based domain decomposition with adaptive natural coarse grid projectors for contact problems. In: Whiteman, J. (ed.) Proceedings of MAFELAP. Elsevier, Amsterdam, 259–270 (2000)
60. Dostál, Z., Gomes, F.A.M., Santos, S.A.: Duality based domain decomposition with natural coarse space for variational inequalities. J. Comput. Appl. Math. **126**, 1–2, 397–415 (2000)
61. Dostál, Z., Gomes, F.A.M., Santos, S.A.: Solution of contact problems by FETI domain decomposition with natural coarse space projection. Comput. Methods Appl. Mech. Eng. **190**, 13–14, 1611–1627 (2000)
62. Dostál, Z., Gomes, F.A.M., Santos, S.A.: FETI domain decomposition for contact 3D problems. In: Neittaanmäki, P., Křížek, M. (eds.) Proceedings of International Conference on Finite Element Method in Three-Dimensional Problems, GAKUTO International Series, Math. Sci. Appl. **15**, Tokyo, 54–65 (2001)
63. Dostál, Z., Haslinger, J., Kučera, R.: Implementation of fixed point method for duality based solution of contact problems with friction. J. Comput. Appl. Math. **140**, 1–2, 245–256 (2002)
64. Dostál, Z., Horák, D.: Scalability and FETI based algorithm for large discretized variational inequalities. Math. Comput. Simul. **61**, 3–6, 347–357 (2003)
65. Dostál, Z., Horák, D.: Scalable FETI with Optimal Dual Penalty for Semicoercive Variational Inequalities. Contemp. Math. **329**, 79–88 (2003)
66. Dostál, Z., Horák, D.: Scalable FETI with Optimal Dual Penalty for a Variational Inequality. Numer. Linear Algebra Appl. **11**, 5–6, 455–472 (2004)

67. Dostál, Z., Horák, D.: Theoretically supported scalable FETI for numerical solution of variational inequalities. SIAM J. Numer. Anal. **45**, 500–513 (2007)
68. Dostál, Z., Horák, D., Kučera, R.: Total FETI - an easier implementable variant of the FETI method for numerical solution of elliptic PDE. Commun. Numer. Methods Eng. **22**, 1155–1162 (2006)
69. Dostál, Z., Horák, D., Kučera, R., Vondrák, V., Haslinger, J., Dobiáš, J., Pták, S.: FETI based algorithms for contact problems: scalability, large displacements, and 3D Coulomb friction. Comput. Methods Appl. Mech. Eng. **194**, 2–5, 395–409 (2005)
70. Dostál, Z., Horák, D., Stefanica, D.: A scalable FETI-DP algorithm for coercive variational inequalities. IMACS J. Appl. Numer. Math. **54**, 3–4, 378–390 (2005)
71. Dostál, Z., Horák, D., Stefanica, D.: A scalable FETI–DP algorithm with non–penetration mortar conditions on contact interface. Submitted (2007)
72. Dostál, Z., Horák, D., Stefanica, D.: Quadratic Programming and Scalable Algorithms for Variational Inequalities. In: de Castro, A.B., Gómez, D., Quintela, P., Salgado, P. (eds.) Numerical Mathematics and Advanced Applications–ENUMATH 2005, Springer–Verlag, New York, 61–76 (2006)
73. Dostál, Z., Horák, D., Stefanica, D.: A Scalable FETI–DP Algorithm for Semi-coercive Variational Inequalities. Comput. Methods Appl. Mech. Eng. **196**, 8, 1369–1379 (2007)
74. Dostál, Z., Schöberl, J.: Minimizing quadratic functions over non-negative cone with the rate of convergence and finite termination. Comput. Optim. Appl., **30**, 1, 23–44 (2005)
75. Dostál, Z., Vondrák, V.: Duality Based Solution of Contact Problems with Coulomb Friction. Arch. Mech. **49**, 3, 453–460 (1997)
76. Dostál, Z., Vondrák, V., Horák, D., Farhat, C., Avery, P.: Scalable FETI algorithms for frictionless contact problems. In: U. Langer et al. (eds.) Domain Methods in Science and Engineering XVII. Springer, Lecture Notes in Computational Science and Engineering (LNCSE) **60**, Berlin, 263–270 (2008)
77. Dostál, Z., Vondrák, V., Rasmussen, J.: FETI based semianalytic sensitivity analysis in contact shape optimization. In: Hoffmann, K.H., Hoppe, R.H.W., Schulz, V. (eds.) Fast Solution of Discretized Optimization Problems. Birkhäuser, International Series of Numerical Mathematics **138**, Basel, 98–106 (2001)
78. Dureisseix, D., Farhat, C.: A numerically scalable domain decomposition method for solution of frictionless contact problems. Int. J. Numer. Methods Eng. **50**, 12, 2643–2666 (2001)
79. Eaves, B.C.: On quadratic programming. Manage. Sci. **17**, 698–711 (1971)
80. Eck, C., Jarůšek, J., Krbec, M.: Unilateral Contact Problems. Variational Methods and Existence Theorems. Chapman & Hall/CRC, Boca Raton (2006)
81. Elman, H.C., Sylvester, D.J., Wathen, A.J.: Finite Elements and Fast Iterative Solvers: with applications in incompressible fluid dynamics. Oxford University Press, Oxford (2005)
82. Farhat, C., Gérardin, M.: On the general solution by a direct method of a large scale singular system of linear equations: application to the analysis of floating structures. Int. J. Numer. Methods Eng. **41**, 675–696 (1998)

83. Farhat, C., Lesoinne, M., LeTallec, P., Pierson, K., Rixen, D.: FETI-DP: A dual-primal unified FETI method – a faster alternative to the two-level FETI method. Int. J. Numer. Methods Eng. **50**, 1523–1544 (2001)
84. Farhat, C., Mandel J.: The two-level FETI method for static and dynamic plate problems – Part I: An optimal iterative solver for biharmonic systems. Comput. Methods Appl. Mech. Eng. **155**, 129–152 (1994)
85. Farhat, C., Mandel, J., Roux, F.-X.: Optimal convergence properties of the FETI domain decomposition method. Comput. Methods Appl. Mech. Eng. **115**, 365–385 (1994)
86. Farhat, C., Roux, F.-X.: A method of finite element tearing and interconnecting and its parallel solution algorithm. Int. J. Numer. Methods Eng. **32**, 1205–1227 (1991)
87. Farhat, C., Roux, F.-X.: An unconventional domain decomposition method for an efficient parallel solution of large-scale finite element systems. SIAM J. Sci. Comput. **13**, 379–396 (1992)
88. Fiedler, M., Pták, V.: On matrices with non-positive off-diagonal elements and positive principal minors. Czech. Math. J. **12**, 123–128 (1962)
89. Fletcher, R.: Practical Methods of Optimization. Wiley, Chichester (1997)
90. Forsgren, A, Gill, Ph.E., Wright, M.H.: Interior methods for augmented nonlinear optimization. SIAM Rev. **44**, 525–597 (2002)
91. Fortin, M., Glowinski, R.: Augmented Lagrangian Methods: Applications to the Numerical Solution of Boundary Value Problems. North-Holland, Amsterdam, (1983)
92. Fox, L., Huskey, H. D., Wilkinson, J. H.: Notes on the solution of algebraic linear simultaneous equations. Quart. J. Mech. Appl. Math. **1**, 149-173 (1948)
93. Frank, M., Wolfe, P.: An algorithm for quadratic programming. Naval Research Logistic Quarterly **3**, 95–110 (1956)
94. Friedlander, A., Martínez, J.M.: On the maximization of a concave quadratic function with box constraints. SIAM J. Optim. **4**, 177–192 (1994)
95. Friedlander, A., Martínez, J.M., Santos, S.A.: A new trust region algorithm for bound constrained minimization. Appl. Math. Optim. **30**, 235–266 (1994)
96. Friedlander, A., Martínez, J.M., Raydan, M.: A new method for large scale box constrained quadratic minimization problems. Optim. Methods Software **5**, 57–74 (1995)
97. Friedlander, M.P., Leyfer, S.: Global and finite termination of a two-phase augmented Lagrangian filter method for general quadratic programs. SIAM J. Sci. Comput. **30**, 4, 1706–1729 (2008)
98. Gass, S.I.: Linear Programming: Methods and Applications. McGraw Hill, New York (1958)
99. Glowinski, R.: Variational Inequalities. Springer–Verlag, Berlin (1980)
100. Glowinski, R., Le Tallec, P.: Augmented Lagrangian and Operator-Splitting Methods in Nonlinear Mechanics. SIAM, Philadelphia (1989)
101. Glowinski, R., Lions, J.-L., Trèmoliéres, R.: Numerical Analysis of Variational Inequalities. North-Holland, Amsterdam (1981)
102. Golub, G.H., O'Leary, D.P.: Some history of the conjugate gradient and Lanczos methods. SIAM Rev. **31**, 50–102 (1989)

103. Golub, G.H., Van Loan, C.F.: Matrix Computations, 2nd ed. Johns Hopkins University Press, Baltimore (1989)
104. Gould, N.I.M., Toint, Ph.L.: A Quadratic Programming Bibliography. Rutherford Appleton Laboratory Numerical Analysis Group Internal Report 2000-1, Chilton Oxfordshire (2003) (see also ftp://ftp.numerical.rl.ac.uk/pub/qpbook/qpbook/bib)
105. Greenbaum, A.: Behaviour of slightly perturbed Lanczos and conjugate gradient algorithms. Linear Algebra Appl. **113**, 7-63 (1989)
106. Greenbaum, A.: Iterative Methods for Solving Linear Systems. SIAM, Philaledelphia (1997)
107. Greenbaum, A., Strakoš, Z.: Behaviour of slightly perturbed Lanczos and conjugate gradient algorithms. SIAM J. Matrix Anal. Appl. **22**, 121-137 (1992)
108. Gwinner, J.: A penalty approximation for a unilateral contact problem in nonlinear elasticity. Math. Methods Appl. Sci. **11**, 4, 447–458 (1989)
109. Hackbusch, W.: Multigrid Methods and Applications. Springer–Verlag, Berlin (1985)
110. Hackbusch, W.: Iterative Solution of Large Sparse Systems. Springer–Verlag, Berlin (1994)
111. Hager, W.W.: Dual techniques for constraint optimization. J. Optim. Theory Appl. **55**, 37–71 (1987)
112. Hager, W.W.: Applied Numerical Linear Algebra. Prentice Hall, Englewood Cliffs (1988)
113. Hager, W.W.: Analysis and implementation of a dual algorithm for constraint optimization. J. Optim. Theory Appl. **79**, 37–71 (1993)
114. Hager, W.W.: Iterative methods for nearly singular linear systems. SIAM J. Sci. Comput. **22**, 2, 747–766 (2000)
115. Hager, W.W., Zhang, H.: A new active set algorithm for box constrained optimization. SIAM J. Optim. **17**, 526–557 (2006)
116. Hestenes, C.: Multiplier and gradient methods. J. Optim. Theory Appl. **4**, 303–320 (1969)
117. Hestenes, M.R., Stiefel, E.: Methods of conjugate gradients for solving linear systems. J. Res. Natl. Bur. Stand. **49**, 409–436 (1952)
118. Hintermüller, M., Ito, K., Kunisch, K.: The primal–dual active set strategy as a semismooth Newton method. SIAM J. Optim. **13**, 865–888 (2003)
119. Hintermüller, M., Kovtumenko, V., Kunisch, K.: The primal–dual active set strategy as a semismooth Newton method. Adv. Math. Sci. Appl. **14**, 513–535 (2004)
120. Hlaváček, I., Haslinger, J., Nečas, J., Lovíšek, J.: Solution of Variational Inequalities in Mechanics. Springer–Verlag, Berlin (1988)
121. Horák, D.: FETI based domain decomposition for variational inequalities. Ph.D. Thesis, FEECS VŠB–Technical University of Ostrava, Ostrava (2007)
122. Hüeber, S., Stadler, G., Wohlmuth, B.I.: A Primal-Dual Active Set Algorithm for Three-Dimensional Contact Problems with Coulomb Friction. SIAM J. Sci. Comput. **30**, 572–596 (2008)
123. Hunek, I.: On a penalty formulation for contact–impact problems. Comput. Struct. **48**, 2, 193–203 (1993)

124. Iontcheva, A.H., Vassilevski, P.S.: Monotone multigrid methods based on element agglomeration coarsening away from the contact boundary for the Signorini's problem. Numer. Linear Algebra Appl. **11**, 2–3, 189–204 (2004)
125. Kanto, Y., Yagawa, G.: A dynamic contact buckling analysis by the penalty finite element method. Int. J. Numer. Methods Eng. **29**, 4, 755–774 (1990)
126. Keller, G., Gould, N.I.M., Wathen, A.J.: Constraint preconditioning for indefinite linear systems. SIAM J. Matrix Anal. Appl. **21**, 1300–1317 (2000)
127. Kikuchi, N., Oden, J.T.: Contact Problems in Elasticity. SIAM, Philadelphia (1988)
128. Kinderlehrer, D., Stampaccia, G.: An Introduction to Variational Inequalities and Their Applications. Academic Press, London (1980)
129. Klawonn, A., Rheinbach, O.: A parallel implementation of dual–primal FETI methods for three dimensional linear elasticity using a transformation of basis. SIAM J. Sci. Comput. **28**, 5, 1886–1906 (2007)
130. Klawonn, A., Widlund, O.B., Dryja, M.: Dual–primal FETI methods for three-dimensional elliptic problems with heterogeneous coefficients. SIAM J. Numer. Anal. **40**, 159–179 (2002)
131. Kornhuber, R.: Monotone multigrid methods for elliptic variational inequalities. II. Numer. Math. **72**, 4, 481–499 (1996)
132. Kornhuber, R.: Adaptive monotone multigrid methods for nonlinear variational problems. Teubner–Verlag, Stuttgart (1997)
133. Kornhuber, R., Krause, R.: Adaptive multigrid methods for Signorini's problem in linear elasticity. Comput. Vis. Sci. **4**, 1, 9–20 (2001)
134. Krause, R.H.: On the Multiscale Solution of Constrained Minimization Problems. In: U. Langer et al. (eds.) Domain Methods in Science and Engineering XVII. Springer, Lecture Notes in Computational Science and Engineering (LNCSE) **60**, Berlin, 93–104 (2008)
135. Krause, R.H., Wohlmuth, B.I.: A Dirichlet-Neumann type algorithm for contact problems with friction. Comput. Vis. Sci. **5**, 3, 139–148 (2002)
136. Kruis, J.: Domain Decomposition Methods for Distributed Computing. Saxe-Coburg Publications, Stirling (2006)
137. Kruis, J., Matouš, K., Dostál, Z.: Solving laminated plates by domain decomposition. Adv. Eng. Software **33**, 7–10, 445–452 (2002)
138. Kučera, R.: Minimizing quadratic functions with separable quadratic constraints. Optim. Methods Software **22**, 3, 453–467 (2007)
139. Kunisch, K., Rendl, F.: An infeasible active set method for quadratic problems with simple bounds. SIAM J. Optim. **14**, 1, 35–52 (2002)
140. Lanczos, C.: Solution of systems of linear equations by minimized iterations. J. Res. Natl. Bur. Stand. **49**, 33–53 (1950)
141. Langer, U., Steinbach, O.: Boundary element tearing and interconnecting methods. Computing **71**, 205–228 (2003)
142. Laursen, T.: Computational Contact and Impact Mechanics. Springer, Berlin (2002)
143. Lions, P.-L., Duvaut, G.: Les Inéquations en Méchanique et en Physique. Dunod, Paris (1972)

144. Lukáš, D., Dostál, Z.: Optimal Multigrid Preconditioned Semi–Monotonic Augmented Lagrangians Applied to the Stokes Problem. Numer. Linear Algebra Appl. **14**, 741–750 (2007)
145. Lukšan, L., Vlček, J.: Indefinitely preconditioned inexact Newton method for large sparse non-linear programming problems. Numer. Linear Algebra Appl. **5**, 157–187 (1999)
146. Luo, Z.Q., Tseng, P.: Error bounds and convergence analysis of feasible descent methods: a general approach. Ann. Oper. Res. **46**, 157–178 (1993)
147. Mandel, J.: Étude algébrique d'une méthode multigrille pour quelques problèmes de frontière libre (French). C. R. Acad. Sci. Ser. I **298**, 469–472 (1984)
148. Mandel, J., Tezaur, R.: On the Convergence of a Dual-Primal Substructuring Method. Numer. Math. **88**, 543–558 (2001)
149. Mandel, J., Tezaur, R., Farhat, C.: A scalable substructuring method by Lagrange multipliers for plate bending problems. SIAM J. Numer. Anal. **36**, 1370–1391 (1999)
150. Marchuk, G.I., Kuznetsov, Yu.A.: Theory and applications of the generalized conjugate gradient method. Adv. Math. Suppl. Stud. **10**, 153–167 (1986)
151. Menšík, M.: Cholesky decomposition of positive semidefinite matrices with known kernel. Bc. Thesis, FEECS VŠB–Technical University of Ostrava, Ostrava (2007) (in Czech)
152. Meurant, G., Strakoš, Z.: The Lanczos and conjugate gradient algorithms in finite precision arithmetic. Acta Numer., 471–542 (2006)
153. Moré, J.J., Toraldo, G.: On the solution of large quadratic programming problems with bound constraints. SIAM J. Optim. **1**, 93–113 (1991)
154. Nicolaides, R.A.: Deflation of conjugate gradients with applications to boundary value problems. SIAM J. Numer. Anal. **24**, 355–365 (1987)
155. Nocedal, J., Wright, S.F.: Numerical Optimization. Springer–Verlag, New York (2000)
156. Of, G.: BETI - Gebietszerlegungsmethoden mit schnellen Randelementverfahren und Anwendungen. Ph.D. Thesis, University of Stuttgart (2006)
157. O'Leary, D.P.: A generalised conjugate gradient algorithm for solving a class of quadratic programming problems. Linear Algebra Appl. **34**, 371–399 (1980)
158. Parlett, B.N., Scott, D.S.: The Lanczos algorithm with selective orthogonalization. Math. Comput. **33**, 217–238 (1979)
159. Polyak, B.T.: The conjugate gradient method in extremal problems. USSR Comput. Math. Math. Phys. **9**, 94–112 (1969)
160. Powell, M.J.D.: A method for nonlinear constraints in minimization problems. In: Fletcher, A. (ed.) Optimization. Academic Press, New York, 283–298 (1969)
161. Pshenichny, B.N., Danilin, Yu.M.: Numerical Methods in Extremal Problems. Mir, Moscow (1982)
162. Rusten, T., Winther, R.: A preconditioned iterative method for saddle point problems. SIAM J. Matrix Anal. Appl. **13**, 887–904 (1992)
163. Saad, Y.: Iterative Methods for Large Linear Systems. SIAM, Philadelphia (2002)

164. Sadowská, M.: Scalable Total BETI for 2D and 3D Contact Problems. Ph.D. Thesis, FEECS VŠB-Technical University of Ostrava (2008)
165. Schöberl, J.: Solving the Signorini problem on the basis of domain decomposition techniques. Computing **60**, 4, 323–344 (1998)
166. Schöberl, J.: Efficient contact solvers based on domain decomposition techniques. Int. J. Comput. Math. Appl. **42**, 1217–1228 (2001)
167. Simo, J.C., Laursen, T.A.: An augmented Lagrangian treatment in contact problems involving friction. Comput. Struct. **42**, 97–116 (1992)
168. Sluis, A. van der, Vorst, H.A. van der: The rate of convergence of the conjugate gradients. Numer. Math. **48**, 543–560 (1986)
169. Stewart, G. W.: Conjugate Direction Methods for Solving Systems of Linear Equations. Numer. Math. **21**, 285–297 (1973)
170. Stewart, G. W.: Error and perturbation bounds for subspace associated with certain eigenvalue problems. SIAM Review, **15**, 4, 727–763 (1973)
171. Strang, G.: Linear Algebra and Its Applications. Academic Press, New York (1976)
172. Szegö, G.: Orthogonal Polynomials. AMS, Providence (1967)
173. Tai, X.-C., Tseng, P.: Convergence rate analysis of an asynchronous space decomposition method for convex minimization. Math. Comput. **71**, 1105–1135 (2001)
174. Tarvainen, P.: Two–level Schwarz method for unilateral variational inequalities. IMA J. Numer. Anal. **14**, 1–18 (1998)
175. Toselli, A., Widlund, O.B.: Domain Decomposition Methods – Algorithms and Theory, Springer Series on Computational Mathematics 34, Springer–Verlag, Berlin 2005.
176. Trottenberg, U., Oosterlee, C., Schüller, A.: Multigrid. Academic Press, New York (2001)
177. Vanderbei, R.J.: Linear Programming. International Series in Operations Research & Management Science, Vol. 37, Springer, New York (2001)
178. Vorst, H. van der: Iterative Krylov Methods for Large Linear Systems. Cambridge University Press, Cambridge (2003)
179. Wohlmuth, B.I.: Discretization Methods and Iterative Solvers Based on Domain Decomposition. Springer, Berlin (2001)
180. Wohlmuth, B., Krause, R.: Monotone methods on nonmatching grids for nonlinear contact problems. SIAM J. Sci. Comput. **25**, 324–347 (2003)
181. Wriggers, P.: Contact Mechanics. Springer, Berlin (2005)
182. Wright, S.J.: Primal-Dual Interior Point Methods. SIAM, Philadelphia (1997)
183. Yang, E.K., Tolle, J.W.: A class of methods for solving large, convex quadratic programs subject to box constraints. Math. Program. **51**, 223–228 (1991)
184. Zeidler, E.: Nonlinear Functional Analysis and Its Applications III: Variational Methods and Optimization. Springer, New York (1984)
185. Zeng, J., Zhou, S.: On monotone and geometric convergence of Schwarz methods for two–sided obstacle problem. SIAM J. Numer. Anal. **35**, 600–616 (1998)

Index